Diseases of Field
and
Horticultural Crops

The Editor

Dr. Pallem Chowdappa received M.Sc. in 1980 from Sri Venkateswara University, Tirupathi, Ph.D in 1985 from Mangalore University, Mangalore, Karnataka and post doctoral research at CABI Bioscience, U.K. He joined as Scientist-SI in 1985 at Central Plantation Crops Research Institute, Kasaragod, Kerala and was elevated to Principal Scientist in 2006 at Indian Institute of Horticultural Research, Bangalore. Dr. Chowdappa served as Scientist-in-Charge, Central Plantation Crops Research Institute Research Centre, Hirehalli and Head, Central Horticultural Experimental Station, Hirehalli from December, 2000 till April, 2006. He became Director,Central Plantation Crops Research Institute, Kasaragod in September, 2014.

Dr. Chowdappa is specialized in molecular plant pathology and has over 30 years of research experience in molecular characterization and management of *Alternaria, Colletotrichum* and *Phytophthora* associated with diseases of horticultural crops. He attended international training program on 'Oomycetes bioinformatics' at Virginia Tech, USA in 2014. Dr. Chowdappa was awarded DFID fellowship for Post-Doctoral research at CABI Bioscience, UK in 1998 .

Dr. Chowdappa has published more than 120 research papers in leading national and international journals, 8 books, 22 technical bulletins, 32 book chapters and 03 experimental manuals.

Diseases of Field
and
Horticultural Crops

Edited by

P. Chowdappa

2016

Daya   Publishing House®

A Division of

Astral International (P) Ltd

New Delhi 110 002

Cataloging in Publication Data—DK
Courtesy: D.K. Agencies (P) Ltd. <docinfo@dkagencies.com>

Diseases of field and horticultural crops / edited by P. Chowdappa.
pages cm
Contributed articles.

ISBN 978-93-5130-956-7 (International Edition)

1. Field crops—Diseases and pests. 2. Horticultural crops—Diseases and pests.I. Chowdappa, P. (Pallem), 1957- editor.

SB601.D57 2016   DDC 632   23

Published by   : **Daya Publishing House®**
A Division of
Astral International Pvt. Ltd.
– ISO 9001:2008 Certified Company –
4760-61/23, Ansari Road, Darya Ganj
New Delhi-110 002
Ph. 011-43549197, 23278134
E-mail: info@astralint.com
Website: www.astralint.com

Laser Typesetting   : **Rajender Vashist,** Delhi - 110 059

Printed at   : **Replika Press Pvt. Ltd.**

Disclaimer

This book is designed to provide helpful information on the subject. Any views or opinions or information presented in chapters are solely those of the authors and do not necessarily represent those of the editors or publishers. The editors or authors or publishers accept no liability what so ever by reason of negligence or otherwise arising from the use, or loss of use, or release of this information or any part of it. Further, the inclusion of product trade names does not indicate any preference or endorsement. Users must always read the label before using the product.

भारत सरकार
कृषि अनुसंधान और शिक्षा विभाग एवं
भारतीय कृषि अनुसंधान परिषद्
कृषि एवं किसान कल्याण मंत्रालय, कृषि भवन, नई दिल्ली 110 001

GOVERNMENT OF INDIA
DEPARTMENT OF AGRICULTURAL RESEARCH & EDUCATION
AND
INDIAN COUNCIL OF AGRICULTURAL RESEARCH
MINISTRY OF AGRICULTURE AND FARMERS WELFARE
KRISHI BHAVAN, NEW DELHI 110 001
Tel.: 23382629; 23386711 Fax: 91-11-23384773
E-mail: dg.icar@nic.in

डा. एस. अय्यप्पन
सचिव एवं महानिदेशक
Dr. S. AYYAPPAN
SECRETARY & DIRECTOR GENERAL

Foreword

Low yields are common in many field and horticultural crops and increased productivity is very important in reducing poverty and ensuring food security. Even though factors for low productivity are complex, one major contributory factor for crop losses is due to plant health problems. Annually 30-40% of a crop is lost due to crop health-related problems at global level. Among the crops health related problems, diseases are major factors contributing to crop losses.

Plant diseases have had an enormous impact on livelihoods throughout human history. The Irish Potato and the great Bengal famines are excellent examples of the effect of a plant disease on food security and changing demographics. There are several current examples too. Ug99, a virulent strain of black stem rust (*Puccinia graminis tritici*) that has originated from Africa, is threatening the high yielding wheat varieties. *Phytophthora infestans* that caused the Great Irish potato famine, still remains the most destructive pathogen of potatoes and tomatoes, causing crop losses of up to $ 6.7 billion annually.

The threat of climate change, coupled with declining land, water and agricultural labour force, besides rising input costs, have necessitated the development of efficient cost effective and sustainable integrated disease management approaches to minimize crop losses and to produce quality and pesticide residue-free produce. Accurate identification of the plant diseases affecting field and horticultural crops is a key factor for bio-security preparedness and adopting eco-friendly and effective control strategies.

The book on "Diseases of Field and Horticultural Crops" focuses primarily on diseases of important cereals, pulses, oilseeds, sugar and fibre yielding, vegetable, fruit and ornamental crops. The history, distribution, crop losses, etiological agents, disease diagnosis, epidemiology and control strategies of diseases of each crop is presented in the book. I take this opportunity to compliment

Dr. P. Chowdappa, Director, ICAR-Central Plantation Crops Research Institute, Kasaragod for his painstaking efforts in compiling and editing this valuable information on diseases of field pathologists, horticulturists, nursery managers, teachers, students, exporters, importers, extension specilists and other stakeholders associated with these crops.

(S. Ayyappan)

Preface

To meet food and nutritional requirements of ever growing global population in era of declining land and other natural resources coupled with climate changes, agricultural and horticultural production is being augmented through breeding new crop varieties and agronomic practices. The spread of existing pathogens to newer areas, the occurrence of new virulent strains of existing pathogens, the occurrence of fungicide resistance, the emerging and rapid spread of many previously unknown pathogens and the hybridization between emerging and existing species are some of the factors that increasing the impact of pathogens on crop plants. The dramatic increase in late blight problems on potato and tomato crops in India since 2008, due to the introduction of the 13_A2 clonal lineage of *Phytophthora infestans* from Europe, is also a matter of serious concern and has showed how bio-security measures have failed.

Despite myriad important discoveries made, plant pathogens continue to cause huge crop losses and need directions to implement cost-effective and durable management strategies. The basic knowledge gained from genome studies and effector biology and GM technology is yet to be explored to counteract the challenges ahead. Under these circumstances, knowledge on disease diagnosis, survival and spread and associated weather conditions for disease development are required for effective disease management. An integrated approach involving all available technologies is to be utilized to ensure food and nutritional security.

The required information for managing diseases of important cereals, pulses, oilseeds, sugar yielding, vegetable, fruits and ornamental crops is presented in this compilation through 18 Chapters. Disease symptoms are presented in coloured photographs for easy diagnosis and special emphasis has been given on management strategies. I hope that this book would cater the needs of research scientists, teachers and students of plant pathology, agriculture and horticulture and seed production agencies and other stakeholders who want to manage health of their crops.

P. Chowdappa
Editor

List of Contributors

Biju, C. N. Indian Institute of Spices, Research Cardamom, Research Centre, Appangala, Heravanadu Post, Madikeri - 571201, Karnataka. Email: bijucn123@rediffmail.com

Chander Rao, S. Directorate of Oilseeds Research, Rajendranagar, Hyderabad-500030, Andhra Pradesh.

Chattopadhyay, C. Directorate of Rapeseed-Mustard Research Bharatpur - 321303, Rajasthan

Chinmay Biswas. Cop Protection Division, Central Research Institute for Jute and Allied Fibres, Barrackpore, Kolkata-700120, West Bengal. Email: mkb.psb@gmail.com

Chowdappa, P. Indian Institute of Horticultural Research, Hesaraghatta, Lake Post, Bangalore-560089, Karnataka. Email: pallem22@gmail.com

Dam, S. K. Central Tobacco Research Institute, Rajahmundry - 533105, Andhra Pradesh. Email: damskd01@yahoo.co.in

Das, I. K. Directorate of Sorghum Research, Rajendranagar, Hyderabad-500030, Andhra Pradesh. Email: das@sorghum.res.in

Dheepa, R. Department of Plant Pathology, Tamil Nadu Agricultural University, Coimbatore – 641003.

Dinesh, D. Department of Plant Pathology, Tamil Nadu Agricultural University, Coimbatore – 641003.

Dubey, S. C. Principal Scientist, Division of Plant Pathology, Indian Agricultural Research Institute, New Delhi. Email: scdubey@iari.res.in

Durga Prashad. Department of Mycology & Plant Pathology, Dr. Y.S. Parmar University of Horticulture & Forestry, Nauni, Solan.

Duttamajumder, S. K. Sugarcane Breeding Institute, Coimbatore – 641007, Tamil Nadu

Jawaharlal, M. Department of Plant Pathology, Tamil Nadu Agricultural University, Coimbatore – 641 003.

Krishnaveni, D. Directorate of Rice Research, Rajendranagar, Hyderabad- 500030, Andhra Pradesh. Email: dkrishnaveni@drricar.org

Kumar, Vinod. Directorate of Rapeseed-Mustard Research Bharatpur - 321303 Rajasthan.

Ladha Lakshmi, D. Directorate of Rice Research, Rajendranagar, Hyderabad- 500030 Andhra Pradesh. Email: lakshmiladha@drricar.org

Laha, G. S. Directorate of Rice Research, Rajendranagar, Hyderabad- 500 030, Andhra Pradesh. Email: gslaha@drricar.org

Lijo Thomas. Directorate of Rapeseed-Mustard Research Bharatpur – 321303 Rajasthan

Loganathan, M. Indian Institute of Vegetable, Research, P.O. Jakhini Sahanshapur, Varanasi - 221305 Uttar Pradesh Email: logumuruga@gmail.com

Meena, P.D. Directorate of Rapeseed-Mustard Research Bharatpur - 321303 Rajasthan, Email: pdmeena@gmail.com

Meena Shekhar. Directorate Maize Research, IARI Campus, Pusa, New Delhi - 110001. Email: shekhar.meena@gmail.com

Nakkeeran, S. Department of Plant Pathology, Tamil Nadu Agricultural University, Coimbatore - 641003. Email: nakkeeransingai@yahoo.com

Negi, H. S. Department of Mycology & Plant Pathology, Dr. Y.S. Parmar University of Horticulture & Forestry, Nauni, Solan.

Pankaj Sharma. Directorate of Rapeseed-Mustard Research Bharatpur - 321303, Rajasthan.

Patil, J. V. Directorate of Sorghum Research, Rajendranagar, Hyderabad-500 030, Andhra Pradesh.

Prakasam, V. Directorate of Rice Research, Rajendranagar, Hyderabad- 500030, Andhra Pradesh. Email: prakasamv@gmail.com

Prasad, R. D. Directorate of Oilseeds, Research, Rajendranagar, Hyderabad-500030, Andhra Pradesh. Email: ravulapalliprasad@gmail.com

Praveena, R. Indian Institute of Spices, Research Cardamom, Research Centre, Appangala, Heravanadu Post, Madikeri - 571201, Karnataka.

Rai, A. B. Indian Institute of Vegetable, Research, P.O. Jakhini Sahanshapur, Varanasi - 221305, Uttar Pradesh.

Rao, G. P. Sugarcane Breeding Institute, Coimbatore - 641007, Tamil Nadu

Raoof, M. A. Directorate of Oilseeds, Research, Rajendranagar, Hyderabad - 500030 Andhra Pradesh.

Renukadevi, P. Department of Plant Pathology, Tamil Nadu Agricultural University, Coimbatore - 641003.

Roy, S. Central Tobacco Research Institute, Rajahmundry - 533105, Andhra Pradesh.

Sangit Kumar. Directorate Maize Research, IARI Campus, Pusa, New Delhi - 110001.

Santha Lakshmi Prasad, M. Directorate of Oilseeds Research, Rajendranagar, Hyderabad - 500 030, Andhra Pradesh. Email: santhalakshmib@gmail.com

Sarkar, S. K. Central Research Institute for Jute and Allied Fibres, Barrackpore, Kolkata - 700120, West Bengal.

Selvarajan, R. National Research Centre for Banana, Thogamalai Road, Thayanur (Post), Tiruchirappalli - 620102, Tamil Nadu. Email: selvarajanr@gmail.com

Sharma, I.M. Department of Mycology & Plant Pathology, Dr. Y.S. Parmar University of Horticulture & Forestry, Nauni, Solan - 173230, Himachal Pradesh. Email: imsharma18@gmail.com

Singh, Birendra . Indian Agricultural Research Institute, New Delhi.

Srinivas Prasad, M. Directorate of Rice Research, Rajendranagar, Hyderabad - 500030, Andhra Pradesh. Email: data.msprasad@gmail.com

Thirumalaisamy, P. P. Directorate of Groundnut Research, PO Box 5, Ivnagar Road Junagadh - 362001, Gujarat. Email: thirumalaisamypp@yahoo.co.in

Ved Ram. Department of Mycology & Plant Pathology, Dr. Y.S. Parmar University of Horticulture & Forestry, Nauni, Solan.

Venkataravanappa, V. Indian Institute of Vegetable, Research, P.O. Jakhini Sahanshapur, Varanasi - 221305, Uttar Pradesh.

Viswanathan, R. Sugarcane Breeding Institute, Coimbatore - 641007, Tamil Nadu Email: rasaviswanathan@yahoo.co.in

Contents

Diseases of Field and Horticultural Crops   *Pages* **1-32**
Editor-in-Chief: **P. Chowdappa**
Published by: **Daya Publishing House, New Delhi**

1

Diseases of Rice

V. Prakasam, D. Ladha Lakshmi, G.S. Laha, D. Krishnaveni and M. Srinivas Prasad

Rice (*Oryza sativa* L.) is one of the major food crops of the world. It is unique among the major food crops by virtue of its extent and adaptability to wide range of climatic, edaphic and cultural conditions. Rice cultivation has spread to many parts of the world due to its versatility. It can grow in the desert conditions of Saudi Arabia, in the wetland deltas of Southeast Asia and in the flooded rice plain as well. Being grown worldwide, it is the staple food for more than one and a half of the world's population. It is a nutritious cereal crop, provides 20 per cent of the calories and 15 per cent of protein consumed by world's population. Besides being the chief source of carbohydrate and protein in Asia, it also provides minerals and fibre. Rice straw and bran are important animal feed in many countries.

India is the largest rice growing country accounting for about one-third of the world acreage under the crop. It is grown in almost all the states of India, covering more than 30 per cent of the total cultivated area. It accounts for about 43 per cent of food grain production and 55 per cent of cereals production in the country. India ranks first in area (44 million hectare) and second in the world in production (102.7 million tons) after China. Its cultivation is mostly concentrated in the river valleys, deltas and low lying coastal areas of northeastern and southern India, especially in the states of Andhra Pradesh, Assam, Bihar, Chhattisgarh, Karnataka, Kerala, Maharashtra, Odisha, Tamil Nadu, Uttar Pradesh and West Bengal, which together contribute about 97 per cent of the country's rice production. Contributing about 42 per cent to country's food grain production, rice not only forms the mainstay of diet for majority of its people (>55 per cent), but also is the livelihood for over 70 per cent of the population in the traditional rice growing regions. In India, rice occupies a pivotal place as the primary source of calories for more than 70 per cent population. To meet the food needs of ever growing population (1.8%) in India, there is a need to increase the rice production around 130 million tons by 2030.

The increase in rice production in India is primarily contributed by adoption of semi-dwarf, early maturing and fertilizer responsive high-yielding rice varieties and hybrids coupled with improved management practices and increased cropping intensity. Such intensive and extensive cultivation systems have however, brought about a shift in pest and disease problems in rice. Rice crop is attacked by number of fungal, bacterial, viral and nematode diseases. Serious incidences of diseases such as blast, sheath blight, false smut, brown spot, bacterial blight and rice tungro disease have been reported from many rice growing areas of India (Table 1.1). All pathogenic organisms together limit the rice productivity to great extent while causing diseases at different stages of the crop (Table 1.2).

Table 1.1. Major Rice diseases and their casual agents in India

Disease	Pathogen/vectors
Fungal diseases	
Blast	*Pyricularia grisea*
Sheath blight	*Rhizoctonia solani*
Brown Spot or Sesame leaf spot	*Helminthosporium oryzae*
Sheath rot	*Sarocladium oryzae*
Udbatta disease	*Ephelis oryzae*
Stackburn disease	*Trichoconis padwickii*
Stem rot	*Sclerotium oryzae*
Foot rot or Bakanae disease	*Fusarium moniliforme*
Leaf scald	*Rhynchosporium oryzae*
Grain discoloration	Complex of Pathogens
Narrow brown spot	*Cercospora oryzae*
Bunt or Kernel Smut	*Neovossia horrid*
Bacterial diseases	
Bacterial blight	*Xanthomonas oryzae* pv. *Oryzae*
Bacterial leaf streak	*Xanthomonas campestris* pv. *Oryzicola*
Viral and other diseases	
Tungro (Rice tungro spherical virus	*Nephotettix virescens, N. nigropictus*
and rice tungro bacilliform virus)	*Recelia dorsalis*
Ragged stunt (Rice ragged stunt virus)	*Nilaparvata lugens*
Grassy stunt (Rice grassy stunt virus)	*Nilaparvata lugens*
Yellow dwarf (Phytoplasma)	*Nephotettix virescens; N. nigropictus*

Table 1.2. Occurrence of rice diseases at different stages of crop growth

Crop Stage	Diseases
Seeds and Seedlings	Seed rot and seedling blight, blast, brown spot, bacterial leaf blight (*kresek* phase) and rice tungro
Tillering	Blast, sheath blight, brown spot, bacterial leaf blight, rice tungro and bakanae
Reproductive stage	Blast, sheath rot, brown spot, sheath blight, bacterial leaf blight, stem rot and narrow brown leaf spot
Maturity stage	Neck blast, false smut, grain discolouration, kernel smut and udbatta disease

1. Rice blast

Rice blast, caused by *Pyricularia grisea* (Cooke) Sacc, remains as a first and foremost important disease which acts as major constraint in rice production. In recent years, this disease has been reported to appear regularly at an alarming intensity in many areas, which were earlier considered non endemic to this disease and the losses are heavy when the climatic conditions favour the disease development.

1.1. Economic impact: The disease is found in approximately 85 countries throughout the world. It was first reported as *'rice fever disease'* in China by Soong ying-shin in 1637; in Japan it was reported as *'Imochi-byo'* by Tsuchiya in 1704. In Italy it was reported as a *'brusone'*, Brugnatilli and in India it was first reported in the Thanjavur delta of Tamil Nadu in 1913 (Padmanabhan, 1965). The blast fungus can attack more than fifty other species of grasses. It causes disease at seedling and adult stages on the leaves, nodes and panicles. The disease often results in a significant yield loss, as high as 70–80 % during an epidemic (Ou, 1985). In India blast epidemics were reported from the Sub-Himalayan regions of Jammu and Kashmir, Andhra Pradesh, Tamil Nadu and Coorg regions of Karnataka and North eastern region comprising the states of Arunachal Pradesh, Manipur, Mizoram, Meghalaya, Assam and uplands of Bihar and Odisha (Table 1.3 and Fig. 1.1).

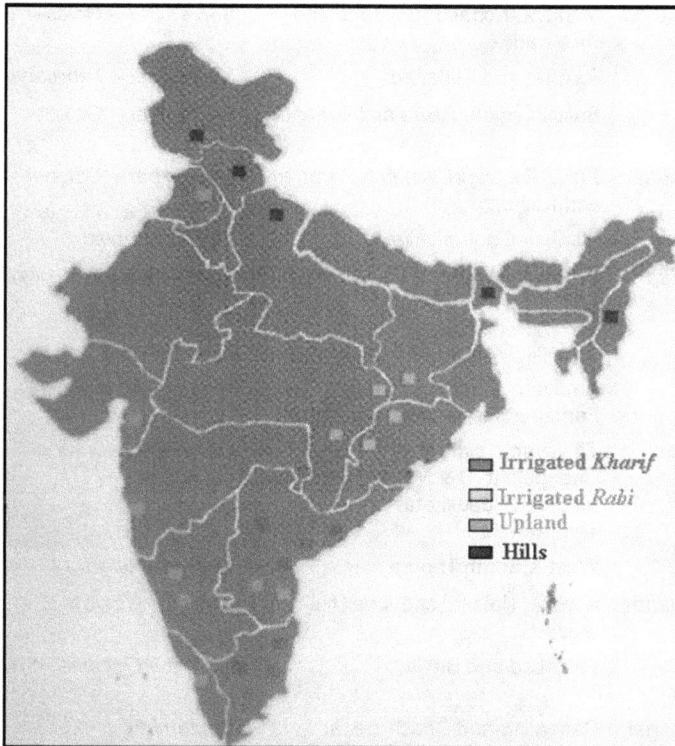

Fig. 1.1. Distribution of blast disease in India

Table 1.3. Distribution of blast disease in India

State	Endemic Districts/Area	Favorable Period
Andhra Pradesh	Srikakulam, Vishakapatnam, Guntur, Nellore, Chittoor, Nizamabad, Medak, Ranga Reddy, Mahboobnagar, East & West Godavari	September –February
Arunachal Pradesh	Arunachal Pradesh	April – July
Assam	Karimganj, Tinsukia, Nowgong, Kamrup, Goalpara and N. Lakhimpur	August – October
Bihar	Ranchi and Hazaribagh	August – October
Chhattisgarh	Northern hill regions	September – October
Gujarat	Kheda	September – October
Haryana	Hissar and Karnal	August – October
Himachal Pradesh	Kangra valley (Malan, Palampur), Kulu and Mandi	August – October
Jammu & Kashmir	Hill zones of Anantnag, Rajouri, Jammu, Udampur and Larnoo	July – September
Karnataka	Mandya, Kodagu, Shimoga and Dharwad	September – October
Kerala	Palghat and Kuttanad	September – February
Madhya Pradesh	Bastar region, Rewa and Bilaspur	September – October
Maharashtra	Pune, Ratnagiri, Kolaba, Parbhani and Kolhapur	September – October
Manipur	Manipur Central valley	July – October
Meghalaya	West Khasi hills	June – October
Mizoram	Mizoram	August – October
Odisha	Cuttack, Ganjam and Koraput	July – August
Punjab	Amritsar, Bhatinda, Patiala, Ferozpur, Ropar and Hoshiarpur	August – October
Tamil Nadu	Tanjavur, Coimbatore, Chengalput, S & N Arcot, Periyar, Madurai, Pudukkotai and Thirunalvelli	October – February
Tripura	West & South Tripura	July – October
Uttarakhand	Almora, Nainital and other hill areas	August – October
Uttar Pradesh	Faizabad and Ballia	August – October
West Bengal	Darjeeling and Cooch Behar	September

1.2. Symptoms

1.2.1. Leaf blast: Lesions are typically spindle-shaped on leaves; wide at the center and pointed toward either end. Large lesions usually develop a diamond shape with greyish center and brown margin. Under favourable conditions, lesions on the leaves expand rapidly and tend to coalesce, leading to complete necrosis of infected leaves giving a burnt appearance from a distance. Hence the name rice blast given to this disease. The symptoms on leaves may vary according to the environmental conditions, the age of the plant, and the levels of resistance of the host cultivars. On susceptible cultivars, lesions may initially appear grey-green and water-soaked with a darker green border and they expand rapidly to several centimeters in length (Fig. 1.2). On susceptible cultivars, older lesions often become light tan in colour with necrotic borders. On resistant cultivars, lesions often remain small in size and brown to dark brown in colour.

Fig. 1.2. Spindle-shaped lesions caused by *Pyricularia oryzae* on leaves

1.2.2. Collar blast/Nodal Blast: The collar of a rice plant refers to the junction of the leaf and the stem sheath. Symptoms of infection of the collars consist of a general area of necrosis at the union of the two tissues. Collar infections can kill the entire leaf and may extend a few millimeters into and around the sheath. The fungus may produce spores on these lesions (Fig. 1.3a).

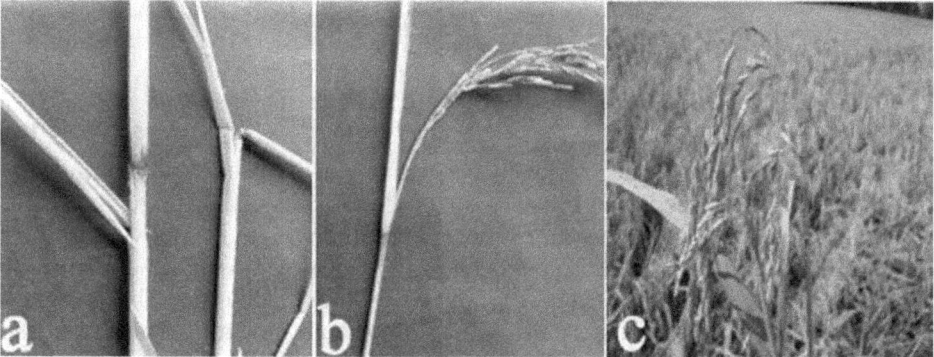

Fig. 1.3. A typical symptom of node blast (a); neck blast (b & c)

1.2.3. Neck Blast: The neck of the rice plant refers to that portion of the stem that rises above the leaves and supports the panicle. Necks are often infected at the node and infection leads to a condition called rotten neck or neck blast.

Infection of the necks can be very destructive, causing failure of the seeds to fill or causing the entire panicle to fall over as if rotted. The pathogen can also infect the panicles. Lesions can be found on the panicle branches, spikes, and spike lets. The lesions are often grey brown and causes discolourations of the branches of the panicle, over the time the branches may break at the lesion. Out of the three symptoms, neck blast is more destructive and directly reduces the economic value of the produce. If the pathogen attacks before grain filling, the entire panicle will be chaffy and if the infection is at the grain filling stage, the grains will not be properly filled and if the pathogen attacks the crop after grain filling, the grains may fall of (Fig. 1.3b & c).

1.3. Causal organism: The fungus *Magnaporthe grisea* (Hebbert) Barr (Anamorph: *Pyricularia grisea* (Cooke) Sacc) is the causal agent of rice blast disease. It belongs to Ascomycetae family. The rice blast fungus pathogen has been known as *Pyricularia grisea*. The telomorph stage, *Magnaporthe grisea*, has not been found in nature, but it has been produced after crossing appropriate compatible isolates in the laboratory. The fungus produces simple, grey conidiophores that bear terminal, pear shaped, mostly two septate conidia (Fig. 1.4a). Mycelium of the fungus is hyaline to olivaceous, septate and highly branched. Conidia are produced in clusters on long septate, olivaceous slender conidiophores. Conidia are obpyriform to ellipsoid, attached at the broader base by a hilum. Conidia are hyaline to pale olive green, usually 3 celled. The perfect state of the fungus is *M. grisea*. It produces perithecia. The ascospores are hyaline, fusiform, 4 celled and slightly curved.

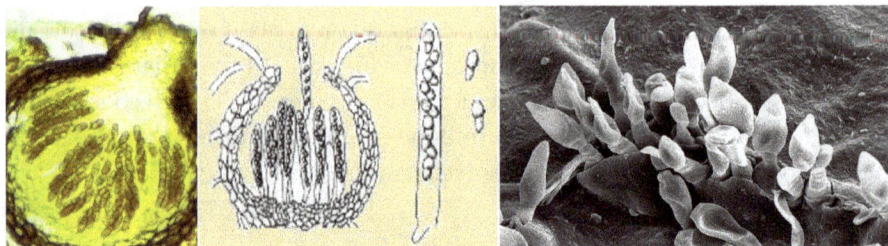

Fig. 1.4. Conidiophores **(a)** and pear shaped conidia **(b)** of *Pyricularia oryzae* and Scanning electron micrograph of conidia and appresorium of *Pyricularia oryzae* **(c)**

1.4. Epidemiology: The pathogen perpetuates as mycelium and conidia on diseased rice straw, seed, and possibly on weed hosts also. The fungus produces conidia and releases in to the atmosphere when there is high relative humidity (>90%). The conidia are air borne and fall on the rice plant and adhere strongly to the leaves through the mucilage produced by them at the tip. These conidia germinate when the rice leaves are wet or have high moisture. They form the appresoria on the leaves through which they adhere to the leaf surface and penetrate the leaf surface by penetration peg from the appresoria or enter through the stomata (Fig. 1.4c). Production and accumulation of melanin in the appresorium cell wall seem to be necessary for successful penetration. Rice seedlings and young or tender tissues are more vulnerable than the older ones. At optimum temperatures, new blast lesions appear within 4-5 days after they fall on leaf surface. In warm and wet weather conditions, new conidia are produced within hours from the appearance of the lesions and this continues for several days. Most of the conidia are released between midnight and sunrise. This is a polycyclic disease and completes several cycles within a season and causes epidemics if conducive weather conditions prevail and are coupled with the critical growth stages of susceptible variety. The critical growth stages are seedling stage, tillering stage and panicle initiation stage of the crop.

1.5. Management: Among several methods available for the management of the disease, chemical control has been very successful and widely practiced in many countries. However, due to concern over the excessive use of chemical pesticides, emphases are being given towards alternative approaches for the management of the disease. Development of an integrated approach by combining the effective cultural, chemical and host plant resistance components would be ideal for the management of this dreaded disease.

1.5.1. Cultural Practices: In endemic area, avoid using susceptible cultivars. Healthy seeds collected from disease free fields should be used. Destruction of weeds, collateral hosts and crop residues can greatly reduce the primary inoculum and terminal disease severity. Seedlings should be raised in the water covered seed beds. Seedlings raised in upland nurseries are more susceptible to blast even after they are transplanted. This is due to lower silicon content in the epidermal cells. In general, the late planted crops suffer more from the blast infection. Balanced application of nitrogenous fertilizers should be practiced. Application of farm yard manures (FYM) should be encouraged. As the disease intensity is more in closely planted crop, wider spacing (20 cm x 15 cm) can reduce the disease severity considerably.

1.5.2. Host Plant Resistance: Many varieties have been screened under artificial condition (Fig. 1.5) and reported to be resistant to this disease. In endemic areas the varieties like Rasi, IR 64, Prasanna, IR-36, Vikas, Tulasi, Sasyasree, Aditya, Krishna Hamsa, Vikas, Mandya Vijaya and VL Dhan 221 may be cultivated and the preference may vary according to the region of cultivation.

Fig. 1.5. Screening of resistant sources against blast under field condition (a) and under UBN (b).

1.5.3. Chemical control: Adopt seed treatment with pyroquilon 50 WP (Fongorene) @ 1 g/kg or tricyclazole 75 WP @ 1 g/kg or carbendazim 50 WP @1-1.5 g/kg. If the disease, appears in the main fields, spray tricyclazole 75 WP @ 0.6 g/l or iprobenphos 48 EC @ 2g/l or isoprothiolane 40 EC @ 1.5 ml/l or carpropamid 30 SC @ 1 ml/l or carbendazim 50 WP @ 1 g/l or kasugamycin 3 SL @ 2.5 g/l or Swing 250 EC® (epoxyconazole 125 g/l + carbendazim 125 g/l) @ 0.5 ml/l. Among the botanicals, Biotos (Plant activator-monoterpenes) in combination with tricyclazole was found very effective against blast. The number of sprays should be decided by carefully monitoring the fields (DRR Progress Report 1975-2010). Under experimental conditions, many combination products like Filia 52.5 SE (Tricyclazole - propiconazole combination) and tricyclazole-mancozeb combination (Tricyclazole + Mancozeb 80 WP), Nativo 75 WG (trifloxystrobin 25% + tebuconazole 50%) @ 0.4 g/litre and RIL 013/F1 35 SC (fenoxalin 5% + isoprothiolane 30%) @ 1.5 and 2 ml/litre, feroxalin + isoprothiolane 35 SC @ 1.5 ml/L and metaminostrobin 20 SC @ 2.0 ml/L were found effective against blast.

2. Sheath blight

Sheath blight is one of the major biotic constraints that affects rice production in India and is considered economically important disease of rice in the world. The disease is caused by *Rhizoctonia solani* Kuhn (teleomorph: *Thanetophorus cucumeris* (Frank) Donk), a fungal pathogen of both rice and soybean. The yield loss due to this disease is reported to range from 5.2-50 per cent depending on the environmental condition, crop stage at which the disease occurs, cultivation practices and cultivars used. The disease was first recorded in Japan, and also found to be widespread in East and South - East Asian countries hence it named as 'Oriental sheath and leaf blight'. Due importance to this disease was not in the past due to senility nature of this disease on rice crop. However, planting of fertilizer responsive, dwarf and short duration varieties, heavy application of nitrogenous fertilizers and change in the weather factors during rice cultivation has resulted in the outbreak of this disease from seedling stage onwards instead of its appearance at older stage of the crop.

2.1. Economic importance: Sheath blight disease has shown a remarkable spread, wherever rice is grown. In India, its incidence is reported throughout country. In some countries like Sri Lanka, China, Taiwan and Japan, sheath blight is considered as a major problem in rice cultivation. The extent of yield loss in rice due to sheath blight is based on its severity and spread of infection. In Arkansas, for example, sheath blight was found to be present in 50-66 per cent of rice fields, causing 5-15 per cent yield reduction in 2001. A yield loss of 30-40 per cent was reported, in case of severe infection of leaf sheath and leaf blade. A yield loss of 25 per cent was observed when the disease extended up to flag leaf and in West Bengal reported yield loss ranged from 5.2 to 50 per cent (Rajan, 1987).

2.2. Occurence and distribution: In order to understand the spread of sheath blight of rice disease in the country, data was collected over the past twenty two years (1990-2011) from production oriented survey (POS) reports and disease distribution map has been generated. Results revealed that occurrence and intensity of sheath blight was moderate before 2000. But the disease has spread widely in terms of both occurrence and intensity over the past twelve years (POS-2001-2012). At present, it is a major production constraint in the states of Punjab, Haryana, Uttarakhand, Eastern UP, Bihar, West Bengal, Odisha, Jharkhand, Chhattisgarh, Andhra Pradesh, Tamil Nadu, Karnataka and Kerala (Fig. 1.6).

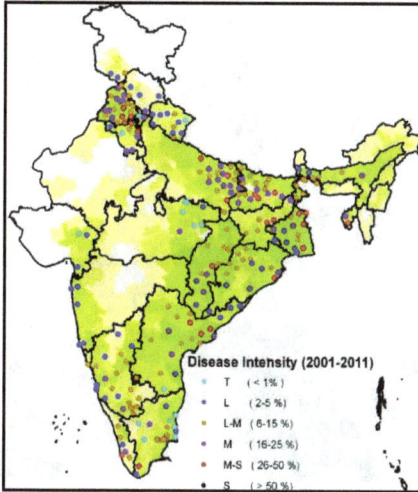

Fig. 1.6. A map showing the distribution and intensity of sheath blight in different rice growing areas of India (2000-2012)

2.3. Symptoms: The disease has been named as 'Sheath blight' because of primary infection on leaf sheath. The fungus affects the crop from tillering to heading stage. Five to six week old leaf sheaths are highly susceptible. The disease attacks the leaf sheath, leaf blades and in severe cases symptoms are also observed on emerging panicles. Initial symptoms usually develop as lesions on sheaths of lower leaves near the water line when plants are in the late tillering or early internode elongation stage of growth. The spots are seen first as ellipsoid or ovoid, somewhat irregular, greenish grey, varying from 1-3 cm in length, gradually enlarging and becoming greyish white with a blackish brown margin. With age, the lesions expand and the center of the lesions may become bleached with an irregular tan to brown border. When humidity exceeds 90 per cent and temperatures are in the range of 28-35 °C, infection spreads rapidly by means of runner hyphae to upper plant parts, including leaf blades, causing extensive, irregularly shaped lesions with brown borders (Fig. 1.7a). Lesions on the upper parts of plants extend rapidly coalescing with each other to cover entire tillers from the water line to the flag leaf (Fig. 1.7b). The presence of several large lesions on a leaf sheath usually causes death of the whole leaf and plant, and in severe cases all the leaves of a plant may be blighted in this way. The infection extends to the inner sheaths resulting in death of the entire plant. Plants heavily infected at these stages produce poorly filled grain, particularly in the lower portion of the panicle. Additional losses result from increased lodging or reduced ratoon production due to infection of the culm and reduced grain filling.

Fig. 1.7. Sheath blight symptoms on tillers (a) and on leaves (b)

2.4. Causal organism: *Rhizoctonia solani* Kuhn. (Anamorph) is the acceptable name for the sheath blight fungus with *Thanatephorus cucumeris* (Frank) Donk. as the perfect stage and AG-1 as anastomosis group comes under Basidiomycota phylum. The genus concept in *Rhizoctonia* was first established in 1815 by De Candolle. The most important species of *Rhizoctonia, R. solani,* was originally described by Julius Kühn on potato in 1858 which is the most widely documented and the most important and destructive species of *Rhizoctonia. R. solani* has a wide host range of host plants and can infect plants belonging to more than 32 plant families and 188 genera. These host plants favouring to establish the inoculum load in the soil to make quorum to cause the disease.

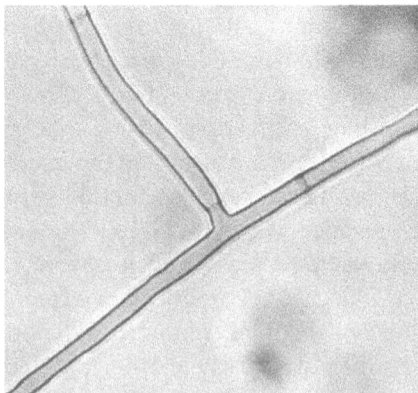

Fig. 1.8. Right angle branching of matured hyphae

Rhizoctonia is considered a genus of basideomycetous imperfectous fungi. According to *R. solani* was characterized by the diameter of vegetative hyphae (8-12 µm), constriction at the point of branching, and right angle branching of matured hyphae (Fig. 1.8). Initially the young hyphae are colourless but become yellow and ultimately brown with age. The fungus produces three different types of mycelia on the host. They are, (a) Runner mycelium: the straight, creeping, tropic and long runner hyphae or sometimes thickened and flattened. (b) Lobate mycelium: originate from the runner hyphae as short, swollen, much branched and single or multiple lobate appresoria. Penetration peg of the fungus will enter through intra cellular space of the host tissue through infection cushion and produces lesion on the rice sheath. Such mycelium can withstand dessication. (c) Moniloid mycelium: involved in the formation of sclerotia in *in vivo*. It is produced on the wall of petriplate or test tube while culturing under *in vitro*.

When the mycelium is 6-8 days old it starts producing sclerotia (Fig. 1.9). The shape of the sclerotia is oblate, kidney, roughly spherical or somewhat flattened and irregular. Young sclerotia are composed of compact masses of hyphal cells about 5-8 µm wide and the cell wall thickness is about 0.9 µm. The size of the sclerotia is varying and composed of inner and outer layer. The outer layer is consisted with 10-30 layers of dead cells which is occupying about 50 per cent of sclerotial aperture. The secretions are dispatched through germination aperture in the sclerotia germination process and mycelia also extend from the aperture.

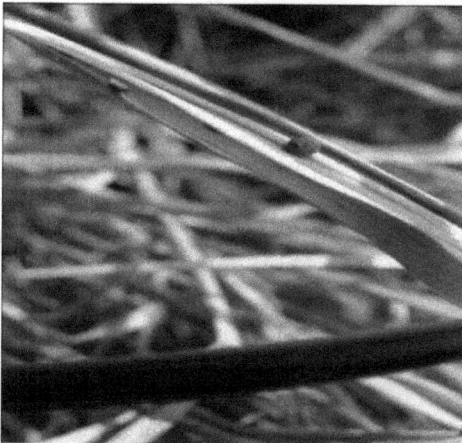

Fig. 1.9. Sclerotium of *R. solani*

Basidial stage is observed on upland rice. It appears in white powdery layer on healthy leaves or healthy areas adjacent to lesions under high humidity content. A Basidium is a unit cell, colourless and oval, obvate or cylindrical with 2-4 small stalks generated on the top of mycelial cell and each stalk has one basidiospores. Basidiospore (2-4) terminal in imperfect cymose or racemose cluster formed by short cells of ascending hyphae with sterigmata. The basidia are usually 11-15x8-9 µm produced at night or during wet day. Basidiospores production is very rare in Indian rice ecosystem.

2.5. Epidemiology

Dynamic nature of temperature, precipitation and other climatic variables were found to have tremendous influence on plant diseases. High temperature (22-35 °C) and high relative humidity (>90%) are favourable for development of sheath blight fungus. The temperature and RH within the crop canopy (microclimate) varies as compared to the surrounding air temperature and RH. Close planting and heavy application of N-fertilizers lead to thick growth of plant canopy and tends to create favourable microclimate to infection by the pathogen. Mycelial growth and sclerotia formation are optimum at pH 6.0-7.0. Disease development progresses very rapidly in the early heading and grain filling stages and during periods of frequent rainfall and overcast skies. It was reported that sheath blight disease severity was positively correlated with sandiness of soil. Further, the disease incidence was highest in wet soils with 50-60 per cent water holding capacity (WHC) and lowest in submerged soils with 100 per cent WHC. Sheath blight occurs in more severe form under the shaded conditions than in the open field.

2.6. Management

2.6.1. Cultural practices: The pathogen has a very wide host range so practice of clean cultivation is effective in minimizing the disease. Burning the infected crop debris and stubbles after harvest, deep ploughing in summer to eliminate the weed hosts, bunds cleaning, and allowing the field for fallow will reduce inoculums. Avoid close planting because sparse planting resulted in lower sheath blight occurrence and greater lodging resistance in rice. Square method of transplantation is contributed to reduced sheath blight intensity and higher grain yield. Planting of rice seedlings far from the bund resulted in reduced sheath blight incidence since bunds harbour weed hosts of *R. solani*. Submergence of the crop and alternate wetting and drying had a negative effect on disease progress and resulted in reduced sheath blight disease development. Avoid flow of irrigation water from infected fields to healthy fields. The application of silicon to complement host resistance to sheath blight appears to be an effective strategy for disease management in rice, especially when the soil is low in silicon. Soil solarization during May month is effective in reducing the soil borne inoculum of the pathogen and helps to produce healthy seedlings for transplanting (IRRI, 2006). However in India soil solarization is not feasible to reduce the sclerotial load in soils among rice farmers due to lack of resources, fragmentation field and most of the regions rice is cultivated throughout the year.

2.6.2. Host Plant Resistance: Host plant resistance is the most important tool for rice disease management and has played a key role in sustaining rice productivity. The major problem in sheath blight resistant breeding is the lack of donors having high degree of resistance to the pathogen. Some identified promising genotypes are Teqing, Tetep, Tadukan, Jasmine 85 and WSS5. Among the land races, few accessions were found to show moderate level of resistance. Despite the fact that the rice variety is having absolute resistance to sheath blight, there are significant differences in tolerance levels among rice cultivars (AICRIP, 1965-

2011). Several wild germplasm accessions with moderate resistance to sheath blight were identified in the recent past. Accessions of *Oryza nivara, O. rufipogon, O. glaberrima* and *O. latifolia* were reported to show tolerance to sheath blight. Targeted resistance breeding for sheath blight was not done in the past primarily because of lack of donor having high level of resistance. Though a number of varieties have been released through AICRIP with moderate level of resistance to sheath blight most of them do not exhibit desired level of resistance under farmers field conditions. Shusk samrat, Warangal samba, Amulya, CR 1002 are some of the moderately varieties released in India.

2.6.3. Biological control: Seed treatment with *Pseudomonas fluorescens* @ of 10 g/kg followed by seedling dip @ of 2.5 kg product (*P. fluorescens*) /ha dissolved in 100 liters of water and dipping for 30 minutes. Soil application of *P. fluorescens* @ of 2.5 kg/ha after 30 days of transplanting (this product should be mixed with 50 kg of FYM/Sand and then applied). Seeds are treated with salted water followed by seed treatment with *Trichoderma harzianum* and *Pseudomonas fluorescens* @ 5g each /kg seed or Pant Bio-agent-3 @ 10 g/kg seed reduces the disease incidence.

2.6.4. Chemical control: New fungicidal formulations were also found effective against rice sheath blight. Among them, Amistar 25 SC @ 1.0 ml/l (30.6%) and RIL-010/ FI 25 SC at 0.75 ml/l (30.1%) showed a high degree of efficacy in reducing the disease severity and increased grain yields and superior over the standard fungicides (validamycin at 2.5 ml/l). Pencycuron (monceren 250 SC) was most effective when sprayed at 35 and 55 days after transplanting in reducing the sheath blight intensity. Strobilurins, a new group of fungicides were very effective both in terms of disease reduction as well as in increasing grain yields (Biswas, 2006). Some of the fungicides are popular among the farmers in India to manage the sheath blight are carbendazim 50 WP (1 ml/l), propiconazole 25 EC (1 ml/l), hexaconazole 5 EC (2 ml/l), validamycin 3 SL (2.5 ml/l) and iprophenphos 48 EC (2 ml/l). In recent time many combination fungicides product like, Filia 52.5 EC (tricyclazole+propiconazole) 2.5 ml/l, Nativo 75 WG trifloxytrobin+tebuconazole) 0.4g/l and Luster 37.5 SE (flusilazole+carbendazim) 300 a.i. g/h are also very effective against sheath blight pathogen.

3. False smut

In the recent years, false smut has become a serious threat to rice cultivation because the disease affects the grain development and causes direct economic loss to farmers. Earlier the disease was considered as farmer's friendly disease and locally known as 'lakshmi disease' because it was always found associated with bumper yields. The reason behind this statement is that the conditions which are favourable for the growth of the fungus particularly high humidity and rainfall are similar to those resulting in good growth of the rice crop. In earlier days, importance was not given to this disease because its occurrence was irregular/ sporadic and the symptoms were mostly restricted to one or two grains per panicle. However, use of high fertilizer-responsive varieties and hybrids, heavy application of nitrogenous fertilizer and changes in climatic conditions have paved the way for the outbreak of this disease and it has been observed that the disease can affect huge number of grains in a panicle which can lead to disease epidemic and heavy yield loss (Fig. 1.10).

Fig. 1.10. Heavy incidence of false smut disease of rice caused by *Ustilaginoidea virens*

3.1. Occurrence and distribution: False smut of rice is caused by *Ustilaginoidea virens*. The disease is also known as orange and green smut which was first reported in India by Cooke in 1878 from Thirunelvelly district of Tamil Nadu. Subsequently, the disease was reported from more than 40 countries including almost all paddy growing regions of the world *viz.*, India, Philippines, Myanmar, Colombia, Peru, Bangladesh, Mauritius, Nigeria, Burma, Sri Lanka, Fiji, Africa (Biswas, 2001), USA (Rush *et al.*, 2000) and Egypt (Atia, 2004). In India the disease has been reported from Assam, Andaman and Nicobar Islands, Andhra Pradesh, Arunachal Pradesh, Haryana, Chhattisgarh, Gujarat, Himachal Pradesh, Jammu and Kashmir, Karnataka, Kerala, Madhya Pradesh, Maharashtra, Manipur, Meghalaya, Mizoram, Nagaland, Odisha, Punjab, Rajasthan, Sikkim, Tamil Nadu, Tripura, Uttar Pradesh, Uttarakhand and West Bengal (Dodan and Ram Singh, 1996). To know the actual spread of the disease in our country, data were collected for the past twenty years (1990-2010) from Production Oriented Survey reports and disease distribution map has been generated. Results revealed that occurrence and intensity of false smut disease was less during 1990-2000, but the disease has spread widely in terms of both occurrence and intensity over the past ten years (2001-2010) (Fig. 1.11).

3.2. Economic importance: In India, the disease has been reported to occur in moderate to severe intensity from the year 2000 onwards. The yield losses in different states of the country have been estimated to vary between 0.2% to 49% depending on the disease intensity and rice varieties grown in those areas (Dodan and Ram Singh, 1996). The pathogen converts the rice grain into ball of mycelial mat covered with powdery mass of pathogen spores and thereby causing both quantitative and qualitative losses. Disease infection on one spikelet may also cause sterility on the neighbouring spikelets. Yield loss occurs due to sterility of the adjacent spikelets, chaffiness, reduction in the number of healthy grains and thereby reduction in thousand grain weight. The per centage of chaffy grains increased with intensity of the disease indicating a direct correlation between disease intensity and yield loss and might contribute to heavy yield loss under high disease pressure. Studies showed that 10% disease incidence could cause about 25 % of chaffiness leading to 9 % reduction in thousand grain weight (Chib *et al.*, 1992). Similarly 44.1 % incidence could cause about 28.5 % reduction in grain weight. Apart from this, experimental results proved that the germination of seeds dusted with chlamydospores was reduced up to 35 % (Baruah *et al.*, 1992). Under field conditions, spreading of spores from smut balls to healthy seeds could cause high abortive grain rate of 75% (Hu, 1985).

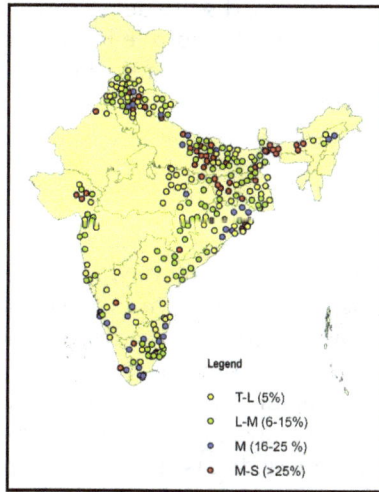

Fig. 1.11. A map showing the distribution and intensity of false smut in different rice growing areas of India (2000-2010)

Cursory perusal of the literature revealed that false smut disease incidence varied widely from region to region depending upon the varieties grown, cultivation practices followed and variation in environmental conditions. In hybrid rice cultivation, incidence of false smut during the rainy season reduced the yield to 3 t/ha as against 6 t/ha (Elazegui *et al.*, 2009). Information from the Production Oriented Survey (POS) reports for the last six years (2005 - 2010) revealed that across the locations, the disease intensity was high in hybrids compared to inbred varieties except in few cases where high incidence was also noticed in inbreds.

3.3. Symptoms: The false smut pathogen, *Ustilaginoidea virens* (Cooke) (Takahashi) affects the young ovary of the individual spikelet and transforms it into large, yellow to velvety green balls (smut ball) and symptoms produced are visible from milky stage onwards. Initially, the smut balls are small in size and remain confined between glumes. They gradually enlarge and enclose the floral parts. Young smut balls are white in colour, gradually change into yellow and both are covered with white colour membrane. Later, the membrane bursts and the colour changes to yellowish orange, olive green and finally greenish black. If infection occurs before fertilization, most of the glumes remain sterile without any visible sign of infection. Typical large, velvety, green smut balls develop when infection occurs after fertilization. The fructifications replacing the grains represent the conidial, pseudosclerotial and sclerotial stages of the pathogen. The pseudosclerotia (green smut balls) consist of mycelial tissue and spore masses, remnants of anthers and portions of palea and lemma. In general only few grains are affected in a panicle but the number may rise up to 100 in case of severe disease incidence.

Wintering fungal structures such as sclerotia and chlamydospores may initiate the disease. Prevalence of conducive environmental factors *viz.*, low temperature (especially low night temperature), high humidity (> 90%) and drizzling condition during flowering time, favours the disease development. The number of rainy days during flowering period have profound role in disease development and is more important compared to the amount of rain fall. In addition to these factors, application of high doses of nitrogen fertilizers during flowering stage of the crop may also lead to outbreak of the disease.

3.4. Management

3.4.1. Cultural practices: Early transplanted rice had higher disease incidence when compared to late planting (Chhottaray, 1991; Dodan and Ram Singh, 1995) while Sanne (1980) reported that false smut can be avoided by early sowing. To escape severe damage, sowing date and heading period could be planned in such a way that flowering should not coincide with rainy period. Use of sclerotia free seeds for sowing and cleaning of bunds may help the farmers to reduce the initial occurrence of the disease. In respect of cultivation practices, furrow irrigated rice cultivation system recorded less disease severity compared to flooded fields. The mechanism behind is the reduction on the survival period of chlamydospores in soil and occurrence of physiological changes in the host plant in response to shift of rice cultivation from anaerobic to aerobic growing conditions.

3.4.2. Host plant resistance: Use of disease resistant varieties is considered as the most effective, cheap and eco-friendly method of controlling the disease. Many workers have evaluated rice varieties/cultures against false smut of rice. A number of varieties have been reported to have moderate to high level of resistance to false smut. Some of the varieties which have been found resistant or tolerant to false smut are Zenith, IR22, IR 26, IR 28, IR 30, Ch 45, Sabrmati, Saket 4, Sona, Vijaya under field condition.

3.4.3. Chemical control: Copper fungicides *viz.*, copper oxy chloride, copper hydroxides are highly effective *U. virens* (Dodan and Ram Singh, 1995; Dodan and Ram Singh, 1996). However, due to the limitations of phytotoxicity, triazole group of fungicides are preferred. Spraying of 0.1% propiconazole (Tilt) (1 gm/lt water) during booting and 50% panicle emergence can satisfactorily reduce the incidence. Application of simeconazole at submerged condition at 3 weeks before heading was also effective against false smut (Tsuda *et al.*, 2006) as per the *in vitro* test conducted against *Ustilaginoidea virens*, propiconazole (Tilt) and Trifloxystrobin 25% + Tebuconazole 50% (Nativo 75WG) are on par in their efficacy and inhibited the fungal growth (100 %) compared to control. The same chemicals are also tested under field conditions at different crop stages *viz.*, booting, 50 % panicle emergence (PE) and 100 % PE across the locations. It is found that application of Tilt and Nativo 75WG at 50 % PE is effective in reducing the disease in terms of per centage of infected panicles/m^2 and per centage of infected spikelets/panicle (unpublished data). False smut of rice is known to occur in almost all the rice growing areas in the world. The pathogen replaces the grains into smut balls and causes substantial quantitative and qualitative losses depnding on the environmental conditions and genetic composition of cultivars and virulence of the pathogens. A considerable progress has been made on various aspects of the disease. However, additional research is required on estimation and prediction of yield losses, pathogenic variability, role of infected seeds in annual reference of the disease, formation of true sclerotia in plains, development of simple and rapid inoculation technique for identifying resistance in rice cultivars, use of biocontrol agents and adjustment of agronomic practices in disease management.

4. Brown Spot or Sesame leaf spot

4.1. Economic impact: This disease was considered to be the major factor contributing to the **Great Bengal Famine** in 1942, resulting in yield losses of 50% to 90% and causing the death of 2 million people. Epidemics in India have resulted in 14-41% losses in higher yielding varieties. Under favourable environment, yield loss estimates ranging from 16 to 40% in Florida, USA were reported.

4.2. Symptoms: The fungus attacks the crop from seedling in nursery to milk stage in main field. Symptoms appear as lesions (spots) on the coleoptile, leaf blade, leaf sheath, and glume, being most prominent on the leaf blade and glumes. The disease appears first as minute brown dots, later becoming cylindrical or oval to circular (Fig. 1.12). The several spots coalesce and the leaf dries up. The seedlings die and affected nurseries can be often recognised from a distance by their brownish scorched appearance. Dark brown or black spots also appear on glumes which contain large number of conidiophores and conidia of the fungus. It causes failure of seed germination, seedling mortality and reduces the grain quality and weight.

4.3. Causal organism: *Helminthosporium oryzae* (Syn: *Drechslera oryzae*) (Sexual stage: *Cochliobolus miyabeanus*) is the causal agent of the disease. *H. oryzae* produces greyish-brown to dark brown septate mycelium. Conidiophores may arise singly or in small groups. They are straight, sometime geniculate, pale to brown in colour. Conidia are usually curved with a bulge in the centre and tapering towards the ends occasionally almost straight, pale olive green to golden brown colour and are 6-14 septate. The perfect stage of the fungus is *C. miyabeanus*. It produces perithecia with asci containing 6-15 septate, filamentous or long cylinderical, hyaline to pale olive green ascospores. It produces C25 terpenoid phytotoxins called ophiobolin A, (or Cochliobolin A), ophiobolin B (or cochliobolin B) and ophiobolin I. Among them Ophiobolin A is most toxic. This breaks down the protein fragment of cell wall resulting in partial disruption of integrity of cell.

Fig. 1.12. Cylindrical or oval shaped brown spots

4.4. Epidemiology: Temperature of 25-30°C with relative humidity above 80 per cent are highly favourable. Excess of nitrogen aggravates the disease incidence. The infected seeds are the most common source of primary infection. The conidia present on infected grain and mycelium in the infected tissue may viable for 2 to 3 years. The fungus may survive in the soil for 28 months at 30°C and 5 months at 35°C. Airborne conidia infect the plants both in nursery and in main field. Maximum flight of conidia takes place at a wind velocity of 4.0 - 8.8 hr. Minimum temperature of 27 -28°C, Relative humidity of 90-99% and rainfall of 0.4 -14.4 mm favoured the dispersal of the conidia to maximum extent. The fungus also survives on collateral hosts like *Leersia hexandra, Arundo donux,* and *Echinochlora colonum.*

4.5. Management: Field sanitation (removal of collateral hosts and infected debris in the field), crop rotation, use disease free seeds, adjustment of planting time and proper fertilization are useful practices to reduce the initial inoculum. Use of slow release nitrogenous fertilizers is advisable. Grow disease tolerant varieties *viz.,* Co44, Cauvery, Bala Bhavani. Treat the seeds with Thiram or Captan at 4 g/kg. Spray the nursery with edifenphos or mancozeb or captafol. Spray the crop in the main field with Edifenphos 500 ml or Mancozeb 1 kg or Captafol 625 g/ha.

5. Sheath rot

Occurrence of sheath rot (ShR) disease became widespread in mid 1970s when the high yielding technology was introduced by planting improved N-responsive cultivars. Today, ShR is a major constraint in rice production in irrigated as well as rainfed, upland rice. The disease has been reported in many rice growing countries and its spread causes concern because of its transmission and dissemination. The causal fungi, *Sarocladium* spp., have biological association with rice tungro virus, mites, stem-borers and mealy bugs and therefore increase the ShR incidence as well as severe yield losses. Scant information is available on the biology of the causal organism both *in vitro* and *in vivo*. Although a few tall cultivars have been identified as resistant, almost all the high-yielding, improved cultivars are susceptible to ShR. Crop losses ranging from 3 to 20-60% were reported in Taiwan in 1980. In India, a yield reduction of 9.6% to 26%, with an average of 14.5%, was reported in 1978.

5.1. Symptom: Initial symptoms are noticed only on the upper most leaf sheath enclosing young panicles. The flag leaf sheath show oblong or irregular greyish brown spots. They enlarge and develop grey centre and brown margins covering major portions of the leaf sheath. The young panicles may remain within the sheath or emerge partially. The panicles rot and abundant whitish powdery fungal growth is formed inside the leaf sheath (Fig 1.13).

Fig 1.13. Oblong or irregular greyish brown spots along with abundant whitish powdery fungal growth

5.2. Causal organism: *Sarocladium oryzae* - (Syn: *Acrocylindrium oryzae)* is the cause of disease: The fungus produces whitish, sparsely branched and septate mycelium. Condiophore is slightly thicker than the vegetative hyphae. Conidia are hyaline, smooth, single celled and cylindrical in shape. It spreads through air and infected seed.

5.3. Epidemiology: Closer planting, high doses of nitrogen, high humidity and temperature around 25-30°C. Injuries made by leaf folder, brown plant hopper and mites aids disease development by retarding panicle emergence and providing entry wounds for the fungus. The disease is most often associated with the presence of stem borers and other forms of injury or insect damage of the flag leaf sheath. The disease appears to be favoured by low nitrogen levels.

5.4. Management: The disease inoculum increases in proportion from year to year, appropriate management of this disease under field conditions with fungicides and biopesticides is essential. If initial symptoms are noticed fungicide Carbendazim 250 g or edifenphos 1 litre may be sprayed to cover a hectare. Application of gypsum 500 kg/ha in equal splits, one at basal and another at active tillering stage also effectively reduces the disease incidence. Alternatively spray neem seed kernal extract (5 %) or neem oil (3 %) and application of *Pseudomonas fluroescens* talc based formulation at 500 g/ha reduces the disease incidence. Since grass hosts, such as *Echinochloa crusgalli* and *Panicum rapens*, in rice fields and bunds, harbour the pathogen, they have to be weeded out.

6. Bakanae

In India, *bakanae* or foot rot is known to cause severe damage in the states of Tamil Nadu, Andhra Pradesh, eastern districts of U.P. and in Haryana. In Punjab, foot rot has become an important constraint to basmati rice production throughout Gurdaspur district 1988. In 1996 and 1997 several reports of the foot rot disease were received and it was observed that the incidence of foot rot varied from 5 to 20 per cent in this region. It has been observed that the basmati rice varieties of basmati being grown in this area vary considerably in their reaction to the disease. In Punjab, Basmati-385 has been found to be more severely infected when compared to Basmati-370, Basmati-386 and Pusa 1.

6.1. Symptoms: In the nursery, the affected seedlings are pale yellowish green, thin and abnormally elongated (several inches taller than the normal plants) (Fig. 1.14). They are normally scattered throughout the seed bed and do not occur in definite patches. Many affected seedlings die before transplantation and those survive produce plants which at the tillering stage are taller than the normal plants and are yellowish. In older plants, the leaves dry up and turn brown. Inside affected stems and lower nodes are discoloured brown. On the outside of dead leaf sheaths, just above water level, a white or pink bloom of fungal mycelium develops. The root system is not affected and the dead plants, when pulled up, tend to snap at the collar. Also typical of the disease is the production of adventitious roots from one or more nodes above water level. In some cases, infected plants flower earlier than the healthy ones. In some cases, the infected plants do not show the typical elongation but are instead severely stunted: this tends to occur at lower temperature and when the soil is rather dry.

Fig. 1.14. Foot rot incidence (thin and abnormally elongated plants) in rice field

6.2. Causal organism: The disease is caused by *Fusarium moniliforme*.

6.3. Epidemiology: The disease is mainly seed-born, infection taking place at flowering. The disease can also be soil-borne. There may also be carryover of the disease in rice straw and stubble. The optimum temperature for the growth of the fungus, *Fusarium moniliforme* is about 27 to 30°C whereas 35°C is the temperature suitable for seedling growth and for infection.

6.4. Management: As the disease is mainly seed-borne, seed treatment with fungicides should be followed. The seed to be used for the sowing of next year crop should be collected from the disease-free fields.

7. Stem rot

Significant climate changes over the past few years have led to 'alarming' increase in the hitherto minor or unknown disease that are causing wilting and other effects on the crops in different parts of the Andhra Pradesh State. For instance, paddy crop in over 45,000 hectares in East Godavari district was infected by stem-rot disease in just three days during the kharif season 2007. These fungi overwinter as small resting structures, the sclerotia, in infected crop residue or free in the soil. In severe cases, the infected plants become weak and fall to the ground, causing incomplete rice grain formation that leads to high losses. The disease also causes decay of the leaf sheath and culm, which contributes to lodging and losses in yield. Incomplete grain formation lowers the grain milling quality due to the light and chalky character of the grain. Estimates of yield loss due to the disease range from 10 to 75%.

7.1. Symptoms: The first symptoms are generally observed in the field after the mid- tillering stage. Initially, the disease appears as a small, blackish, irregular lesion on the outer leaf sheath near the water line. The lesion enlarges as the disease progresses with the fungus penetrating into the inner leaf sheaths. Eventually, the fungus penetrates and rots the culm while the leaf sheath is partially or entirely rotten. Infection of the culm may result in lodging, unfilled panicles, chalky grains, and in severe cases, death of the tiller. Brownish-black lesions appear and finally one or two internodes of the stem rot and collapse. Upon opening the infected stem, dark greyish mycelium may be found within the hollow stem and numerous tiny, black sclerotia are embedded all over the diseased leaf sheath tissues (Fig 1.15). Sclerotia and mycelium of the fungus are generally present inside the infected culms. The presence of sclerotia is usually a positive and easy way of diagnosing the disease. Attack on the stems increases in intensity as the plants approach maturity and reaches its peak during harvest time. Weakened stalks break during this stage and plants lodge making harvest difficult. Plants infected early produce lower yields.

7.2. Causal organism: *Sclerotium oryzae* (Catt.) is the causal agent of the disease

7.3. Epidemiology: The fungus survives between crops as sclerotia in crop debris or in soil. After flooding, sclerotia float to the surface of the paddy water where they infect rice sheaths at the water line. Sclerotia are produced abundantly on diseased tissue as the disease progresses and the rice approaches maturity. The fungus also produces conidia and ascospores on infected plants that may serve as an additional source of inoculums.

Fig 1.15. Presence of tiny, black sclerotia found

7.4. Management: Application of total amount of nitrogen in split doses and avoids excess nitrogen application. In stem rot infected fields, final nitrogen application should be avoided. Brown plant hopper damage to the crop should be checked which in turn reduces the stem rot incidence. Destruction of infected stubbles, infected plants, weeds, etc. should be followed. By spraying of Iprobenphos 48 EC @ 2ml/l or carbendazim 50 WP @ 1g or Thiophanate-Methyl 70 WP @ 1g may check the disease.

8. Grain discolouration

Grain discolouration is a serious problem in hybrid rice seed production plots. This might be due to the clipping-off of flag leaf and foliar application of gibberellic acid to encourage pollination and obtain a higher hybrid seed-set in cytoplasmic male sterile female parental lines. High atmospheric humidity and rainfall accompanied by cloudy days during flowering favours grain discolouration. When infection is by field fungi, the main effects are reduced viability and grain quality. Seedling blights and other diseases may occur when the seeds are planted. When the infection is by storage fungi, besides a reduction in viability and grain quality, there may also be production of toxins. Injuries from insects in the field, or from wind and rainstorms during storage, may also increase grain discolouration. Spikelet sterility and grain discolouration in Andhra Pradesh was suspected to be associated with mite, saprophytic fungal species and white-tip nematode.

8.1. Management: The fields can be protected from grain dicolouration by application of propiconazole 25 EC (1ml/l) in the evening hours around panicle initiation to flowering period (DRR, 2002). This application can be limited to cover only emerging panicles.

9. Bacterial Leaf Blight

Bacterial blight (BB) of rice caused by *Xanthomonas oryzae* pv. *oryzae*, remains a major production constraint in rice cultivation. In Punjab and Haryana states of India, major epidemics occurred in 1979 and 1980; severe kresek was observed and total crop failure was reported (POS, 1979 & 1980). The disease was again reported in epidemic form during 1998 in Pallakad district of Kerala and since then it has become endemic in that region. This disease is a major problem in *kharif* season (wet season) crop.

8.1. Economic impact: The extent of yield loss depends on the growth stages of the crop at which it is infected, the level of susceptibility of the cultivar, season, climatic condition and level of nitrogen fertilizer applied which lead to rapid buildup of the disease. Damage may be due to partial or total blighting of the leaves (leaf blight phase) or due to complete wilting of the affected tillers (Kresek phase) leading to unfilled grains. This disease is a problem of rainy season (kharif season). In general, late infection results in only slight reduction in yield, but the early infection leading to 'kresek' (wilt) causes far heavier losses and sometimes, the entire crop may be lost. Generally, the stage between maximum tillering and booting is highly sensitive to disease infection, as it affects the yield significantly in terms of filled grain weight per hill and total yield. In a study, it was observed that when the disease occurred at the maximum tillering stage and onwards, the yield reduction approached 70% in highly susceptible cultivars like TN1 and upto 20% in field tolerant cultivars like IR20 whereas post-flowering occurrence of the disease resulted in 22% and 9% yield reduction in TN1 and IR20, respectively (Reddy, 1974).

In a study on extent of yield loss in 19 different varieties, the yield loss varied from 6.12% in CR-44-35 to 74.20% in the cultivar Bala. In general, the reduction in yield is due to increase in number of chaffy/ unfilled grains and reduction in grain weight and partly due to reduction in panicle number/m^2 and reduction in panicle length. It was also found that different yield components were affected to varying degrees in different cultivars.Depending on the stage of infection and severity of the disease under natural condition, the extent of loss has been reported to vary from 6-60%. Some other workers have reported yield loss up to 50% depending on the variety, severity and stage of infection.

In India, bacterial blight is considered as a serious production constraint especially in irrigated and rainfed lowland ecosystem. In Punjab and Haryana states of India, major epidemics occurred in 1979 and 1980; severe kresek was observed and total crop failure was reported. The disease was again reported in epidemic form during 1998 in Pallakad district of Kerala and since then it has become endemic in that region. This disease is a major problem in *kharif* season (wet season) crop in rice growing regions of Punjab, Haryana, Uttarakhand, Bihar, West Bengal, Tripura, Assam, Tamil Nadu, eastern Uttar Pradesh and Andaman and Nicobar islands; coastal areas of Andhra Pradesh and Kerala and parts of Maharashtra, Chhattisgarh, Gujarat, Himachal Pradesh and Karnataka (Fig. 1.16).

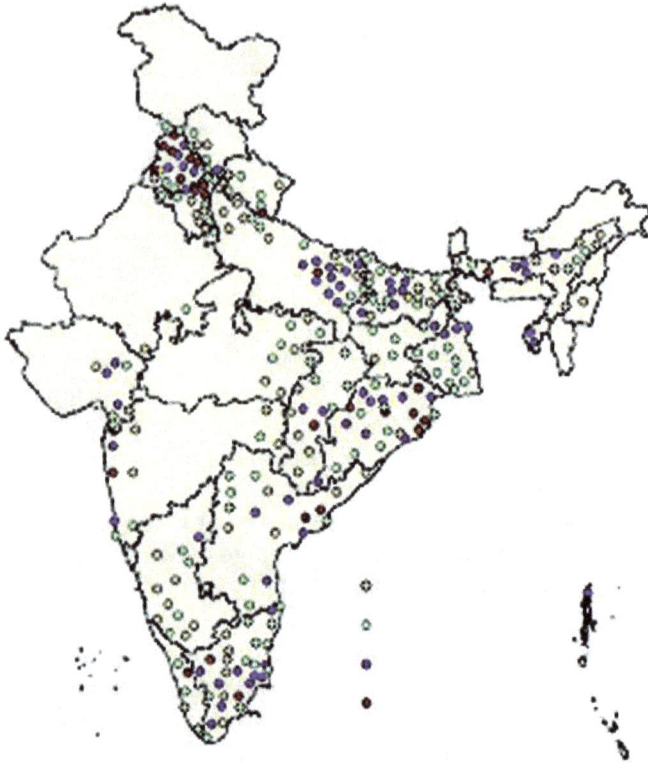

Fig. 1.16. A map showing the distribution and intensity of bacterial blight of rice in different rice growing areas of India

8.2. Symptoms: This is a typical vascular disease and has three distinct phases of symptoms.

8.2.1. Leaf blight Phase: Leaf blight phase is most common. The symptom starts as water soaked lesions on the tip of the leaves and increases in length downwards. Initially, the lesions are pale green in colour and later turn into yellow to straw coloured stripes with wavy margins (Fig. 1.17). The lesions adjoining the healthy part show water soaking. Lesions may start at one or both the edges of the leaves. Occasionally, the linear stripes may develop anywhere on the leaf lamina or along the midrib with or without marginal stripes. As the disease advances, the lesion covers the entire leaf blade, turns white and later becomes grayish or blackish due to growth of various saprophytic fungi. In humid areas, on the surface of the young lesions, yellowish, opaque and turbid drops of bacterial ooze may be observed during early morning. They dry up to form small, yellowish, spherical beads on the lesions.

Fig. 1.17. Leaves showing typical bacterial blight symptoms

8.2.2. Kresek Phase: The most destructive phase of the disease in the tropics is 'kresek' or wilt phase (Fig. 1.18) resulting from early systemic infection in the nursery or from seed infection. The leaves roll completely, droop, turn yellow or grey and ultimately the tillers wither away. In severe cases, the affected hills may be completely killed.

Fig. 1.18. Kresek or wilt phase of the disease

8.2.3. Pale yellow leaf phase: This phase of the disease has been reported from Philippines. Some of the youngest leaves in a clump may become pale yellow or whitish. The diseased leaves later wither, turn yellowish brown and dry up. The pale yellow phase has not been reported from other countries

8.3. Causal organism: Bacterial blight of rice is caused by *Xanthomonas oryzae* pv. *oryzae* (Ishiyama) Swings *et al*. According to the new classification system, the bacterium has been placed in the family Xanthomonadaceae, order Xanthomonadales, class Gammaproteobacteria and phylum proteobacteria in the domain Bacteria. *X. oryzae* pv. *oryzae* (*Xoo*) is gram negative, non-spore forming and rod shaped bacterium which is motile with a single polar flagellum (Fig. 1.19). Individual cells vary in length from approximately 1-2 m and in width from 0.4-0.7 m. Colonies on solid culture medium are round, convex, mucoid and yellow in colour (Fig. 1.20) due to production of a non-diffusible yellow pigment called xanthomonadin (a brominated, aryl polyene pigment), characteristics of the genus. Very rarely the pigment deficient albino form of the pathogen is detected. The bacterium produces copious capsular extra-polysaccharide (EPS), which is important in the formation of droplets of bacterial exudates from the infected leaves, providing protection from desiccation and aiding in wind- and rain-borne dispersal. The bacterium grows slowly in the culture medium. Among the different carbon sources, sucrose is most favourable followed by glucose, mannose, galactose and maltose. Among the nitrogen sources, L-glutamic acid is most favourable followed by L-cysteine.

Fig. 1.19. Transmission electron microscopic picture showing single cell of *X. oryzae pv. oryzae* with single polar flagellum

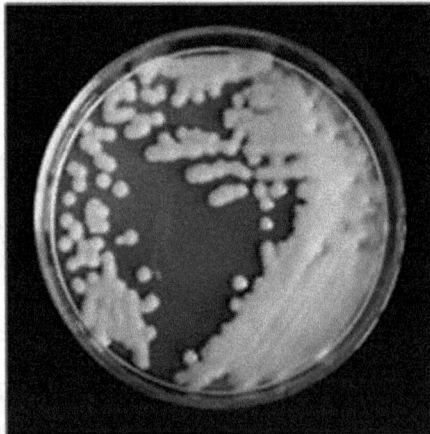

Fig. 1.20. Characteristic colonies of *X. oryzae pv. oryzae* on culture plate

8.4. Epidemiology: A number of factors influence the development of the disease. This is mainly a disease of monsoon season. Flooding and water logging conditions encourage the disease development. Strong winds or cyclone accelerates the disease development especially when accompanied with rains. Strong winds not only cause wounds that hasten up infection and enhance severity of the disease but also aid in dissemination of the pathogen. High humidity (>80%), rain and cloudy condition accelerate the disease development. A moderate amount of rainfall evenly distributed during the crop season can bring about an epidemic. Reduced duration of sunshine or cloudy condition favours the disease development. The intensity of the disease is more in that part of the field which is under shade. A moderate temperature of 28-30°C coupled with high humidity helps in the build-up of the disease. It has also been found that alternate irrigation and drainage aggravates the disease in the fields. In India, combination of cloudy and rainy weather or drizzling condition, floods, cyclone or strong winds and moderate temperature of 28-30°C favour the rapid build-up of the disease. Excess of nitrogen, especially in organic form and late top dressing increases the disease. The effect of nitrogen on disease is perhaps mainly due to the enhanced vegetative growth of the plants, which influences the humidity and dissemination of the pathogen.

8.5. Management

8.5.1. Cultural practices: The importance of cultural practices is being reemphasized as an essential component in the integrated disease management practice. The following practices have been found to reduce the bacterial blight incidence. Keep the field free from weeds (collateral hosts). Species of *Cyperus* and *Leersia* must be removed from the field. Infected plant debris, self-sown rice plants and ratoons have to be ploughed down and the field is to be irrigated a month before sowing and transplanting in order to bring down the inoculum potential. Infected straw or chaff should not be left *in situ* or applied to the field. Rather, it should be burnt. Several wild rice species occurring in ponds, ditches and canals around rice fields serve as potential source of infection to the cultivated crop. Eradication of wild rice near the field will minimize the inoculum potential for the cultivated crop. The time of sowing and transplanting should be adjusted as far as possible in such a way that the period of tillering to maximum tillering do not coincide with peak rainfall period. Pruning of leaves either at the time of transplanting or subsequently should be avoided. Field to field irrigation should be discouraged. Growing of rice crops under the shade should be avoided. Healthy seeds collected from the disease free fields should be used for raising the nursery. If such seeds are not available, disinfection of seeds should be done with antibiotics (mentioned earlier) or with hot water treatment. Soaking of seeds in water for 12 hours followed by exposure to hot water at 53°C for 30 minutes is enough to eradicate the seed borne inoculum. Seedlings should be raised in upland nurseries as seedlings from raised seedbed tend to reduce the inoculum potential in the transplanted field. Inundation of nursery or main field up to maximum tillering stage is to be avoided by giving only shallow irrigation. Water from infected fields should not be let into fields, which are free from infection. Controlling insects to avoid physical injury will reduce the spread of the disease.

8.5.2. Use of resistant cultivars: Use of resistant cultivars is the most economic and environment safe strategy for the management of the disease. Although several BB resistance genes have been identified and characterized, the effectiveness of these resistance genes varies due to difference in virulence profile of the pathogen in different geographical regions. Moreover, wide spread use of few resistant genes has resulted in narrow genetic base and as a result new and more virulent forms of the pathogen have appeared. For example, large scale and long term cultivation of varieties carrying resistance gene *Xa4* has resulted in significant changes in pathogen population of *X. oryzae* pv. *Oryzae* and in many places, rice varieties with only *Xa4* have become susceptible to BB. In India also, a large number of rice varieties with some level of resistance to BB have been released and some of these varieties are still grown at several places in the country. Many of these varieties are with the BB resistance gene *Xa4*. However, due to changes in the pathogen population structure, majority of these varieties have become moderately to highly susceptible to BB. Similarly, there are reports that the near isogenic line carrying resistant gene *Xa21* showed susceptibility to some isolates of BB pathogen. However, the resistant variety Ajaya (IET 8585) which was developed at AICRIP and released in 1992 still exhibits moderate to high level resistance to most of the strains of the pathogen.

Nutrition management: Increased application of nitrogen is accompanied by a higher incidence of bacterial blight. Therefore, a judicious level of fertilization should be aimed at without sacrificing the yield. An economic level of 60-80 kg N/ha with required level of potassium may be recommended in endemic areas during the wet season. The nitrogen should be applied in 3-4 splits. The application of nitrogen at tillering stage should be avoided as it aggravates the disease.

8.5.3. Chemical control: Several chemicals including antibiotics have been tested and used for control of bacterial blight disease but under Indian conditions none has proved highly satisfactory. The chemicals like Sankel (Nickel dimethyl-dithiocarbamate), Phenazine (Phenazine 5-oxyde), Cellomate (Acetylene-dicarboxa-mide or cellocidin), Streptomycin, Chloramphenicol etc., recommended in Japan, have been found unsatisfactory in India. However, some of these chemicals may offer partial control of the disease especially when disease pressure is low. Two chemicals, ATDA (2-amino-1,3,4- thiadiazole) and TF-130 were found highly effective against the disease in Japan and other countries and the problem of bacterial blight control seemed to have been solved by these chemicals. But it was found that the subacute mammalian toxicity of these chemicals was unexpectedly high and as a result these chemicals were withdrawn. However, the following measures have been tested in many places and can partially reduce the BB severity. Up to 95% of the seed infection can be eradicated by soaking the seeds for 12 hours in 0.025% solution of Agrimycin-100 (an antibiotic containing 15% streptomycin and 1.5% terramycin) plus 0.05% wettable ceresan and then transferring the seed to hot water at 52-53 0C for 30 minutes. Overnight soaking of infected seeds in 100 ppm streptocycline (streptomycin 12% + chlorotetracycline hydrochloride 1.5%) solution can also effectively eradicate the seed infection. Spraying twice with 250 ppm of Agrimycin- 100 can effectively reduce the disease intensity and has been advocated for checking secondary spread of the disease.

9. Rice tungro disease

Rice tungro disease (RTD) caused by rice tungro virus (RTV) is one of the major destructive diseases that cause huge damage to the rice crop. Leaf yellowing symptoms in rice has been reported from several countries such as Philippines, Malaysia, Bangladesh, Indonesia and India, in south East Asia. The viral nature of tungro and its transmissibility with leaf hoppers was demonstrated in Philippines in 1963. Tungro is a Philippino word meaning degenerated growth. Rice tungro virus (RTV) can appear at any time on rice plant right from seedling stage. During the last four decades tungro is increasingly noticed in several rice growing states in India. During 1975–2007, tungro disease occurrence caused considerable damage to rice production in 61 districts. All these tungro-affected districts were under-irrigated and rain fed lowland ecosystems. Severe tungro damage was reported only from Andhra Pradesh, Bihar, Punjab and Tamil Nadu (Fig. 1.21). Tungro incidence was recorded in farmers' fields for four years in Nellore (1984, 1988, 1990 and 1992); three-years each in East Godavari (1986, 1990 and 1995), Telangana area (2003, 2005 and 2007), Puducherry (1991, 1999 and 2000-2003 and 2007), and South Arcot and Thanjavur (1984, 1987 and 1988); and two years each in Burdwan (1999 and 2001), Chengal pattu (1998 and 2000), Kanyakumari and Tirunelvelli (2005-2006), Chittoor (1984 and 1992), Krishna (1977 and 1985), North Arcot (1987 and 1988) and West Godavari (1990 and 1995), Gurdaspur and Amritsar (1998) districts.

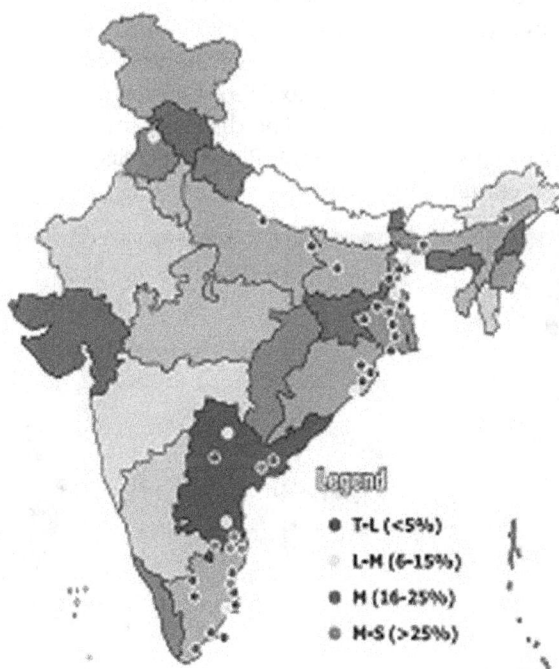

Fig. 1.21. Rice tungro disease distribution in India

9.1. Economic impact: The loss from tungro epidemics steadily increased during 1979- 1980. Three major epidemics in farmers' fields (1984, 1988 and 1990) caused severe quantitative and monetary losses. Each of the other two epidemics (1987 and 1998) led to a similar loss of more than a million tonnes in rice production but showed a steady increase in loss in terms of real value. An epidemic outbreak of tungro during 2001 in only three districts of West Bengal caused a paddy or unmilled rice production loss of 0.5 million tonne that amounted to Rs. 2911 millions at current prices. This study clearly demonstrated that tungro epidemics could cause a maximum yield loss of 53% in a district, 23% in any one state and 2% in all-India rice production.

9.2. Symptoms: In the initial stages of tungro virus infection interveinal chlorosis and twisting of infected leaves are noticed. The type of discolouration of leaves mostly depends on the variety. Yellowing of infected leaves normally starts from the tip of the lower leaves with the development of intense colouration. Generally, infected leaves in japonica cultivars show yellow shade, while indica cultivars show orange shade. Small, rusty, necrotic spots may also be seen on the discoloured areas of the older leaves. Tungro causes stunting in rice plant. The extent of stunting is dependent on the variety and age of the plant at the time of infection. Younger plants get infected more than the older plants during early growth period. Tillering is reduced in the infected plants. Occasionally profuse tillering may be seen in some cultivars infected by tungro virus. Leaves are discoloured from green to yellow to orange (Fig. 1.22a &b) or yellowish-brown colour. Infected plants exhibit poor root growth and delayed flowering and maturity. The panicles are often small, sterile and incompletely exerted. Grains are also covered with dark blotches and grain weight is reduced. The disease reduces number and length of panicles, number of spikelets, and yield; it also decreases the grains filling, grain weight and starch content.

Fig. 1.22. (a) Rice tungro disease infected filed; (b) Interveinal chlorosis caused by Rice tungro virus

9.3. Causal organism: At present, rice tungro considered to be caused by two viral particles namely, spherical (RTSV) and bacilliform (RTBV) particles. RTSV is believed to be responsible for the transmission of RTBV by green leafhoppers.

9.4. Disease Management

9.4.1. Host plant resistance: Host plant resistance is the most effective and economical method to control plant diseases and insect pests. The resistant varieties create adverse effect on the survival and development of the nymphs and fecundity of females and hatching of the eggs. Ever since the identification and confirmation of tungro virus incidence in India, attempts were made to identify donors and develop breeding lines resistant to tungro disease. Several field evaluation methods were developed to rapidly screen breeding lines. The cultivars like Latisal, Kataribhog, Ambemohar 102, Pankhari 203 and Tjina were found to be resistant to both virus and vector. Vikramarya (IET 7302), a medium duration variety from the cross between RPW 6-13 and Ptb 2 having long bold grains was found to be resistant to rice tungro disease and tolerant to brown plant hopper and sheath blight disease. The variety Nidhi (IET 9994) was also found highly resistant to rice tungro disease. These varieties have been released for cultivation in tungro endemic area. A number of high yielding cultures with resistance or tolerance to tungro and/or green leafhoppers have been released for commercial cultivation in specific rice growing areas.

9.4.2. Sowing time: The seedlings are susceptible at the time of transplanting. Protection of seedlings with insecticides in seed bed as well as in early stages of transplanting is crucial in the management of tungro virus disease. In a rice-based cropping pattern, especially in tungroendemic areas, inclusion of a non-host crops like legume, oilseed, fibre or tuber crop in the crop rotation will help to interrupt the disease perpetuation and prevent vector build-up. Diseased plants serve as a source of inoculum. Thorough and periodic rouging and removal of affected plants reduce the inoculum source and effectively contain the spread of the disease. Ratoons and stubbles in the affected rice fields also serve as the potential source of virus inoculum and the vector build up. Ratoon cropping may be discouraged in endemic areas. Destruction of stubbles by ploughing the affected fields checks the spread and severity of the disease.

9.4.3. Chemical control: The spread of rice tungro disease can be checked indirectly by controlling the vector by suitable pesticide application. As the plants are more vulnerable to tungro virus infection during early stages of growth, chemical protection of the nursery effectively reduces green leafhopper population and there by minimizes the build-up of virus inoculum as well as the pace of transmission. Incorporate carbofuran 3 G @ 30 to 35 kg/ ha or phorate 10 G @12 to 15 kg / ha of nursery in top 2-5 cm layer of the soil before sowing sprouted seeds. If such incorporation is not possible, broadcast the recommended insecticides 4 to 5 days after sowing in a thin film of water and allow this water to seep completely. To protect the transplanted crop, foliar spray application may be made with monocrotophos 36 EC @ 1.0 L/ ha or carbaryl 50 WP @ 2.0 kg/ha or phosphamidon 85 WSC @ 0.65 L/ha on 15th and 25th day after sowing, depending on green leafhopper population. Usually gall midge and stem borer also appear in the field along with green leafhoppers as a complex during later crop growth stages. In such a situation, carbofuran 3 G @ 25 kg /ha or phorate 10 G @ 7.5 kg /ha may be applied 10 days after planting.

REFERENCES

Atia, M. M. M. 2004. Rice false smut (*Ustilaginoidea virens*) in Egypt. Jour. of Plant Dis. and Prot. 111: 71-82.

Baruah, B. P., Senapoty, D., and Ali, M. S. 1992. False smut: A threat to rice growers in Assam. Indian J. Mycol. Pl. Pathol. 22: 274-277.

Biswas, A. 2001. False smut disease of rice: A review. Environ and Biol.19: 67-83.

Chhottaray, P. K. 1991. Doctor of Philosophy Thesis, Orissa University of Agriculture and Technology, Bhubaneswar, Orissa, 175.

Dodan, D. S., and Ram Singh. 1995. Effect of planting time on the incidence of blast and false smut of rice in Haryana. Indian Phytopathol. 48: 185-186.

Dodan, D. S., and Ram Singh. 1996. False smut of rice: Present status. Agri Rev. 17: 227-240.

DRR (1975-2010). Production Oriented Survey (POS), Directorate of Rice Research (ICAR), Rajendranagar, Hyderabad-500 030, Andhra Pradesh, India.

Elazegui, F. A., Castilla, N. P., Nieva, L. P., and Vera Cruz, C. M. 2009. Notes on Rice diseases. http://www.knowledgebank.irri.org/IPM/index.php/false-smut/economic-importance.

Hu, D. J. 1985. Damage of false smut to rice and effect of the spores of *Ustilaginoidea virens* on germination of rice seeds. Zhejiang Agril. Sci., 4: 164-167.

Ou, S. H. 1985. Rice Diseases. 2nd Edition, Commonwealth Mycological Institute, Kew, Surrey, England, 380.

Padmanabhan, S. Y. 1965. Estimating losses from rice blast in India. In the rice blast disease: Johan Hopkins Press, Baltinoie, Maryland. 203-221.

Rajan, C. P. D. 1987. Estimation of yield losses due to sheath blight disease of rice. Indian Phytopath. 40: 174-177.

Reddy, A. P. K. 1974. In: International Rice Research Conference, IRRI, Los Banos, Philippines, April 22-25, 1974.

Rush, M. C., Shahjahan, A. K. M., and Jones, J. P. 2000. Outbreak of false smut of rice in Louisiana. Plant Dis. 84: 100

Sanne, G. 1980. Studies on false smut disease caused by *Ustilaginoidea virens* on Paddy in Karnataka, India. Intern. Rice Res. Newsl. 5: 4-5.

Tsuda, M., Sasahara, M., Ohara, T., and Kato, S. 2006. Optimal application timing of simeconazole granules for control of rice kernel smut and false smut. J. Gen. Plant Path. 72: 301-304.

Diseases of Field and Horticultural Crops
Editor-in-Chief: **P. Chowdappa**
Published by: **Daya Publishing House, New Delhi**

2

Diseases of Maize

Meena Shekhar and Sangit Kumar

Maize is the third largest planted crop after wheat and rice in India. It is a leading feed crop but is also an important staple food in some pockets of the country. Diversified uses of maize for starch industry, corn oil production, baby corns, popcorns, etc., and potential for exports has added to the demand of maize all over world. In India, maize is grown in a wide range of environments. The crop is very popular in the low and mid-hill areas of the western and northeastern regions. The maize potential states are Andhra Pradesh, Gujarat, Karnataka, Madhya Pradesh, Maharashtra, Rajasthan, Tamil Nadu, Uttar Pradesh, Bihar and West Bengal. While international trade accounts for only 12 per cent of world maize production, it represents over one-third of total cereal trade. Global trade in maize has increased significantly over the past two decades, from 55 million tonnes to around 80 million tonnes, with the fastest expansion taking place more in recent years. The structure of the world maize market can be characterized as one with a high level of concentration in terms of exports but very low concentration on the import side. The main reason for this development is the fact that those countries which usually have significant maize surpluses for exports are relatively few in number, while those relying on international markets to meet their needs for domestic animal feeding purposes by importing maize (as a primary feed ingredient) are many.

In order to increase the production and productivity of maize, the government adopted the new approach for area expansion for maize in view of serious competition from food and cereal crops. The program envisages transfer of improved technology through demonstration on improved crop production technology and Integrated Pest Management training programs, seeds production programs and distribution of insecticides, pesticides, weedicides and other inputs, etc. A vast number of plant pathogens from bacteria and fungi cause diseases in maize crops in India. Their effects range from mild symptoms to catastrophes in which large areas planted to crops are destroyed. Plant pathogens are difficult to control because their populations are variable in time, space, and genotype. Most insidiously, they evolve, often overcoming the resistance that may have been the hard-won achievement of the plant breeder. In order to combat the losses they cause, it is necessary to define the problem and seek remedies. At the biological level, the requirements are for the speedy and accurate identification of the causal organism, accurate estimation of the severity of disease and its effect on yield, and identification of its virulence mechanisms is necessary. Disease may be minimized by the reduction of the pathogen's inoculum, inhibition of its virulence mechanisms, and promotion of genetic diversity in the crop (Richard *et al.*, 2005). Conventional plant breeding for resistance has an important role to play that can now be facilitated by marker-assisted selection. At the political level, there is a need to acknowledge that plant diseases threaten our food supplies and to devote adequate resources for their control for food security.

The problem of plant disease, particularly in developing countries, is exacerbated by the paucity of resources devoted to their study. In part, this may be a result of difficulty of quantifying plant disease and relating this to the failure of crops to reach achievable yields. Other difficulties faced by plant pathologists are the reliable identification of the causal agents of a disease upto levels-species, formae species and races-that are appropriate to their properties as plant pathogens and therefore relevant to their control; identification of sources of inoculum; the wide host range of some pathogens; lack of availability of sources of durable genetic resistance; and the slow acceptance in some cultures of the value of transgenic traits in plant improvement. Tropical climates, prevalent in most part of the country, cause particular difficulties in that they allow continuous cropping with the consequent build-up of inoculum. In contrast, the winters of temperate climates reduce inoculum levels owing to the reduced availability of hosts and inclement weather.

Table 2.1. Estimation of losses in due to major diseases of maize in India

Diseases	Per cent loss	Annual grain loss (tones)
Seed and seedling blights	0.1	13,510
Downy mildews	2.1	2,83,710
Foliar diseases (Turcicum leaf blight, Maydis leaf blight, Phaeosphoria leaf spot, Rusts, Brown stripe downy mildew)	5.0	6,75,500
Stalk rots, root rots and ear rots	5.0	6,75,500
Sheath blights, smuts, Viruses and others	1.0	1,35,100
Total	13.2	17,83,320

Based on research efforts for the last four decades, under the aegis of All India Coordinated Maize Improvement Project, 16 out of 62 diseases adversely affecting this crop have been identified as a major constraint. Banded Leaf and Sheath Blight, Pythium stalk rot, Bacterial stalk rot, Post Flowering Stalk Rots, Polysora rust, common rust and downy mildews are the major threat to the potential yield of maize. The total loss in economic products of the crop due to diseases has been estimated to the tune of Rs. 17, 83, 320 (Table 2.1.) and in terms of per cent losses is 13.2% (Payak and Sharma, 1985). Some economically important diseases of maize, their distribution, symptoms and pathogen biology management practices are described in this chapter(Table 2.2).

Table 2.2. Major diseases affecting the maize crop

S.No	Name of Disease	Affected part	Pathogen	Losses
1.	Seed and Seedling Blights	Seedlings	Species of *Fusarium, Rhizoctonia, Pythium Penicillium, Aspergillus, Acremonium, Cephalosporium* etc.	By reducing plant stand
2.	Turcicum Leaf Blight	Foliar part at knee high stage	*Exserohilum turcicum*(Pass) Leon. & Sugs	Severe infection causes a premature death of plant 83 % yield reductions
3.	Maydis Leaf Blight	Foliar part at knee high stage	*Drechslera maydis*Niskado Syn. *H. maydis*	Severe infection causes a premature death of plant
4.	Common Rust	Foliar part	*Pucciniasorghi* Schw	Leaves may turn chlorotic and senesce prematurely when severe disease. Yield losses as high as 50% have been recorded
5.	Polysora Rust	Foliar part	*Puccinia polysora* Underw.	Yield losses in excess of 45% have been recorded
6.	Brown Stripe Downy Mildew	Foliar part	*Sclerophthora rayssiae* var. *zeae* Payak and Renfro	Losses ranging from 20-100% depending on the species and host cultivars in maize.
7.	Sorghum Downy Mildew	Foliar part & in severe case tassels.	*Peronosclerosporasorghi* (Weston &Uppal) Shaw	Severe outbreaks have occurred in all over world including India, The yield loss as high as 90% has been reported.
8.	Rajasthan Downy Mildew	Foliar part	*Peronosclerospora hetropogoni* Siradhana *et.al*	Very serious in Rajasthan & Yield losses in excess of 70% have been recorded.
9.	Brown Spot	Leaf & mid rib	*Physoderma maydis* Miyake	In India it is not so important, however, when conditions are favorable the disease can develop and greater incidence and cause alarm situations.
10.	Banded Leaf and Sheath Blight	Started from lower leaf & whole plant including cob	*Rhizoctonia solani* f. sp. *Sasakii* Exner	Yield losses close to 40% have been attributed to BLSB
11.	Pythium Stalk Rot	Stalk at pre flowering stage	*Pythium aphanidermatum* (Eds) Fitz	The yield reduction in susceptible genotypes has been reported to the tune of 100%
12.	Bacterial stalk rot	Stalk at pre flowering stage	*Erwinia chrysanthemi* p.v. *zeae* (Sabet) Victoria, Arboleda & Munoz *Fusarium moniliforme* Sheld	Early infection led to complete death of plant, infection at flowering stage resulted in 92.2%, yield
13.	Post Flowering Stalk Rots	Stalk at post flowering stages	*Macrophomina phaseolina* (Goid) Tassi*Cephalosporium maydis*	loss and late infection in 57.0 and 36.3% yield losses
(a)	Charcoal Rots	Stalk at post flowering stages	Samara, Sabeti & Hingorani *Macrophomin aphaseolina* (Goid) Tassi	
(b)	Fusarium Stalk Rots	Stalk at post flowering stages	*Fusarium moniliforme*Sheld	
(c)	Late Wilt	Stalk at post flowering stages	*Cephalosporium maydis* Samara, Sabeti & Hingorani	

1. Estimates of yield losses in experimental field: The experiment conducted on avoidable yield losses against major diseases were conducted in different 'hot spot' locations of the country during 2007 – 11. Genotypic variation on the extent of the avoidable yield losses was observed due to BLSB, PFSR and TLB (Table 2.3). In BLSB the losses was from 29.02 to 36.23% in Pant Sankul Makka -3 at Pant Nagar, in PFSR caused by Charcoal rot the losses was from 19-42% at Hyderabad & Delhi and losses due to Fusarium rot was from 36.23 – 38.93% in Mahi Dhawal at Udaipur, in TLB the avoidable losses was 11.00-29.84% in different genotypes at Arabhavi. The extent of losses in TLB was from 10.58 – 14.03% in HM 8 and Bio 9681 at Delhi whereas the yield loss due to SDM was observed 54095% in genotype COH (M) -5 at Coimbatore.

Table 2.3: Avoidable yield losses in experimental field

Genotype	Disease	Location	Yield loss (%)	Range of losses
Pant Sankul Makka – 3	BLSB	Pant Nagar	29.02	BLSB -
Pant sankul makka – 3	BLSB	Pant Nagar	36.23	29.02 – 36.23
30V92	PFSR(*M. phaseolina*)	Hyderabad	20.74	PFSR–
HM 9	PFSR(*M. phaseolina*)	Delhi	31.40	C. Rot
30V92	PFSR (*M. phaseolina*)	Hyderabad	19.26 %	19- 42%
30V92	PFSR (*M. phaseolina*)	Hyderabad	20.11	
Pro 311	PFSR(*M. phaseolina*)	Delhi	42 %	
Mahi Dhawal	PFSR(*F. moniliforme*)	Udaipur	36.23	PFSR (F. Rot)
Mahi Dhawal	PFSR(*F. moniliforme*)	Udaipur	38.93 %	36.23-38.93%
DHM – 2	TLB	Arabhavi	14.26	
EH 434042(Arjun)	TLB	Arabhavi	11.97	TLB –
Bio 9681	TLB	Arabhavi	29.84	11.00 – 9.84%
DHM – 2	TLB	Arabhavi	20.84 %	
EH 434042 (Arjun)	TLB	Arabhavi	11.00 %	
Bio 9681	TLB	Arabhavi	16.55	
HM 8	MLB	Delhi	14.03	MLB –
Bio 9681	MLB	Delhi	10.58	10.58-14.03%
COH(M)-5	SDM	Coimbatore	54.95	SDM - 54.95%

BLSB – Banded leaf and sheath blight; PFSR – Post flowering stalk rot; TLB- Turcicum leaf blight; MLB-Maydis leaf blight; SDM – Sorghum downey mildew.

2. Seed and Seedling Blights

A variety of pathogens are associated with seed rots and seedling blights including species *Pythium, Fusarium, Acremonium, Penicillum, Rhizoctonia, Macrophomina, Sclerotium* etc. Species of *Fusarium, Rhizoctonia, Penicillium, Aspergillus, Acremonium, Pythium,* and *Cephalosporium* etc. are also known to be associated with disease. The disease is prevalent in compacted and poorly drained and wet soils when temperature exists between 10-15°C at planting time. Most of seed rots and seedling blight on maize are more severe in wet soil, in low lying areas in a field & in soil that have remain wet for an extended period of time favours seed rots

& seedling blight. Disease severity is affected by planting depth, soil type, seed quality, mechanical injury to seed etc. Germinating maize seedlings are attacked by number of soil borne or seed borne fungi that causes seed rot & seedling blight consequently plant stand is reduced.

Symptom appears as brown sunken lesions on mesocotyl, rotting at collar region leading to wilting & toppling of seedlings (Fig. 2.1). The disease poses a serious problem in temperate areas by reducing plant stand. However, they are not a serious threat in the major tropical environments of India because of rapid emergence of seedlings. Low quality seed also produce seedlings that are weak and survive poorly in cold and wet soils. The problem becomes severe with the use of old seeds stored under high temperature and humidity and sown during winter (Dey *et. al* 1992). Control is obtained by using high quality seed which have been treated with protective fungicides. Residues left on soil surface may influence the incidence & severity of seedling blight.

Fig. 2.1. Symptoma of seedling blight

3. Turcicum Leaf Blight

Exserohilum turcicum (Pass) Leon.&Sugs. (Teleomorph: *Setophariaturcica*) is the responsible for the cause of disease. The disease prevalent in areas where cooler condition with high humidity prevails and maize is planted in high lands. Disease is also common in winter planting in the plains as the cool/moderate humid conditions (18-27°C) favours disease development. In India the disease is common in Jammu & Kashmir, Himachal Pradesh, Sikkim, West Bengal, Meghalaya, Tripura, Assam, Rajasthan, Uttar Pradesh, Uttarakhand, Bihar (in winter crop), Madhya Pradesh, Gujarat, Maharashtra, Andhra Pradesh, Karnataka and Tamil Nadu. Turcicum leaf blight was first reported by Passerini (1876) in Perma, Italy. In India, the disease was for the first time reported by Butler during 1907 from Bihar. Later it was reported from many parts of the country, *viz.*, Lalmardi, Srinagar, Punjab, Himachal Pradesh and Kashmir valley (Payak and Renfro, 1968). The sexual stage of the fungus, *Trichometasphaeria turcica* Luttrel rarely occurs in nature. Disease intensity was moderate to severe on commercially cultivated maize hybrids in southern Karnataka and disease intensity depends on cultivar susceptibility weather factors and location. It was reported an epidemic of northern corn leaf blight from Texas. They estimated the loss in grain yield of susceptible hybrids ranging from 40 to 50%. Turcicum leaf blight incidence on maize was very severe during 2000 at Almora (Uttarakhand) attaining epidemic proportion which resulted in 83% yield reductions (Babu *et al.*, 2004).

Yield loss is caused predominantly through loss of photosynthetic leaf area due to blighting. Under severe infestation, sugars of the plant can be diverted from the stalks for grain filling leading to crop lodging. If disease establishes before silking and spreads to upper leaves during grain filling, severe yield losses can occur. Crop lodging is a particular concern where maize is mechanically harvested as the lodged cob cannot be harvested. The local cultivars recorded 66 % reduction in yield due to TLB (Payak and Sharma, 1985).Yield losses as high as 70% have been recorded due to Turcicum leaf blight. Slightly oval water-soaked, small spots are produced on leaves which grow into long, elliptical, grayish green or tan lesions ranging from 2.5 to 15 cm. in length (Fig. 2.2a) develop first on the lower leaves and later on the disease progresses upward on the plant. The disease can develop rapidly after anthesis resulting in complete blighting of leaves. In damp weather, large numbers of grayish black spores are produced on the lesions; lesions may form on the outer husks. Severe infection causes a premature death and gray appearance that resembles frost or drought injury (Fig. 2.2b). Pant *et al.* (2001), who reported about 91% reduction in the rate of photosynthesis when severity of TLB in maize exceeded 50%.

Fig. 2.2 (a) Symptoms of TLB   **Fig. 2.2 (b).** Field view of susceptible plant to TLB

In addition to maize, *E. turcicum* is known to infect sorghum, Johnson grass, gama grass, teosinthe and Sudan grass. However, isolates recovered from these species do not infect maize. Pathogen *E. turcicum* overwinters as mycelium and chlamydospores in infected crop debris. At the onset of the subsequent/favourable season, begin to sporulate. Levy (1984) reported that *E. turcicum* overwintered on sorghum and maize plant debris. Spores (conidia) are then disseminated by wind and rain splash to freshly planted maize. They germinate in temperatures ranging from 17-27°C. Secondary infection through conidia and disseminated by rain splash and wind.

4. Maydis Leaf Blight

This disease is caused by *Drechslera maydis*Niskado Syn. *H. Maydis* (Teleomorph: *Cochliobolus heterostrophus*). Disease is prevalent in warm humid temperate to tropical climate where the temperature ranges from 20-30°C during cropping period. In 1970, the disease was first reported in February from southern Florida, near Belle Glade. In India it is distributed in Jammu & Kashmir, Himachal Pradesh, Sikkim, Meghalaya, Punjab, Haryana, Rajasthan, Delhi, Uttar Pradesh, Bihar, Madhya Pradesh, Gujarat, Maharashtra, Andhra Pradesh, Karnataka and Tamil Nadu. Damage is caused by loss of photosynthetic leaf area, due to foliar lesions which reduce photosynthetic production for grain filling. Further damage is caused by lodging, which occurs when plants divert sugars from the stalks for grain filling during severe disease pressure. Race T pathotypes are also able to infect stems, sheaths and ears, which can result in ear rot, ear drop, and lodging. Damage is most critical if infection occurs prior to silking and if weather conditions are favourable for disease development during the reproductive growth stages. When the cultivating varieties is having the Texas source of male sterility then the infection by Race T pathotypes, is extensive and stalk and ear rots is very common. Increased application of nitrogen fertilizer and increased crop density are associated with increased disease severity 83 per cent yield reductions (Babu *et al.*, 2004), due to *D. maydis* in susceptible varieties.

Plants are affected at knee height stage. Young lesions are small and diamond shaped, become elongate (2-3 cm) at maturity. Lesions may coalesce, producing a complete burning of large areas of the leaves. Symptoms may vary according to the causal race and host germplasm. Race O produces lesions that are initially small and diamond-shaped. These lesions elongate as they mature, although growth of lesions is restricted by leaf veins (Fig. 2.3). Final lesions are rectangular (2-6 × 3-22 mm), restricted by leaf veins, and tan in colour. Lesions caused by isolates of Race O are restricted to leaves. Lesion produced by race 'T' is tan, 0.6-12 x 0.6-2.7 cm. elliptical with yellow green. Later, the race 'T' lesion have dark, reddish brown borders and may occur on the leaves, stalks, leaf sheath, ear husk, ear; and cob rot can also occur with substantial losses in harvesting and shelling. Seedling from infected kernels (Race 'T') may wilt and die within three to four weeks after planting. Severe blighting of leaves caused by either race predisposes plant to stalk rot. Race T and Race O are morphologically similar, although Race T is pathogenic to maize where the Texas male sterility has been incorporated. Additionally, Race T produces larger lesions than Race O.

Fig. 2.3. Symptoms of MLB

In addition to maize, *D. maydis* is also known to infect sorghum. The fungi overwinter as mycelium in infected crop debris that remains on the soil surface between growing seasons. At the onset of the subsequent growing season, in response to favourable condition begin sporulating and produce condia which disseminated through wind and rain splash. Conidia infects plants through stomata, giving rise to characteristic lesions within which conidia are produced, which leads to secondary infection.

5. Common Rust

The causal organism first reported by Schweinitz (1832). Barclay (1891) made the first record of the existence of *Puccinia sorghi* Schw in India from the maize field near Shimla. It was further collected by Butler in 1905 near Poona (Butler, *et. al.* (1931). The disease is prevalent in a cool weather and occurs worldwide in subtropical and temperate and high land environments with mild temperatures (16 to 20°C), high relative humidity, and high moisture. Disease is severe during the ear filling period and can spread rapidly if favourable condition prevails. In India it is common in Jammu & Kashmir, Himachal Pradesh, Sikkim, Meghalaya, West Bengal, Punjab (Rabi), Haryana (Rabi), Rajasthan, Uttar Pradesh, Bihar (Rabi), Madhya Pradesh, Maharashtra, Andhra Pradesh, Karnataka and Tamil Nadu. Damage is caused by loss of photosynthetic leaf area, chlorosis, and premature leaf senescence, leading to incomplete grain filling and poor yields. Damage is most critical when environmental factors favours the disease development (temperature of 16 to 25°C, relative humidity above 95%, and over 6 hours of leaf wetness) and susceptible varieties/hybrids are cultivated. It has been estimated that reduction in yield as high as 7% is possible for each 10% of total leaf area infected. Yield losses from Diara region in Bihar ranges from 11.2 – 33.6% reported by Gupta (1981) and Sharma *et al* (1982) reported up to 32.18% losses due to this disease. In some cases the yield losses as high as 50% have been recorded. Sweet corn varieties are particularly susceptible to common rust compared to dent maize varieties. Damage is most serious in early infected plants causes more damage than the late (during flowering time) and when disease spreads to leaves above the ear, which contribute most to grain filling.

Disease is common at Knee high stage. Circular to elongate (0.2 to 2 mm), with dark brown pustule (uredinia) scattered over both leaf surfaces giving the leaf a rusty appearance (Fig. 2.4). Pustules may emerge in circular bands due to infection that occurred in the whorl. Pustules break through the leaf epidermis and release powdery reddish-brown spores (uredinospores). As pustules mature, they release brownish-black spores (teliospores) which are the overwintering spores. Under severe disease pressure, leaves may turn chlorotic and senesce prematurely. *Oxalis* sp. of plants are (alternate host) frequently infected with light orange coloured pustules. The life cycle of *P. sorghi* involves two hosts (maize and *Oxalis* species) and five spore stages (teliospores, basidiospores, spermatia, aeciospores and urediniospores). In tropical or subtropical regions uredionospore overwinters and serve as the primary source of inoculum. Urediniospores are disseminated by wind over vast distances (hundreds of kilometers) and frequently spread from tropical/subtropical regions to temperate regions in spring and summer when maize is cultivated. The sexual stage of the life cycle occurs predominantly in tropical and subtropical regions. Teliospores are source of secondary inoculum.

Uredial pustules

Fig. 2.4. Symptoms of Common rust

6. Polysora Rust

Southern corn rust caused by *Puccinia polysora* Underw was identified in Alabama in 1981 on *Tripiacum dactyloides* L. (Underwood, 1897). In India the disease was first noticed in maize cultivars from Mysore district (Payak, 1992). Polysora rust (Southern rust) is a major disease of maize in tropical and subtropical regions worldwide. In India earlier it was considered as less economically important disease and emerging as a threat in coastal areas of A.P. and Karnataka. The incidence of P. rust has taken a heavy toll in majority of the cultivars grown in Southern districts of Karnataka *viz* Mysore, Mandya, Hassan, Part of Coorg, Kolar, Shimoga and Chitradurga district (Anon. 2002). The disease is favoured by mild temperature (27°C) and high relative humidity.

Damage is caused by loss of photosynthetic leaf area, chlorosis, and premature leaf senescence leading to incomplete grain filling and poor yields. Under severe disease pressure, sugars are diverted from the stalks for grain filling leading to plant lodging. Damage is severe when environmental factor favours disease development (temperatures above 24°C, high relative humidity and leaf wetness) in susceptible host. It has been estimated that reductions in yield as high as 8% are possible for each 10% of total leaf area infected. Damage is most serious in early infected plants and when disease spreads to leaves above the ear which contribute most to grain filling. Late planted maize is most vulnerable and prone to this disease. Polysora rust is an important disease in many tropical regions where maize is cultivated continuously. Melching (1975) reported yield reduction up to 37% whereas Rodriguez Ardon *et al.* (1980) recorded yield losses up to 45% especially on late planted maize recorded. There were reports of 60% losses in grain yield.

Polysora rust are very similar to common rust (*P. sorghi*). However the two can be distinguished, as pustules of Polysora rust occur predominantly on the upper leaf surface, whereas pustules of common rust occur abundantly on both leaf surfaces. Pustules (uredinia) are golden - orange in colour, circular to oval (0.2 to 2 mm long), raised, and most prominent on the leaf, but also occur on the stems and sheaths (Fig. 2.5). When the pustules rupture, uredinospores are released, resulting in the secondary cycles of the disease, giving the leaves a rusty appearance. Towards plant maturity, pustules become dark brown to black and release teliospores. Under severe disease pressure, leaves turned in to chlorotic and plant senesce prematurely. Rupturing of pustules, results in the release of powdery brown spores. An alternate host for *P. polysora* has not been identified so far. Uredinospores of *P. polysora* are known to infect maize and various grasses. Uredinospores are source of both primary and secondary inoculum. At the onset of the growing season, uredinospores germinate and infected corn debris is disseminated by wind and rain splash. They germinate and infect maize plants through the stomata and spread rapidly under favourable condition.

Fig. 2.5. Symptoms of Polysora rust

7. Downy Mildews

Downy mildews are caused by up to ten different species of Oomycetes fungi in the genera *Peronosclerospora, Scleropthora* and *Sclerospora*. This group of the pathogens constitutes one of the most important factors limiting maize production in India. The important downy mildew of maize in India are the Sorghum downy mildew, Brown stripe downy mildew and Rajasthan downy mildew. It has been determined that the crop is most vulnerable to downy mildew infection at seedling stage (15 to 20 days after planting). Downy mildews are very significant maize diseases in tropical/subtropical regions of India and Asia, where prolonged periods of leaf wetness and cultivation of alternate hosts are prevalent during the growing season. Cool, wet and humid conditions are optimal for disease development. In favourable conditions, disease cycles are rapid, leading to severe infection and spread of disease is very fast.

Systemic infection of young seedlings leads to stunted growth, chlorosis and premature death, limiting yields. Under severe infection, tassels are malformed and ears are barren, leading to extensive yield loss. Damage is critical when systemic infection of young seedlings occurs, which can result in premature death of the plant and barren ears. Disease causes losses ranging from 20 to 100% depending on the species and host cultivars in maize. Symptoms of downy mildew on maize caused by the various pathogenic species are similar, although symptoms can vary depending on plant age, prevailing climatic conditions, and host germplasm. Infection of maize plants at the seedling stage (less than 4 weeks old) results in stunted and chlorotic plants and premature plant death. Leaves on older plants display characteristic symptoms of downy mildews which include mottling, chlorotic streaking and lesions, and white striped leaves that eventually shred. 'Downy' growth is often observed on both leaf surfaces, but is more common on the lower leaf surface. Infected plants have leaves that are narrower and more erect as compared to healthy leaves. Infected plants are often stunted, tiller excessively and have malformed reproductive organs (tassels and ears). Infected plants may not seed, while tassels may exhibit 'bushy' growth called phyllody.

At the onset of the growing season, at soil temperatures above 20°C, oospores in the soil germinate in response to root exudates from susceptible maize seedlings. The germ tube infects the underground sections of maize plants leading to characteristic symptoms of systemic infection including extensive chlorosis and stunted growth. When oospores initiate infection, the first leaf generally remains disease free as it is able to outgrow the fungi. However, the whole plant will show disease symptoms if the pathogen was seed-borne. Oospores are reported to survive in nature for up to 10 years. Once the fungi has colonised host tissue, sporangiophores (conidiophores) emerge from stomata and produce sporangia (conidia) which are wind and rain splash disseminated and initiate secondary infections. Depending on the species, sporangia germinate directly or release zoospores that initiate infection. Sporangia are always produced in the night. They are fragile and cannot be disseminated more than a few hundred meters and do not remain viable for more than a few hours. Germination of sporangia is dependent on the availability of free water on the leaf surface. If sufficient water is available, sporangia germinate and infect the plant through stomata on the leaf, sheaths, or stems in a couple of hours. Initial symptoms of disease occur in 3 days. Conidia are produced profusely during the growing season. As the crop approaches senescence, oospores are produced in large numbers. Perenation of infected crop debris serves as a source of inoculum in subsequent seasons. Cultivation of alternate hosts in rotation or simultaneously will build pathogen pressure. Moist soils favour oospores germination and therefore damp soil as a result of irrigation or reduced tillage techniques will encourage disease development.

8. Brown Stripe Downy Mildew

The disease is caused by Sclerophthora rayssiae var. zeae Payak and Renfro. The disease is most common in the Himalayan areas of northern India and the disease is limited to location below 1500 masl. Cool, wet and humid conditions are optimal for disease development. In India it is commonly distributed in Himachal Pradesh, Sikkim, West Bengal, Meghalaya and Uttarakhand. Initially, lesions develop on the leaves as narrow, chlorotic or yellowish stripes, 3-7 mm wide with well-defined margins and delimited by the veins. The stripes later become reddish to purple (Fig. 2.6). Lateral development of lesions causes severe striping and blotching occurs prior to flowering. Downy or woolly cottony whitish growth occurs in early morning hours on lower surfaces of the lesions.

Fig. 2.6. Symptoms of Brown stripe downy mildew

9. Sorghum Downy Mildew

Peronosclerospora sorghi (Weston & Uppal) Shaw is responsible for this disease .The disease prevails in cool, wet and humid conditions. It is commonly distributed in Gujarat, Maharashtra, Andhra Pradesh, Karnataka, and Tamil Nadu. Infected plants are chlorotic which includes the base of the blade with transverse margin and easily defined between diseased and healthy tissue (Fig. 2.7). Leaves of infected plants tend to be narrower and more erect than these healthy plants. A white downy growth may appear on lower surfaces of infected leaves (Fig. 2.7). In severe cases the tassels of diseased plants may exhibit phyllody (Fig. 2.7). There is no seed set in such plants. In tolerant varieties, the plant show symptoms of infection but have normal seed setting. The sporangiophore is erect, dichotomously branched, 180 to 300µm in length. Emerge singly or in groups from stomata. The sporangia/ conidia are oval, borne on sterigmata (about 13µm long) and oospores are spherical (36µm in diameter on average), light yellow or brown in colour.

Fig. 2.7. A Field view of Resistant & Susceptible plant to SDM

Both conidia and oospores are sources of disease inoculum. Overwintering spores produced between the leaf veins exist in the soil for long periods. The oospores can survive in the soil for several seasons. The germ tube infects the underground sections of maize plants leading to characteristic symptoms of systemic infection including extensive chlorosis and stunted growth. Oospores cannot produce symptoms if the seedlings emerge in cool soils. Conidia are produced profusely during the growing season on the leaves and are disseminated by the wind, providing secondary inoculum, which can induce systemic infection in plants up to four weeks old. Oospores are produced in large numbers only when plant systemically infected. The pathogen can also be seed borne and seed-transmitted in cases where seeds are fresh and have high moisture content.

10. Rajasthan Downy Mildew

Peronosclerospora hetropogoni Siradhana *et al.* is the causal agent of the disease. Disease prevails in hot and humid conditions in Rajasthan and surrounding areas. It is a major yield constraint in 11 districts of major maize growing belt of Rajasthan. Several outbreaks due to this disease have been recorded in 3 districts – Udaipur, Rajsamand, and Bhilwara during 1973-1980. Rajasthan downy mildew causes heavy losses under favourable condition. The disease severity increased with the increase of inoculum density that resulted in 88.7% reduction of grain yield of maize.

The systemic symptoms in early seedling stage (2-3 leaves) are characterized by the pale appearance of bases of second & third diseased leaves of the seedling giving a complete chlorosis or chlorotic strips (Fig. 2.8a). On infected leaves, yellow stripes observed at the base also extend up to upper green portion. Severely infected plants give yellowish appearance even from a distance. In severe infection plants die at about knee-high stage. The secondary symptoms start when plant approaches at 2-3 leaf stage until tassels and silks are formed. Under humid conditions, whitish fluffy growth due to abundant fructification of the fungus can be observed on the lower and upper leaf surfaces. The symptoms of this disease are similar to *P. sorghi* except one difference that is *P. heteropogoni* also infects the grass, *Heteropogon* spp. Oospores are produced abundantly on *H. contorts* but not on maize. Conidia produced on *Hetropogon contortus* or *H. melonocarpus* (Fig. 2.8b) are the primary source of inoculum to infect maize. Tassels may be malformed producing less pollen while ears may be aborted resulting in partial or complete sterility in severe cases. In early symptoms plants are stunted and may die. Conidia and oospores are sources of disease inoculum. Oospores are produced abundantly on alternate host, a grass *H. contortus* but not on maize. Conidia produced on *H. contortus* or *H. melonocarpus* are the primary source of inoculum to infect maize.

Fig. 2.8. A Symptoms of RDM

11. Brown Spot

This disease is not considered to be important in India, however, when conditions are more conducive for disease development, the disease can develop and greater incidence and cause alarm because its symptoms are similar to some other more serious diseases. Brown spot is caused by the chytridiomycete fungus, *Physoderma maydis* (syn. *P. zeae-maydis*), which is closely related to the oomycete or water mold fungi, such as the downy mildews. The disease is favoured by high temperatures and high humidity. It attacks leaf blades, sheaths, and stalks, producing small, reddish-brown to purplish-brown spots which may merge together to form large brown blotches. This disease is prevalent in subtropical areas with abundant rainfall with moderate temperature. Commonly the disease is distributed in Jammu & Kashmir, Himachal Pradesh, Sikkim, West Bengal, Punjab, Rajasthan, Madhya Pradesh and Karnataka.

Plants are most susceptible at 50 to 60 days after germination. Lesions are small and round to oblong, yellowish to brown in colour, and can develop on the leaf blade, stalk, sheath, and husks. On the leaf blade, these young lesions can resemble the symptoms those caused by rusts, such as early southern rust. The brown spot lesions frequently develop in distinct bands across the leaf, particularly at the base of the leaf. These lesions appear different in the midrib than on the remainder of the leaf blade. On the leaf midrib these lesions tend to be darker in colour and sometimes larger, so the difference in appearance in this area from the surrounding leaf blade is a clue to the identity of this disease. However, white lesions on the lamina continue as chlorotic spots (Fig. 2.9). Such brown lesions appear on nodes and internodes also. As the disease progresses, the lesions expand in size, coalesce with neighboring lesions into larger lesions and darken in colour ranging in colour from chocolate to reddish brown or purple. In severe infection, the lesions increase in size and coalesce with adjoining ones into larger lesions and darken in colour (chocolate to reddish brown or purple) which may induce stalk rotting.

Fig. 2.9. Symptoms of Brown Spot

The pathogen overwinters in infested tissue or soil and produces brown sporangia that are packed inside infected cells. Conidia/sporangia are oval to cylindrical (18-26 x 29-67μm) and oospores are spherical (29-37μm in diameter), brown in colour. Each sporangium releases up to 50 motile zoospores that require both light and water to germinate and infect the plant. Infection most commonly occurs in the whorl where water tends to accumulate during periods of rain and irrigation as a result lesions tend to develop in bands across the leaf. This pathogen survives in crop debris and may be more common in continuous corn and fields with abundant residue, such as where reduced tillage practices are employed.

12. Banded Leaf and Sheath Blight

Banded leaf and sheath blight (BLSB) is caused by the basidiomycete fungi *Rhizoctonia solani* f. sp. *sasakii* Exner (Teleomorph: *Corticium sasakii* Syn. *Thanatephoruscucumeris*). BLSB is a significant impediment to maize production in many hot and humid environments in the tropics and subtropics. In particular, BLSB is recognized as a serious constraint to maize production in China, South Asia and Southeast Asia. The disease was first recorded from Sri Lanka (Bertus, 1927). In India, the disease was first recorded from Tarai region of Uttar Pradesh, in 1960 (Payak and Renfro 1966) Now it has become increasingly severe and assumed epidemic proportions in the next two decades. Disease severity is often closely associated with prevailing climatic conditions and farming practices. Humid conditions and irrigated fields are highly favourable for the disease. Additionally, high crop densities impact disease severity. *R. solani* is widespread in tropical and subtropical regions of the world and once established in a field the fungus often remains indefinitely. It is common in Jammu & Kashmir, Himachal Pradesh, Uttarakhand, Sikkim, Meghalaya, Assam, Nagaland Punjab, Haryana, Rajasthan, Madhya Pradesh, Delhi, Uttar Pradesh and Bihar. The disease is prevalent in hot humid foothill region in Himalayas and in plains.

Maximum damage is caused when ears are infected. In addition to ear rots, kernels are often wrinkled, dry, chaffy and light in weight. The disease causes direct losses, resulting in premature death, stalk breakage and ear rot indirect losses by reducing the grain yield. In India, losses in grain yield have been estimated in the range of 23.9 to 31.9% (Lal *et al.* 1980). However, the magnitude of grain loss may reach as high as 100% if the ear rot phase of the disease predominates. The disease caused drastic reduction in grain yield-to the tune of 97 per cent (Butchaiah, 1977) and exhibited a direct correlation with other yield parameters (Barua, 1979). In India, yield losses as high as 40% have been recorded.

The disease appears on leaves and sheaths on 40-50 days old plants and later on spread to the ears. The characteristic lesions are first seen on lower leaves and sheaths (first and second) in the form of concentric bands and rings (Fig. 2.10a). The disease can spread to the ears. Ear rot is characterized by light brown, cottony mycelium on the ear and the presence of small, round, black sclerotia (compact mass of hyphae that can survive in unfavourable conditions). The developing ear is completely damaged and dried up prematurely with cracking of the husk leaves. Brown rotting of the ears may develop which show conspicuous light brown cottony mold with small, round black sclerotia (Fig. 2.10b). *R. solani* is also pathogenic to a wide range of cultivated crops. In addition to maize, anastomosis group AG1-IA is also pathogenic to rice, wheat, sorghum, bean (*Phaseolus* species), and soybean. *R. solani* survives in the soil and on infected crop debris as sclerotia or mycelium for several years in the soil. The fungi spread by water (flooding), irrigation, movement of contaminated soil, and plant debris. At the onset of the growing season, in response to favourable humidity and temperatures (15 to 35°C), fungal growth is attracted to freshly planted host crops by chemical stimulants released by growing plant cells.

Fig. 2.10. A Symptoms of Banded leaf & Sheath blight

Fig. 2.10 B. Plant showing sclerotia of *Rhizoctonia soloni* f. sp. sasakii

13. Pythium Stalk Rot

Pythium stalk rot is caused by the Oomycete *Pythium aphanidermatum* (Eds) Fitz. In India the diseases was recorded for the first time by Srivastava and Rao (1964) from Delhi. The disease is prevalent in some hot and humid tropical and subtropical zones and in some temperate areas. The disease is more prevalent in wet areas and in poorly drained soils. The disease occurs in depressed areas of the field or in river bottom fields that are characterized by wet soils. High temperatures (25-35°C), with relative humidity of 80-100%, high crop density, and high levels of nitrogen fertilizer all favour disease severity. The disease is common in Sikkim, Himachal Pradesh, West Bengal, Punjab, Haryana, Rajasthan, Delhi, Uttar Pradesh and Bihar

Pythium stalk rot results in lodging of the plant at pre-flowering stage. Plants are therefore killed prematurely, leading to yield loss. Often infection occurs prior to tasseling. Damage is most prevalent in warm and humid regions where soils are particularly wet. Valley areas and river bottom fields are prone to infection. However dispersal of the pathogen by natural means is often limited and hence infection can be limited to particular areas of a field. Pythium stalk rot is generally confined to hot and humid regions where maize is cultivated in poorly drained soils. Damage is usually localized in a field. Occasionally entire parts of a field may lodge prematurely, although extensive damage in the field is rare as the pathogen dispersal by natural means is often limited. The yield reduction in susceptible genotypes has been reported to the tune of 100%.

Pythium stalk rot is usually confined to a single internode just above the soil line. The diseased area of the stalk is brown, water-soaked, and soft and the stalk is collapsed. The affected plants topple but do not die up to two weeks after attack. The plant gets twisted (Fig. 2.11) due to rotting at infected portion resulting in lodging, though, they do not break off completely. Infected plants remain green and turgid up to several weeks because the vascular bundles remain intact.

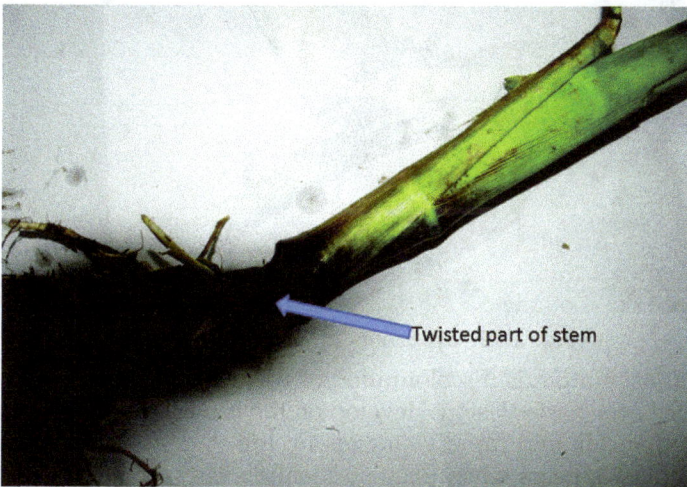

Fig. 2.11. Symptoms of Pythium stalk rot

Pythium aphanidermatum is known to infect a wide range of cultivated crops, including cereals and grasses, cucurbits, horticultural crops, and cotton. *P. aphanidermatum* is also an important greenhouse pathogen. *P. aphanidermatum* survives in the soil as mycelium or oospores for several years. They produce sporangia which release motile zoospores. Zoospores swim towards plant or root tissue in response to exudates which serve as nutrient or chemotrophic stimulants. Zoospores infect the plant directly leading to the development of characteristic lesions. Mycelium within the infected host tissue gives rise to oogonium and antheridium, the sexual stage of the life cycle. The antheridium fertilizes the oogonium producing oospore, which is able to overwinter in crop debris or in the soil. The pathogen is dispersed with either the movement of infected crop debris or with flooding or excess wetness, which transports oospores and enables zoospores to swim freely. However the pathogen can survive on the turf grass (*Agrotis palustris*) throughout the year (Saladini *et al.*, 1983)

14. Bacterial stalk rot

The bacterial stalk rot is caused by *Erwinia chrysanthemi* p.v.*zeae* (Sabet) Victoria, Arboleda & Munoz. It is a motile, gram-negative, rod shaped bacterium. The disease is prevalent in high temperature (32-35°C) and high relative humidity and flooding. It is commonly distributed in Himachal Pradesh, Sikkim, West Bengal, Punjab, Haryana, Rajasthan, Delhi, Uttar Pradesh, Uttarakhand, Bihar and Andhra Pradesh. Early infection led to complete death of plant, infection at flowering stage resulted in 92.2%, yield loss and late infection in 57.0 and 36.3% yield losses (Saxena and Lal, 1980).

Fig. 2.12. Sypmtoms of Bacterial stalk rot

The initial symptom is discolouration of the leaf sheath and stalk at a node. As the disease progresses, lesions develop on the leaves and sheath. Disease then develops in the stalk and rapidly spreads up the stalk and into the leaves. The disease generally appears when plant suddenly falls over and are seen scattered in the field (Fig. 2.12). Splitting of stalk exposes internal discolouration and soft slimy rot at the nodes. Bacterial stalk rot can affect the plant at any node from the soil surface up to the ear leaves and tassels. Infections that occur high on the plant may impair normal tasseling and affect subsequent pollination. Although it may spread along the plant to infect additional nodes, the bacteria do not usually spread to neighboring plants unless vectored by an insect. Splitting the stalk reveals internal discolouration and soft slimy rot mostly initiating at the nodes. In advance stage of infection, a foul odour can be sensed from macerated tissues and the top of such plants can be very easily removed from the rest of the plant. Affected plants may remain green for several days. Because the bacteria usually do not spread from plant to plant, diseased plants are quite often found scattered throughout the field.

The bacterium has a wide host range. It can attack tubers of potato and sweet potato, onion bulbs, bean pods, roots of carrot, turnip, radish and sugarbeet, fruit of tomato, brinjal, tobacco, cabbage (Thind, 1970; Rangarajan and Chakravorty, 1971; Hingorani *et al.*, 1959). These bacteria survive in corn, sorghum stalks and residue. The bacteria enter the plants through natural openings; wounds from hail, high winds, or insect feeding (e.g., stalk borers) can provide additional entry sites into the plant. Saxena and Lal (1984) reported that the pathogen survives for 120 days in a field soil containing infected debris.

15. Fusarium Stalk Rot

Fusarium verticilloides is the cause of the fusarium stalk rot. The disease is distributed worldwide and most common in dry, warm regions and cause extensive crop damage by premature drying and resulting plant death. The disease is most common in Rajasthan, Uttar Pradesh, Bihar and Andhra Pradesh. The infected plants typically wilt, leaves turn dull greyish-green and symptoms become conspicuous when the crop enters senescence phase. Damage is caused by premature plant death, lodging, and interference with translocation of water and nutrients during grain filling, leading to poor yields. Damage is most severe if infection occurs early in the season. In these cases yield loss is directly affected by premature plant death or reduced kernel filling due to interference with translocation of water and nutrients in the stem. When crops are infected later in the season and lodging occurs, losses can be incurred particularly where maize is machine harvested. *F. verticillioides* is also seed borne, in which case the majority of plants may suffer from stalk rot leading to severe crop loss. Precise yield loss data for most stalk rots is difficult to ascertain. However, extensive losses are possible under severe disease infestation. Yield losses influenced by various factors like climatic conditions, crop density, fertilization rates & cultural practices estimating precise yield loss due to maize stalk rot is often complicated by number of factors involved The yield losses in susceptible maize cv from experimental field in Udaipur was reported 38.93 % (Meena Shekhar *et al.* 2012).

The leaves of infected plant turn to dull green instead of dark green colour and the lower stalk becomes yellowed/ straw-colour and whole plant is wilted (Fig. 2.13a). The internal pith of the lower nodes gets disintegrated and softened which can be easily pressed between thumb and finger. Fungal mycelium can often be seen at such nodes. The stalks show pink-purple discolouration on splitting (Fig. 2.13b). Often black perithecia or mycelium growth can be observed at lower stalk nodes where the plant is infected. The pathogen commonly affects the crown regions of roots and lower internodes. The pathogen has Broad host range: infects many crops, including maize, sorghum (*Sorghum bicolour*), sugarcane (*Saccharum officinarum*), wheat, cotton (*Gossypium* spp.), banana (*Musa* spp.), pineapple (*Ananascomosus*), and tomato (*Solanum lycopersicum*). Mycelium in infected crop debris on the soil surface produces macro conidia and micro conidia, which are wind and rain splash disseminated to freshly planted maize. *F. verticillioides* is also seed borne, in which case the pathogen may be present during the entire life cycle of the plant. Disease is more severe when crop density and nitrogen fertilizer rate is increased. Stressed crop/plants are more prone to stalk rots; stalks are already vulnerable at the time of grain filling stages as sugars are diverted for grain filling.

Fig. 2.13(a). Symptoms of Fusarium stalk rot

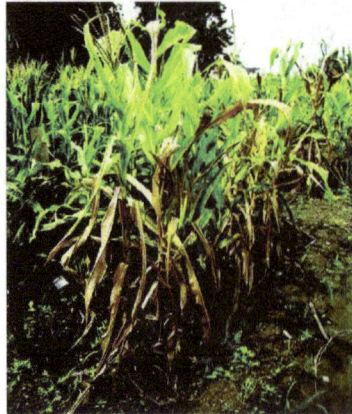

Fig. 2.13(b). Field view of Fusarium stalk rot infected plants

16. Charcoal Rot

Charcoal stalk rot is a common disease of maize and most prevalent in hot, dry environments and is caused by the fungi *Macrophomina phaseolina* (Goid) Tassi. Drought and high temperature encourage disease severity. The disease is strongly associated with drought conditions. Soil temperatures of 30-42°C increase disease severity. Charcoal stalk rot is characterized by the stalk turning grey-black. The name of the disease is derived from its symptoms. In India it is commonly distributed in Jammu & Kashmir, West Bengal, Punjab, Haryana, Rajasthan, Delhi, Uttar Pradesh, Madhya Pradesh, Andhra Pradesh, Karnataka and Tamil Nadu.

Infection of the stalk can lead to premature senescing of the crop, lodging, and interference with translocation of water and nutrients, which all result in yield loss. The disease is particularly prevalent in drought years and in arid regions where maize is regularly cultivated in rotation with other host crops. The disease is heat and stress (drought) driven and is therefore rare in cooler climes and irrigated fields. Increased losses may be experienced where maize is machine harvested due to lodging. Charcoal stalk rot is recorded as a serious disease of maize in several arid maize growing regions where extensive yield loss has been documented when the crop is infected early.The disease incidence was recorded in Karnataka ranged from 10 to 42% (Desai *et al.*, 1991). Whereas the yield losses in susceptible maize cv from experimental field in Delhi due to C. rot was reported 42.9% (Meena Shekhar *et al.*, 2012).

This disease is prevalent in comparatively drier maize growing areas. Symptoms of this disease observed once the crop/plant approaches maturity. Affected plants dry prematurely (Fig. 2.14a). The affected internodes become disintegrated and show black discolouration. Presence of numerous, minute black sclerotia on the vascular bundles and inside the rind of the stalks is a distinguishing character (Fig. 2.14b & 14c). Vascular bundles within the stem often shred. The disease is usually confined to first or second internode above soil level. Water stress at or after flowering period has been found to predispose the plant to infection. Evidence of the carbohydrate sink size of the grain also influencing the occurrence of stalk rot supported the photosynthetic stress-translocation balance concept of stalk rot (Dodd, 1980).

Fig. 2.14 (a) Toppled plant susceptible to C.rot showing toothpick

Fig. 2.14 (b). Susceptible reaction to M. phaseolina

Fig. 2.14 (c). Microsclerotia on the cortical region

M. phaseolina infects an extremely wide range of hosts, including sorghum, soybean, cucurbits and various weed species. Over 500 hosts have been documented for *M. phaseolina*. It overwinters as sclerotia in the soil and can remain viable for several years. In dry and hot conditions fungi infect the roots of maize plants and colonize the lower stalk, eventually giving rise to characteristic symptoms (abundant, minute, black sclerotia and charring and shredding of the pith tissue). Isolates that are pathogenic to maize are not known to produce conidia. Charcoal rot is a soil borne disease.

17. Late Wilt

Late wilt, or black bundle disease, is a vascular wilt disease of corn that is caused by the soilborne fungus, *Harpophora maydis* (Samra *et al.*, 1966) W. Gams with synonyms: *Cephalosporium maydis* Samra, Sabet, & Hingorani and *Acremonium maydis* (*CABI*, 1999; E l- *Shafey and Claflin*, 1999). Late wilt was first reported as a vascular wilt disease of corn in Egypt in 1960 and is now considered endemic throughout Egypt. Late wilt occurs in Andhra Pradesh, Uttar Pradesh, Bihar, and Rajasthan provinces of India (Payak *et al.*, 1970). It is prevalent in Andhra Pradesh, Uttar Pradesh and Rajasthan. Yield losses of up to 40 per cent have been reported (El-Shafey and Claflin 1999). The fungus is one of the most important pathogens of maize in some parts of India (Payak *et al.* 1970, Singh and Siradhana 1987b), causing yield losses of up to 100 per cent (Satyanarayana 1995). Late wilt, or black bundle disease, poses a moderate to severe threat to corn production in India where it occurs endemically, with incidence as high as 70% and economic losses up to 51% (Johal *et al.*, 2004).

The first symptoms observed as moderately rapid wilting of the leaves beginning at tasseling time. It steadily progressed from the lower to upper leaves; the leaf tissues between the veins changing first to a pale green colour then the whole leaf rolling inward lengthwise. Some leaves dried up and became brittle. Vascular bundles in the stalk are discoloured (Fig. 2.15a). As leaf wilting advanced, yellowish

Fig. 2.15(a). Resistant & susceptible plant to Late wilt

Discoloration of V. B.

Fig. 2.15(b). Symptoms of Late wilt

or reddish brown streaks appeared on the basal internodes of the stalk, which dried up and became shrunken (Fig. 2.15b) and hollow with or without wrinkling turn purple to dark brown which is more prominent on lower 1-3 internodes. When the stalk was split, a brown discolouration extended along the internodes. Some secondary organisms also develop on stalk rots that cause wet rot with some typical sweetish smell. It is primarily soil borne and may infect young maize plants more readily than other plants through roots or mesocotyl. Infected seed, biological culture, or a small amount of infected crop residue could provide pathways for introduction. Secondary dissemination could occur through contaminated seed and the movement of agricultural equipment. Disease symptoms appear late in the season, infected seeds remain symptomless.

REFERENCES

Anonymous. 2002. Proceedings of the Annual Discipline Technical Progamme meeting of Plant Pathology 2003-04, Department of Plant Pathology, UAS, Bangalore 1-49.

Babu, R., Mani., Pandey, A. K., Pant, S. K., and Rajesh Singh. 2004. Maize Research at Vivekanand Parvatiya Krishi Anusandhan Sansthan, An Overview, Techn. Bull, Vivekanand Parvatiya Krishi Anusandhan Sansthan, Almora, 21: 31.

Barclay, A. 1891. Additional uredineae from the neighborhood of Simla, Jour. Asiatic soc., Bengal 60: 214.

Barua, P. 1979. Studies on Some Aspect of Banded Sclerotial Disease of Maize Caused by *Rhizoctonia solani f. sp. sasakii*. MSc. Thesis. G.B. Pant Univ. of Agric. & Tech., Pantnagar, 56.

Bertus, L. 1927. A sclerotial disease of maize (*Zea mays* L.) due to *Rhizoctonia solani* Kuhn. Yeal Book, Deptt. Agric., Ceylon 46-48.

Butchaiah, K. 1977. Studies on Banded Sclerotia! Disease of Maize. MSc. Thesis. G.B. Pant Univ. of Agric. & Tech., Pantnagar, 65.

Butler, E. J., and Bisby, G. R. 1931. The Fungi of India. Scientific Monograph No. 1. The Imperial Council of Agricultural Research. Government of India Central Publication Branch, Calcutta, India.

Chenulu, V. V., and Hora, T. S. 1962. Studies on losses due to *Helminthosporium* blight of maize. Indian phytopath. 15: 235-237.

Desai., Suseelendra., and Hegde, R. K. 1991. A preliminary survey of incidence of stalk rot complex of maize in two districts of Karnataka. Indian Phytopath. 43: 575-576.

Dey S. K., Dhillon, B. S., and Khera. 1992. Diseases of maize in Asia-Problem and Progress. In Progress of Plant Pathological Research. (Eds. S. S. Sokhi; *et. al.*) 88-108.

Dodd, J. L. 1980. Grain sink size and predisposition of *Zea mays* to stalk rot. Phytopathology 70: 534-535.

El-Shafey, H. A., and Claflin L. E. 1999. Late Wilt. *In*: D. G. White, (ed.), Compendium of Corn Diseases (3rd). St. Paul: APS Press, 43-44.

Gupta, U. 1981. Assessment of losses in grain yield due to maize rust (P. sorghi Schw.)Seventh Annual Progress Report of Pathological Research on Diseases of maize presented at Rabi Workshop, New Delhi.

Hingorani, M. K., Grant, U. J., and Singh, N. 1959. Erwinia carotovora f.sp.zeae, destructive pathogen of maize of India. Indian phytopath. 12: 151-157

Johal, L., Huber, D. M., and Martyn, R.. 2004. Late wilt of corn (maize) pathway analysis: intentional introduction of *Cephalosporium maydis*. *In*: Pathways Analysis for the Introduction to the U.S. of Plant Pathogens of Economic Importance. United States Department of Agriculture, Animal and Plant Health Inspection Service, Technical Report No. 503025.

Lal, S., Barauh, P., and Butchaiah, K. 1980. Assessment of yield losses in maize cultivars due to banded sclerotial diseases. Indian Phytopath. 33: 440-443.

Levy, Y. 1984. Overwintering of *Exserohilum turcicum* in Israel. Phytoparasitica 12: 177-182.

Meena Shekhar, Sangit Kumar and S. S. Sharma (2012) Avoidable yield losses due to Post Flowering Stalk Rots in Maize Paper ID No. IAC2012/Sym-V Third International Agronomy Congress on *"Agriculture Diversification, Climate Change Management and Livelihoods"*, at IARI, Pusa Campus, New Delhi during November 26 -30, 2012.

Melching, J. S. 1975. Corn Rusts; Types, races; and destructive potential. In. Proc. 30th Annu. Corn. and Sorghum. Res. Conf. 90-115.

Pant, S. K., Pramod Kumar., and Chauan, V. S. 2001. Effect of Turcicum leaf blight on photosynthesis in maize. Indian Phytopath. 54: 251-252.

Passerini. 1876. Lanebbia Delgranotur Co. Bol. Comiz, Agriculture. Parmense, 10: 3.

Payak, M. M., and Renfro, B. L. 1968. Combating maize diseases. Indian Farmer Dig. 1: 53- 58.

Payak, M. M., and Sharma, R. C. 1985. Maize diseases and their approach to their management. Tropical Pest Management 31: 302-310.

Payak, M. M., Lal, S., Lilaramani J., and Renfro B. L. 1970. *Cephalosporium maydis*-a new threat to maize in India. Indian Phytopath. 23: 562-569.

Payak, M. M. 1992 Interception of Puccina Polysora Southern Rust of Maize in India, NBPGR, IARI, New Delhi. 1-5.

Rangarajan, M., and Chakravarti, B. P. 1971. *Erwinia carotovora* (Jones) Holland, the inciting agent of corn stalk rot in India. Phytopathologia Mediterranea 10: 41-45.

Richard, N., Strange., and Scott, P. R. 2005. Plant Disease: A Threat to Global Food Security. Annu. Rev. of Phytopath. 43: 83-116

Rodriguez, A. R., Scott, G. E., and King, S. B. 1980. Maize yield losses caused by Southern Corn rust. Crop Sci. 20: 812-814.

Saladini, J. L., Schmitthenner, A. F., and Larsen, P. O. 1983. Prevalence of *Pythium* species associated with cottony-blighted and healthy turfgrasses in Ohio. Plant Dis. 67: 517-519.

Samra, A. S., Sabet, K. A., Abdel Rahim, M. F., El Shafey, H. A., Mensour, I. M., Foll, F. A., Dawood, N. A., and Jhalil, H. I. 1966. Investigation on stalk rots diseases of maize in UAR. Maize and Sugarcane Dis Cont Plant Prot Dept. UAR 204.

Satyanarayana, E. 1995. Genetic studies of late wilt and turcicum leaf blight resistance in maize. Madras Agricultural Journal 82: 608-609.

Saxena, S. C., and Lal, S. 1984. Use of meterological factors in prediction of Erwinia stalk rots of maize. Tropical Pest Management 30: 82-85.

Saxena, S. C., and Lal, S. 1980. Assessment of yield losses due to bacterial stalk rot of maize (abstract) Indian Jour. of Mycol. and Pl. Pathol. 10.

Schweinitz, L. D. 1832. Synopsis fungorum in America Boreali media degentium trans. Am. Phil. Soc., N.S., 4: 141-318.

Sharma, R. C., Payak, M. M., Shankerlingam, S., and Laxminarayana, C. 1982. A comparison of two methods of estimating yield losses in maize caused by Common rust. Indian Phytopathol. 35: 18-20.

Singh, S. D., and Siradhana B. S. 1987. Survival of *Cephalosporium maydis*, incitant of late wilt of maize. Indian J. Mycol. Pl. Pathol. 17: 83-85.

Thind, B. S. 1970. Investigations on bacterial stalk rot of maize (Erwinia carotovora var. zeae Sabet). Ph.D thesis, Indian Agricultural Research Institute, New Delhi, 113.

Underwood, L. M., 1897. Some new fungi chiefly from Alabama Bull. Torrey. Bot. Club. 24: 81-86.

Diseases of Field and Horticultural Crops   *Pages* **61-88**
Editor-in-Chief: **P. Chowdappa**
Published by: **Daya Publishing House, New Delhi**

3

Diseases of Sorghum

I.K. Das and J.V. Patil

Sorghum (*Sorghum bicolour* L. Moench) is one of the most important cereal crops grown extensively in the semi-arid tropics for its nutritious food, juicy stalk and precious fodder. It is main staple food for the world's poorest and most food-insecure people across the semi-arid tropics. With exceptions in some regions, it is mainly produced and consumed by rural people. India is the largest sorghum grower in the world with an area of 8.6 million ha and contributes about 17% of the world production. It is the fourth most important cereal crop in the country. However, productivity level is very low (1tonne/ha) in India. Prominent sorghum growing states in India are Maharashtra, Karnataka, Gujarat, Rajasthan, Madhya Pradesh, Andhra Pradesh, Tamil Nadu, Uttar Pradesh and Uttarakhand (Fig. 3.1). Maharashtra holds the highest area (around 50%) followed by Karnataka (around 18%). The major challenge facing sorghum in India is to evolve technologies that will enable transformation of subsistence farming into commercial and profitable production.

There are three types of sorghum based on end product utilization such as grain sorghum, sweet sorghum and forage sorghum (Fig. 3.2). They are grown either in kharif (rainy season) or in rabi (post-rainy season). The utilization of kharif sorghum grain as a raw material in various industries is increasing, given the limited prospects of rainy season (kharif) sorghum for human consumption. Post-rainy season sorghum is a highly valued food grain, and too expensive to be used as industrial raw material. The main industries currently using sorghum in India are the poultry feed, animal feed and alcohol distilleries while its usage is quite sporadic in starch industry. At present poultry feed sector is using approximately 1.3 million tonnes annually; animal feed sector uses about 0.45 million tonnes followed by alcohol distillers (about 0.092 million tonnes). The major user poultry feed sector largely depends on maize which constitute 30-35% of poultry ration. The estimation for the future demand shows that poultry feed industry is going to be the major industry which will absorb huge quantity of sorghum (4.0 million tonnes), followed by dairy feed industry (0.6 million tonnes).

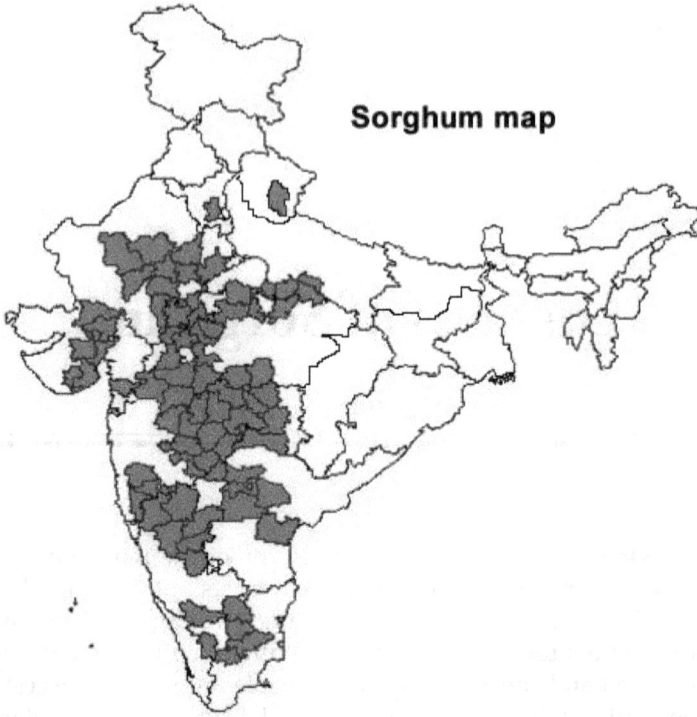

Fig. 3.1. Distribution of sorghum cultivation in India

Fig. 3.2. Sorghum crop. (a) Grain sorghum, (b) Sweet sorghum, and (c) Forage sorghum

Sorghum is a wonder crop from the physiological point of view. No cereal can outbeat sorghum in its productivity under rainfed conditions. It has efficiency to accumulate high levels of sugars in the stalk. These unique attributes of sorghum is considered highly favourable for its eventual emergence as a bioenergy crop.

Active research is now underway in several countries on sweet sorghum. This crop could also be useful to produce about 3-4 tonnes jaggery and 4 tones syrup/ha. Kharif grain has also good potential as a raw material for production of good quality potable alcohol with a competitive cost over molasses. This may provide new opportunities for higher profitability, for agri-business and employment generation in the sorghum growing semi-arid regions and can offer as a supplement to sugarcane molasses as an alternative feed stock for bioenergy production for blending with petrol.

Table 3.1. List of common diseases on sorghum in India

Sr. No.	Disease	Causal organisms
Foliar diseases		
1	Anthracnose	*Colletotrichum sublineolum*
2	Downy mildew	*Peronosclerospora sorghi*
3	Leaf blight	*Bipolaris turcica*
4	Rust	*Puccinia purpurea*
5	Sooty stripe	*Ramulispora sorghi*
6	Zonate leaf spot	*Gleocercospora sorghi*
7	Grey leaf spot	*Cercospora sorghi*
8	Bacterial leaf stripe	*Pseudomonas sorghicola*
9	Bacterial leaf streak	*Pseudomonas rubrilineans*
10	Bacterial leaf spots	*Pseudomonas syringae*
11	Viral diseases	Maize stripe virus, Maize mosaic virus
Root and stalk diseases		
12	Charcoal rot	*Macrophomina phaseolina*
13	Fusarium stalk rot	*Fusarium moniliforme*
14	Pythium root rot	*Pythium arrhenomanes*
15	Pokkah boeng	*Fusarium moniliforme* var. *subglutinans*
Panicle diseases		
16	Grain mold	*Fusarium, Curvularia, Alternaria, Phoma* spp.
17	Ergot/ Sugary disease	*Sphacelia sorghi*
18	Loose smut	*Sporisorium cruenta*
19	Covered smut	*Sporisorium sorghi*
20	Head smut	*Sporisorium reilianum*
21	Long smut	*Tolyposporium ehrenbergii*

Sorghum diseases heavily come in the way of increasing productivity, production of quality grain in grain sorghum, fodder yield and quality in forage sorghum and sugar yield in sweet sorghum. Fungal diseases are of major significance on sorghum followed by viral diseases (Frederiksen, 1986). Bacterial diseases though reported are not of economic significance in India. A list of common diseases of sorghum in India is given in the Table 3.1. In grain sorghum, grain mold, downy mildew, anthracnose and ergot are major diseases during *kharif* season whereas; root and stalk rot and viral diseases are common during *rabi* season. Other diseases like leaf spots, pokkah boeng and smuts occur sporadically and assume economic significance under specific environments depending on relative humidity and temperature during crop growth period. In forage sorghum diseases contribute negatively on fodder yield and quality. Diseases that are of economic significance on forage are foliar diseases *viz.*, leaf spots, sooty stripes, leaf blight, downy mildew, anthracnose and rust. Others like viral diseases (maize stripe virus), sugary disease and head mold assume significance under specific environments depending on relative humidity and temperature.

Leaf diseases destroy active leaf area required for photosynthesis, adversely affect accumulation of sugar in stalk and thus interfere with the quantity and quality of fodder. Sweet sorghum or high energy sorghum has immense potential as bio-energy producer and has great relevance in the national renewable energy security. In addition to the grain, the juice from stalk, the stover and the bagasse can be used to produce alcohol (ethanol) in an environment friendly manner. Most of the diseases of grain sorghum also occur in sweet sorghum depending on growing conditions and environment. Leaf anthracnose, red stalk rot, leaf blight, downy mildew, rust, sugary disease, head mold and virus diseases are common in sweet sorghum. As sweet sorghum is a crop having high commercial value, cost intensive management practices can be advocated for this crop.

Major diseases of economic importance

Many fungal, viral and bacterial diseases are reported on sorghum but all are not economically important. Economic significance of a particular disease varies with location, prevailing environmental conditions, cropping season (*kharif* or *rabi*) and type of sorghum grown (grain, forage and sweet sorghum). The diseases like grain mold, anthracnose, downy mildew, ergot, maize stripe virus, leaf blight and leaf spots, and charcoal rot occurs almost every year on different types of sorghum either in moderate or severe form in various parts of India causing considerable economic losses (Thakur *et al.*, 2003).

1. Grain mold

1.1. Occurrence and distribution: Grain mold is a major biotic constraint in the way of production, marketing and utilization of quality sorghum grain (Indira and Rana, 1987). It is one of the most important diseases of sorghum in many countries in Asia, Africa, North America and South America. The disease is particularly important on improved, short- and medium-duration sorghum cultivars that mature during the rainy season in humid, tropical and subtropical climates. The disease is more severe on high yielding kharif hybrids that are grown for grain (Audilakshmi *et al.*, 2011). In India grain mold is very common on kharif sorghum growing areas in the states of Maharashtra, Karnataka, Andhra Pradesh, Tamil Nadu and parts of Gujarat and Madhya Pradesh (Fig. 3.3).

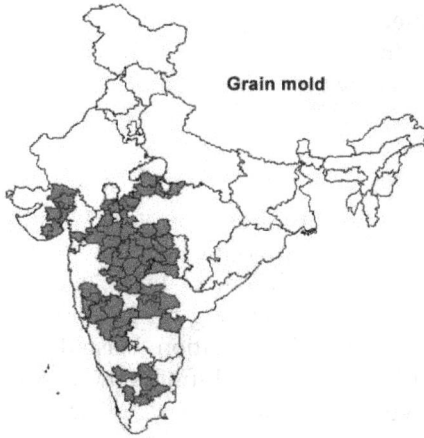

Fig. 3.3. Map showing distribution of sorghum grain mold in India

1.2. Economic significance: Grain mold is very severe on kharif sorghum and affects both hybrids and varieties. On an average around one-third of the kharif grains produced become molded every year. Early infection of flower causes loss in grain formation or production of light-weight grain. The disease sometimes induces premature sprouting of grain in the panicle. Loss of final produce (i.e., grain) is more because of discolouration of grain and less due to yield loss (yield loss occurs mostly due to reduction in seed size and weight of grain). Moldy grain fetches around 20-40% less market price than the normal one. Seed value of the produce gets affected because of loss of germination in molded seed. Monetary loss due to grain mold varies from year to year and state to state depending on rainfall and humidity conditions during crop maturity stage. On current production and price structure an approximate loss of rupees 300 to 500 crores occurs every year in India (Das and Patil, 2013). Cost of protection using agro-chemical is around Rs 1000-1200/ acre. Low cost of sorghum grain acts as a hindrance towards chemical control. Moreover, because of rain and tall nature of the crop spraying is not always feasible, and use of tolerant variety is thought to be the only way out. The disease has very significant impact on human and livestock population and crop pattern. Deterioration of grain quality by mold is one of the main reasons for less acceptability of kharif grain and decline in areas of kharif sorghum from 11.29 million ha (QE 1970) to 3.58 million ha (QE 2007). Molded grains become discoloured and loose appeal in the market and fetches less price. Nutritive value of food and feed, the cooking quality of the grain get deteriorated.

Molded grains often contain mycotoxins which are harmful to human, animal and poultry bird. When stored under inappropriate (humid) conditions grains get further deteriorated due to growth of storage fungi like *Aspergillus, Penicillium* etc. Mycotoxins contamination levels are relatively more on such grains. In India, fumonisin levels between 0.1-2.7 mg/kg grains have been reported in sorghum. Fumonisin has been implicated as a possible cause of an acute disease outbreak in human beings in several areas of India (Bhat *et al.*, 1997). An outbreak of food poisoning characterized by abdominal pain and diarrhea, attributed to the ingestion of fumonisin-contaminated moldy sorghum and maize had been reported from several villages in South India. Major share of kharif sorghum grains are utilized as feed for animal and poultry birds. Because of presence of mycotoxins, there is health risk if molded grains are continuously feed to these live stocks including porcine pulmonary edema, liver toxicity and liver cancer in rats, atherosclerosis in monkeys, and immunosuppression in poultry.

1.3. Symptoms: Symptoms of grain mold vary with the severity of infection and grain development stages. First visible symptom of infection is pigmentation of the spikelet tissues including sterile lemma, palea, lodicules and glumes. Anthers and filaments are also infected depending on severity of infection. Infection at anthesis results in loss of caryopsis formation, blasted florets, poor seed set and small, shriveled grains (Fig. 3.4). Under humid conditions infected grain may be covered by fungal growth even before physiological maturity and such grain

Fig. 3.4. Symptoms of sorghum mold

disintegrate under slight pressure. Such disintegration of molded grain is termed as 'pre-mature seed rotting'. Grain discolouration varies from whitish, pinkish, grayish, to shiny black depending on fungus species. *Fusarium* infection generally produces pinkish white mycelium; powdery in appearance during early stages which later becomes pinkish fluffy and fluffy white. Gray or dull grayish fungal bloom is generally produced by *Alternaria* spp. while black to shiny black colour is produced by *Curvularia* spp. *Curvularia lunata* appears as a shiny, velvety black, fluffy growth on the grain surface. *Phoma sorghina* produces small, round, black pinheads like pycnidia in the grain surface giving it a rough appearance. Internal colonization of grain sometimes leads to sprouting of grains in the field under wet conditions. Such sprouted grains become soft due to digestion of parts of the endosperm by α-amylase and are predisposed to extensive colonization by mold fungi, primarily species of *Fusarium* and *Curvularia*. The most obvious sign of mold infection on mature grain is the appearance of pink, orange, gray, white, or black mycelium on the grain surface. Discolouration of grain is more prominent on white-grain than on brown/red- grain sorghums (Fig. 3.5). Fungal growth first occurs at the hilar end of the grain, and subsequently extends on the pericarp surface. Sometimes visibly normal grain may not show external symptoms but develops fungal growth under inappropriate storage conditions.

Fig. 3.5. Symptoms of grain mold at post-maturity stage on white (left) and red grain (right) cultivar

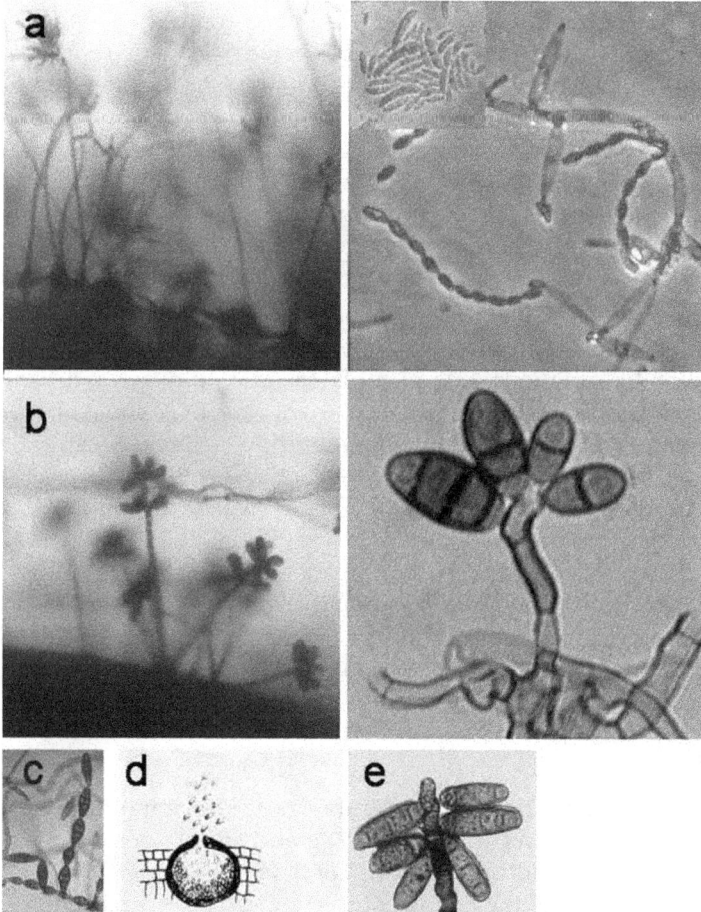

Fig. 3.6. Fungi causing grain mold in sorghum

1.4. Causal organisms: Usually many fungi are involved in grain mold disease complex. A few of them are pathogen and majority is saprophytes. Among the pathogenic fungi *Fusarium* spp. and *Curvularia* spp. are predominant in India and in some locations *Alternaria* (e.g., *A. alternata*) and *Phoma* spp. (e.g., *P. sorghina*) may also be observed (Fig. 3.6). Many saprophytic fungi may colonize mature sorghum grain under the condition of high humidity and rainfall. Frequently encountered fungi are *Bipolaris* spp., *Colletotrichum* spp., *Aspergillus* spp., *Cladosporium* spp., *Exerohilum* spp. etc. Among the *Fusarium* spp. predominant one is *F. moniliforme* (Das *et al.*, 2012) (presently comes under *F. andiyazi*, *F. nygamai*, *F. proliferatum*, *F. thapsinum* and *F. verticillioides*) and among *Curvularia* spp. *C. lunata* are the most frequently occurred species on molded sorghum grain. *F. moniliforme* produces microconidia in chain (Fig. 3.6a).

1.5. Epidemiology: Early infection occurs at flowering by air-borne inocula of the mold fungi. Moderate temperature (25-35 °C) and high humidity (>90% RH) favours infection and subsequent disease development. Frequent rain during grain development and maturity stage provide congenial conditions for saprophytic mold development. Plant residues and soil debris containing fungal hyphae and conidia are the primary sources of inoculums in the field. Crop residue buried deep survives longer than the surface residue. Under moist conditions senescent lower leaves of sorghum plants also produce abundant spores. Spores are readily disseminated by wind and rain splash for both primary and secondary spread. The natural inocula present over sorghum field during rainy season are suggested to be sufficient for grain mold development under favourable conditions. Cool and dry weather are not favourable for grain mold. Long duration cultivar that matures during rain free periods generally escapes the disease.

1.6. Management: Best option for management of grain mold is to use cultivars that mature during a period of no rains. But this is hardly practicable under changing climatic situations. Use of mold tolerant cultivars and harvesting the crop at physiological maturity followed by drying of grain is the second best option to avoid grain deterioration due to saprophytic mold. Chemical sprays are recommended and works if not washed out or disrupted by rains. Three sprays on the ear heads with Captan (0.3%) + Dithane M 45 (0.3%) at 10 days interval from flowering period onward; or spray of Tilt 25% EC @ 0.2% starting at flowering and another spray after 10 days are recommended. However, it is hardly economical because of low price of sorghum grain and high price of chemicals and human labours. For high value seed production plot spraying panicles with fluorescent pseudomonad may be undertaken that significantly reduces grain mold severity and improve seed quality.

2. Anthracnose

2.1. Occurrence and distribution: Anthracnose is an important disease causing substantial economic losses to grain, forage and sweet sorghum (Mathur *et al.*, 2002). It was first reported in 1902 from Togo, West Africa in 1980, and has since been observed in most of the regions of the world where sorghum is grown. The disease appears on several plant parts causing seedling blight, leaf blight, stalk rot and head blight. It occurs almost every year on forage sorghum in northern India in moderate to severe form and sporadically on grain sorghum in central India. The disease is common on forage sorghum growing regions in Rajasthan, Gujarat, Haryana, Uttarakhand, Uttar Pradesh and Madhya Pradesh and on grain sorghum areas in the states of Maharashtra, Karnataka and Andhra Pradesh (Fig. 3.7).

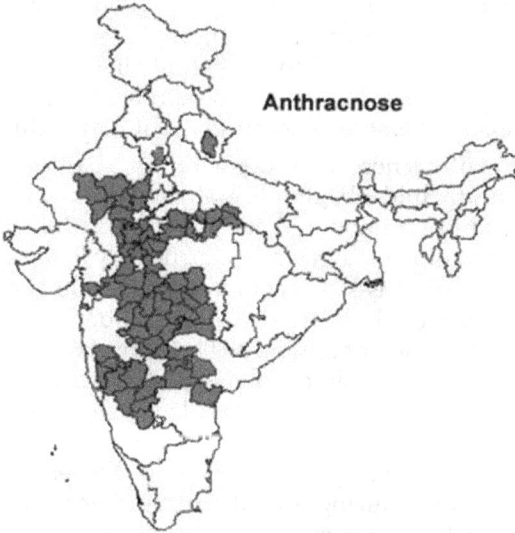

Fig. 3.7. Distribution of sorghum anthracnose in India

2.2. Economic impact: Anthracnose occurs in mild to moderate from on sorghum both during kharif and rabi seasons. Incidence varies over locations and seasons. Disease severity ranges from 1.5 to 4.0 (on a 1-5 scale) depending on environmental conditions in a year. Average disease grade of 2.5 is common on forge and grain sorghum. Leaf anthracnose affects photosynthetic areas and thus reduce amount of current photosynthesis. Red rot phase, on the other hand, damages stalk tissue and may affect movement of sap/ nutrients to the ear head. The pathogen may produce phytoalexins in plant. Financial loss varies with incidence, environment, and resistance in cultivars. Seedling blight phase of the disease may sometimes affect final plant stand in the field. Damages on leaf areas result in reduction of grain and fodder yield. Grain yield losses of 50% or more can occur under severe epidemics which may go up to 80% on highly susceptible cultivars. At national level it is a major disease on forage sorghum. However, seedling blight stage of this disease is of lesser importance because of high seed used for growing forage crop. Major economic loss occurs due to anthracnose leaf blight which has impact on grain and fodder yield. Estimation of monetary loss is complex as quantifiable effect of this disease on fodder loss in not available and grains are hardly considered for forage crop. However in grain sorghum it might cause considerable economic loss. Clean cultivation, elimination of crop residues and grasses on which the fungus can survive, and enhancement of the conditions that hasten decomposition of host residues are practices used to control the disease. Cost of disease management may be around Rs. 800 per acre.

The disease has impacts on human and cattle population and on quality of the produce. It has profound effect on grain and stover yield, stover quality and also sugar accumulation in sweet sorghum. Loss of production thus directly impacts sorghum farmers. Severely infected plants produce neither grain nor quality fodder and thus have significant effect on yield. Reduction in forage yield is particularly important and can cause significant losses in income from crop-livestock production system in rural India especially in northern states since cattle population in these regions heavily depends on sorghum fodder. Leaf anthracnose severely affects forage quality. On grain sorghum it may reduce seed mass.

2.3. Symptoms: Anthracnose may occur on any parts of a sorghum plant including leaf, leaf sheath, mid-rib, glume, grain and even on stalk. Among these, leaf anthracnose is the most pronounced and devastating. Initial symptoms on the leaves appear as small, elliptic to circular spots, usually 5 mm or less in diameter (Fig. 3.8a). Centre of the spot is straw-coloured with wide margin. colour of the margin may be red, orange, blackish purple, or tan, depending on the pigment present in the cultivar (purple or tan)(Fig. 3.8b). Under favourable conditions, the spots increase in number and coalesce to give a blighted appearance on the leaf (Fig. 3.8c). Examination of lesions with a hand lens reveals small, black, hair like structures (setae) protruding from the acervuli. In suitable environments, creamy to pinkish masses of spores are produced among the setae. Anthracnose may defoliate plants markedly, reducing growth and further development. In severe cases, plants die before they reach maturity. Infected mature plants may develop red rot symptoms. Such stalk when split open show discolouration, which may be continuous or discontinuous giving the stem a marbled appearance (Fig. 3.8d). Nodal tissues are rarely discoloured. If infection occurs at seedling stage plants become stunted and yellow and often do not produce any tiller. In severe cases, the seedlings wilt and die.

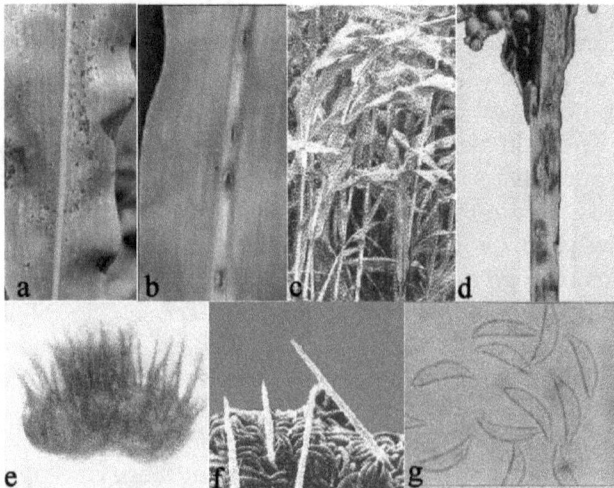

Fig. 3.8. Sorghum anthracnose

2.4. Causal organisms: The disease is caused by *Colletotrichum graminicola* (Ces.) Wilson (Synonym: *C. sublineolum* (Henn.) Kabat & Bub, Perfect state: *Glomerella graminicola* Politis). Mycelium of *C. graminicola* is gray to olivaceous, septate, and sparingly branched when grown on culture medium. The fungus produces dark brown to black colour acervulus on infected host tissue that appears as black speck on the tissue. Stromata are 70-300 µm in dia. and have prominent, dark, septate setae (up to 100 µm long) (Fig. 3.8e & f). Acervuli produce numerous erect, hyaline, nonseptate conidiophores on which conidia are borne terminally among the setae. The conidia (5 x 30 µm) are hyaline, nonseptate, and cylindric to obclavate but become sickle-shaped with age (Fig. 3.8g). Germination can occur from any area of the conidium. More than 40 races / pathotypes have been reported from different geographical areas of the world, using different sets of putative host differentials. Nine pathotypes have been reported from India.

2.5. Epidemiology: The fungus may infect below or just above the ground. If the infection is early and severe, pre emergence damping-off may occur. Conidia from wild sorghum species or crop residues serve as the primary inoculums which are spread by wind or rains. On coming in contact with water films conidia germinate producing germ tubes. Germ tube develops into an appressoria that penetrates directly through the epidermis or stomata. The disease is most severe during extended periods of cloudy, warm, humid, and wet weather, especially when these conditions occur during the early grain-filling period. Conidia are produced on lesions under conditions of high humidity or high moisture and require about 14 hr to mature at 22°C. The fungus can survive as mycelium in host residue, wild sorghum species, and some weeds and as conidia or mycelium on seed. It can persist up to 18 months in diseased residues on the soil surface. Mycelia survive for only a few days in the absence of residues. Sorghum isolates of *C. graminicola* can colonize leaves of senescent sorghums, other grasses, and some other grains.

2.6. Management: The best control for anthracnose is the use of resistant cultivars. Anthracnose can also be controlled by cultural practices. One-year crop rotation with a species other than a host offers reasonable control. Clean cultivation, elimination of crop residues and grasses on which the fungus can survive, and enhancement of the conditions that hasten decomposition of host residues have also been used to control the disease.

3. Downy mildew

3.1. Occurrence and distribution: Sorghum downy mildew (SDM) has been reported from many countries in the tropical and sub-tropical zones. It was initially originated in Africa and Asia and subsequently spread to Americas in the late 1950s. It is a common and potentially destructive disease in peninsular India. The disease infects both sorghum and maize and cause significant loss in both grain and fodder yield. SDM is quite common in Karnataka, Tamil Nadu, Andhra Pradesh and Maharashtra (Fig. 3.9) and reported from Rajasthan. The disease was in check for quite long time and occurred mainly during kharif season. Recently it is showing increasing trends and infecting both kharif as well as rabi crops in India (Pande *et al.*, 1987).

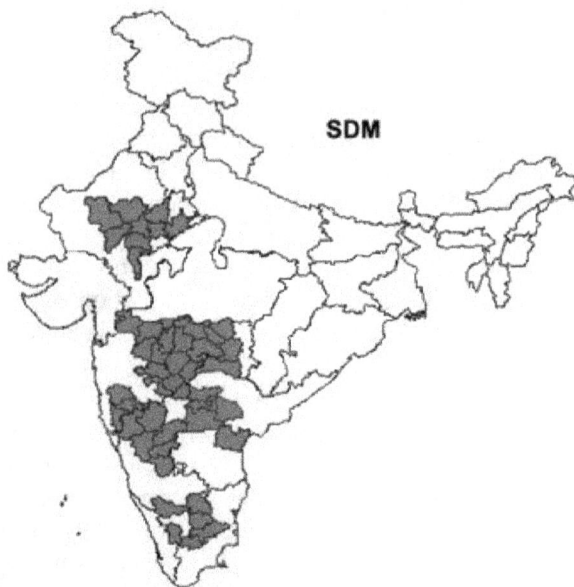

Fig. 3.9. Distribution of sorghum downy mildew in India

3.2. Economic impact: Downy mildew is a serious disease on sorghum especially in peninsular India. Systemic infection of the plant results in a barren inflorescence. There is a linear relationship between the incidence of systemic SDM and yield loss at normal sowing densities. Under favourable conditions the disease may outbreak as an epidemic. Several epidemic of downy mildew has occurred in the past. Annual yield loss due to SDM was at least 0.1 million tons in parts of India. In a single season in the USA, a SDM epidemic in grain sorghum in Texas caused an estimated loss of US$ 2.5 million. The disease affects both grain and stover yield. Crop losses estimated due systemic infection were found to vary from 2-20% depending on the time of infection and prevalence of favourable conditions and the crop cultivars grown. Monetary loss due to this disease varies from year to year and state to state depending on cultivars and prevailing weather conditions. Karnataka and Tamil Nadu generally experience more damage than other states. Recently it is fast spreading in other states. On current production and price structure of grain and stover a combined annual loss may figure around Rs 350 crores during kharif and another 780 crores during rabi. Cost of chemical management maybe around Rs. 1000-1200 per acre. Use of disease resistant cultivars is the best way to manage this disease and many cultivars have moderate level of field resistance.

Downy mildew has severe impact on human and live-stock population. Sorghum crop is grown by the marginal and poor farmers, who do not have adequate resources to invest in agriculture like water and other inputs. Saving this crop from the ravages of downy mildew would alleviate 0.3 million tones grains and 1.4 million tones green fodder, and also enhance the level of income of the poor farming community, who are mainly dependent on subsistence agriculture.

Systemically infected plants produce neither grain nor fodder and thus have significant effect on yield. Loss of stover yield is particularly important and can cause significant losses in income from crop-livestock production system in rural India since sorghum crop residue is an important feed for cattle in this system. Local infection during grain development stages results in poor grain filling and yield loss.

3.3. Symptoms: Sorghum downy mildew occurs as either systemic or localized infection. The systemic infection occurs when the pathogen colonizes the apical meristematic tissues. Systemically infected seedlings are pale yellow or have light-coloured streaking or mottling on the leaf, chlorotic and stunted and may die prematurely (Fig. 3.10a). Usually, the first symptoms are exhibited on the lower part of the leaf blade. Leaves that exhibit symptoms later show progressively more chlorosis. In cool, humid weather, the lower surfaces of chlorotic leaves become covered by a white, downy growth consisting of conidia and conidiophores of the pathogen (Fig. 3.10b). During later stages of the disease, leaves emerging from the whorl exhibit parallel stripes of vivid green and white tissue (Fig. 3.10c). The infected striped areas die, turn brown, and disintegrate, resulting in a shredded appearance of the leaf (Fig. 3.10d). Systemically infected plants do not produce any earhead. Conidia produce in the infected plants cause local infection on leaves. Local lesions are rectangular in shape, initially pale yellow in colour that later turn brown (Fig. 3.10e).

Fig. 3.10. Symptoms of sorghum downy mildew

3.4. Causal organisms: *Peronosclerospora sorghi* (Weston & Uppal.) Shaw. (Synonym: *Sclerospora sorghi* (Kulk.) Weston & Uppal.) is causal agent of the downy mildew. *P. sorghi* is an obligate parasite. However, it has been successfully grown in dual culture with host tissue on a modified White's medium. In cool, humid weather, the abaxial surfaces of chlorotic leaves become covered by a white, downy growth consisting of conidia and conidiophores of the pathogen. Conidiophores are erect, hyaline, and dichotomously branched and emerge through stomata in the lower surface of leaf (Fig. 3.11a). Conidia are 15-29 x 15-27 μm in size, hyaline, obovate, nonpapillate, and nonporoid. Oospores are produced in the mesophyll tissue between the fibrovascular bundles (Fig. 3.11b). They are spherical (25-12 μm in dia.), hyaline to yellow, and enclosed in an irregularly thickened, brown oogonial wall. Conidia and oospores both germinate by a germ tube. Nine pathotypes / races of *P. sorghi* have been reported in countries such as US, Brasil and Honduras.

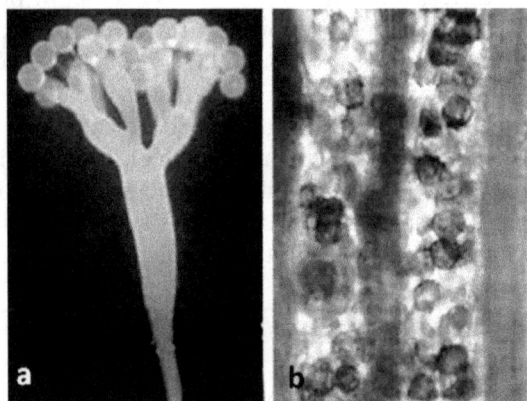

Fig. 3.11. (a) Dichotomously branched conidiophores of P. sorghi bearing conidia, and (b) oospores produced in mesophyll tissues

3.5. Epidemiology: Systemic infection of young seedlings occurs either from oospores in soil or from conidia produced on early infected plants. Such infection normally occurs a week after emergence. The minimum soil temperature for infection by oospores is 10°C. Oospore infection of sorghum is favoured by low soil moisture. Production of conidia and infection process are favoured by a cool environment and high humidity. The optimum temperature for conidia production is 18°C. The pathogen spreads through oospores in the glumes of sorghum seeds and in plant debris mixed with the seed. Conidia are fragile and play no role in the long distance dissemination of the fungus. The pathogen may be present as mycelium in infected seeds, but mycelium is inactivated when seeds are dried and thus is seemed to be of negligible importance in spreading the disease. *P. sorghi* survives as oospore in soil and plant debris. It can also infect grasses such as *Euchlaena, Panicum, Pennisetum,* and *Zea* spp. on which it can produce conidia and oospores. Occurrence of this disease even during rabi season recently may be an indication of change in pathogen population. The collateral hosts serve as source of inoculums for the sorghum crop.

3.6. Management: Deep ploughing, crop rotations, rouging of infected plants, adjust the sowing dates are some toe recommended practices for management of the disease. Most of the present day cultivars have moderate level of field resistance.

However, chemical spray becomes necessary under favourable conditions for disease. Seed treatment with Ridomyl-MZ @ 6g/kg seed + followed by one spray of Ridomyl-MZ @ 3g/liter reduces SDM incidence. Three sprays of Ridomil MZ-72 (metalaxyl and mancozeb @0.25%) at 45 days intervals beginning 40 days after sowing provided maximum disease control (82%) and increase in yield up to 49%. Use of disease resistant cultivars is recommended.

4. Charcoal rot

4.1. Occurrence and distribution: *Macrophomina* stalk rot, popularly known as charcoal rot, is a widespread disease and is reported from all the diverse areas of sorghum culture in the tropics, subtropics, and temperate regions. It is a stress related disease which is prevalent in entire rabi sorghum growing tracts of Maharashtra, Karnataka and Andhra Pradesh (Fig. 3.12). Though it is predominant disease during rabi season sometimes it may also occur on kharif crop if the crop is exposed to post-flowering drought. Moisture stress particularly post-flowering drought acts as inducing factor for the disease.

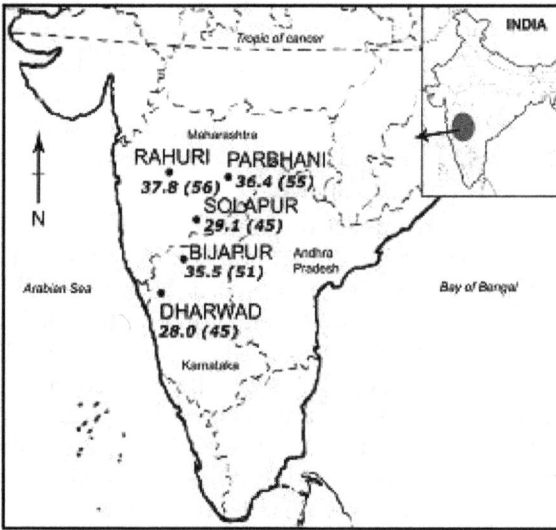

Fig. 3.12. Map showing distribution of sorghum charcoal rot in India. The locations shown in the states of Maharashtra and Karnataka represent areas prone to the disease

4.2. Economic impact: Charcoal rot occurs in mild to moderate from on rabi sorghum and mainly affects high yielding cultivars. Land races are less affected due to this disease. On an average around 20-35% plants show charcoal rot symptoms depending on cultivars and stress conditions. The pathogen infects root, destroy cortical tissues and may block water movement through vascular bundles and thus physiologically weaken the plants. Rotting and breaking of the basal internodes cause lodging of the crop which in turn facilitates further loss of water from the basal cracks. Grain filling process in such plants is greatly affected. In a diseased plant about 30-100 cm basal stalks may be damaged due to rotting and such stalks become unfit for consumption by cattle. Yield losses vary, depending on weather and the growth stage of cultivars at the time of infection and colonization by the fungus. In general, the disease becomes conspicuous near crop maturity. In some fields, particularly of hybrid sorghums, more than 50% of the plants may break over at the base. There may be four different types of crop losses that include (a) poor crop stands due to seedling blight, (b) loss in yield and quality of grain due to premature lodging of plant and improper grain filling, (c) Post production loss of grain yield in lodged plants due to destruction of such plants by termites and rodents in the field, and (d) loss in quality and quantity of fodder due to infection and destruction of the stalk. Grain yield loss depends on stage of the crop when lodging takes place and extent of lodging. Crop lodging of 100% could cause 23-64% loss in grain yield under experimental conditions. Loss in grain weight may vary from 15-55%. Financial loss due to charcoal rot is approximately Rs 450 to 650 per acre at current productivity. Monetary loss due to this disease varies from year to year and state to state depending on cultivars and prevailing stress conditions during grain filling stage. On current production and price structure of grain and stover a combined loss of around Rs 200 to 400 crores may occur in the states of Maharashtra, Karnataka, Andhra Pradesh and Tamil Nadu. Of the total loss around 55% is due to loss in grain yield and remaining due to loss in fodder yield. It is difficult to estimate cost of management of the disease directly as the disease is mainly managed through by practicing clean cultivation and cultural management. Chemicals are not generally used. Adequate level of genetic resistance is not available for this disease and drought tolerant, non-senescing sorghum genotypes generally show some resistance.

The disease has profound impact on human and livestock population and quality of the produce. The disease affects both grain and stover yield and also stover quality. Sorghum grain and stover produced during rabi are valued equally. A large section of rural population in Maharashtra and Karnataka consume rabi sorghum grain as food by making roti. Loss of production thus directly impacts sorghum farmers. The disease cause rotting of basal portion of stalk and thus reduces stover quality. In India as in many semi-arid tropical countries sorghum stover is nearly as valuable as grain. In the diseased stalk, the pathogen produces many toxins. The toxin phaseolinone, produced by *M. phaseolina*, can cause anemia in mice (LD_{50} 0.98 g/ kg/ body wt.). Thus, the disease has indirect implications on animal health. Grains produced on charcoal rot infected plant loose luster and normal size and thus fetches less price. Diseases stalks cannot be stored for long time as it loses moisture through cracks and becomes unfit for consumption by cattle.

4.3. Symptoms: Lodging of the crop and poor grain filling indicate that the crop is infected by charcoal rot pathogen (Fig. 3.13a). Infected roots and lower stem of the infected plant show water-soaked lesions that slowly turn brown or black (Fig. 3.13b). Affected stalks become soft at the base and often lodge even due to moderate wind or by bending the plants. Thus pre-mature lodging is the most apparent symptom of charcoal rot (Fig. 3.13b). When infected stalk is split open, the pith is found disintegrated across several nodes. The cortical tissues are disintegrated and vascular bundles get separated from one another (Fig. 3.13c). The vascular tubes contain numerous minute, dark, charcoal-coloured sclerotia of the pathogen which gives the disease the name charcoal rot. Normally the disease appears during post-flowering stage, but in some cases seedlings can be infected (Fig. 3.13d).

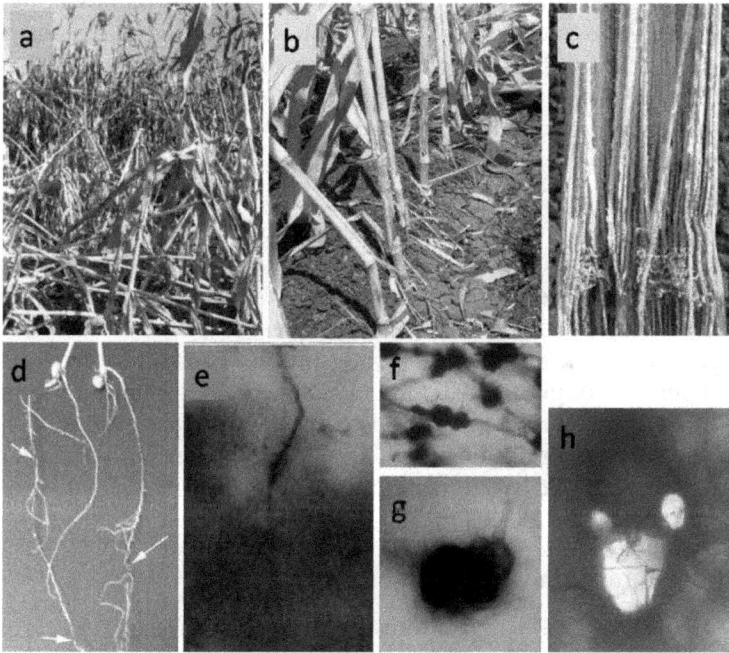

Fig. 3.13. Sorghum charcoal rot

4.4. Causal organism: *Macrophomina phaseolina* (Tassi.) Goid belongs to Deuteromycotina and its perfect state is called *Sclerotium bataticola* Taub. The mycelium is aerial, superficial or immersed, hyaline to brown, septate, profusely branched or dendroid. The fungus is highly variable in size of sclerotia and presence or absence of pycnidia. Sclerotia are called microsclerotia which are loose type, brown to black, irregular in shape and size, and are highly variable within an isolate (Fig. 3.13f). Pycnidia stage is widespread in jute and garden beans but uncommon in soybean, maize and sorghum. *M. phaseolina* is present in most cultivated soils and can infect about 500 plant species worldwide, including a wide range of agriculturally important crops.

4.5. Epidemiology: The pathogen is a soil-borne fungus that survives as sclerotia on the infected plant debris. These sclerotia serve as primary source of inoculum, which can survive for 2-4 years in the soil. Dry weather, high temperature and low soil moisture are the important factors predisposing sorghum plants to infection by *M. phaseolina*. Germination of sclerotia present in the soil is triggered by root exudates from sorghum seedlings. Germinated sclerotia can infect the primary root and cause seedling blight. If infection occurs before the emergence of secondary roots, the plants die. Less severely infected seedlings, however, survive and establish secondary roots and grow to mature plants. Fungal mycelium colonizes the xylem vessels blocking the translocation of water and carbohydrate to the upper plant parts (Fig. 3.13h). The prevalence of high soil temperature (35-38°C) and low soil moisture during the post flowering stage are the most important factors predisposing sorghum plants to charcoal rot. Under the condition of stress (moisture, temperature and photosynthesis) that often coincide with the onset of flowering, the host-defense-system is weakened and activity of *M. phaseolina* increases many folds leading to rapid and extensive rotting of roots and stalks that result in lodging of the crop. The pathogen produces at least six phytotoxins of which phaseolinone is the major product. Phytotoxins injure cell protoplast and often prime the infected plants towards more severe disease in later stages.

4.6. Management: Deep sowing, conservation of soil moisture, optimum plant density, wheat straw mulching and mixed cropping with pigeonpea are some of the practices recommended for management of charcoal rot. Early maturing varieties generally escape disease. High level of genetic resistance is not available. Strong relation of the disease with yield and environmental stresses, particularly moisture and temperature, makes the task of evaluating host resistance more challenging. Drought tolerant, lodging resistant and non-senescing sorghum genotypes are supposed to have good tolerance to charcoal rot. However, finding such genotypes with high grain yield under desirable agronomic background are often not easy. Seed treatment with talc based formulation of *Pseudomonas chlororaphis* SRB127 is reported to reduce disease severity and increase seed weight (*Das* et al., 2008).

5. Ergot or sugary disease

5.1. Occurrence and distribution: Ergot or sugary disease is a serious disease that affects hybrid seed production. It was first reported from Kenya in 1915. Two years later (1917) it was reported from India. Subsequently the disease has been reported from many countries in Africa, South America, North America and Australia. The disease is predominantly observed in hybrid seed production plot where there is lack of synchrony in flowering between male and female parents. The pathogen infects unfertilized ovaries in flowering panicles, thus preventing fertilization and seed set. Rapid pollination and fertilization of ovaries prevents ergot infection in most sorghum lines. The disease is mostly seen in Maharashtra, Andhra Pradesh, Karnataka and Gujarat where seed production for sorghum hybrid is undertaken. Ergot has implications for quarantine and seeds harvested from infected fields are often rejected in trading.

5.2. Economic impact: Ergot is a serious limiting factor in production of hybrid seed, particularly if seed-set in male sterile lines is delayed due to lack of viable pollen (Tonapi *et al.*, 2003). Disease incidence ranges from 2-12% over seasons with an average incidence of around 5% in India. There is two-fold damage; directly through loss in seed yield, and indirectly through rejection of ergot- sclerotia contaminated seed lots due to poor quality. Ergot can also cause wide spread damage of male-fertile cultivars in farmers' fields when environmental conditions favourable to pathogen occur at flowering. Hybrid seed production blocks are at great risk from ergot. Losses have been estimated at 10-80% in India. Sclerotia of *C. africana* contain toxic chemicals, in particular the alkaloid dihydroergosine. The disease has great impact on livestock's. Feeding sclerotia contaminated sorghum can affect milk production in cows and sows, and weight gain in cattle. There is a stock feed limit of 0.3% sclerotia by weight in sorghum grain (approximately 1 sclerotium per 100 seeds). Seed lot containing sorghum with sclerotia levels higher than 0.3% is rejected by grain merchants. Floatation techniques can be useful for quickly counting sclerotia.

5.3. Symptoms: Ergot is predominantly a disease of ovary. The infected flowers do not produce grain, reduce grain and seed quality, germination, seedling emergence and predisposes seedlings to other diseases. The pathogen colonizes unfertilized ovary both internally and externally in about 2-3 days converts it into white a fungal mass. The fungus attack unfertilized ovaries, resulting in exudation of honeydew a thin viscous, pinkish to brownish liquid drops that are sweet in test and sticky (Fig. 3.14a). That gives the name sugary or honeydew disease malady. Infected panicles appear as white with fresh colourless honeydew or black if saprophytically colonized and can be recognized from a distance (Fig. 3.14b). Severe incidence results in discoloured panicles without any grain (Fig. 3.14c). Under warm dry conditions sphacelia harden and form solid dense sclerotia known as ergot (Fig. 3.14d).

Fig. 3.14. Ergot or sugary disease

5.4. Causal organisms: *Sphacelia sorghi* McRae (sphacelial/conidial stage), *Claviceps sorghi* Kulkarni *et al.* (in Asia) (Sclerotial stage), *C. africana* Frederickson, Mantle & de Milliano (in Africa), *C. sorghicola* Tsukiboshi, Shimanuki & Uematsu (in Japan) were reported as causal agents of the disease. Among these three strains, *C. africana* is the most widespread and is found throughout Asia, Africa, the

Americas, and Australia. It can produce macroconidia, microcondia and secondary conidia. Macroconidia are oval to oblong (5-8×9-17 µm), slightly constricted at the center, and have two polar vacuoles; microconidia are spherical (2-3 µm dia); and secondary conidia are borne on sterigmata like structures with a distinct protruding hilum. Sclerotia germinate to form stromata that appear as pale with globose proliferations with fully extended pigmented stipes; capitula are subglobose and internally purple and contain perithecia. Perithecia contain asci and each mature ascus produces eight filiform and hyaline ascospores of 1.2×45 µm size.

5.5. Epidemiology: Sorghum flowers are susceptible to ergot infection when stigma is receptive. Primary infection takes place either by ascospores or by conidia. Ascospores are produced in perithecia from germinating sclerotia. Conidia are produced on collateral grass weeds, wild sorghums, and infected panicles or on plant debris in the soil. Honeydew produced on the infected florets contains numerous macroconidia and secondary conidia that help in secondary spread of the disease. Several conidial cycles are completed in a season and conidia are spread by wind, rain and insects. Low night temperatures (<12^{0}C) during the period between 3-4 weeks before flowering to 5 days after flowering, high RH (>90%), cloudy weather or panicle wetness after stigma emergence, favour ergot disease development. The pathogen survives off-seasons via infected panicles left in the field or via sclerotia that are mixed with the seed during threshing and processing. Collateral hosts (*Penniseum typhoides, Ischaemum pilosum* and *Panicum maximum*) also play role in survival.

5.6. Management: Early sowing avoids the occurrence of the sugary disease. Removal of collateral host plants from the field bunds helps to reduce pathogen inoculum and disease. Mechanical removal of sclerotia from seeds, by washing in 30% salt water followed by 3 rinsings in plain water before sowing reduces seed contaminated infection. In seed production plots, ensuring synchrony of flowering between A and R lines avoids the occurrence of disease. Spraying panicles with fungicides (0.1% Bavistin/ 0.2% Thiram/ 0.2% Tilt/ 0.2% Mancozeb) minimizes disease and its subsequent spread. First spray should be done at 50% flowering stage and rest two sprays at 10 days interval.

6. Leaf blight and Leaf spots

6.1. Occurrence and distribution: Leaf blight and leaf spots are important diseases causing substantial economic losses to forage and dual sorghum. Many types of leaf spots are reported in sorghum. Few are economically important while others assume significance under specific conditions. Sooty stripes, zonate leaf spot, gray leaf spot and target leaf spot are frequently encountered on forage cultivars grown in central and northern part of India. Disease incidence is more on purple or red pigment genotypes and relatively less on tan cultivars. The leaf diseases occur almost every year in moderate to severe form on forage sorghum in Rajasthan, Haryana, Uttarakhand and Uttar Pradesh and on dual sorghum in the states of Maharashtra, Karnataka and Andhra Pradesh.

6.2. Economic impact: Leaf blight and leaf spots diseases occur in mild to moderate from on sorghum both during kharif and rabi seasons. Incidence varies over locations and seasons and occasionally it may assume epidemic form. Disease severity ranges from 1.0 to 4.0 (on a 1-5 scale) depending on environmental conditions. Average disease grade of 2.2 is common on forge and grain sorghum. They affect photosynthetic areas and thus reduce amount of current photosynthesis. Financial loss varies with incidence, environment, and resistance in cultivars. Seedling blight phase of the disease may sometimes affect final plant stand in the field. Damages on leaf areas result in reduced yield of grain as well as fodder. Grain yield losses of 50% or more can occur under severe epidemics. At national they are important mainly on forage sorghum. Major economic loss occurs due to leaf damage which has impact on grain and fodder yield. Estimation of monetary loss is complex as quantifiable effect of these diseases on fodder loss in not known and grains are hardly considered for forage crop. However, in grain sorghum they may cause considerable economic loss. Cost of disease management through chemical may be around Rs. 800 per acre.

These diseases have impacts on human and cattle population and on quality of the produce. They affect grain and stover yield, stover quality and also sugar accumulation in sweet sorghum. Loss of production thus directly impacts sorghum farmers. Severely infected plants produce neither grain nor quality fodder and thus have significant effect on yield. Reduction in forage yield is particularly important and can cause significant losses in income from crop-livestock production system in rural India especially in northern states since cattle population in these regions heavily depends on sorghum fodder.

6.3. Symptoms: The typical symptoms of leaf blight are presence of long elliptical necrotic lesions which are straw coloured in the centre with dark brown margins, especially in older plants. Long, elliptic, reddish purple or yellowish tan lesions, up to 12mm wide and 2.5-15 cm long, develop first on lower leaves, and then progresses to upper leaves (Fig. 3.15a). These spots later enlarge and coalesce, so that leaves wilt, turn purplish gray and may die. The lesions vary in morphology according to different levels of host resistance. Grains are not infected. Numerous sporulation of the pathogen causes surfaces of necrotic lesions to appear dark gray, olive, or black. In damp weather, large numbers of grayish black spores are produced on the lesions, often in concentric zones. Very long lesions may develop and coalesce destroying large areas of leaf tissue giving the crop a distinctly burnt or blasted appearance (Fig. 3.15b).

Fig. 3.15. Sorghum leaf blight

Zonate leaf spot is conspicuous on sorghum leaves as circular, reddish purple bands alternating with straw-coloured or tan areas, which form a zonate, pattern with irregular borders (Fig. 3.16a). The spots often occur in semicircular patterns along the margins of leaves (Fig. 3.16b). Leaf sheaths can also become infected. During warm, wet weather, pink to salmon coloured gelatinous spore masses form above stomata. High incidence during seedling stage may result in severe defoliation and even death of infected plants. Sooty stripe is characterized by small, circular to elongated reddish brown spots on leaf with distinct yellow haloes (Fig.3.16c). Lesions coalesce to form necrotic areas with blackish or sooty centre. Numerous sclerotia are formed on the surface of the lesion giving it a rough appearance (Fig. 3.16d).

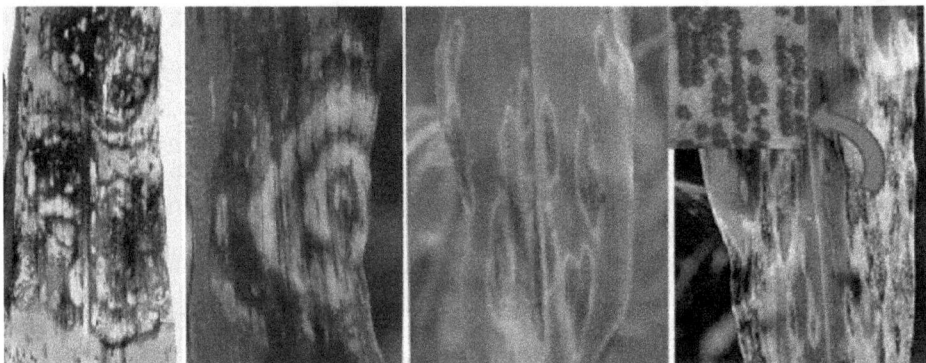

Fig.3.16. Sorghum leaf spots

6.4. Causal organisms: Leaf blight is caused by *Exserohilum turcicum* (Pass.) Leonard & Suggs. (Synnonym: *Helminthosporium turcicum* Pass, *Bipolaris turcica* (Pass.) Shoemaker, *Dreschslera turcica* (Pass.) Subramanian & Jain. Zonate leaf spot is caused by Gleocercospora *sorghi* D. Bain & Edg and Sooty stripe is caused by *Ramulispora sorghi* (*Ell. & EV.*) *L. S. Olive & Lefebvre.* (Synonym: *R. andropogonis* Miura). *E. turcicum* produces conidiophores in group of 2-6. Conidia are produced acropleurogenously on the conidiophores. Conidia are light gray, straight or spindle-shaped or curved with rounded ends. They are 3-11 septate and have a protruding basal hilum, and measure 10-20 × 28-153 μm. They germinate by polar germ tubes. The pathogen can also infect maize, Johnsongrass, teosinte and Sudangrass. G. *sorghi* produces fruiting bodies known as sporodochia. Pathogen hyphae that are emerged through stomata produce these sporodochia. Conidia are borne in a pinkish to salmon-coloured slimy matrix, are hyaline, needle-shaped, straight or slightly curved, 4-14 septate and of variable length (20-195 x 1.4-3.2 μm). Black sclerotia are usually produced within the necrotic tissues on sorghum. Mycelia of *R. sorghi* get densely packed inside substomatal cavity forming sporodochia. Conidia are borne singly at the tips of the conidiophores that emerged through stomata. Many conidia aggregate into a gelatinous mass. Conidia (36-90 x 2-3 μm) are filiform, curved, tapering at the ends, hyaline, and 3-8 septate and have one to three lateral, septate or nonseptate branches. The sclerotia are attached to the substomatal stroma by narrow columns of sclerotized hyphae that pass through the stomata. Often, the sclerotia germinate by producing sporodochia and conidia. The fungus grows slowly in culture, with optimum growth at 28°C, forming a circular, compact colony with an entire margin. After several weeks, sporulation occurs, and a light pinkish, cone shaped, gelatinous mass forms over the rough surface of the black fungal growth. Sclerotia are not known to be produced in culture.

6.5. Epidemiology: Foliar pathogens favour moderate temperature (20-27°C), extended periods of cloudy weather, high humidity, heavy dews, warm and humid wet weather. Intermittent rains and cool winds favour secondary spread with in fields and also for long distance dissemination of pathogen by wind. Leaf blight and leaf spots are economically important and widespread disease of sorghum especially in humid areas. Leaf blight pathogen persists as mycelia and conidia in infected crop residues on or in the soil and on the glumes of sudangrass seed. Conidial cells can be transformed into chlamydospores. Air-borne conidia are responsible for secondary spread within and between fields. At least two types of resistance are known: polygenic resistance, characterized by few and small lesions, and monogenic resistance, characterized by a hypersensitive fleck and little or no lesion development. Zonate leaf spot pathogen overseasons as sclerotia formed within the dead tissue of old leaf lesions, where they appear as small, raised bodies in lines parallel with the veins. They are also produced abundantly on millet and other grasses, and sclerotia from these hosts may also function in the survival of the pathogen. The sclerotia germinate and form conidia, which infect the next crop. During wet weather, conidia are produced on the new lesions and cause further spread of the disease. The fungus may also be carried on seed. Sooty stripe pathogen survives as sclerotia in leaf residue on or below the soil surface. When conditions become favourable, sclerotia and sclerotial sporodochia produce abundant conidia. Wind and rain disseminate the conidia and sclerotia to new foliage and to other plants. The fungus survives on some perennial hosts, such as *S. bicolour* subsp. *bicolour*, *S. haleppnse*, and *S. purpureosericeum*.

6.6. Management: Use of good quality and healthy seeds, crop rotation with non-hosts once in two years, clean cultivation before taking up planting and after planting, cultural practices like adjusting dates of sowing, proper tillage and intercropping with non-hosts reduces foliar disease incidence. Destruction of weeds, volunteer, wild sorghum and alternate hosts help to reduce primary inoculum. Need based use of fungicides (metalaxyl, captan, thiram) with right dosage and at right time is beneficial. However, use of disease resistant cultivars is the best option considering the cost of chemical treatments.

7. Viral diseases

7.1. Occurrence and distribution: In India virus diseases on sorghum are distributed in all sorghum growing regions. Three groups of plant viruses namely, potyvirus, tenuivirus and rhabdovirus infect sorghum: Among potyviruses sugarcane mosaic virus (SCMV), maize dwarf mosaic virus (MDMV), johnson grass mosaic virus (JGMV) and sorghum mosaic virus (SrMV); among tenuiviruses maize stripe virus or MStV (formerly MStpV); and among rhabdoviruses maize mosaic virus (MMV) have been reported to naturally infect sorghum crop in different countries. However, MStV, MMV and SRSV are economically important on sorghum in India. In recent times there is an increasing trend in incidence of these viruses on sorghum. This may be due to change in agricultural practice, intensive cultivation and introduction of high yielding hybrids and varieties. Sorghum stripe has assumed economic significance particularly in *kharif* and irrigated *rabi* crops in Maharashtra, Karnataka and Andhra Pradesh.

7.2. Economic impact: Viral diseases occur in mild to moderate from on sorghum both during kharif and rabi seasons. Early infected plants generally do not produce any ear head and thus loss is directly related to per cent disease incidence. Sometimes late infected plats may produce earhead but with chaffy grains. Disease incidence may vary from 4 to 14%. On an average yield loss of 2-4% for grain and 2-8% for fodder is common particularly in peninsular India. Yield losses vary with the stages of infection. Infection at early stages results in higher grain and fodder losses in comparison to that at later stages. Around 2-4% yield losses is common in India. Reduction of plant height, ear head weight and grain mass to the extent of 73, 93 and 25 per cent respectively have been reported in the variety CSV15. Financial loss due to viral diseases is approximately Rs 550 to 600 per acre. Total monetary loss due to this disease varies from year to year and state to state depending on cultivars and prevailing weather conditions. On current production and price structure of grain and stover an annual loss of around Rs 890 crores may occur in the states of Maharashtra, Karnataka, and Andhra Pradesh. Around 70% of the total monetary loss occurs due to loss on rabi crop. Clean cultivation and cultural management keep the disease in check. Resistant sources are not known however, the disease can be managed by control of insect vector through application of insecticides. Cost of management will be around Rs. 1200 per acre.

The disease has considerable impact on human and live-stock population. It affects sorghum during both the growing seasons and has profound effect on yield and quality. A large section of rural population in Maharashtra and Karnataka consume rabi sorghum grain as food by making roti. Loss of production thus directly impacts sorghum farmers. Symptomatic plants produce neither grain nor fodder and thus have significant effect on yield. Reduction in stover yield is particularly important and can cause significant losses in income from crop-livestock production system in rural India since sorghum crop residue is an important feed for cattle in this system. Apart from effects on grain viral diseases have severe adverse effect on forage quality. In forage sorghum leaf protein content was reduced by 16.98 and 37.58 per cent in MKV Chari-1 and SSG 59-3, while the total soluble solid content reduction was 3.25 and 3.19 per cent respectively. Infection with SRSV significantly reduced plant height, leaf area, juice yield and chlorophyll content at all stages of growth.

7.3. Symptoms

7.3.1. Sorghum stripe (MStV): The characteristic external symptoms on sorghum include appearance of continuous chlorotic stripes/ bands between the veins of the infected leaf (Fig. 3.17a). The width of chlorotic stripe varies depending of stages of disease development. Leaves on the infected plant appear as yellow with continuous stripes progressing from the base towards the tip of the leaves. The infection is systemic and subsequent leaves appear with yellow stripes on them. Affected plants appear stunted in growth. Early infected plant dies sooner or later without emergence of earhead (Fig. 3.17b). Plants infected at later stages appear dwarf with short internodes, show partial exertion of earhead having few or no seed formations (Fig. 3.17c). The disease is also known as chlorotic stripe stunt or sorghum stripe disease (SStD), the name derived from its characteristic stripe symptoms on leaf. In an infected plant, expression of first chlorotic symptom can take place on any leaf starting from 4th to 11th leaf. Symptom expression frequency increases from 4th to 7th leaf, reaches peak on 7th or on 8th leaf and thereafter gradually decreased on subsequent upper leaves.

Fig. 3.17. Viral disease of sorghum

7.3.2. Sorghum mosaic (MMV): Sorghum mosaic is characterized by fine discontinuous chlorotic streaks between the veins on leaf (Fig. 3.17d). The lesions become necrotic as the disease progress. Infected plants become stunted in growth with short internodes. Early infected plant dies sooner or later without emergence of earhead. Plants infected at later stages may develop earhead with or without grain. It should be noted that there is chance of multiple virus infection in a plant under natural field conditions. This chance increases when a single vector is able to transmit more than one virus (e.g., *P. maidis* can transmit both MStV and MMV). In such case the symptoms will vary accordingly creating more confusion at field level identification.

7.3.3. Sorghum red stripe (SRSV): The disease is characterized by systemic symptoms of mosaic followed by necrotic red stripe and temperature dependent red leaf (Fig. 3.17e). Severely infected crops in a field may develop general necrosis and such crop produces burning appearance when viewed from a distant (Fig.

3.17f). Recently there was an outbreak of red stripe disease in Solapur and adjoining regions in Maharashtra during 2010.

7.4. Causal organisms: Occurrence of sorghum stripe was first reported in India during nineties. The sorghum isolates was found to be the variants of MStV and designated as MStV-Sorg (or MStV-S) to distinguish it from MStV which readily infects maize. Sorghum mosaic characterized by fine discontinuous chlorotic streaks between the veins was first observed on sorghum in peninsular India during 1988. In immuno-double diffusion tests, the virus reacted positively with antisera to MMV from Reunion (MMV-RN) and Hawaii (MMV-HI), and was designated as MMV-S isolate. One of the earliest reports of the occurrence of sorghum red stripe in India was from Maharashtra. A potyvirus naturally infecting sorghum grown in the proximity of sugarcane in Maharashtra was termed as sugarcane mosaic virus-Jg (SCMV-Jg). Later the virus has been named as sorghum red stripe virus-Indian isolate (SRSV-Ind). Sorghum red stripe virus (SRSV) was reported to be related to potyvirus SCMV but was distinct from MDMV-A, MDMV-B and SrMV.

7.5. Viral transmission: MStV-S and MMV-S are transmitted by insect vector *Peregrinus maidis*, the delphacid plant hopper on sorghum. *P. maidis* is known as shoot bug and is also a major pest on sorghum in India (Fig. 3.18a). Shoot bug sucks sap from the leaves, leaf sheaths and stem during exploratory feeding and in the process transmit virus from healthy to diseased plants. Figure 18 shows a colony of *P. maidis* with different stages of the vector on the sorghum leaf. *P. maidis*

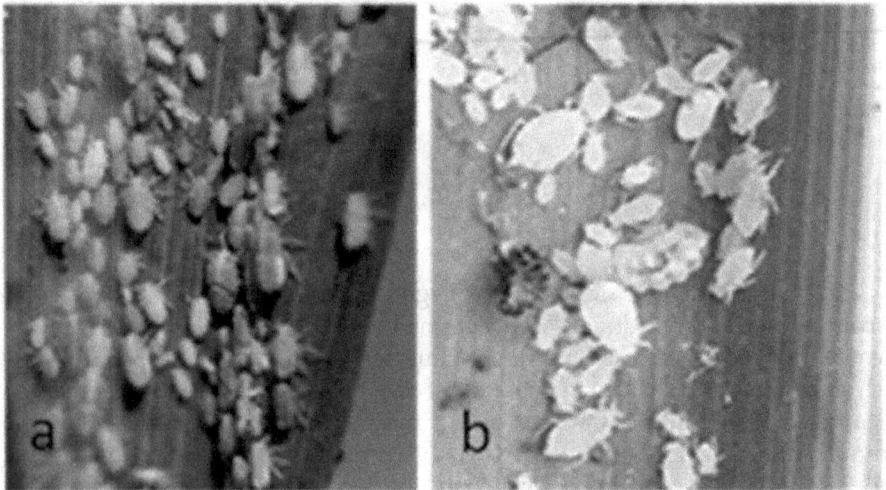

Fig. 3.18. Different stages of the vector (a) Plant hopper (P. maidis), and (b) Sugarcane aphid

requires feeding on the infected plant for at least four hours to acquire the virus. After the virus is acquired, the vector needs another 8-22 days to be ready for infection of other plant. The vector requires a minimum of 1 hour feeding to transmit the virus in a plant. A viruliferous vector retains the virus until its death and transmits it from one generation to another through eggs. The virus multiplies in the vector and is transmitted in a persistent manner (Transovarial transmission). The nymphs and macropterous females are more efficient transmitters of MStV and MMV than the males. Neither of these two viruses (MStV-S and MMV-S) is transmitted by seed or mechanically by plant sap. Mode of transmission of sorghum red stripe virus (SRSV) is not yet known. The potyvirus SCMV which is related to SRSV may be transmitted mechanically by plant sap and by various aphids in non-persistent manner.

7.6. Epidemiology: Plant growth stages between 36 to 65 days after emergence have been identified as highly susceptible for the development of chlorotic stripe virus in rabi sorghum. MStV-S and MMV-S are transmitted by a shoot bug (*P. maidis*) in a persistent manner. Bothe these viruses and the vector can infect wild graminaceous hosts like Johnsongrass. Such wild grasses on the field bund and ratooned sorghum crop in the field serve as important sources for primary inoculums at the beginning of the season. The persistent and transovarial nature of virus transmission across insect generations contributes greatly to the perpetuation of the disease. Early shown crop develops more incidence compare to late sown one.

7.7. Management: The disease can be managed or its incidence can be reduced by practices like clean cultivation, vector control and adjustment of sowing time (Narayana *et al.*, 2011). The practice of uprooting and burning of the infected plants help to reduce source of inoculum for the vector and thus reduce spread of the disease in the field. Spraying of Imidachlorpid @ 1.5 ml/L water effectively reduces vector population and the disease. Disease incidence is greatly reduced as sowing of *rabi* sorghum is shifted from September to October. When sowing is delayed from 1st week of September to 1st or 3rd week of October disease incidence is reduced by 40 and 65 per cent respectively. Therefore, early sowing of *rabi* sorghum should be avoided to reduce crop loss by this viral disease.

REFERENCES

Audilakshmi, S., Das, I. K., Ghorade, R. B., Mane, P. N., Kamatar, M. Y., Narayana, Y. D., and Seetharama, N. 2011. Genetic improvement of sorghum for grain mould resistance: I. Performance of sorghum recombinant inbred lines for grain mould reactions across environments. Crop Prot. 30: 753-758.

Bhat, R. V., Shetty, P. H., Amruth, R. P., and Sudershan, R. V. A. 1997. Food borne disease outbreak due to the consumption of moldy sorghum and maize containing fumonisins mycotoxins. Journal of Toxicol. Clin. Toxic. 35: 249-255

Das, I. K., Audilakshmi, S., and Patil, J. V. 2012. Fusarium Grain Mold: The Major Component of Grain Mold in Sorghum (*Sorghum bicolour* L. Moench). The Eur. Jour. of Plant Sci. and Biotech. 6: 45-55.

Das, I. K., Indira, S., Annapurna, A., Prabhakar., and Seetharama, N. 2008. Biocontrol of charcoal rot in sorghum by fluorescent pseudomonads associated with rhizosphere. Crop Prot. 27: 1407-1414.

Das, I.K., and Patil, J.V. 2013. Assessment of economic loss due to grain mold of sorghum in India, Pages 59-63 In: Compendium of Papers and Abstracts: Global Consultation on millets promotion for health and nutritional security, 18-20 Dec., 2013 (editors: S. Rakshit, JK Das, G. Shyamprasad, JS Mishra, CV

Ratnavathi, RR Chapke, Vilas A Tomar, B Dayakar Rao and JV Patil). Society for Millet Research, Directorate of Sorghum- Research, Hyderabad-500030, AP, India pp. 356.

Frederiksen, R. A. 1986. Compendium of sorghum diseases. Amer. Phytopathol. Soc. Texas. P82.

Indira, S., and Rana, B. S. 1997. Variations in physical seed characters significant in grain mould resistance of sorghum. In: Proceedings of the International Conference on Genetic Improvement on Sorghum and Pearl Millet, 23e27 September 1996. INSTORMIL and ICRISAT, Lubbock, USA, pp. 652-653.

Mathur, K., Thakur, R. P, Neya, A, Marley, P. S., and Casela, C. R. 2002. Sorghum anthracnose- problem and management strategy. Pages 211-220 *in* Sorghum and Millets Diseases (Leslie JF, ed.). Ames, Iowa, USA: Iowa State Press.

Narayana, Y. D., Das, I. K., Bhagwat, V. R., Tonapi, V. A., and Patil, J. V. 2011. Viral disease of sorghum in India, Directorate of Sorghum Research, Rajendranagar, Hyderabad 500030, Andhra Pradesh, India. 30pp. ISBN: 81-89-335-35-9

Pande, S., Bock, C. H., Bandyopadhyay, R., Narayana, Y. D., Reddy, B. V. S., Lenné, J. M., and Jeger, M. J. 1997. Downy mildew of sorghum. Information Bulletin no. 51. Patancheru 502 324, Andhra Pradesh, India: International Crops Research Institute for the Semi-Arid Tropics. 32pp.

Thakur, R. P., Reddy, B. V. S., and Mathur, K. 2007. Screening Techniques for Sorghum Diseases. Information Bulletin No. 76. ICRISAT, Patancheru 502 324, Andhra Pradesh, India, ISBN 978-92-9066-504-5, 92 pp.

Tonapi, V. A, Wirojwattanakul, K., Vinh, D. V., Thein, M. M., Navi, S. S., and Tooley, P. W. 2003. Prevalence of Sorghum Ergot in Southeast Asia. International sorghum and millets newsletter 44: 95-97.

Diseases of Field and Horticultural Crops
Editor-in-Chief: **P. Chowdappa**
Published by: **Daya Publishing House, New Delhi**

Pages **89-112**

4

Diseases of Chickpea

S.C. Dubey and Birendra Singh

Chickpea (*Cicer arietinum* L.) known as Bengal gram, gram, Spanish pea in English and chana in Hindi, is a high protein crop cultivated in many areas of the world. It is one of the first grain legumes to be domesticated by humans (Vander Maesen, 1972). Chickpea contain 20.4-28.1% protein, which is 3 times greater than the cereals. The chemical score of protein (biological value) improves greatly when wheat and rice is combined with one of the pulses, because of the complementary relationship of the essential amino-acids. While cereals as a group are relatively deficient in lysine and rich in the sulphur containing amino-acids, the reverse is true in pulses. Chickpea contains 6.5% lysine as compared to 2.5% in cereals. The other important reason for such a significant place for pulses in agriculture is to their capability to fix nitrogen from the atmosphere and improving soil health.

Unlike other pulses which are primarily used as *Dhal*, the chickpea has multiple uses as *besan* for preparation of sweets and *namkeen*, consumed as whole grain, as salad, parched and roasted health food. India is the largest producer of chickpea in the world sharing 65% of area and 69% of production. In India it is grown in 8.17 mha with total production of 7.48 m tonnes with average productivity of 915 kg ha^{-1} (Agricultural Statistics at a Glance, 2011). Madhya Pradesh, Uttar Pradesh, Rajasthan, Maharashtra, Gujarat, Andhra Pradesh and Karnataka are the major chickpea producing states sharing over 93% area. The area under chickpea cultivation in the northern part of India has been reduced by 29% during 1975 to 1990 due to increase in irrigation facilities and non-availability of input responsive varieties of chickpea. This reduction in area has been compensated by the increase in area in central and southern states (Table 4.1). The increase in production has mainly been due to increase in its productivity level from 642 kg ha^{-1} in 1971-72 to 915 kg ha^{-1} in 2009-10. Production of chickpea is insufficient relative to the needs of human nutrition, particularly in the developing countries like India, where the green revolution has occurred only in cereals production. Amongst the

Table 4.1. Area, production and yield of chickpea in different states

State	Area (000'ha)						Production (000'tonnes)						Yield (kg/ha)					
	1971-72	1991-92	2001-02	2006-07	2008-09	2009-10	1971-72	1991-92	2001-02	2006-07	2008-09	2009-10	1971-72	1991-92	2001-02	2006-07	2008-09	2009-10
Andhra Pradesh	66	64	285	602	609	650	25	47	363	653	857	850	379	734	1274	1085	1410	1308
Bihar	241	150	68	133	61	60	170	148	65	117	57	60	705	987	957	879	920	1014
Gujarat	57	67	49	246	175	130	40	40	27	214	177	130	702	597	554	870	1010	947
Haryana	1119	305	143	108	123	80	647	201	122	91	128	60	578	659	853	843	1040	738
Karnataka	152	196	480	651	726	970	62	77	282	308	401	570	408	393	587	473	550	591
Madhya Pradesh	1686	2138	2554	2463	2841	3090	1148	1715	2408	2413	2786	3300	681	802	943	980	980	1071
Maharashtra	433	434	756	1308	1143	1290	133	206	450	924	774	1110	307	475	595	706	680	863
Punjab	335	25	7	4	3	-	282	18	6	4	3	-	842	720	873	1000	1170	-
Rajasthan	1642	1029	970	1011	1260	880	885	679	736	873	981	530	539	660	759	863	780	604
Uttar Pradesh	1989	1105	841	675	554	620	1566	943	817	501	562	510	296	853	971	742	1010	824
Sub total	7720	5513	6153	7201	7495	7770	4958	4074	5276	6098	6726	7120	-	-	-	-	-	-
Other States	190	67	267	429	398	400	122	46	194	232	334	360	-	-	-	-	-	-
Total	7910	5580	6420	7630	7893	8170	5080	4120	5470	6330	7060	7480	642	739	853	830	890	915

factors responsible for low yield are diseases, pests and non-availability of input responsive varieties of chickpea. In India 30 fungi, 1 bacterium, 7 viruses, 1 mycoplasma and 44 nematodes have been reported on chickpea (Nene *et al.*, 1996). Economically important diseases of chickpea are wilt, ascochyta blight, dry root rot, grey mould, wet root rot and stunt (Table 4.2) and their ranks in different states of India varied (Table 4.3). The chickpea diseases have been reviewed earlier (Nene and Reddy, 1987; Haware, 1998; Dubey *et al.*, 2007; Dubey and Singh, 2010). Present chapter is focused on distribution and losses along with management options available for the major chickpea diseases in India.

Table 4.2: The important diseases of chickpea and economic losses

Disease	Causal organism	Extent of loss (%)	References
Major diseases			
Wilt	*Fusarium oxysporum* f. sp. *ciceris* (Padwick) Matuo and K. Sato	10 10-20 14.1- 32.0	Singh and Dahiya, 1973 Dubey *et al.*, 2010
Ascochyta blight	*Ascochyta rabioi* (Pace.) Labr.	30-50 5-75	Bedi and Athwal, 1962 Grewal, 1982
Dry root-rot	*Rhizoctonia bataticola* (Taub.) Butler = *Macrophomina phaseolina* (Tassi.) Goid	10-20 5-35	Vishwadhar and Chaudhary, 2001 Anonymous, 2010
Grey mould	*Botrytis cinerea* (Pers. ex Fr.)	70-100	Grewal and Laha, 1983
Wet root rot Stunt	*Rhizoctonia solani* Kuhn Luteoviruses *1) Beet western yellows virus* *2) Chickpea stunt disease associated virus*	8-10 80-95	Vishwadhar and Chaudhary, 2001 Kotasthane and Gupta, 1978
	Geminivirus*Chickpea chlorotic dwarf virus*	75-100	Horn *et al.*, 1993
Minor diseases			
Rust	*Uromyces ciceris arietini* (Grogn.) Jacz and Boyer	32-40	Anonymous, 2010
Stemphylium blight	*Stemphylium sarciniforme* (Cav.) Wilts.	10-30	Anonymous, 2010

Table 4.3. Diseases of chickpea in different states of India in order of their importance

State	Diseases
Uttar Pradesh, Bihar, Jharkhand, West Bengal and Assam	1. Wilt 2. Dry root rot 3. Grey mould 4. Wet root rot
Punjab, Haryana, Himachal Pradesh, Jammu & Kashmir and Uttarakhand	1. Wilt 2. Ascochyta blight 3. Grey mould 4. Dry root rot 5. Wet root rot
Gujarat, Maharashtra, Madhya Pradesh, Chhattisgarh and Rajasthan	1. Wilt 2. Dry root rot 3. Stunt virus
Andhra Pradesh, Karnataka and Tamil Nadu	1. Dry root rot 2. Stunt virus 3. Wilt

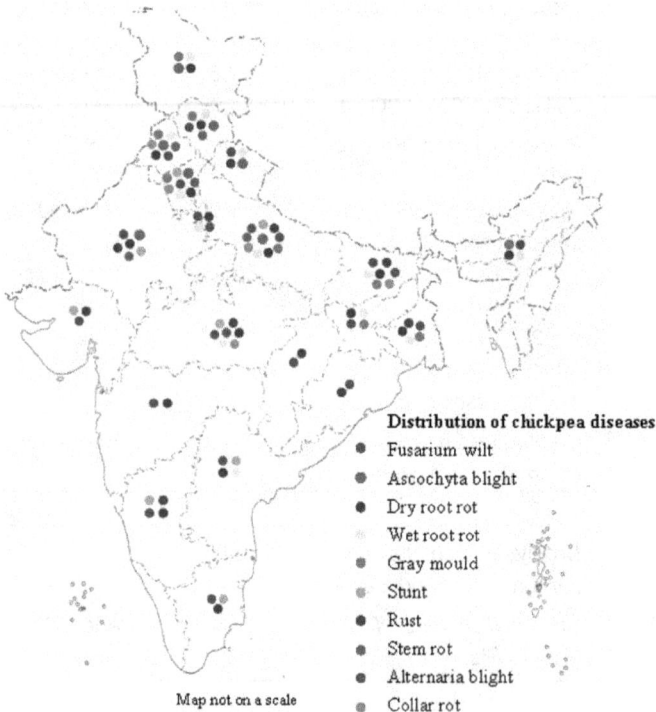

Distribution of chickpea diseases
- Fusarium wilt
- Ascochyta blight
- Dry root rot
- Wet root rot
- Gray mould
- Stunt
- Rust
- Stem rot
- Alternaria blight
- Collar rot

Map not on a scale

Fig 4.1. A map of India showing distribution patterns of the major diseases of chickpea

1. Wilt

1.1. Occurrence and distribution: Among the diseases affecting chickpea, wilt caused by *Fusarium oxysporum* f. sp. *ciceris* is one of the most severe diseases throughout the world (Gupta *et al.*, 2009). In India, it is prevalent in all chickpea growing states (Fig. 4.1).

1.2. Economic impact: Wilt causes an annual loss of 10% (Singh and Dahiya, 1973). However, it was observed that early wilting causes 77-94% losses, while late wilting causes 24- 65% losses (Haware and Nene, 1980). A survey was conducted for the occurrence of chickpea wilt during 2009-2010 and the highest wilt incidence was recorded in Karnataka (2-48%) and the lowest (in traces) in Andhra Pradesh. In Bihar, Rajasthan and Maharashtra, the wilt incidences were 2-20%, 8% and 20%, respectively (Anonymous, 2010). Recently, Dubey *et al.*, (2010) has made an extensive survey in the major chickpea growing states of India and observed that the incidence of chickpea wilt varied from 14.1 to 32.0% with the highest in the state of Rajasthan, followed by Jharkhand and the lowest in Punjab.

1.3. Symptoms: The disease can affect the crop at any stage of growth. Early wilt symptoms were recorded on susceptible cultivars within 25 days of sowing. The infected seedlings showed drooping of leaves, change of colour to a dull green and uneven shrinkage at the collar region. Whole seedling collapse and lie on the ground. Late wilt also occurred during reproductive growth, but affected plants did not show external signs of root-rot. Such wilted plants when cut transversely in the colour region or split the stem vertically from the colour region downwards, brown to black discolouration of both pith and xylem can be seen. Partial wilting also occurs occasionally (Fig. 4.2).

Fig. 4.2. Symptoms of wilt (a) field view of chickpea wilt, (b) drooping of leaves (c) internal blackening of the stem.

1.4. Causal organisms: Chickpea wilt was first reported from India (Butler, 1918). Earlier it was believed that two different causes were associated with wilt, one was the species of *Rhizoctonia* resembling to *R. bataticola* while others was of physiological disorder (Dastur, 1935). Finally, the causal agent was compared a *F. oxysporum* f. sp. *ciceri* (Chattopadhyay and Sen Gupta, 1967).The mycelium of *F. oxysporum* f. sp. *ciceris* is delicate, white, cottony, becoming felted and wrinkled with age. Hyphae are septate and profusely branched. Microconidia are oval to cylindrical, straight to curved, 2.5-3.5x5-11 µm on simple short conidiophores. Macroconidia develop on the same conidiophores; thin walled, 3-5 septate, fusioid, pointed at both ends and measure 3.5-4.5x25-65 µm in size. Chlamydospores are smooth or rough walled, terminal or intercalary and may develop singly in pairs or in a chain. Recently, Dubey *et al.* (2010) observed high level of variability in morphological characters of 112 isolates of *F. oxysporum* f. sp. *ciceris.*

Fusarium wilt is one of the most important disease, limiting chickpea production worldwide, has been the target of breeding for resistance (Kumar *et al.,* 1985). Cultivation of resistant cultivars is one of the most practical and cost efficient strategies for managing this disease. However, the efficiency of resistant cultivars in disease management is limited by high pathogenic variability in *F. oxysporum* f. sp. *ciceris.* The pathogen has extreme genotypic and phenotypic variability and can adapt a wide range of environmental conditions.

Internationally 8 races are reported. Races 1A, 2, 3 and 4 only from India (Haware and Nene, 1982), whereas 0,1B/C, 5 and 6 are found mainly in the Mediterranean region and United states (Jimenez-Gasco *et al.,* 2001). Unlike the other races, race 1A is more widespread and has been reported in India, California (USA) and the Mediterranean region. Recent, study of Dubey and Singh (2008) and Dubey *et al.* (2010) on the virulence analysis of 64 isolates collected from major chickpea growing states of India on 14 varieties including 10 international differentials revealed that more than one races are prevalent in one state. Majority of the isolates were not matched with the race specific reactions, therefore, it was suggested that the cultivars GPF 2, DCP 92-3 and KWR 108 should be included in the set of differentials to get clear cut differential reactions. Subsequently, Dubey *et al.* (2012) identified 8 races of the pathogen on the basis of virulence analysis on 10 new set of differential cultivars.

The Fusarium wilt is soil and internally seed borne disease. However, the chlamydospores like structure were detected in the helium region of the seed (Haware *et al.,* 1978). The disease development is faster at 24-27⁰C soil and air temperature. Below 17⁰C, infection remains restricted in the root without any wilt symptoms. The importance of the soil temperature has also been substantiated by the observation that late sowing of the crop reduces the incidence of the disease. The polymerase chain reaction (PCR) based methods were developed for the detection of pathogen (Pedrajas *et al.,* 1999; Dubey *et al.,* 2010; Durai *et al.,* 2012).

1.5. Management: Deep ploughing during summer and removal of host debris from the field reduces the inoculum levels in soil. The soil inoculum can be reduced by addition of 15-20 tonnes of farmyard manure amended with *Trichoderma* sp. at 4-5 kg/ha before sowing (Singh and Dubey, 2007). The disease can be managed by seed treatment with bavistin + thiram (1: 1) at 2.5g kg^{-1} seed before sowing. It decreased seedling mortality 7% and increased seed germination 11.2% and grain yield 22.3% (Pal and Singh, 1993).

Use of Resistant cultivars and adjustment of sowing dates are important measures for management of wilt of chickpea. The chickpea cultivar Ayala was moderately resistant to *Foc* when inoculated plants were maintained at a day/ night temperature regime of 24/21°C but was highly susceptible to the pathogen at 27/25°C (Landa *et al.*, 2006). The sowing time may be considered as an important component for expression of level of resistance in chickpea cultivars. Therefore, proper sowing time may be recommended for the management of the diseases. Chickpea crop sown on 9th November recorded higher yield 10.6 q ha^{-1} and the lowest wilt incidence 14.9% as compared to 10th October and 25th October sown crop, yielded 6.3 and 7.4 q ha^{-1} grain yield and 32.2% and 25.9% disease incidence, respectively (Andrabi *et al.*, 2011). The best time for sowing of chickpea in the northern India is the first fortnight of November for higher yield and low disease incidence.

The identification and use of host plant resistance has the great potential in the long term management of wilt. The genotypes H 99-9, Pusa 212, JG 315, JG 322, H 01-36 and PCS 8 (Sel. H 82-2) were found resistant (Dubey and Singh, 2004; 2008). In addition to these a large number of wilt resistant cultivars namely, GPF 2, KWR 108, GNG 469 (Samrat), Haryana chana 3 (H86-10), Jawahar gram16 (SAKI 9516), Haryana kabuli 1 (HK 89-131), Rajas (Phule G 9425-9), Pusa Shubhra (BGD 128), Pusa 547 (BGM 547), Gujarat gram 2 (GCP 107), Gujarat gram 4 (GCP 105) were identified. Two *kabuli* accessions ICC 14194 and ICC 17109 from Mexico showed complete resistance to wilt (Gaur *et al.*, 2006).

Ten isolates belonging to three species of *Trichoderma* i.e. *T. viride, T. harzianum* and *T. virens* were evaluated against four isolates of the pathogen representing 4 different races, Dharwad (race 1), Kanpur (race 2), Ludhiana (race 3) and Delhi (race 4) commonly prevalent in India. *T. viride* (IARI P-1) followed by *T. harzianum* (IARI P-4) and *T. viride* (IARI P-19) inhibited maximum mycelial growth of the pathogens. These bioagents also enhanced seed germination, root and shoot length and decreased wilt incidence under greenhouse condition. The isolates proved potential *in vitro* tests were evaluated individually and in combination with carboxin under field condition. The efficacy of *Trichoderma* species was enhanced in combination with carboxin. The integration of *T. harzianum* (10^6 spores/ml/10g seed) and carboxin (2.0 g/kg seed) enhanced seed germination by 12.0-14.0% and grain yield by 42.6-72.9% and reduced wilt incidence (44.1-60.3%) during 3 years (2002-05) of field experimentations under wilt sick field (Dubey *et al.*, 2007).

2. Ascochyta blight

2.1. Occurrence and distribution: The occurrence of blight (*Ascochyta rabiei* (Pass.) Labr.) was first reported in North West Frontier Province of undivided India by Butler in 1911 (Butler, 1918). Now it has been reported causing losses to chickpea in 35 different countries (Nene *et al.*, 1996). It is particularly serious in India, Pakistan and the countries around the Mediterranean Sea.

2.2. Economic impact: In India the disease is very serious in North-Western region, in the states of Punjab, Haryana, Himachal Pradesh, Jammu and Kashmir, Uttar Pradesh and Rajasthan (Grewal and Pal, 1986). Several epidemics of this disease resulting in total loss in North India have been reported (Butler, 1918; Mitra, 1936; Sandhu *et al.*, 1984). Bedi and Athwal (1962) observed that 30 to 50 % chickpea crop may be damaged by blight. In February 1982, blight appeared in epiphytotic proportions again in these states as well as Bikaner district of Rajasthan causing 5-75% loss (Grewal, 1982). In recent years, the sporadic occurrence of the disease has been observed in Himachal Pradesh and Jammu and Kashmir (Fig. 4.1) and losses were recorded up to 5% (Anonymous, 2010).

2.3. Symptoms: The pathogen infects all aerial parts of the plants. Brown to dark brown circular lesions appears on the leaves, branches and pods, whereas those on petioles and stems are usually elongated, coalesce and affected parts become girdled (Fig. 4.3). In severe infection, the entire plant look blighted. The infected seeds are small, wrinkled and lesions are prominent in *kabuli* type.

Fig 4.3. Symptoms of Ascochyta blight

2.4. Casual organisms: The causal organism was first described as *Phyllosticta rabiei* (Pass.) Trot. by Trotter (1918). Labrousse (1931) transferred the fungus to the genus *Ascochyta* under the species name as *Ascochyta rabiei* (Pass.) Labr. due to the presence of few bicelled spores in the culture media. Pycnidia develop in concentric rings are immersed erumpent, globose and 65-245 μm in diameter. The pycnidiospores are mostly aseptate but 2-5% spores have septum. They are hyaline, oval to oblong, straight or slightly curved at one or both ends, 0-1 seplate, and 10-16x3.5-5.2 μm. The perfect state of the fungus *Mycosphaerella rabiei* (Syn. *Didymella* *rabiei* (Kovachevski) Arx.) was observed on over wintering chickpea debris in southern Bulgaria and USA (Kovachevski, 1936; Kaiser *et al.*, 1987). In Spain fungus grew saprophytically on the soil surface and remained viable for 2 years, however when buried, viability is lost after 2-5 months (Navas-Cortes *et al.*, 1995). Infected plant debris left over in the field after harvest have negligible role in the perpetuation of the disease in the plains in India, where summer temperature are very high.

The disease usually appears in epiphytotic form at flowering stage onwards in the presence of 9-24°C temperature with 10h or more wetness, wet and windy weather. Luthra and Bedi (1932) were the first to demonstrate the seed borne nature of the pathogen. They showed that the seed coat and cotyledons of infected seeds contained mycelium. The fungus has been reported to the host specific and attack species belonging to the genus *Cicer*, however Kaiser (1973) observed that the fungus could infect cowpea (*Vigna sinensis*) and bean (*Phaseolus vulgaris*) when inoculated artificially. The fungus successfully re-isolated from infected peas, beans and cowpea in greenhouse (ICARDA, 1982).

Ascochyta rabiei showed variation in morphological, physiological and pathological characters (Luthra *et al.*, 1939; Vir and Grewal, 1974a). Genetic diversity has been studied at molecular level in Indian, American, Syrian and Pakistani isolates and grouped Indian isolated in 2 categories (Santra *et al.*, 2001). A composite linkage map was constructed on two interspecific recombinant inbred line derived between *C. arietinum* (ILC 72 and ICCL 81001) and *C. reticulatum* (cr 5-10 or cr 5-9). Seven resistance gene analogs (RGAs) PCR-based markers (5 CAPS and 2 d CAPS) were successfully genotyped in the two progenies. Six of them have major quantitative trait loci conferring resistance to ascochyta blight. These six RGAs were novel sequences (Palomino *et al.*, 2009).

2.5. Management: The use of resistant cultivars is the best methods, since the other methods are either impractical or uneconomical. A large number of high yielding ascochyta blight resistant varieties have been developed through collaborative efforts under ambit of All India Coordinated Research Project on Chickpea by various state agriculture universities (SAU's), Indian Institute of Pulses Research (IIPR) and Indian Agricultural Research Institute (IARI) during last 20-25 years. Some of the prominent varieties are PBG 5, GNG 469, PBG 1, Pusa 413, Pusa 408, Pusa 417, L 551 and Himanchal chana 1. Out of 233 genotypes screened, only two, namely, H 00-108 and GL 92024 were found resistant (Dubey and Singh, 2003). A large number of chickpea genotypes were evaluated for multiple resistance and reported that GL 90168, GL 91137, GL 92015 and PGL 167 were free from wilt and root rot diseases and resistant to ascochyta blight (Kaur *et al.*, 2008).

The spread of the disease can be prevented by burying of the diseased debris at 10 cm/deeper and sowing of infected seed at 15 cm/deeper. Disease free seed is a pre-requisite for disease management. Grewal (1982) recommended seed treatment with combination of carbendazim (bavistin) + tetramethyalthiuram disulphide (thiram) 1: 2 ratio at 2.5g kg^{-1} of seed for eradication of internally and externally seed borne infection of *A. rabiei*. Folier applications of zineb, maneb, captan and daconil also reduced the disease (Vir and Grewal, 1974b). Reddy and Singh (1983) reported that one application of chlorothalonil at the early podding stage in tolerant cultivar ILC 482 prevented yield loss.

The biocontrol potential of fungal antagonists, *Chaetomium globosum, Trichoderma viride* and *Acremonium implicatum* were explored under *in-vitro* and *in-vivo* conditions. *C. globosum* reduced the growth and pycnidiospores germination 48.6 % and 70.9%, respectively under *in-vitro* conditions, whereas its post inoculation, spray reduced 73.1% disease (Rajakumar *et al.*, 2005).

3. Dry root rot

3.1. Occurrence and distribution: The dry root-rot of chickpea caused by *Rhizoctonia bataticola* (Taub.) Butler was first reported from India by Mitra (1931). Since then besides India, it has also been reported from Iran, the USA, Australia, Ethiopia, Sudan and Pakistan (Nene *et al.*, 1996).

3.2. Economic impact: In India severe root-rot was reported from non-irrigated areas of Tamil Nadu, Andhra Pradesh and Karnataka causing 10-20% losses (Vishwadhar and Chaudhary, 2001). In the last three decades the climate is changing largely as a result of human activity. Rainfall patterns are changing; both floods and droughts are becoming more frequent and more severe. The impact will be felt worldwide, but nowhere more actually than in the Indian sub-continents and Mediterranean regions. In India, *R. bataticola* causing severe damage to both *Kharif* and *Rabi* seasons pulse crops in all parts of the country (Fig. 4.1) (Dubey *et al.*, 2009). Recently, survey conducted in the states of Bihar, Andhra Pradesh, Maharashtra, Karnataka and Rajasthan and observed that the dry root rot is causing more damage even greater than the wilt and it was ranging from 5-35% (Anonymous, 2010). The higher incidence was due to exposure of chickpea plants to moisture stress conditions, which ultimately led to more production of sclerotia of *R. bataticola* on chickpea plants roots (Manjunatha *et al.*, 2011). The pathogen is both soil and seed borne. The crop is grown under rain fed conditions and moisture stress, predispose the crop to *R. bataticola* infection.

Fig. 4.4. Symptoms of dry root rot

3.3. Symptoms: Drying of plants, suddenly in the field under dry and hot conditions and affected plants could be pulled out easily due to rot of lateral and finer roots. The leaves and stems of affected plants are usually straw coloured. The tap root becomes dark and brittle towards tip and shows shredding of the bark. Some times at the tap root tip, a grayish mycelial coating can be seen. The minute and dark brown sclerotia can be seen on the exposed root as well as on the inner side of the bark. The pathogen also invades the vascular tissues resulting in death of the plants (Fig. 4.4).

3.4. Causal organisms: Butler (1918) identified the culture of *Sclerotium bataticola* provided by Brinton-Jones as *Rhizoctonia bataticola* and consequently it was designated as *Rhizoctonia bataticola* (Taub.) Butler. The mycelium is hyaline when young and turns olive brown. Pycnidia do not develop on chickpea, but are formed in culture (Haware, 1998). Sclerotia are 80-174 μm, black, round to irregular and develop abundantly in the bark and pith. The populations of *R. bataticola* are highly variable in respect of morphological, cultural and pathological parameters. Twenty seven isolates of *R. bataticola* isolated from chickpea (23), mungbean (1), urdbean (1) and groundnut (1) were variable in respect of growth rate, size of sclerotia, pycnidia and pycnidiospores, and they were grouped into 4 categories on the basis of growth diameter. The size of sclerotia varied from 40-600 μm. Out of 23 chickpea isolates, pycnidia production in culture was encountered only in 3 isolates. The size of pycnidiospores ranged from 5-10x14-30 μm in these isolates (Aghakhani and Dubey, 2009a).

A set of differential cultivars for pathotyping of the pathogen was standardized. The Indian populations of *R. bataticola* represting 11 different states were grouped in 6 pahtotypes based on differential reaction on a set of 10 differential cultivars of chickpea (Aghakhani and Dubey, 2009a). Genetic diversity of the pathogen was also determined by using random amplified polymorphic DNA, internal transcribed spacer restriction fragment length polymorphism and ITS sequencing (Aghakhani and Dubey, 2009b).

The pathogen perpetuates on diseased debris and persists in soil as facultative parasite. The fungus has a wide host range. Gupta and Chohan (1970) reported that the pathogen can attack plants at any growth stage, i.e. immediately after sowing of seed till the maturity of crop. Low soil moisture and temperature 26-35°C are conductive for the disease development. Dry root rot has been found associated at the post flowering stage, increased with crop age and showed a negative correlation with soil moisture (Patel and Anahosur, 2000). Cultivation of chickpea under rain fed predisposes the crop to dry root rot development. The use of sorghum-chickpea in rotation favours disease development.

3.5. Management: *R. bataticola* is diverse plant pathogen, usually soil borne, but seed borne in many pulse crops. Deep ploughing during summer and removal of host debris from the field reduces inoculum levels. Soil solarization alone and in combination with millet residues or paunch contents amendments was assessed in a naturally infested soil. Solarization increased the temperature to 50°C for at least 4h per day during June leading to a significant reduction (44%) in soil inoculum of the pathogen. Paunch contents or millet residues amendments (3 tonnes ha^{-1}) caused 16% or 35% reduction of initial inoculum density, respectively (Ndiaya *et al.*, 2007). They also suggested that the combination of solarization and organic amendments can be a credible alternative to pesticides for managing root-rot disease of pulse crops. Seed treatment with carbendazim + TMTD (1: 1) or carboxin + TMTD (vitavax power) at 2.0 g kg^{-1} of seed reduced incidence of root rot. Seed treatment with *T. viride* or *T. harzianum* reduced dry root-rot and increased the phenolic and carbohydrate contents of chickpea (Singh *et al.*, 1998). Recently, Dubey *et al.* (2011) studied the efficacy of newly developed seed dressing and soil application formulations of *T. viride*, *T.virens* and *T. harzianum* and reported that a combination of soil application of *T. harzianum* based Pusa biopellet 10G at 5kg ha^{-1} and seed treatment of *T. harzianum* based Pusa 5SD at 4g kg^{-1} + carboxin at 1g kg^{-1} provided the highest seed germination, shoot and root lengths and grain yield and the lowest dry root rot incidence in chickpea. Dry root rot pathogen has a wide host range and survives in soil for longer periods. The use of host plant resistance is the most economical approach for management of dry root-rot disease in chickpea. Some of the resistant cultivars are PBG 5, GNG 1365, H 99-9, H 00-108 and Haryana chana 1.

4. Grey mould

4.1. Occurrence and distribution: Grey mould of chickpea caused by *Botrytis cinerea* (Pers. ex Fr.) was first reported by Shaw and Ajrekar (1915) from India. Besides Indian sub-continent grey mould has also been reported from Argentina, Australia, Canada, Colombia, Spain, Hungary, Mexico, Turkey, Myanmar, Vietnam, and United States of America.

4.2. Economic impact: The disease was responsible for heavy losses in Indo-Gangetic plains of India during 1979-1982 (Grewal and Laha, 1983). In January 1979 it appeared in epiphytotic form and destroyed chickpea crop over an area of 20 thousand hectares in Barahiya- Mokamah *Tal* area of Bihar. Again in 1981 the disease caused 70-100% losses at the State Farm Corporation of India, Hisar and in parts of Punjab (Grewal and Laha, 1983). The disease was observed in sporadic during 1982-86 at Indian Agricultural Research Institute, New Delhi (Fig. 4.1). However, its appearance in Tarai region of Uttar Pradesh (now Uttarakhand) is regular feature since its report during 1967- 1968 (Joshi and Singh, 1969). The pathogen caused 50-100% loss of chickpea crop in Bangladesh in 1989-90 and in Nepal during 1987-88 crop seasons. Recently, survey made under All India Coordinated Research Project on Chickpea recorded 10-30% disease incidence along with stemphylium blight in Bihar (Anonymous, 2010).

4.3. Symptoms: The grey mould symptoms appear on stem, leaves, flowers and pods as grey or dark brown lesions covered with mouldy sporophores. Growing tips and flowers are particularly susceptible to the fungus attack (Fig. 4.5). The infected flowers drop off. Stem lesions are 10-30 mm long and girdle the stem. On thick stems the grey mould is gradually transformed in to a dirty gray mass containing dark green to black sporodochia. If the pods infected during early stages of development, no seed formation takes place. Sometime small shriveled seeds are formed. Grayish while mycelium may be seen on immature seeds.

Γig. 4.6. Grey mould symptoms

4.4. Causal organism: *Botrytis cinerea* (Pers. ex Fr.) is the anamorph of the fungus *Botryotinia fuckeliana* Groves and Loveland. The mycelium is septate and brown. Hyphae are thin, hyaline and 8 to 16 μm wide. Conidiophores are light brown with a hyaline tip, septate and 8-24 μm wide. The conidia are hyaline, one celled, oval or globose measuring 4-24x4-18 μm (average 14.9x8.4 μm), borne in clusters on short sterigmata. The fungus has a wide host range and the inoculum is almost always present in the environment waiting for the right weather to become active. On chickpea the pathogen is seed borne and it can survive both internally as well as externally and such seeds become the potential carrier of pathogen to new areas (Grewal and Laha, 1983). Infected plant debris also plays an important role in the survival of the pathogen (Grewal, 1988).The pathogen *B. cinerea* is reported to have extreme variability in morphological, cultural and pathological characters. Existence of 5 pathotypes of *B. cinerea* has been reported from northern India (Rewal and Grewal, 1989a). Free moisture, high humidity, and optimum temperature of 20-25°C are congenial for the infection and disease development. Cloudiness and intermittent rains increased disease severity and the pathogen completes the entire disease cycle in 7 days (Singh *et al.*, 2007).

4.5. Management: Use of pathogen free seed, low seed-rate and wider row and plant spacing are helpful in reducing the disease incidence. Tall (BGM 267) and compact (H 86-143) genotypes has less disease than bushy types (Singh *et al.*, 2007). More than ten thousand germplasm and advanced lines has been screened at G.B. Pant University of Agriculture and Technology, Pantnagar during 1986-1997 and none of the genotypes showed high level of resistance to grey mould (Tripathi and Rathi, 1999). Gray mould tolerant cv. Avrodhi, soil application of di-ammonium phosphate, wider row spacing, seed treatment with carbendazim + TMTD at 2g kg^{-1} seed and need based foliar application of carbendazim , has been devised as an integrated disease management package (Pande *et al.,* 2006). Seed treatment with carbendazim as well as carbendazim + TMTD as (1: 1) at 2.5 g kg[-1] of seed not only eliminated externally and internally seed borne inoculum but also protected the seedlings up to 8 weeks after sowing against aerial infection by *B. cinerea* (Laha and Grewal, 1983; Grewal, 1988). Three genotypes, ICC 1069, ICC 4936 and ICC 5035 were found to be resistant to gray mould under artificially inoculated conditions. Later, Rewal and Grewal (1989b) studied the inheritance of grey mould resistance in the resistant genotypes ICC 1069. When ICC 1069 was crossed with BGM 413 and BG 256 monogenic dominance conferred resistance, but when ICC 1069 was crossed with BGM 419 and 408, a ratio of 13 susceptible: 3 resistant was obtained indicating the presence of epistatic interactions.

5. Wet root rot

5.1. Occurrence and distribution: Wet root rot caused by *Rhizoctonia solani* Kühn is also considered as one of the factors for low productivity of chickpea. Chaturvedi and Nadarajan (2010) observed that the wet root rot has the major threat to early plant establishment in the state of Uttar Pradesh, Bihar, Jharkhand, West Bengal, Assam, Punjab and Haryana. The disease has also been reported from Argentina, Chile, Iran, Mexico, Pakistan, Bangladesh, Canada and the USA. Wet root rot has been commonly observed in Nebraska (USA) in irrigated field (Harveson, 2011). *Rhizoctonia solani* has the potential to be an important constraint to chickpea production in the western Canada (Hwang *et al.,* 2003). The Disease is wide spread in nature and affects many important agricultural and horticultural crops worldwide, causing several diseases (Gonzalez Garcia *et al.,* 2006).

5.2. Economic impact: In India it is prevalent in north and north eastern states (Fig. 4.1) and causes an annual loss around 8-10% (Vishwadhar and chaudhary, 2001).

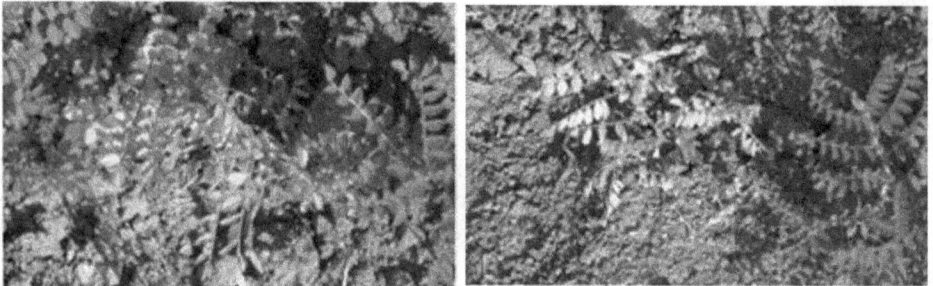

Fig. 4.6. Wet root rot symptoms

5.3. Symptoms: Root rotting often originating at the distal tip of the young root and gradual yellowing and wilting of foliage are the major characteristic symptoms (Fig. 4.6). The finer roots are completely rotten. The rotted, discoloured tissues are soft and wet. Colour of the leaves of infected plants gradually changes from dull-green to yellowish brown. The diseases is most commonly seen early in the season, when soil moisture content is often high, however, in irrigated chickpea the disease may occur at any time. In India it is more frequent in chickpea planted after the rice harvest when the soil is wet (Nene, 1980).

5.4. Causal organism: *Rhizoctonia solani* Kühn, the anamorph of *Thanatephorus cucumeris* (Frank) Donk, is a seed and soil borne pathogen of chickpea. The *R. solani* occurs on a wide range of hosts (Anderson, 1982; Nelson *et al.*, 1996). The sclerotia of *R. solani* are superficially borne on a mycelium either scattered or aggregated at centre or edge of Petriplate. Sclerotia are variable in size, 0.5-2 mm in diameter, often covered by loose mycelium and appear rough and warted. The mycelium is septate and measures 3.5-7.2 μm width. Branches arises near distal end of cell are constricted at point of origin and septate shortly above. The branches may arise at right or acute angle at various points along the cell length. The soil environment has a great influence on the severity of root rot. Chang *et al.* (2004) studied the effect of soil temperature, seeding depth and seeding date on root rot of chickpea and reported that the host responded to warm soils by increasing its growth rate, and the pathogen by increasing its virulence. The *kabuli* cultivar, showed greater susceptibility to root rot caused by *R. solani* than the *desi* culivar. Seedling emergence and seedling dry weight were greater at a seeding depth of 2 cm than at 5 cm in infested soil. Seeding date did not affect the occurrence of root-rot. Nene and Reddy (1987) observed that the temperature range of 18-30°C, in a soil moisture range of 30-80%, and at high nitrogen levels favour disease development. Wet root- rot reduced nodulation (Hwang *et al.*, 2003).

5.5. Management: Management of *R. solani* is difficult because of wide host range and its ability to survive through sclerotia under adverse environmental conditions. Cultural measures include deep ploughing in summer, destruction of infected residues and grass weeds help in reducing the inoculum of *R. solani*. Excessive dose of nitrogenous fertilizer should be avoided. Authors observed that the seed treatment with vitavax power (Carboxin 37.5% + Thiram 37.5%) at 2g kg⁻¹ or carbendazim 50% + TMTD 75% (1: 1) at 2g kg⁻¹ seed reduced the wet root rot, dry root rot and Fusarium wilt disease of chickpea at early stages of plant growth as well as enhanced the emergence and grain yield. Soil application of kaolin-based powder formulation of *T. harzianum* at 5g in one kg of farmyard manure in 3m² area at one week before sowing caused 4.9 % and 1.2 % wet root rot at 30 and 60 days after sowing, respectively (Prasad *et al.*, 2002). They also observed that the soil application is more effective than seed treatment. Whereas, Dubey *et al.* (2012) observed that a combination of soil application of Pusa Biopellet 16G (*T. virens*) at 5 kg ha⁻¹ at the time of sowing and seed treatment with Pusa 5 SD (*T.virens*) + carboxin (4: 1) at 5 g kg⁻¹ of seed increased seed emergence, shoot and root lengths and grain yield and reduced the wet root rot in chickpea. In the field, seed treatment was more effective than soil application. They also observed that the both the formulations enhanced the growth of the plants indicating growth promoting ability of the isolates used for the development of the formulations.

6. Collar rot

Collar rot caused by *Sclerotium rolfsii* Sacc. was noted from Uttar Pradesh, Madhya Pradesh, Haryana, Punjab and Bihar (Fig. 4.1). The seeding showed rotting in the collar region downwards and produced rapeseed like sclerotia. The incidence of collar rot is associated with high soil moisture (wet), presence of undecomposed organic matter and temperatures of 28 to 30°C. Chickpea following paddy shows a higher incidence (Nene and Reddy, 1987).

7. Stem rot

Stem rot (*Sclerotinia sclerotiorum* (Lib.) de Bary) was recorded from Punjab, Haryana, Himachal Pradesh and Uttar Pradesh (Fig. 4.1). The disease appears mostly on adult plants when thick canopy and the soil remain wet for a long period. The symptoms are dropping of petioles and leaflets without turning yellow and later on leaves turn dry straw coloured prematurely (Fig. 4.7). The disease perpetuates through sclerotia in the soil and spreads mainly by ascospores. The variability of the pathogen in India has been described (Mandal and Dubey, 2012). The fungus favoured by a combination of cool and wet weathers, excessive vegetables growth and heavy dew.

Fig. 4.7. Stem rot symptoms

8. Rust

Rust (*Uromyces ciceris -arietini* [Grogn.] Jacz. and Boyer) was reported from west Bengal (Butler, 1918) and subsequently from Bihar, Madhya Pradesh, Uttar Pradesh, Rajasthan and Karnataka (Fig. 4.1). The rust epiphytotics were reported in 1957 and 1985 from Uttar Pradesh and Karnataka, respectively (Asthana, 1957; Grewal, 1988). Recently in 2010 it was observed on cultivars JG 11 and the incidence ranged from 32-40% (Anonymous, 2010). Rust over summer in the uredial stage in hilly areas and affect the crop in plains. High humidity cloudy weather and temperature 20 to 30°C are favourable for disease development.

9. Stemphylium blight

Stemphylium blight (*Stemphylium sarciniforme* (Cav.) Wilts.) of chickpea was first reported form West Bengal. The occurrence of the disease was reported from Uttar Pradesh, Bihar, West Bengal and Madhya Pradesh. In 2010 the incidence of the disease was recorded 10-30% in Bihar (Anonymnous, 2010). The fungus is seed borne and also survives in infected plant debris. Wet weather and temperature 20 to 25°C are favourable for disease development.

10. Alternaria blight

Alternaria blight (*Alternaria alternata* (Fr.) Kiessler] was reported from India, Bangladesh and Nepal (Nene *et al.*, 1996). In India the occurrence of the disease in sporadic form occurs in states of Bihar, Madhya Pradesh, Delhi and Uttar Pradesh (Fig. 4.1). The fungus affects all the aerial parts, mostly in the adult stage and sometimes causes floral blight. The fungus is seed borne. The perfect state of *A. alternata* is *Pleospora infectoria* and this was also reported to cause blight in chickpea. Under favourable weather the entire foliage appears blighted and high humidity cause rapid withering of the individual leaflets. On pods the lesion are circular, slightly sunken and dirty black in colour. Seeds in infected pods are discoloured and shriveled. Foliar sprays of mancozeb (dithane M-45) at 0.25% reduces disease incidence.

11. Colletotrichum blight

This disease (*Colletotrichum dematium* Pers. ex Fr.) was reported from central India (Mishra *et al.*, 1975). The fungus is seed borne and has a wide host range (Nene and Reddy, 1987). The disease appears at any stage of plant growth and cause severe damage to chickpea depending on the climatic conditions. On leaves and pods, lesion are circular to elongate, sunken in the middle and surrounded by yellow margins. Lesions on stem are elongated and black.

These minor diseases (6–11) may not cause serious losses in normal years but could be a potential threat to the crop in suitable weather conditions.

12. Viral diseases

12.1. Occurrence and distribution: Several viruses and mycoplasma has been reported to infect chickpea (Nene and Reddy, 1987) in which the major and most economically important viruses worldwide are *Alfalfa mosaic virus, Bean leaf roll virus, Beat western yellows virus, Chickpea stunt disease-associated virus, Chickpea chlorotic dwarf virus, Cucumber mosaic virus* and *Faba bean necrotic yellows virus* (Kumar *et al.*, 2008).

12.2. Economic impact: In India the chickpea stunt is the most important virus disease and cause 80-95% yield loss in experimental plot at Jabalpur, Madhya Pradesh (Kotasthane and Gupta, 1978). Horn *et al.* (1993) reported a new leaf hopper transmitted *Chickpea chlorotic dwarf geminivirus* (CCDV) provoking symptoms characteristic of chickpea stunt caused 75-100% yield losses, if infection occurs early in the season. Survey was conducted in the states of Rajasthan, Madhya Pradesh and Gujarat during 1991 to 1992 crop seasons and observed the stunt incidence in farmer's fields of Gujarat ranged from 0-45% (average 12%), Haryana 0-29% (average 4%) and Rajasthan 0-5.2 % (Horn *et al.*, 1996). Nene and Reddy (1987) reported its seriousness only in northern India, whereas, Reddy *et al.* (2004)

observed its occurrence throughout India in sporadic form and it was most common in Gujarat and Haryana. Presently its occurrence is increasing in almost all major chickpea growing parts of the country irregularly in different years (Fig. 4.1).

12.3. Symptoms: virus infected plants can be easily spotted in the field by their yellow, orange or brown discoloured foliage and stunted growth. The stunting is due to shortened internodes. The leaflets are smaller. The tips and margins of leaflets are chlorotic before turning brown. The leaflets are stiffer and thicker than normal ones. The leaf reddening in case of *desi* types and yellowing in *kabuli* types are very common (Nene and Reddy, 1987). A horizontal cut through the collar region reveals a brown ring of discoloured phloem and a shallow vertical knife cut at the collar region in variably reveals phloem browning. Plants infected at an early stage show poor growth and often die prematurely.

12.4. Causal organism: Stunt virus of chickpea has been studied extensively and characterized in India. Two different Luteoviruses, *Beet western yellows virus* (BWYV) and *chickpea stunt disease associated virus* (CpSDaV) of genus *Polerovirus*, family *Luteoviridae;* and the leaf hopper transmitted *Chickpea chlorotic dwarf virus* (CCDV), of genus *Cutovirus* , family *Geminiviridae* are found to be associated with the disease (Reddy and Kumar, 2004). The BWYV found at certain locations and only a small proportion of stunt disease plants, whereas a luteovirus (CpSDaV) and geminivirus (CCDV) are predominantly associated with chickpea stunt disease affected plants in India. Both the viruses cause similar symptoms and are difficult to distinguish by symptoms alone (Reddy *et al.,* 2004). Earlier to this the causal virus of chickpea stunt was described as *Pea leaf roll virus* (Nene and Reddy, 1976). Luteoviruses have isometric particles of size 24 nm, single coat protein of size 26-28 kDa and a liner, single stranded RNA genome with positive polarity. Luteoviruses are limited to phloem tissues. Particles of geminivirus are geminate 15x25 nm diameter, with single standard closed circular DNA of size 2.9kb and single coat protein of size 32 kDa (Reddy *et al.,* 2004).

The viruses associated with stunt disease of chickpea neither mechanically transmitted nor seed-borne. Non-vector transmission is by grafting and natural insect vectors transmission is by aphids *Aphis craccivora* (CpSDaV), *Acyrthosiphon pisum* and *Myzus persicae* (BWYV) and by leafhopper *Orosius orientalis* (CCDV). The aphids and leafhoppers are transmitting in a circulative and non-propagative manner (Reddy *et al.,* 2004). Earlier, an aphid (*Aphis craccivora*) has been reported to transmit the stunt virus from Iran and India (Kaiser and Danesh, 1971; Nene and Reddy, 1976). The stunt virus is confined to beans, clovers, fababean, groundnut, pea and lentil.

12.5. Management: Cultural practices such as rouging, alteration in sowing dates and use of early maturing cultivars and sprays of insecticide to control vectors, at early stage of the crop are effective in reducing disease incidence. Identification of resistance to stunt disease of chickpea has been done on large scale at ICRISAT sub-centre-Hisar and reported that chickpea lines ICC 403, ICC 2385, ICC 10466, ICC 10596 and ICC 11155 showed field resistance (Singh and Reddy, 1991). *Desi* types are comparatively less susceptible than *kabuli* types. Delay in the date of sowing reduces the disease in northern India (Nene and Reddy, 1976).

REFERENCES

Aghakhani., Maryam., and Dubey, S. C. 2009a. Morphological and pathogenic variation among isolates of *Rhizoctonia bataticola* causing dry root rot of chickpea. Indian Phytopath. 62: 183-189.

Aghakhani., Maryam., and Dubey, S. C. 2009b. Determination of genetic diversity among Indian isolates of *Rhizoctonia bataticola* causing dry root rot of chickpea. Antonie van Leeuwenhoek 96: 607-619.

Agricultural Statistics at a Glance 2011. Directorate of Economics and Statistics, Department of Agriculture, Ministry of Agriculture, Goverment of India.

Anderson, N. A. 1982. The genetics and pathology of *Rhizoctonia solani*. Annu. Rev. Phytopathol. 20: 329-347.

Andrabi, M., Vaid, M., and Razdan, V. K. 2011. Evaluation of different measures to control wilt causing pathogens in chickpea. J. Plant Protec. Res. 51: 55-59.

Anonymous. 2010. Project coordinator report.2009-10, All India Coordinated Research Project on Chickpea, Kanpur, India pp.45.

Asthana, R. P. 1957. Some observations on the incidence of *Uromyces ciceris-arietini* on *Cicer arietinum*. Nagpur agric. Colle. Magn. 31: 20A-20B.

Bedi, K. S., and Athwal, D. S. 1962. C-235 is the answer to blight. Indian Fmg. 12: 20-22.

Butler, E. J. 1918. Fungi and Disease in Plants. Thacker Spink and Co.,Calcutta, 547pp.

Chang, K. F., Hwang, S. F., Gossen, B. D., Turnbull, G. D., Howard, R. J., and Blade, S. F. 2004. Effects of soil temperature, seeding depth and seeding date on Rhizoctonia seedling blight and root rot of chickpea. Can. J. Plant Sci. 84: 901-907.

Chattopadhyay, S. B., and Sen Gupta, P. K. 1967. Studies on wilt diseases of pulses, I. variation and taxonomy of *Fusarium* spp. associated with wilt diseases of pulses. Ind. J. Mycol. Res. 5: 45.

Chaturvedi, S. K., and Nadarajan, S. 2010. Genetic enhancement for grain yield in chickpea- accomplishments and resetting research agenda. Electronic J. Plant Breeding. 1: 611-615.

Dastur, J. F. 1935. Gram wilts in central provinces. Agric. Liv-stk., India 5: 615-627.

Dubey, S. C., Bhavani, R., and Singh, B. 2009. Development of Pusa 5SD for seed dressing and Pusa Biopellet 10G for soil application formulations of *Trichoderma harzianum* and their evaluation for integrated management of dry root rot of mungbean (*Vigna radiata*). Biol. Control. 50: 231-242.

Dubey, S. C., Priyanka, K., Singh, V., and Singh, B. 2012. Race profiling and molecular diversity analysis of *Fusarium oxysporum* f sp. *ciceris* causing wilt in chickpea. J. Phytopathol. doi: 10.1111/j.1439-0434.2012.01954.x

Dubey, S. C., and Singh, B. 2003. Evaluation of chickpea genotypes against ascochyta blight. Indian Phytopath. 56: 505.

Dubey, S. C., and Singh, B. 2004. Reaction of chickpea genotypes against *Fusarium oxysporum* f. sp. *ciceris* causing vascular wilt. Indian Phytopath. 57: 233.

Dubey, S. C., and Singh, B. 2008. Evaluation of chickpea genotypes against *Fusarium oxysporum* f. sp. *ciceris*. Indian Phytopath. 61: 280-281.

Dubey, S. C., and Singh, B. 2010. Dry root rot of pulses and its management. In Plant diseases and its mangement (P.C. Trivedi, eds): 93-111. Pointer Publisher, Jaipur.

Dubey, S. C., Singh, B., and Bahadur, P. 2007. Diseases of pulse crops and their ecofriendly management. In Ecofriendly management of plant diseases. (S. Ahamad and U. Narain eds): 16–44. Daya Publishing House, Tri Nagar, Delhi.

Dubey, S. C., and Singh, S. R. 2008. Virulence analysis and oligonucleotide fingerprinting to detect diversity among Indian isolates of *Fusarium oxysporum* f. sp. *ciceris* causing chickpea wilt. Mycopathologia. 165: 389-406.

Dubey, S. C., Singh, S. R., and Singh, B. 2010. Morphological and pathogenic variability of Indian isolates of *Fusarium oxysporum* f. sp. *ciceris* causing chickpea wilt. Arch. Phytopathol. Plant Prot. 43: 174-189.

Dubey, S. C., Suresh, M.., and Singh, B. 2007. Evaluation of *Trichoderma* species against *Fusarium oxysporum* f. sp. *ciceris* for integrated management of chickpea wilt. Biol Control 40: 118–127.

Dubey, S. C., Tripathi, A., Bhavani, R., and Singh, B. 2011. Evaluation of seed dressing and soil application formulations of *Trichoderma* species for integrated management of dry root rot of chickpea. Biocontrol Sci. Techn. 21: 93-100.

Dubey, S. C., Tripathi, A., and Singh, B. 2012. Combination of soil application and seed treatment formulations of *Trichoderma* species for integrated management of wet root rot caused by *Rhizoctonia solani* in chickpea (*Cicer arietinum*). Indian J. agric. Sci. 82: 357-364.

Dubey, S. C., Tripathi, A., and Singh, S.R. 2010. ITS-RFLP fingerprinting and molecular marker for detection of *Fusarium oxysporum* f.sp. *ciceris*. Folia Microbiol. 55: 629-634.

Durai, M., Dubey, S. C., and Tripathi, A. 2012. Genetic diversity analysis and development of SCAR marker for detection of Indian populations of *Fusarium oxysporum* f. sp *ciceris* causing chickpea wilt. Folia Microbiol. (DOI 10.1007/s12223-012-0118-5).

Gaur, P. M., Pande, S., Upadhyaya, H. D., and Rao, B. V. 2006. Extra large *kabuli* chickpea with high resistance to Fusarium wilt. *Int.* Chickpea and Pigeonpea Newsl. 13: 5-7.

Gonzalez-Garcia, V. G., Portal, O. M. A., and Rubio, S. V. 2006. Review. Biology and systematics of the form genus *Rhizoctonia*. Spanish J. Agri. Res. 4: 55-79.

Grewal, J. S. 1982. Control of important seed borne pathogens of chickpea by seed treatment. Indian J. Genet. 42: 393-398.

Grewal, J. S. 1988. Diseases of pulse crops an overview. Indian Phytopath. 41: 1-14.

Grewal, J. S., and Laha, S. K. 1983. Chemical control of Botrytis blight of chickpea. Indian Phytopath. 36: 516-520.

Grewal, J. S., and Pal, M. 1986. Fungal disease problems in chickpea. In Vistas in plant pathology (A.Varma and J.P. Verma eds): 157-170. Malhotra Pubishing House, New Delhi.

Gupta, S., Chakraborti, D., Rangi, R.K., Basu, D., and Das, S. 2009. A molecular insight into the early events of chickpea and *Fusarium oxysporum* f. sp. *ciceris* (race 1) interaction through cDNA-AFLP analysis. Phytopathology 99: 1245-1257.

Gupta, V. K., and Chohan, J. S. 1970. Seed borne fungi and seed health testing in relation to seedling diseases of groundnut. Indian Phytopath. 23: 622-625.

Harveson, R. M. 2011. Soilborne root diseases of chickpea in Nebraska. University of Nebraska-Lincoln, Extension publication, available online at http: // extension. Unl. edu/publications.

Haware, M. P. 1998. Diseases of chickpea. In The pathology of food and pasture legumes (D.J. Allen and J.M. Lenne, eds): 473-506. CAB International, Wallingford, U.K.

Haware, M. P., and Nene, Y. L. 1982. Races of *Fusarium oxysporum* f. sp. *ciceri*. Plant Dis. 66: 809-810.

Haware, M. P., and Nene, Y. L. 1980. Influence of wilt at different growth stages on yield loss of chickpea. Trop. Grain Legume Bull. 19: 38-40.

Haware, M. P., Nene, Y. L., and Rajeshwari, R. 1978. Eradication of *Fusarium oxysporum* f. sp. *ciceri* transmitted in chickpea seeds. Phytopathology. 68: 1364-1367.

Horn, N. M., Reddy, S. V., Roberts, I. M., and Reddy, D. V. R. 1993. Chickpea chlorotic dwarf virus, a new leafhopper-transmitted germinivirus of chickpea in India. Ann. Appl. Biol. 122: 467-479.

Horn, N. M., Reddy, S. V., van den Heuvel, J. F. J. M., and Reddy, D. V. R. 1996. Survey of chickpea (*Cicer arietinum* L.) for chickpea stunt disease and associated viruses in India and Pakistan. Plant Dis. 80: 286-290.

Hwang, S. F., Gossen, B. D., Chang, K. F., Turnbull, G. D., Howard, R. J., and Blade, S. F. 2003. Etiology, impact and control of Rhizoctonia seedling blight and root rot of chickpea on the Canadian Prairies. Can. J. Plant Sci. 83: 959-967.

ICARDA 1982. Chickpea Pathology, Progress Report 1981-82. Food legume improvement programme, International Centre for Agricultural Research in Dry Areas, Aleppo, Syria 75pp.

Jimenez-Gasco, M. M., Perez-Artes, E., and Jimenez-Diaz, R. M. 2001. Identification of pathogenic races 0, 1B/C, 5 and 6 of *Fusarium oxysporum* f. sp. *ciceri* with Random Amplified Polymorphic DNA (RAPD). Euro. J. Pl. Pathol. 107: 237-248.

Joshi, M. M., and Singh, R. S. 1969. A Botrytis gray mould of gram. Indian Phytopath. 22: 125-128.

Kaiser, W. J. 1973. Factors affecting growth, sporulation, pathogenicity and survival of *Ascochyta rabiei*. Mycologia 65: 444-457.

Kaiser, W. J., and Danesh, D. 1971. Etiology of virus induced wilt of *Cicer arietinum*. Phytopathology 61: 453-457.

Kaiser, W. J., Hannan, R. M., and Trapero–Casas, A. 1987. Survival of *Ascochyta rabiei* in chickpea debris. Phytopathology 77: 1240.

Kaur, L., Sandhu, J. S., and Gupta, S. K. 2008. Multiple disease resistance – A solution for encouraging chickpea cultivation in various climatic zones of India. Indian J. Agric. Sci. 78: 1067- 1070.

Kotasthane, S. R., and Gupta, O. 1978. Yield losses due to chickpea stunt. Trop. Grain Leg. Bull. 11/12: 38-39.

Kovachevski, I. C. 1936. The blight of chickpea, *Mycosphaeralla rabiei* Nov. sp. Min. Agric. Nat. Domains Sofia, 80pp.

Kumar, J., Haware, M. P., and Smithson, J. B. 1985. Registration of four short duration *Fusarium* wilt resistant kabuli chickpea germplasm. Crop Sci. 25: 576-577.

Kumar, P. L., Kumari, S. M. G., and Waliyar, F. 2008. Virus diseases of chickpea. In Characterization, diagnosis and management of plant viruses. Vol.3 Vegetable and pulse crops (G.P. Rao., P.L. Kumar and R.J. Holguin-Penna eds): 213-234. Studium Press LLC, Texas, USA.

Labrousse, F. 1931. Anthracnose of chickpea. Rev. Path. Veg.et.Ent. Agric. 18: 226-231.

Laha, S. K., and Grewal, J. S. 1983. Botrytis blight of chickpea and its perpetuation through seed. Indian Phytopath. 36: 630- 634.

Landa, B. B., Navas-Cortes, J. A., Jimenez-Gasco, M., Del, M., Katan, J., Retig, B., and Jimenez-Diaz, R. M. 2006. Temperature response of chickpea cultivars to races of *Fusarium oxysporum* f. sp. *ciceris*, causal agent of Fusarium wilt. Pl. Dis. 90: 365-374.

Luthra, J. C., and Bedi, K. S. 1932. Some preliminary studies on gram- blight with reference to its cause and mode of perennation. Indian J. Agric Sci. 2: 499-515.

Luthra, J. C., Sattar, A., and Bedi, K. S. 1939. Variation in *Ascochyta rabiei* the causal fungus of blight of gram. Indian J. Agric. Sci. 9: 791-805.

Mandal, A. K., and Dubey, S. C. 2012. Genetic diversity analysis of *Sclerotinia sclerotiorum* causing stem rot in chickpea using RAPD, ITS-RFLP, ITS sequencing and mycelial compatibility grouping. World J. Microbiol. Biotechnol. 28: 1849-1855.

Manjunatha, S. V., Naik, M. K., Patil, M. B., Devika Rani, G. S., and Sudha, S. 2011. Prevalence of dry root rot of chickpea in north-eastern Karnataka. Karnataka J. Agric. Sci. 24: 404-405.

Mishra, R. P., Sharma, N. D., and Joshi, L. K. 1975. A new disease of gram in India (*Colletotrichum dematium*). Curr. Sci. 44: 621-622.

Mitra, M. 1931. Report of the imperial mycologist. Science Report of the Agricultural Research Institute, Pusa 1929-1930, pp 8-71.

Mitra, M. 1936. Report of the imperial mycologist. Science Report of the Agricultural Research Institute, Pusa 1933. 34: 139-167.

Navas-Cortes, J. A., Trapero-Casas, A., and Jimenez-Diaz, M. 1995. Survival of *Didymella rabiei* in chickpea straw debris in Spain. Plant Pathol. 44: 332-339.

Ndiaya, M., Termorshuizen, A. J., and Bruggen, A. H. C. V.. 2007. Combined effects of solarization and organic amendment on charcoal rot caused by *Macrophomina phaseolina* in the Sahel. Phytoparasitica 35: 392-400.

Nene, Y. L. 1980. Diseases of chickpea proceedings, International workshop on chickpea improvement . ICRISAT, 28 Feb-2 Mar 1979, Hyderabad, India. pp.171-178.

Nene, Y. L., and Reddy, M. V. 1976. Preliminary information on chickpea stunt. Trop. Grain Leg. Bull. 5: 31-32.

Nene, Y. L., and Reddy, M. V. 1987. Chickpea diseases and their control. In The chickpea (M.C. Saxena and K.B. Singh, eds): 233-270. CAB International, Wellingford, U.K.

Nene, Y. L., Sheila, V. K., and Sharma, S. B. 1996. A world list of chickpea and pigeonpea pathogens. 5th Edition, ICRISAT, Patancheru, India 27pp.

Pal, M., and Singh, B. 1993. Channe koo uktha rog se bachayen. Kheti 47: 24-25.

Palomino, C., Fernandez-Romero, M. D., Rubio, J., Torres, A., Moreno, M. T., and Millan, T. 2009. Integration of new CAPS and dCAPS-RGA markers into a composite chickpea genetic map and their association with disease resistance. Thro. Appl. Genet. 118: 671-682.

Pande, S., Galloway, J., Gaur, P. M., Siddique, K. H. M., Tripathi, H. S., Taylor, P., MacLeod, M. W. J., Basandrai, A. K., Bakr, A., Joshi, S., Krishna Kishore, G., Isenegger, D. A., Narayana Rao, J., and Sharma, M. 2006. Botrytis grey mould of chickpea: A review of biology, epidemiology and disease management. Aust. J. Agr. Res. 57: 1137-1150.

Patel, S. T., and Anahosur, K. H. 2000. Association of soilborne fungi infecting chickpea and correlation coefficient between frequency of fungi and other factors. J. Mycol. Plant Pathol. 30: 50-52.

Pedrajas, M. D. G., Bainbridge, B. W., Heale, J. B., Artes. E. P., and Diaz, R. M. J. 1999. A simple PCR-based method for the detection of the chickpea wilt pathogen *F. oxysporum* f. sp. *ciceris* in artificial and natural soils. Eur. J. Plant Pathol. 105: 251-259.

Prasad, R. D., Rangeshwaran, R., Anuroop, C. P., and Rashmi, H. J. 2002. Biological control of wilt and root rot (wet) of chickpea under field conditions. Ann. Plant Protec. Sci. 10: 72-75.

Rajakumar, E., Aggarwal, R., and Singh, B. 2005. Fungal antagonists for the biological control of Ascochyta blight of chickpea. Acta Phytopathol. Entomol. Hun. 40: 35-42

Reddy, M. V., and Singh, K. B. 1983. Foliar application of Bravo 500 for Ascochyta blight control. Int. chickpea Newsl. 8: 25-26.

Reddy, S. V., and Kumar, P. L. 2004. Transmission and properties of a new luteovirus associated with chickpea stunt disease in India. Curr. Sci. 86: 1157-1161.

Reddy, S. V., Kumar, P. L., and Waliyar, F. 2004. Chickpea stunt disease. In Serological and nucleic acid based methods for the detection of plant viruses (P.L. Kumar, A.T. Jones and F. Waliyar eds): 30-32. Manual, ICRISAT, Patancheru, India.

Rewal, N., and Grewal, J. S. 1989a. Differential response of chickpea to grey mould. Indian Phytopath. 42: 265-268.

Rewal, N., and Grewal, J. S. 1989b. Inheritance of resistance to *Botrytis cinerea* Pers. in *Cicer arietinum* L. Euphytica 44: 61-63.

Sandhu, T. S., Bhullar, B. S., Brar, H. S., and Sandhu, S. S. 1984. Ascochyta blight and chickpea production in India. In Proceedings of the workshop on Ascochyta blight and winter sowing of chickpea (M.C. Saxena and K.B. Singh, eds) ICARDA, 4-7 May 1981, Aleppo, Syria: 259-269.

Santra, D. K., Singh, G., Kaiser, W. J., Gupta, V. S., Ranjekar, P. K., and Muehlbaur, F. J. 2001. Molecular analysis of *Ascochyta rabiei* the pathogen of ascochyta blight in chickpea. Theor. Appl. Gene. 102: 676-682.

Shaw, F. J. F., and Ajrekar 1915. The genus *Rhizoctonia* in India. Memories of the Department of Agriculture India, Botany Series 7: 177.

Singh, B., and Dubey, S. C. 2007. Channe kaa mallyni rog - bachaw ke uppaya. Kheti 60: 18-20.

Singh, G., Chen, W., Rubiales, D., Moore, K., Sharma, Y. R., and Gan, Y. 2007. Diseases and their management. In Chickpea breeding and management (S.S. Yadav, R. Redden, W. Chen and B. Sharma, eds) CAB International, Wellingford, UK: 520-537.

Singh, K. B., and Dahiya, B. S. 1973. Breeding for wilt resistance in chickpea. Symposium on wilt problems and breeding for wilt resistance in Bengal gram. Sept. 1973 at IARI, New Delhi, India, pp. 13-14.

Singh, K. B., and Reddy, M. V. 1991. Advances in disease-resistance breeding in chickpea. In Advance in agronomy. Academic Press, Inc. 45: 191-222.

Singh, R., Sindhan, G. S., Parashar, R. D., and Hooda, I. 1998. Application of antagonist in relation to dry root rot and biochemical status of chickpea plants. Plant Dis. Rep. 13: 35-37.

Tripathi, H. S., and Rathi, Y. P. S. 1999. Current research status of botrytis grey mould of chickpea -A review. Agric. Rev. 20: 135-140.

Trotter, A. 1918. The rabia or anthracnosis of chickpea and its products. Rev. Path. Veg et. Ent. Agric. 9: 7.

Vander Maesen, L. J. G. 1972. *Cicer* L. A monograph of the genus with special reference to the chickpea (*Cicer arietinum* L.), its ecology and cultivation. Thesis Agric. Univ., Wageningen.

Vir, S., and Grewal, J. S. 1974a. Physiological specialization in gram blight. Indian Phytopath. 27: 355-360.

Vir, S., and Grewal, J. S. 1974b. Evaluation of fungicides for the control of gram blight. Indian Phytopath. 27: 641-643.

Vishwadhar., and Chaudhary, R. G. 2001. Disease resistance in pulse crop-current status and future approaches. In The role of resistance in intensive agriculture (S.Nagarajan and D.P. Singh eds) Kalyani Publisher, New Delhi: 144-157.

Diseases of Field and Horticultural Crops
Editor-in-Chief: **P. Chowdappa**
Published by: **Daya Publishing House, New Delhi**

Pages **113-166**

5

Diseases of Sugarcane

S.K. Duttamajumder, R. Viswanathan and G.P. Rao

Sugarcane is a native to India and in cultivation since pre-historic times and one finds mention of use of sugar in the Vedas. In the current agriculture scenario of India sugarcane occupies a prized position in the agrarian economy. About 6 million sugarcane farmers, their dependants and a large mass of agricultural labourers are involved in cane cultivation. Besides, about a half a million skilled and semi skilled workers mostly from rural areas are engaged in the cane industry, the second largest agro-based industry of India. During the 2009-10 crop season, sugarcane was cultivated in 4.17 million hectares with a production of 292 million tonnes of cane with a productivity of 70 tonnes/ha and 490 sugar mills crushing about 186 million tonnes of cane produced 18.91 million tonnes sugar with a sugar recovery of 10.20%. Thus, white sugar accounted for nearly 63% of the cane; *gur* and *khandsari* for 25% and rest 12% accounted for other purposes including seed. Commercial sugarcane is essentially a captive crop and monocultured in and around sugar mill command area for obvious reasons.

In India, there are two distinct sugarcane growing zones viz., subtropical north (which accounts for 60% of the total acreage and only 50% of the total production) is characterized by the harsher climate, thin canes, short active growing span, and low yield and the tropical south is characterized with higher yield, thick canes and little restriction of growing span. The climatological and edaphic factors have also influenced the pathogen scenario of sugarcane in these two distinct growing regions.

In India more than 60 diseases have been reported, but only seven diseases viz., red rot, wilt, smut (fungal disease), leaf scald, ratoon stunting disease (bacterial disease), grassy shoot disease (Phytoplasma), and mosaic (viral disease) cause substantial damage to the crop. Among the fungal diseases red rot is the most notorious one with several epidemics to its credit. It hit hard the Indian Sugar Industry during 1938-39. Smut is mainly the problem of the tropical belt, occasionally it becomes problematic in the subtropical north also (in localized patches). Wilt is present throughout the length and breadth of the country but its impact is more pronounced in the subtropical India limiting the cultivation of thick *'officinarum'* type canes and high sugar genotypes. Ratoon stunting disease, leaf scald, grassy shoot disease and mosaic are prevalent throughout the country but seldom assumed epidemic proportion. Pineapple disease (sett rot) sometimes causes concern in tropical area and bacterial top rot occasionally flare up in certain pockets. Recently, yellow leaf disease (viral disease) is spreading and causing concern. Leaf diseases are of less economic importance except rust, ring spot (*Leptosphaeria*) and Pokkah boeng which in localized pockets cause some concern. In India, overall loss due to various diseases vary from 15-25% in sugarcane but in certain pockets the disease may cause complete crop failure hitting the farmers hard and being an industrial crop the mills suffer badly due to the shortage of raw material.

As of now control measures in sugarcane pathology are predominantly preventive; once the disease appears in the field there is hardly any measure available that can cure the affected plant(s). In sugarcane, heat therapy is quite effective and, in many cases, is the only recommendation that can put a number of diseases under check. Due to the industrial nature of the crop, it is difficult to practice the principles of avoidance for all practical purposes; sugarcane being the raw material of the industry it has to be grown in the command area of sugar mills to ensure a regular supply. Similarly, in flood prone areas, where no crop sustains, sugarcane is the ray of hope for the farmers, even if this condition is very favourable for the flare up of red rot. All these practical considerations pose substantial difficulty in the management of sugarcane diseases.

Sugarcane is a highly heterozygous polyploid crop plant and thus it is extremely difficult to breed desired sugarcane genotypes. But it is bestowed with vegetative propagation; ones a desired variability (hybrid vigour) is obtained, it can be maintained over generations without any genetic contamination/change. However, this also has become an inherent weakness; if a pathogen gets an entry in stalk, it can carry over from one generation to other generation without any hitch. Thus, it becomes very difficult to eradicate the pathogen from the plant system. The major constraints of sugarcane disease management as outlined by Agnihotri and Duttamajumder, (1994) are

(1) Vegetative nature of the crop and it needs immediate planting after harvest.

(2) Damaging diseases are essentially sett-borne and the primary inoculum present in the sett is the major source of the pathogen.

(3) The setts are bulky with tender buds on the surface and thus are difficult to handle.

(4) These tender buds are quite vulnerable to any harsh physical or chemical treatment for the eradication of the pathogen(s).

(5) The crop is virtually available to the pathogen throughout the year, and thus there is no dead season. Pathogens usually do not venture out for survival.

(6) The practice of ratooning (raising of a crop from the stubble after harvest of

upper ground portion), which is a must for profitable sugarcane cultivation, provides congenial niche to the pathogen(s) to proliferate and attack from the very beginning of the season (high initial inoculum load and that too available on/in the host itself).

The details of fungal diseases (red rot, wilt, smut, pineapple disease and rust), bacterial diseases (ratoon stunting and leaf scald) and viral and mycoplasmal diseases (mosaic, yellow leaf and grassy shoot) are described below.

A. Fungal Diseases

1. Red rot

1.1. Economic impact: Red rot is the number one disease of sugarcane in India with considerable economic repercussions. It is a problem throughout the length and breadth of the country and considered a disease of national importance. In fact, even today red rot is the major bottleneck in the profitable cultivation of sugarcane in India, and as a statutory practice, today no sugarcane variety is released for general cultivation unless it is tested resistant against the prevalent red rot race(s) of the particular zone for which the variety is targeted/proposed.

Red rot was first reported to the scientific world from Java (now Indonesia) by Went (1893) and subsequently from India by Barber (1901). One also finds its mention in the Buddhist literature (Deerr, 1949; Daniels and Daniels, 1976) indicating that red was prevalent in the India much before it was known to the scientific fraternity (Duttamajumder and Angnihotri, 1992). The 1895-1900 epidemic of red rot in the Godavari delta is the milestone in sugarcane research and development in India, as it paved the way for the establishment of Sugarcane Breeding Institute, Coimbatore in 1912 and ushered a new era of hybrid cane. In fact, first green revolution in India was witnessed in sugarcane with the release of Co 205 in 1918. The high yield of new hybrid sugarcane along with tariff protection provided the requisite boost to the sugar mills and from 29 mills in 1931, the number rose to 111 by 1934. Thus, it necessitated production of sufficient sugarcane to feed the mills. Every sugar mills tried to expand the cane acreage with the new hybrid canes subverting all the seed selection norms and any available canes were planted. Red rot struck in a big way and devastated the new sugarcane plantations of Co 213 in eastern Uttar Pradesh and Bihar in 1938-39. The total cane crushed was only one third to one half of the normal cane crop season (Chona and Padwick, 1942). The red rot epidemic of 1938-39 was again acted as another turning point, as the epidemics prior to this were mostly restricted to thick the 'noble' cane (*Saccharum officinarum*). During this epidemic, it not only affected the interspecific *Saccharum* hybrid but also produced a more virulent *'light race'* of the pathogen. This red rot epidemic (1938-39) along with the brown spot of rice epidemic (1942-43) paved the way for plant pathology to come out from the fold of botany and became a separate university subject. This also led to the establishment of another central Institute (IISR) in sugarcane heartland of Uttar Pradesh to work on all aspects of sugarcane cultivation except breeding. Even though, with the passage of time more and more resistant genotypes have been pumped in these areas, damage due to red rot still continues (Chona, 1980; Agnihotri, 1990; Duttamajumder, 2008).

1.2. Symptoms: *C. falcatum* can attack any part of the sugarcane plant - from root to leaf and completes the life cycle on leaf. However, damage to the leaf does not cause much harm to the plant or cause any significant economic loss. The most damaging phase of this disease occurs when the pathogen attacks the stalk. The infected cane setts, if used for planting, carry the primary infection to the field and takes toll by killing the bud cause germination failure. This poor germination leads to gappy crop stand and reduction in yield. The damage continues further as tiller mortality, a few tillers are affected. Hereafter, the first diagnostic symptom, the spindle infection appears. Its frequency is more during the pre-monsoon and early part of monsoon and gradually the frequency tapers as monsoon advances. Spindle infection contributes significantly in the flaring up of the disease in the crop (Duttamajumder, 2008). The spindle infection is characterized by the infection of lower side of the mid-rib of unfolded leaves of crown. Generally, infection traverses from the sett to the stalk without producing any conspicuous damage or reddening of the stalk tissues and finally reaches the crown where it expresses. Only one or two odd discoloured vascular bundles indicate the path which *C. falcatum* traversed. On the crown leaf, it starts from the leaf base as water-soaked elongated (oval, eye shaped) lesion at the lower side of the mid-rib (which remains exposed in the unfolded crown leaves). On the leaf lamina, it also produces elongated lesion surrounded by a yellow halo. In three to four days time, these water soaked lesions turn light brown and eventually become dark coloured due to formation of acervuli bearing abundant setae and spore masses. Sporulation on this spindle infection is very fast, and in a week's time, millions of spores (conidia) are formed, whereas in case of mid-rib infection it takes more than a month to sporulate. In the damaged unfolded leaves of the crown, abundant chlamydospores are also formed. The spindle infection is a very characteristic of this disease and is a confirmatory symptom of red rot, like internal rotting with white spot (Duttamajumder, 2008).

The characteristic symptoms as described in textbooks (stalk rot phase) appear as the grand growth phase progresses and sufficient stalk formation takes place. This starts with the yellowing of one or two leaves of the crown, yellowing as well as withering of leaves starts from the margin. The third or fourth open leaf being the most prominent leaf of the crown attracts one's attention from a distance (which is incorrectly written as the yellowing of 3rd or 4th leaf as the first appearance of the disease in most of the writings). Later on, the entire crown becomes light orange yellowish and dries. At this time, if the stalk is split open longitudinally, the typical red rot symptoms become visible in the internodal tissues. Reddening of the internal tissues interspersed with white patches (white spots), which are usually at right angle to the stalk axis, along with the presence of sour alcoholic smell confirm the disease. Probably, these white spots in the internodal tissues of cane stalk act as the inn, where the pathogen safeguards itself from the host defences. In fact, in the white spot area, tissue damage is most prominent and the tissues are usually devoid of cell contents and filled with air and web of fungal hyphae. When the stalk tissues with prominent white spots are allowed to dry, the white spot area becomes sunken, indicating greater moisture loss and tissue degradation of this particular area (Duttamajumder, 2008). With the progress of the disease, cavities are formed in the intermodal tissue and in which the fungus often sporulates. The other characteristic of red rot is nodal rotting. In red rot, invariably the node is badly damaged and the affected cane can be broken easily. This nodal rotting quickly distinguishes it from wilt where no nodal rotting takes place. *C. falcatum* proliferates within the cane tissue and emerges out to sporulate when the cane starts drying. Usually, it sporulates on nodal region, especially on the root primordia. On dried affected cane acervuli formation is also very common on the rind. Due to formation of setae, these appear as small dark projections which give prickly feeling on touching (Abbott, 1938; Edgerton, 1959, Singh and Singh, 1989; Agnihotri, 1990; Duttamajumder, 2008). It is also to be pointed out here that sporulation is blazingly fast on cut or damaged portion of the diseased cane. Once the diseased cane is cut or damaged exposing the affected intermodal tissues, a good amount of sporulation appears on the exposed surface even in overnight incubation, provided the cut surface does not dry up (Duttamajumder, 2008).

The other most important phase of the disease is the infection on leaf mid-rib and leaf lamina. *C. falcatum* produces reddish lesions on the upper surface of the mid-rib. The lesions gradually increase in size and may elongate up to the entire leaf mid-rib. With the age of the spot, it develops ashy white centre where the pathogen produces acervuli laden with conidia and setae. The mid-rib lesions, for a long time have been taken as the chief source of inoculum for causing infection in the standing cane stalk. The spindle infection was taken as mid-rib infection and thus resulted in conflicting claims. Nevertheless, the mid- rib infection never puts any discernible effect on the cane plant in terms of juice quality and yield. Several trials with different fungicides as well as removal of affected leaves by various workers did not yield any dividend and therefore, it has been concluded that the mid-rib population of *C. falcatum* does not pose any serious threat to cane cultivation directly.

The third type of infection that affects the leaf is termed laminar infection. On the lamina, numerous tiny reddish lesions on the upper surface of the leaves appear and sometimes lesions are devoid of any red or dark colouration (Singh and Alexander, 1970). Later on, at the infection site acervuli are formed especially on the dying/dead cane leaves. Laminar infection is often encountered in fields of sugarcane breeding experiments where large number of cane genotypes with varying resistance/susceptibility to *C. falcatum* is usually grown.

1.3. Causal organism: The fungus causing red rot of sugarcane is commonly known by its imperfect state, i.e., *Colletotrichum falcatum* Went. The disease was first described to the scientific world in 1893 by Went (Went, 1893) from Java (now Indonesia). He called the malady "Het rood snot" meaning red smut. Butler (1906), from India, gave the name"Red rot" to this disease due to the characteristic reddening of the internal tissues of the stalk., The perfect state was first recorded by Spegazzini from Argentina in 1896 (Spegazzini, 1896) but it took almost 50 years to establish the association between *Colletotrichum falcatum* Went and *Physalospora tucumanensis* Speg. (Carvajal and Edgerton, 1944). In 1954, von Arx and Muller placed it in genus *Glomerella* and named it *Glomerella tucumanensis* (Speg.) von Arx and Muller (syn. *Physalospora tucumanensis* Speg.). von Arx (1957) included most of the described species of *C. falcatum* on Gramineae as synonyms of *Colletotrichum graminicola* (Ces.) Wils. and accepted *Glomerella tucumanensis* (Speg.) von Arx and Muller (syn. *Physalospora tucumanensis* Speg.) as the name of the perfect state of this species. However, due to the strong and specific pathogenic reaction of *C. falcatum* to sugarcane, it is considered different from *C. graminicola* (Sutton, 1980; Abbott and Hughes, 1961).

The *Colletotrichum falcatum* colony is greyish white with sparse aerial mycelium and small dense felty patches, reverse white to grey, conidial masses salmon pink(light race); some cultures have abundant greyish white aerial mycelium with poor sporulation and no distinct acervuli (dark race). Sclerotia not produced by both the races, setae sparse, conidia falcate (but not markedly so), fusoid, apices obtuse, 15.5 (25-26.5) 48 µm x 4 (5-6) 8 µm and contents are granular and sometime contain oil globules. Appressoria medium brown, clavate or circular, edge entire, 12.5-14.5 µm x 9.5-12 µm (6-21 x 6-17 µm). In the acervulus, dark brown setae measuring 100 - 200 µm also develop (Abbott, 1938, Agnihotri, 1990, Singh and Singh, 1989). These are septate stout structures with bulbous base tapering toward the tip. These under favourable condition, when tip remains hyaline, also produce conidia (Edgerton, 1959; Duttamajumder, 2008).The fungus forms chlamydospores on the surface of the stalk/ leaf or within the cane tissues, and in culture media. (Edgerton and Carvajal, 1944).

Perithecia are inconspicuous and are almost entirely embedded in the cane tissues. They are found in abundance on dead or dying leaf blades, mid-ribs, and sheaths. In the leaf tissue they are crowded between the vascular strands, and measure 100-260 µm in width by 80- 250 µm in height. Asci are clavate, measuring 70-90 µm x 7.5 (13)-18 µm, and the hyaline, straight to slightly fusoid, one-celled ascospores measuring 18-22 µm x 7-8 µm are arranged in a biseriate pattern (Carvajal and Edgerton, 1944; Chona and Bajaj, 1953 ; Chona and Srivastava, 1952; Duttamajumder, 2008).

1.4. Race and molecular characeraiztion: For successful breeding of disease resistant genotypes and their deployment, a clear picture of race flora of the pathogen and their distribution is essential. With the advancement in sugarcane breeding, emphasis was given for breeding resistant varieties against the pathogen. The erratic behaviour of pathogen in artificial culture, frequent loss of sporulation as well as pathogenicity always put the pathologists on the back foot. Time ans again workers resorted to collect available variability each year and use the most virulent one from the collection for the screening purpose.

The concept of physiologic specialization in red rot pathogen dates back to 1935 when Abbott (1938) from Louisiana reported that red rot isolates form two morphologically distinct groups - one dark velvety non sporulating and the other lighter and sporulating and usually capable of infection on inoculation. Similar observations were also made in India. Indicated that the epidemic of 1938-39 developed a new highly virulent light type, which was responsible for the failure of Co 213. Chona (1956) and others also have reported the variation in the pathogen in terms of morphology and pathogenicity on different cane genotypes. Kirtikar (from Shajahanpur) designated the variation in R series, like R 141, R 200 etc., whereas Rafay (from Lucknow), Chona (from New Delhi) and others grouped them as A, B, C, D, E, F, G, H, and I stains of *C. falcatum*. There was no consensus among the workers in this regard, and due to the overlapping morphological groupings it was extremely difficult to put a *C. falcatum* isolate in the above categories.

In 1992, in the meeting of AICRP (Sugarcane) under the direction of the then ADG (Crop Protection) Dr. S. Nagarajan and K. C. Alexander (PI, Pathology) a workable system of virulence mapping was chalked out. The group identified differentials and designated races for validating the pathogenicity data from different locations and environment. Accordingly, three subtropical isolates and three tropical isolates were selected for their testing at all India level for their stability in different environment. An isolate from Co 1148 was named as Cf 01, isolate from Co 7717 got the designation Cf 02 and isolate from CoJ 64 was designated Cf 03. Similarly, isolate from Co 419, received the designation Cf 04, isolate of Co 997 got Cf 05 and the isolate from CoC 671 was designated Cf 06. The six isolates typically produced the HS reaction with the matching genotype. These six representative cane genotypes were taken as the maintainers of virulence. For the differentials three species representative viz., SES 594 (*S. spontaneum*), Khakai (*S. sinense*) and Baragua (*S. officinarum*) were taken. For the differentiation of subtropical isolates Co 975 and for tropical isolates Co 62399 were taken. BO 91 and CoS 767 were taken as the next vulnerable genotypes, as any attack on these genotypes had the potential to affect major chunk of cane area. Of late, five more races have been designated (Cf 07, Cf 08 and Cf 11 from CoJ 64, Cf 09 from CoS 767 and Cf 10 from the clone 85A261).

It was quite natural to explore the variability of *C. falcatum* using molecular tools. Around the same time at IISR, Lucknow Dr. Madan and his associate tried with the six designated races to establish the variation at molecular level (RAPD) but no clear cut picture came out which could explain the variation (Madan *et al.*, 2000); later on Suman *et al.*, (2005) also expanded the RAPD profiling work with the available races and could broadly divided the races in to two groups. The molecular work was further expanded by Satyavir from Haryana (Kumar *et al.*, 2010) and in addition to RAPD and ISSR markers they also used URP (universal rice primers) for the characterization of races and isolates of *C. falcatum*. However, due to high non-structured molecular diversity among the isolates a good correlation with the pathogenic behavior could not be established.

1.5. Host range: In addition to *Saccharum officinarum, S. barberi, S. sinense, S. robustum, S. spontaneum, C. falcatum* has a number of collateral hosts like *Sorghum vulgare, Sorghum halepense, Leptochloa filiformis, Miscanthus* (Edgerton, 1959, Singh and Singh, 1989). However, the role of non *Saccharum* host is very limited.

1.6. Epidemiology: *C. falcatum* can attack any part of the sugarcane plant from root to leaf. The infected cane setts, if used for planting, carry the primary infection to the field. This poor germination leads to gappy crop stand and reduction in yield. If, at all, the buds of the infected setts are able to germinate and grow, then above ground symptoms appear. The type of symptoms varies depending on the prevailing weather condition. At first, the symptoms appear as the death of young and emerging shoots without any conspicuous identifiable symptom (in March-April-May in north Indian condition of spring planting) and it continues further as tiller mortality. Hereafter, the spindle infection appears and its frequency is more during the onset of monsoon and gradually the frequency tapers. Sporulation on this spindle infection is very fast, and in a week's time, millions of spores (conidia) are formed. The spores formed in spindle infection act as the major source of secondary inoculua in the field. These gets spread to the adjoining canes through, wind blown rains, rain splash, running water or get piggy backed with the borer (Duttamajumder, 2008). During the monsoon, the disease remains in peak and can cause wholesale damage to the crop, if weather condition favours the spread and establishment in the new cane. With the cessation of monsoon rains and drop in temperature in the ambient environment the pathogen gets restricted. New infections causing death of the plants become few. Even if the infections are there, *C. falcatum* go latent and the infection is not discernible with common eye. This type incipient infection is the major source of inoculum that goes along with the sett and helps in the perpetuation of the disease.

The pathogen is a transient visitor to the soil and remains viable as long as the debris lie undecomposed in soil (Singh, *et al.*, 1977; Agnihotri *et al.*, 1979). Due to its high sensitivity to microbial antagonism, *C. falcatum* cannot withstand the stiff competition with the saprophytes in soil (Chona and Nariani, 1952).

1.7. Management: Clean cultivation is the primary prerequisite for growing a healthy sugarcane crop. This starts with the selection and planting of healthy setts, preferably in a healthy field. This practice reduces the load of initial inoculum and helps in checking the red rot disease. This practice of sett selection should be religiously followed if one really desires to check the damage from red rot. Removal of diseased plants, as and when detected, also helps in reducing the inoculum available for the secondary spread. It has been observed that the early inoculum developed as spindle infection is the key secondary spread of the disease in the crop season (Duttamajumder, 2008)

Sugarcane, being available around the year, it is the environment that modulates the disease incidence and intensity. Therefore, a resistant variety to the prevalent races/ pathotypes of the pathogen in a specific geographical area becomes the primary prerequisite to raise a healthy crop. Secondly, sugarcane is commercially propagated through stem cuttings called "setts"; these setts should be made free of red rot infection to prevent the ingress of the initial inoculum through the planting material. Thirdly, fields having immediate history of red rot should preferably be avoided for cane planting. Although, the soil inoculum does not pose an immediate threat to the crop, this surely produces contaminated planting material which shows its impact later on. Irrigation or rain/flood water plays a very important role in the dissemination of the red rot pathogen. Moreover, the damaging phase of the crop occurs only during the late monsoon emphasizing the importance of rain, cloudy weather, etc. in the spread of the disease. In fact, this is a nagging problem in eastern Uttar Pradesh and Bihar where flooding or waterlogging is common.

In India, the prevalent legislation is unable to contain the movement of canes from one zone to other and this unrestricted movement of the cane from one district or state to another has largely been responsible for the migration of *C. falcatum*. Earlier, pathogens had spread to the new areas and now virulent pathotypes are moving to new pastures (Duttamajumder and Agnihotri, 1992). Therefore, legislation as well as its strict compliance is essential to curb the spread of the disease through the agency of man.

1.7.1. Physical Treatment: Sugarcane, a vegetatively propagated crop, responds well to the use of physical energy in the form of heat therapy/treatment. In this treatment, setts are exposed to a critical temperature that is lethal to the pathogen but not to healthy buds and thus effectuates a differential killing of the pathogen is achieved. It has been tried in various forms with varying degrees of success against red rot (Singh, 1973; Singh, *et al.*, 1980; Natarajan and Muthuswamy, 1981). Heat treatment can root out fresh superficial infections very well, but it becomes largely ineffective against the deep seated dormant infections present in the nodal region of seed cane (Srinivasan, 1971; Singh and Agnihotri, 1987). Although complete elimination is not achieved by the MHAT treatment at 54°C for 2½ hours it significantly reduces the number of surviving propagule/inoculum causing infections later in the season (Singh, *et al.*, 1980). Curative property of hot water treatment may be increased by adding the desired fungicide in the water tank or by using a post fungicidal treatment (Singh and Agnihotri, 1987; Agnihotri, 1990).

1.7.2. Chemical Treatment: Several fungicides, both systemic and non-systemic (e.g., Bavistin, Vitavax, Topsin, Aretan, Blitox, Dithane Z-78, etc. are effective in killing *C. falcatum in vitro* but they don't provide satisfactory control of the disease under field conditions (Chand *et al.*, 1974; Agnihotri, 1990).

1.7.3. Resistant Varieties: Breeding red rot resistant variety/genotype has become the mainstay in the perpetual fight against red rot. However, over the years only a few spontaneum clones have been used as a source of resistance gene(s). Thus, the genetic base for the red rot resistance has become quite narrow in the present commercial varieties. Over the years, several excellent varieties have been bred and released for general cultivation but with the passage of time most of them have fallen prey to the red rot in due course. Varieties like Co 213, Co 312, Co 419, Co 453, Co 997, CoC 671, Co 1148, Co 7717, CoJ 64, CoS 767, CoS 8436, CoLk 8001, CoLk 8102, etc. have succumbed to red rot in due course and the list is increasing in every passing year (Duttamajumder, 2008).

The task of breeding sugarcane genotypes with good degrees of resistance has become cumbersome due to the non-availability of high degree of resistance or immunity in donor parents (Srinivasan, 1965). It is not that scientists have not tried to understand the genetics of resistance against red rot pathogen. Rather, scientists have spent their life time to decipher the underlying mystery but no definitive answers could be given. Even for the trait 'sugar', which has high heritability, scientists are yet to put a definitive conclusion. Thus, the 'proven crosses' rarely yields an excellent variety. Most of the sugarcane varieties that sustained for a longer period in the field and served sugar industry bear this testimony. The inheritance is complex one.

Information on the inheritance of red rot resistance in sugarcane is rather sketchy. There is no clear-cut pattern of inheritance that one can embark upon. The inheritance, as known today, is indiscriminate as crosses between resistance and susceptible clones in possible combinations (RxR, SxR , SxS and reciprocals) give rise to some resistant plants. Degree of resistance of parents make little difference to the per centage of resistant progenies, however, variation do occur in certain cross combinations. In sugarcane, the best-known source of resistance to red rot is *S. spontaneum* and most of the present day commercial hybrids owe their red rot resistance to *S. spontaneum*. However, except the initial involvement of a few *spontaneum* clones in the crossing programme, very little has been done to tap this potential resistant donor. Of late, a programme (ISH development) to broaden the genetic base of the parental pool was run at SBI, Coimbatore and a few clones have come up with new gene/gene combinations. However, the major constraint in introducing *spontaneum* genome/genes in order to further broaden the genetic base of red rot resistance is the requirement of time and dedicated effort in this regard. It is hoped that a structured programme at the national level will run to introduce an array of genes from the wild relatives like *S. spontaneum, S. robustum* and *Erianthus* sp. to tackle the problems of pests and diseases as well as the problems of drought and waterlogging.

A close look at the data supplied and clubbing the *robustum, sinense* and *barberi* with hybrid canes on the non-tenability of these species reveals an interesting picture. It became apparent that out of 104 old hybrids (*robustum, barberi* and *sinense*) only 5 (4.81%) were R, 17(16.34%) were MR, 15 (14.42%) were MS, 16 (15.38%) were S and 51 (49.04) were HS (Duttamajumder, 2008). The susceptibility picture is also no different in freshly developed ISH hybrids. The corresponding figures are quite similar. However, there is a definite shift in resistance reaction in the time tested old hybrids (recognised as *barberi* and *sinense*) that withstood the rigor of red rot in the sub-tropical India over the millennium. More of MR type was encountered, with the significant reduction of MS type. It should be noted that the pattern of high susceptibility was around 50% in both the time tested and the new ISH hybrids. It indicated the role of selection pressure that has gradually shifted towards more of tolerance in the intermediate category (MS and MR) rather than imparting any major change in resistance or susceptible pattern (R and S category). Probably physical resistance/entry point resistance has evolved along with as has been observed by Duttamajumder and Misra (unpublished) with Khakai. *C. falcatum* isolates that produced HS reaction on plug method of inoculation failed completely to get an entry in the cane through nodal method.

1.7.4. Three Tier Seed Programme: Based on the success of MHAT in checking sett borne diseases of sugarcane, including red rot, a programme was launched to produce healthy seed (Singh, 1983) by eliminating/reducing the level of initial inoculum. The programme has three well defined stages, viz., (i) Breeders seed, (ii) Foundation seed, and (iii) Certified seed. Breeder seed is raised by treating apparently healthy canes using moist hot air (54°C for 2½ hours) and subsequent generations are raised with inspection. There should be total freedom from red rot in seed nurseries. As the crop raised from heat treated cane may get reinfected with the diseases, a strict vigil is necessary to guard against the possibilities of reinfection and to maintain the freedom from diseases. The crop raised from breeders seed is called foundation seed and from foundation seed commercial seed is produced. This is a continuous process, and every third year, healthy seed is provided to the farmers. This method is not only used in India but also in the neighbouring countries for raising disease free nurseries.

1.7.5. Biological control: *Trichoderma* is one of the most important biocontrol agent used for management of different diseases including red rot. Use of biological agent is more important, as chemicals treatments are not very effective. Work on *Trichoderma* was started in sugarcane to control seedling diseases (raised from fluff) at SBI, Coimbatore by Alexander and his co-workers. At IISR, Lucknow potent *Trichoderma* cultures obtained from Pantnagar were tried to control red rot with sett, soil as well spray treatment in standing cane. Results of *Trichoderma* were positive but the efficiency of control was far below what was achieved by the comparable fungicide treatment. Similarly, *Trichoderma* was also tried by Satyavir and associate from HAU, Hisar with positive results in growth promotion and disease control. From IISR, Lucknow Vijay Singh and associates conducted elaborate experiments with *Trichoderma* for growth promotion and red rot management through increasing the efficacy of *Trichoderma* and came out with technique of mass multiplication of *Trichoderma* in pressmud (a sugar mill waste byproduct) and procedures of soil application at the time of planting and application at the time of tillering (Kumar and Satyavir, 1998, 1999, Singh and Duttamajumder, 1995, Singh *et al*, 2008, Singh *et al.*, 2010, Srivastava, *et al.*, 2006,Yadav *et al.*, 2008). *Pseudomonas fluorescence* is another biocontrol agent that has also shown promise against red rot. The group led by Vidyasekharan and later by Samiyappan at TNAU, Coimbatore lead to the development of talc formulation. Subsequently, Viswanathan and co-workers tested the efficacy of *P. fluorescens* against red rot and got positive response in relation to germination, growth and reduction in the intensity of red rot.

1.7.6. Integrated disease management: The research carried out at the Indian Institute of Sugarcane Research, Lucknow and elsewhere suggest that there is no single method available that can control red rot disease to a high degree of satisfaction (cent per cent). Therefore, an integrated approach is essential to contain this pathogen. Resistant variety remains the most practical option to combat this disease, even though red rot pathogen is breaching the introduced resistance with impunity in a regular fashion. The integrated approach currently recommended to achieve this goal is as follows:

 (i) Selection of field having no immediate history of red rot. Preferably, the field should be well drained/ upland type.
 (ii) Selection and use of healthy planting material.
(iii) Whereever possible, the planting material should be subjected to MHAT.
(iv) Chemical treatments of seed material with a fungicide like Dithane M-45, Bavistin, etc. to protect against seed as well as soil borne inoculum and also to protect the setts from secondary infections after MHAT.

(v) Application of bio-agents like *Trichoderma* as sett- as well as soil- treatment at the time of planting, and also drenching around clump before the onset of monsoon.

(vi) Roguing of infected stools (entire clump) and spot application of fungicide as and when disease is detected (pre monsoon) is necessary to reduce the level of available inoculum in the field and to remove the focus of infection.

(vii) Discouraging the practice of ratooning, if any incidence of red rot is observed in the plant crop.

2. Wilt

2.1. Economic impact: The wilt disease of sugarcane is of considerable importance and probably it is the number one disease of sugarcane in the subtropical India, as it imposes severe limitations in the selection of thick canes and high sugar genotypes. In fact, most of the thick high sugar genotypes are culled out due to the wilt during the selection process. Probably it is due to wilt, moderate thick cane along with moderate sugar dominates the varietal scenario of subtropical India. This disease is also has all India foot print but is more pronounced in the subtropical sugarcane growing belt.

The wilt disease of sugarcane was first reported from India from Pusa, Bihar (Butler and Khan, 1913). Subsequently the disease has been reported from several countries like Bangladesh, Philippines, Uganda, South Africa, West Indies, Mexico, U.S.A., and some South American countries (Ganguly, 1964). In India, many excellent varieties like Co 527, Co 951, Co 1007, Co 1223, CoS 245, CoS 321 were phased out of cultivation due to wilt (Kirtikar *et al.*, 1972) and subsequently it affected many varieties like Co 419, Co 453, Co 527, Co 775, Co 975, Co 997, CoC 671, Co 86032 and the list is increasing day by day and in the immediate past varieties like CoLk 8001, Co 7717, Co 89003, CoS 88230, CoH 92 have fall prey to wilt. This disease has a penchant for the early high sugar genotypes. In fact, it has been ascribed as the bottleneck in the development of early high sugar cane genotypes in subtropical India where the propensity of this disease is more (Duttamajumder *et al.*, 2004). This is evidenced from the fact that the tropical cane genotypes usually catch wilt infection when grown in the subtropical region.

The wilt is a killer disease and depending on the susceptibility of a cane genotype and availability of soil inoculum the disease sometimes becomes annihilating, causing cent per cent crop loss. The loss in the plant crop increases over period of time; least at the starting of the season i.e. October and increases several fold in the following March. However, the loss is more pronounced in the ratoon crop as wilt affected clumps usually fail to germinate and rarely give rise to any harvestable ratoon. Besides the occasional heavy toll the disease it appears in most of the cane varieties in a milder way and certain per centage of yield is lost every year due to wilt. The wilt is often associated with red rot (Butler and Khan, 1913, Agnihotri, 1990) in northern India and with pineapple disease in peninsular India (Srinivasan, 1964) aggravating the crop damage. Associations of root borer with wilt and nematodes have been implicated to predispose the plant for taking of wilt infection.

2.2. Symptoms: The 'wilt' indicates some suddenness in the death of the plant-whether pathological or physiological. Once permanent wilting point is reached, it cannot be reversed. However, the 'wilt' as it is known in sugarcane has no suddenness. It is gradual death of the plant over a period of time due to diminishing supply of water in the stalk tissues. Obviously the so called 'wilt' in sugarcane is a misnomer. The disease is basically a dry rot and took 2-4 months to dry-up the cane. The impact of wilt is felt only after the rains when a moisture stress in the field aggravates the situation. Usually the disease surfaces at the end of grand growth phase (September-October) which coincides with the withdrawal of monsoon in subtropical India. The affected cane looses water content in stalk (juice) and become light. The crown of the affected plant initially shows yellowing of leaves (older leaves); later on, these leaves droop and dry out. The drying starts from lower leaves and gradually upper leaves get affected. During this slow drying of leaves a host of opportunist fungi grow on the debilitated leaves imparting a dirty appearance. At the advanced stage of the disease only 3-4 crown leaves remain alive (pale green / pale yellow) before dying out. The cane gradually loose moisture, and eventually the epidermis caves in resulting in a bobbin shaped nodes and internodes. At the advanced stage of drying of cane, large cracks in the internode exposing the internal cavity form; sometimes crack extends in the opposite directions and providing passage to light passes through. The cavity formation starts from the basal internodes and gradually extends to upper internode.

On splitting of the cane the internal hollowness becomes inevitable. Depending on the time and duration of the disease, the nature of hollowness differs. At the beginning of the wilt affected stalk, if split open water-soaking of the central core of the internodal tissues is observed. Gradually water soaking spreads laterally to the periphery. Concomitant with water-soaking, a significant drop in the brix% juice also takes place. As the disease further advances internodal tissues gradually lose moisture and gets pressed towards epidermis and forms a central cavity in the internode. Gradually the cavity increases and entire internode become hollow. Typically the cavity does not extend beyond the internodal tissues. Initially the intermodal tissues remain white, but with the progression of the disease, the internal tissues gets different shades of colour, from a muddy brown to reddish brown depending on the secondary invader and varietal response to this disease. The wilt affected canes do not emit any foul smell as in red rot.

2.3. Causal organism: Conflicting claims have been made regarding the causal organisms of wilt (Ganguly, 1964). A perusal of the literature indicates that two fungi either singly or in combination are mainly responsible for the wilt disease. Butler and Khan (1913) named the causal organism as *Cephalosporim sacchari* Butler and others have accepted it (Subramaniam and Chona, 1938; Ganguly, 1964; Srinivasan, 1964). Subsequently several investigators reported *Fusarium moniliforme* var. *subglutinans* as the causal organism of this disease as authenticated wilt samples yielded only this organism (Booth, 1971). Gams (1971) coined a new name for this fungus and now it is recognised as *Fusarium sacchari* (Butler) Gams to which both *Cephalosporium sacchari* and *Fusarium moniliforme* var. *subglutinans* were made synonyms. Subsequently Nirenberg (1976) divided *F. sacchari* into two varieties, viz. var. *sacchari* with unseptate conidia mostly in aerial mycelium and var. *subglutinans* with 1-3 septate microconidia. However, serious difficulties were encountered by subsequent workers in reproducing typical wilt in artificial inoculations. Of late, Viswanathan *et al.* (2011) also claimed *Fusarium sacchari* as the actual causal organism of wilt.

Singh *et al.* (1975) identified *Acremonium implicatum* as the major causal organism of this disease. They were in the view that *Fusarium sacchari* is chiefly the pathogen of parenchymatous tissue and is responsible for the damage in the internodes whereas *Acremonium* is the vascular pathogen and mostly manifests in the fibrous node and responsible for slow choking of the cane to death.

In a series of experiments conducted at IISR, Lucknow during the last 10 years by Duttamajumder and Misra (unpublished) have indicated that wilt is not a sett-borne disease as thought earlier. With adequate nutrition supply, a healthy crop can be raised from even badly wilted canes. The so called claimed pathogen, *Fusarium sacchari* is an endophyte and is intimately associated with sugarcane system. Isolation from healthy canes, even from the stalk of seedlings raised from fluff (true seed) always yielded *Fusarium sacchari*. Isolation efficiency usually varies in the range from 40-60%. The *Fusarium sacchari*, thus, fulfills all the requirement of Koch's postulate and molecular detection except the specific pathogenicity. Duttamajumder and Misra (unpublished observation) concluded that wilt is probably a soil- borne disease which selectively degenerates the roots and arrests the development of new roots and thus impairs the water availability in the stalk and plant gradually dries up. For the survival of the young (the axial bud of the stalk is the young of the cane through which cane propagates) it channelizes all the stored nutrients of the internode to the node and thus nodes remain healthy till the end.

2.4. Epidemiology: Primary transmission of the pathogen(s) is thought to be via infected sett and soil (Ganguly, 1964, Kirtikar *et al.*, 1972) while the secondary transmission is via wind, rain and irrigation water (Ganguly, 1964). The wilt fungus survives in soil for 27 to 31 months (Ganguly and Chand, 1963). In subtropical area (Bihar) neutral to slightly alkaline pH favours the rapid proliferation (Ganguly and Khanna, 1955) whereas in the tropical area (Anakapalle) it is favoured high nitrogen and acidic soil pH. The disease is basically a soil-borne one and is influenced by soil conditions such as pH, soil texture, soil moisture, etc. In Bihar, high disease incidence was observed in soil having neutral or slight alkaline pH (7.0-8.0) whereas, in Andhra Pradesh the disease is prevalent in acidic soil. High C: N ratio in soil also predisposes cane to wilt (Ganguly, 1964). A direct correlation between wilt incidence and moisture regime in soil has been reported (Sarma, 1976). Negligible wilt occurred in fields having 8.1 to 9.1% soil moisture, while it increased appreciably when the moisture level depleted to 3.5%.

2.5. Management: In the absence of proper identity of the pathogen it is difficult to target out the real cause. No effective method is available to check the disease once it surfaces in the field. Therefore, a concerted approach has to be taken to put a check on this disease. The reported pathogens are well known soil inhabitants. However, experiments for their elimination from the soil with fungitoxicants have met with varied success. Use of resistant genotypes along with field sanitation has become the mainstay to check this menace.

3. Smut

3.1. Economic impact: The whip smut of sugarcane is the most easily identifiable disease. It was first recorded from 1870's from Natal, South Africa and identified as *Ustilago sascchari* changed the name to *Ustilago scitaminea*. In 2002, due to presence of columella, peridia and strile cell among the teliospores in the sori as well as molecular rDNA data. It was proposed a new combination *Sporisorium scitamineum* for the sugarcane smut pathogen.

In India the disease was first reported by Butler (1906) and it continues to be a major problem in sugarcane cultivation in all the sugarcane growing states. Many prominent varieties like Co 419, Co 453, Co 740, Co 975, Co 1158, CoS 510, BO 11, etc., have incurred heavy losses due to this disease. An unprecedented epidemic broke out during 1942-43 in Bihar affecting large area (Chona, 1956). In 1947-48, it affected Co 419 in Karnataka and in 1950-52 affected Co 513, BO 11 in Bihar. This is a widely distributed disease of sugarcane and it has been a major constraint to sugarcane production at one time or another in almost all the sugarcane growing countries. The disease caused severe yield loss to sugarcane for long time in Maharashtra and Northern Karnataka regions till Co 740 was under cultivation. Replacement of Co 740 with Co 86032 and CoC 671 reduced the smut severity in these regions. Currently the disease occurs in varying intensities in different states and heavy incidence of smut was found in CoV 05356 (99V30), CoA 92081 in different parts of Andhra Pradesh. The area under popular varieties like Co 86002 and CoSi 95071 has started reducing in Gujarat due to smut. Moderate to severe occurrences of smut were found on Co 92012, Co 97009, CoSi 95071, CoSi 6, CoT 8201, Co 86002, Co 86032, Co 8014, CoC 671, Co 94012, Co 419, Co 7219, Co 1305, Co 6507, Co 8011 (tropical region). CoSe 01424, CoSe 92423, CoS 767, CoS 88230, CoS 95255, BO 91, Co 1158, (subtropical region) in the country. In addition to loss in cane tonnage due to reduced number of millable canes, the disease infection reduces sugar recovery in the mills. The disease can cause yield losses of up to 50 per cent. In general, ratoon crops face severe crop loss as compared to plant crop.

3.2. Symptoms: The disease can be noticed in all the crop stages after germination, but it follows a bimodal pattern. First peak in the incidence occurs after germination of the sett, mostly during April –May and second peak appears in October, after the cessation of monsoon rains. The disease is characterized by the production of a black whip-like structure from central core of the meristematic tissue. Whips are also produced on the lateral buds, mainly due to secondary infection. The whips vary in size widely; from a few centimetres to large whips about 1.5 m long extending high above the crop canopy may be observed. The whips are composed of a central core of host tissue surrounded by a layer of black spores that is covered by a thin silvery white membrane. This membrane quickly disintegrates exposing the spores and central core of host tissue. Older exposed whips that have been weathered for a long time usually are devoid of spores, as the spores are quickly get dispersed by wind and rain. Sometimes lateral buds of infected stalks may show small whips. Infected clumps may appear grassy/bushy with an abnormally high number of lanky stalks with terminal whips. Diseased plants can be spotted even before the appearance of whips as leaves become thin, stiff and remains at an acute angle, as if the plant is going to flowe.

3.3. Host Range: Natural infection of *Imperata arundinacea* and *Erianthus saccharoides* and artificial inoculation of *Sorghum bicolour* and *Rottboellia exaltata* have been reported but are unimportant as far as sugarcane smut epidemiology is concerned.

3.4. Causal organim: Identified the sugarcane whip smut pathogen as *Ustilago scitaminea*. Mundkur (1939) made an exhaustive study of sugarcane smut reviewed the taxonomic status and identity of the pathogen. In 2002, due to presence of columella, peridia and strile cell among the teliospores in the sori as well as molecular rDNA data proposed a new combination *Sporisorium scitamineum* (Syd.) Piepenbr., Stoll & Oberw. The teliospores are round, dark brown in colour, minutely punctuate and measure 5.5 to 9.5 µm. Teliospores germinate readily under moist conditions, each giving rise to a promycelium of variable dimensions averaging 16µm long and 3-4 µm wide and usually divided transversely into three or four cells. Each of these cells produces single cellular hyaline sporidia of 6 x 2 µm in size. The sporidia in turn germinate and produce long septate hyphae. Hyphal or sporidial fusions result in dikaryon formation. This dikaryotic mycelium infects sugarcane. Genetic factors controlling sexual compatibility segregate at meiosis so that sporidia and fusion hyphae are of different strains. The hyphae penetrate the basal portion of bud scales and invade the mesitematic region of the bud. Mycelia become established in germinating shoots and when the stalk matures systemic infection proceeds. Finally whip production is initiated by converting the apex region.

When cells containing the two mating type alleles are brought together, developmental changes occur, shifting from the saprophytic, budding yeast-like appearance to a mycelial form characteristic of the parasitic phase. The promycelium can also germinate to give hyphae capable of fusion, or the promycelial cells may fuse to form the dikaryon (Alexander and Ramakrishnan 1977; Lee-Lovick, 1978)

3.5. Molecular characterization: Xu *et al.* (2004) studied genetic diversity of *S. scitamineum* using RAPD marker and results suggested that molecular variation and differentiation was associated in some degree with geographical origin, but not related to the host origin. Singh *et al* (2005) studied intraspecies diversity within *S. scitamineum* isolates from South Africa (SA), Reunion Island, Hawaii and Guadeloupe using RAPDs, bE mating-type gene detection, rDNA sequence analysis, microscopy and germination and morphological studies. No polymorphism was found in sporidial types of *S. scitamineum*. DNA fingerprinting of the smut isolates collected from Western Australia and Indonesia showed that the isolates were identical (Croft and Braithwaite, 2006).

Using micro satellites Raboin *et al.*, (2007) have shown that genetic diversity of either American or African *S. scitamineum* populations was extremely low and all strains belong to a single lineage. The observed worldwide lineage (WL) is a combination of alleles mostly detected in strains from Asia. The results strongly suggested that *S. scitamineum* WL isolates have originated from a single isolate/ genotype (WG) from Asia that has spread the disease around the world.

3.6. Epidemiology: Primary transmission of the disease takes place through infected setts. Systemically infected plants do not always show whip symptoms but most of the buds in them will be infected by the pathogen. If such symptomless shoots are used as seed, it would serve as a potential source for primary transmission. From the infected plant crop, ratoon gets the disease. The disease is very well adapted to aerial dispersal and spread. The large terminal whips, often protruding above the crop canopy or smaller whips produced in the clump from the infection of basal buds serve as sources of inoculum within a field. From the infected clumps, more than a million teliospores per whip per day are liberated, become air borne, and get dispersed (Agnihotri, 1990; Ferreira and Comstock, 1989). In affected standing cane lateral buds are also catch infection, however, most of these infected buds remain dormant unless the stalks are cut and planted. In addition, spores present in or on the soil surface are also carried to different fields via rain or irrigation water.

Smut is generally favoured by hot dry weather conditions. Plant stress increases frequency of whip development. Under high stress conditions the cultivars may show symptoms otherwise they normally do not produce whips.Teliospore survival is decreased rapidly by soil moisture. Similarly, high rainfall reduces the severity of smut development. In subtropical areas, after severe winter the level of smut decreases probably due to death of smut infected plants.

3.7. Management

Since the disease is mainly transmitted through infected seed canes, care should be taken to obtain seed from a disease free field. Field hygienic practices like removal and destruction of the whips as and when emerges will reduce chances of teliospores release to air and disease spread in the field. Hence, continuous monitoring for disease and roguing is to be done in the disease prone areas to restrict pathogen spread. As smut spores are air-borne and can lodge after travelling long distance, roguing is not always effective in commercial fields.

Heat therapy is very effective in controlling seed borne infection of smut. Two types of heat treatment viz., hot water and moist hot air are in vogue. Hot water treatment is done either at 50 °C for 2 hours or at 52 for 30 min. Moist hot air treatment (MHAT) is carried out at 54 for 2 ½ hours at 95-99% RH (Singh, *et al.* 1980) . Hot water treatment at 52°C for 30 minutes can give 98% control and the long hot water treatment of 50°C for 3 hours is also effective. Softening of the buds during hot water treatment can render the buds more susceptible to reinfection from spores in the soil. Hot-water treated cane resulted in the production of thicker and heavier canes and in an increased number of millable canes per clump. Hence a three-tier seed programme to raise smut-free seed canes has to be followed. A hot water treatment at 52°C for 30 min is routinely used at SBI, Coimbatore to treat seed canes during germplasm exchange and for quarantine. Mixing of 0.1% Triadimefon in the hot water tank has been found to be more effective to eliminate the pathogen in the setts.

3.7.1. Host of resistance: The disease can be easily managed by cultivating disease resistant varieties. Most of the varieties under cultivation are resistant to smut.Sources of resistance to smut have been reported (Alexander and Rao, 1976, 1981 a & b). It has been reported that among the different species of *Saccharum, S. spontaneum* clones were found to possess higher levels of smut resistance. Burner *et al.* (1993) evaluated clones of *Saccharum* spp., *Erianthus* spp., *S. barberi, S. sinensi, S. officinarum, S. robustum, S. spontaneum* and *Saccharum* interspecific for smut resistance. It was observed that clones of *Erianthus* spp. were the most resistant and clones of *S. robustum* and S. *officinarum* were the most susceptible.

3.7.2. Chemical control: Sharififar and Kazemi (1999) evaluated five different fungicides on control of sugarcane smut in Iran and observed that propiconazole gave complete disease control and also recorded that the effect of fungicide and the interaction between fungicide and soaking time on incidence of smut was significant. The combination of thermotherapy and chemotherapy has completely eliminated sett-borne infection of smut. The fungicides which have been found effective are triadimefon and propiconazole. Cold fungicide treatments have not been effective (Croft and Braithwaithe, 2006). This fungicide was reported to be environmentally safe and provided broad-spectrum disease control with three different modes of action for effective resistance management.

4. Pineapple disease

4.1. Economic impact: Like red rot, this disease was also first reported by Went (1893) from Java (now Indonesia) and named this sett rot as pineapple disease because affected setts emit a odour resembling that of over ripe pineapple. The odour is due to production of ethyl acetate in affected setts. The pathogen is soil-borne and mainly affects the setts on planting and cause germination failure. In India, the disease is mainly takes its toll in the tropical sugarcane belt. However, damage of standing cane is often observed in subtropical belt, especially after a flush flood. The disease occurs throughout the country. However the disease occurrence is seasonal. Only during the germination phase in causes infection. Occasionally it can cause damage to standing canes as secondary invader if they are infected by red rot or other stalk pathogens. The disease is rarely a problem when seed cane germinates rapidly after planting. However, under favourable conditions, the disease can cause substantial losses in yield. Death of young shoots of infected setts results in a gappy stand. The disease is widespread in occurrence in most of the sugarcane growing countries.

4.2. Symptoms: Typical disease symptoms are detected in setts after 2-3 weeks of planting. The badly affected setts fail to germinate or settlings die prematurely. When the affected setts are split open, typical red discoluration of internodal tissues with sweet odour of mature pineapple is encountered. With passage of time, indernodal tissues changes to black due to sporulation of the fungus. In most of the cases rotting of setts or slow drying of emerging shoots (6-15 cm) take place. If germination is faster, and good early root development takes place, the plant survives but growth is drasticslly affected due to lack of food supply from the sett.The affected settlings show either partial withering or complete death. Initially the leaves of the affected plant show reddish pink discolouration from leaf tip downwards. Later all the 2 or 3 leaves in the settlings become discoloured and dry. The pathogen enters mainly through the cut ends and proliferates rapidly in parenchymatous tissues of the internode.. Fibrovascular bundles do not disintegrate. Cavities are formed inside the severely affected internodes and the fungus sporulates luxuriantly in them. The disease infection prevents rooting of setts.

The fungus causes considerable damage to standing crop if lodging of canes occurs during maturity phase especially under flooded conditions. Conditions of borer damage, injuries caused by animals, internodal cracks, etc. aggravate the disease in such situation. The pathogen also infects canes in the field as secondary invader following red rot or wilt infection. In standing canes, the pathogen causes reddening and blackening of the internal tissues of several internodes from the base.

4.3. Causal organism: *Ceratocystis paradoxa* (Dade) C. Moreau, an ascomycetous fungus; anamorph = *Chalara paradoxa* (de Seynes) Sacc. *Thielaviopsis paradoxa* (de Seynes) Hohnel. The fungus produces two types of imperfect spores viz., conidia (microspores) and chlamydospores (macrospores). Conidia are cylindrical to somewhat oval, thin walled, hyaline at first, changes to brown, measure 6-24 µm (mean 13 µm) x 2-5.5 µm. The conidia are formed endogenously and in chains from the open ends of the conidiophore. The conidiophores arise laterally from hyphae, are slender, tapering, septate with a long terminal cell and are up to 200 µm. Chlamydospores are produced terminally and in chains from short, lateral hyphal branches. These are obovate to oval, thick-walled, brown, measure 10-25 µm x 7.5-20 µm. The chlamydospores are black and impart black appearance of the internodal tissue of the rotted setts (Wismer and Bailey, 1989). The fungus is heterothallic and readily forms peritheria in culture and perithecia remain immersed in the medium. Perithecia are light brown, globose, ornamented and measure 190-350 µm. The neck of the perithecium is very long and measures up to 1400 µm. The ascospores are ellipsoid and measures 7-10 µm x 2.5-4 µm and ooze out from the ostiole and appear as small white drop on the neck.

4.4. Host Range: *Ceratocystis paradoxa* is a cosmopolitan pathogen and attacks many crop plants including sugarcane. Crops range from filed crops like maize, sugarcane to horticultural crops like banana, pineapple cocoa, coconut, arecanut, oil palm etc. (Wismer, 1961).

4.5. Epidemiology: The disease is essentially soil-borne and conidia and chlamydospores present in the soil readily infect the planted setts through cut-ends. Infection of standing cane stalks occurs through air borne or rain-splashed spores gaining entry through damaged tissue. The inoculum in the soil moves from field to field through irrigation water or run-off water. Factors which slow down germination of setts favour infection by the pathogen, such as cool temperatures, excessive soil moisture, drought or an inability of buds to germinate readily. Hot water treatments render the setts more susceptible to pineapple disease, if they are not associated with a fungicide treatment (Wismer, 1961).

C. paradoxa survives in the soil for many years. Any factor that delays germination of the buds on the setts increases the likelihood of infection and rotting by the pathogen. Factors which slow down germination of setts favour the pathogen, such as cool temperatures, very deep planting, excessive soil moisture, drought or an inability of buds to germinate readily. Hot water treatments render the setts more susceptible to pineapple disease if they are not treated with fungicide. Setts from matured cane stalk germinates less vigorously than those from younger parts, hence old seed cane is more prone to sett rot.

4.6. Management: Although potentially highly destructive, pineapple disease can be efficiently avoided or controlled if a range of precautions or control measures are taken. A general recommendation is to use healthy setts of an appropriate physiological age to ensure rapid germination, setts with at least three nodes to increase the likelihood that the buds towards the centre will germinate before the fungus invades all the tissues, and crop management practices that promote germination and rooting (drainage, irrigation, etc.) (Wismer and Bailey, 1989).

A method aimed at the physical protection of setts from the pathogen was proposed by Croft (1998): he demonstrated that polyethylene coating of short, hot water treated setts significantly improved the control of pineapple disease, especially when the setts are also treated with a fungicide. Some promising experimental results were obtained when different species of *Trichoderma* and two species of *Gliocladium* were used as biological control agents for sugarcane pineapple disease (Guevarra, 1990; Sampang, 1991). The efficiency of these fungi on a larger scale has, however, yet to be proven. Fungicide treatments at planting may be necessary, particularly after hot water treatment. Benzimidazoles like benomyl, or triazoles like propiconazole, can be used as fungicide sprays in the furrow at planting or, even better, as a fungicide bath for the setts before planting (Wismer and Bailey, 1989). Raid *et al.* (1991) felt that the treatment of sugarcane setts with a fungicide may enable sugarcane growers to reduce planting density while maintaining stalk population and yield. Almost all the varieties are susceptible to the disease. Planting optimum sized setts of 6-8 months age in well drained soil, prophylactic fungicidal treatment, avoiding deep planting and providing better drainage will take care of the disease in the country.

5. Rust

5.1. Economic impact: As described earlier, in sugarcane leaf diseases never became threatening to cane cultivation in India except some localized damage. Rust is one of the important leaf diseases which sometimes flare up especially in the tropical sugarcane belt. Sugarcane rust was first described from Java (now Indonesia) and identified as *Uromyces kuehnii* Butler (1918) transferred it to the genus *Puccinia*. Observed variation in teliospores and the rust was appeared different from what reported and named it *Puccinia sacchari*. This rust is currently recognized as *Puccinia melanocepahala* Egan (1979) (syn= *P. sacchari* and *P. erianthi*) and is most commonly encountered in India. The disease is found to occur in serious proportion during post monsoon or cloudy seasons in the country. Occasionally the disease assumes epidemic form as noticed in parts of Karnataka, Maharashtra and Andhra Pradesh. Now rust of sugarcane is categorized into common (brown) and orange rusts based characteristic symptoms and causative pathogens.

5.2. Symptoms: The first symptoms of common rust (*P. melanocephala*) infection are small, elongate yellow flecks that are visible on both leaf surfaces. These flecks increase in size mainly in length and turn reddish-brown in colour within 3-4 days, and in 10-14 days spoulation takes place in uredinia. The size of pustules varies and is in the range of 2-20 mm in length and 1-3 mm in width. Pustules are formed parallel to the veins. At maturity the pustules erupt, exposing orange to reddish brown masses of uridiniospores. Severely infected leaves have large numbers of pustules that coalesce, causing large areas of leaves to become necrotic.

When the rust is severe, numerous lesions occur affecting most of the leaves and impart brown or rusty appearance from a distance. In such situation, if one enters in the field, garment gets coloured with dislodged spores. Premature death of leaves occur due to necrosis. Instances of epidemics after monsoon floods and subsequent water logging covering few lakh hectares were reported in the country. Such situation would show burnt up appearance till new leaves emerge.The initial symptoms of orange rust (*P. kuehnii*) are minute, elongated yellow spots which take on a pale yellow-green halo as they increase in size. As the lesion grows, an orange to orange-brown or yellow brown colour develops depending on the cane variety. In orange rust, the pustules never get dark brown colour. Pustules of orange rust tend to occur 'in groups' on the affected leaf surface.

5.3. Causal organism: Orange rust: *Puccinia kuehnii* E.J. Butler, Brown rust or common rust: *Puccinia melanocephala* H. & P. Sydow, The urediniospores of *P. kuehnii* have an apical thickening, are paler in colour, with generally few paraphyses and teliospores are not abundant. Urediniospores of *P. melanocephala* are smaller, brown with more prominent pores, no apical thickening, abundant paraphyses and teliospores. The surface features of the urediniospores are also different with those of *P. melanocephala* having regularly placed spines 1.0-1.5 µm apart, while in *P. kuehnii* the spines are spaced further apart and are irregularly spaced. Urediniospores of *P. melanocephala* are smaller (21-40 µm x 17-27 µm) than those of *P. kuehnii* (25-57 µm x 17-34 µm) (Ryan and Egan, 1989). Moreover, the urediniospores are darker with more prominent pores with no apical thickening. The spines on the urediniospores can also be used to distinguish the two organisms. The spines on *P. melanocephala* urediniospores are closer, spaced 1-1.5 µm apart in a regular pattern whereas those of *P. kuehnii* are 3-4 µm apart and are not as regularly spaced.

5.4. Host range: Species of *Saccharum* are the main hosts of *P. melanocephala*, although sporulating pustules also have been observed on *Erianthus fulvus* and *Narenga porphyrocoma*. Resistant-type symptoms, in the form of red flecks, have been reported on *Bambusa* sp. and additional *Erianthus* sp. *Puccinia kuehnii* has been reported to infect *Saccharum spontaneum, S. officinarum, S. robustum, S. edule*, commercial hybrid varieties, *Erianthus arundinaceus* and *Sclerostachya fuscum* by Butler (1918).

5.5. Epidemiology: Rust fungi are usually transmitted by wind and water splash. Sreeramulu and Vittal (1970) monitored urediniospores above a rust susceptible variety and concluded that spore numbers were greatest on dry rather than wet days, and incidence peaked in the middle of the day (10.00-14.00 h).

Movement of orange rust has not been as well characterized as common rust, where it has been suggested that air mass movements spread the disease around the world in the late 1970s. Egan (1964) noted that *P. kuehnii* moves more readily during periods of hot humid weather in summer and warm to cool periods in autumn (fall) in Queensland. Under Indian conditions, *P. kuehnii* is more prevalent in the cooler months (Srinivasan and Chenulu, 1956). Urediniospores produced within pustules, typically during periods of high humidity and/or leaf wetness, are increasingly subject to passive release as relative humidity decreases during early daylight hours. Spores dislodged from the pustule by gravity, wind currents or foliar movement, become air-borne. Once aloft, rust spores may travel distances ranging from a few centimetres to hundreds of kilometres. There is even evidence that sugarcane rust may have spread from Africa to the Western Hemisphere by trans-oceanic high altitude air currents (Purdy *et al.*, 1985). The wind-disseminated nature of rust spores accounts for the disease's ability to spread great distances over a relatively short period of time. Since the rust pathogen is not systemic within sugarcane stalks, it is not spread via seed. Rust severity can rapidly increase within a short time period because of the short reproductive cycle. A rust urediniospore can land on a leaf, infect and develop into a sporulating pustule within 14 days. Within 6 weeks, a field planted with a susceptible cultivar may appear reddish-brown when observed from a distance.

5.6. Management: Disease resistant variety is the only economic means of control (Comstock and Raid, 1994). Cultivar diversification is also recommended due to the possible presence of rust variants. Although fungicides have proven effective in controlling rust in research studies, their commercial use is not economically feasible.

B. Bacterial diseases

6. Ratoon stunting disease

6.1. Economic impact: Ratoon stunting disease (RSD) is an important seed-piece transmissible bacterial disease of sugarcane (*Saccharum* interspecific hybrid) with worldwide distribution. It was first recognized as a specific disorder of sugarcane during the summer of 1944-45, in Queensland, Australia when following a dry spring, a number of fields of Q28 produced extremely poor ratoon crops compared with adjacent ratoons of the same variety and this attracted the attention of Dr. Steindl who named it as Q 28 syndrome as no apparent cause was easily discernible. However, by 1949 he was able to show that Q 28 syndrome can be transmitted by artificial inoculations and used the term *"ratoon stunting disease"* to denote Q 28 syndrome (Steindl, 1949, 1950). Even though, the name ratoon stunting continues, in the present day understanding, it is certainly a misnomer as it affects both plant and ratoon crops alike. The disease is an insidious one, and it is difficult to diagnose the disease at field level. Due to this nature, it is often ascribed as one of the major causes of yield decline/varietal deterioration in sugarcane. This disease is now recognised as the most important bacterial disease of sugarcane in India causing substantial loss in cane yield. In the absence of proper diagnostic symptoms of this disease, its presence and impact on the cane growth and yield is less appreciated. Due to the blockage in xylem water supply gets affected and as a result cane shows false maturity as indicated by higher brix (%) value (less extractable juice from the cane). Recovery performance increases due to less water in juice but sugar yield per unit area diminishes.

In India, the presence of the disease was felt only when Prof. Chilton, who visited India as a delegate to attend IX ISSCT Congress, detected its presence in the cane genotype CoS 510 during his visit to Jamnabad farm, Golagokarnath (Chona, 1956,). Reported it possible presence on Co 419 in Bombay state. However, the question of its presence in India was actively debated as most of the experiments conducted subsequently in India was not of affirmative nature (Khanna, *et al.*, 1958) and the nodal symptoms were elusive/ephemeral. Later on, Singh (1966) confirmed the presence of the RSD from the deteriorating cane genotype Co 290, obtained from Jaora Sugar Mill area, Jaora, M.P., by curing it with hot air treatment at 54°C for 8 hours. Dr. Singh dispelled the myth of the presence of RSD in India by his ingenious way. He split the affected canes in two halves – used the heat therapy to treat one half and left other half as control. The excellent performance of cane plants developed from the treated half driven home his point to all and sundry. Hereafter, a number of sugarcane genotypes including the wonder cane Co 419, Co 453, Co 213, Co 312, Co 740, Co 997, CoL 9, BO 22 etc., were reported to be inflicted with this disease (Singh, 1969, 1973). Of late, it has been observed that most of the popular cane genotypes are in cultivation are carrying substantial amount of RSD bacterium. The predominant varieties of subtropical India like BO 91, Co 1148, Co 1158, Co 62399, CoH 56, CoH 72, CoJ 64, CoJ 84291, CoS 767, CoS 88216, CoS 90269, CoS 95255, CoLk 7901, CoPant 84211, CoPant 84212, CoPant 90222, CoSe 92423, etc. are in the clutch of RSD (Duttamajumder, 2001). In fact, no cane area in India is free from the RSD menace; wherever a conscious search is made, its presence was immediately apparent.

6.2. Symptoms: General stunting and the decline in the yield are the only perceptible effects of this disease. To a great extent an uneven growth of the crop of the field is often observed. Various workers have reported that a field fully affected by RSD does not display uniform stunting; rather the crop displays a characteristic "up and down" appearance. No abnormality is seen either in the root system or in the underground portion of the cane. The only definite, obvious symptom of ratoon stunting is a marked reduction in growth, particularly in the ratoon crop. The loss in yield in the plant crop is generally not discernible under commercial conditions since, if the crop is not quite up to expectations, the farmer can usually recall some other factor that may have been responsible. Diseased plant crops usually have fewer canes per stool than healthy ones, though this is not so obvious, and they are the first to show symptoms of distress, such as wilting, scorching of the tips and edges of the leaves, and premature death of the older leaves, when dry conditions are experienced. In ratoon crop, losses are obvious, although there is seldom any death of complete stools, the diseased shoots are very much slower to regenerate and differences between healthy and diseased are usually apparent. In mature cane, the affected stools have fewer canes and the canes are much reduced in length and thickness. Dry weather accentuates the disease and, conversely, adequate water from properly controlled irrigation masks the disease to such an extent that almost normal crops of ratoon are harvested.

Nodal symptoms can be observed only after splitting the cane through the basal nodal area with a sharp knife so that vascular bundles are cut through exposing the coloured deposits within. The symptoms should be observed immediately after splitting the cane, because symptoms become inconspicuous as the tissue dries. Vascular bundles are in the shades of yellow, orange, pink-red, and reddish brown appear as dots, commas or as short lines and may be seen just below the region of attachment of the leaf sheath. This colouration is due to the plugging of xylem vessels with coloured gummy substances (pectinaceous). Such symptoms have been observed in many varieties like Q 28, Q 47, CoL 9, Co 290, Co 975, Co 6611, NCo 310, CP 36-11, etc. Many varieties do not display conspicuous symptoms, yet they carry the bacterium, and these are generally referred to as symptomless carriers. On the contrary, many varieties (NCo 310, Co 975) do display conspicuous symptoms without any association of the bacterium (Duttamajumder, 2001). Several factors affect the expression of the disease symptoms, such as, varietal characteristics, climatic conditions, physiological status of the plant and the presence of other diseases like chlorotic streak, leaf scald, etc. Nodal symptoms, similar to RSD but distinguishable also develop when stalks are damaged mechanically or by insect-pests.

The juvenile symptoms can be observed by cutting longitudinally a 1-2-month-old stalk just above the point of attachment to the seed piece. Salmon pink colour is seen below the meristematic area (Hughes, 1955).

Before the establishment of bacterial etiology, isolation and identification of the causal bacterium *Leifsonia xyli* subsp. *xyli*, RSD was mostly diagnosed on the basis of nodal symptoms. Nowadays, not much credence is given because not all clones display nodal symptoms and not all nodal symptoms are due to bacterial infection. To overcome this, biological assay using indicator plants was developed. Indicator varieties like CP-36-105 (Schexnayder, 1960) and Co 421 (Singh, 1969) and elephant grass (*Pennisetum purpureum*) were used to detect RSD pathogen. Several biochemical tests based on polyphenol oxidase, sieve tube lignifications, amino acid contents of the foliage and coloured reaction of nodal tissues have been developed from time to time. These methods did not appeal to the researchers, as usually the results were elusive. Nowadays, detection is based either on the direct observation of the bacterium in the xylem fluid through the microscope or indirectly on the serological/immunological techniques like ELISA, Dot-Blot, Tissue blot, EIA, etc.

Due to the peculiar morphology of the bacterium, it can be detected through phase contrast and darkfield microscopy (Kao and Damann, 1980; Davis, 1985; Roach and Jackson, 1992, Duttamajumder, 2001) from the other bacteria adapted in the specific ecological niche of sugarcane xylem. Dark field microscopy was found superior to phase contrast microscopy. Identification of the bacterium was fairly accurate due to its atypical shape and non-motility. Alkaline induced autofluorescence is also indicative to the presence of RSD but it needs confirmation of the presence of the bacterium. Damaged or distressed canes also showed such autofluorescence (Damann, 1988).

6.3. Causal organism: In the beginning due to the non-detection of any known pathogenic organisms through conventional isolations and its juice transmissible nature, the disease was thought to be of viral etiology and most of the researches spanning two and a half decades were carried out assuming it a viral disease. However, the non-crystallizing nature of this pathogen like other known viruses haunted the pathologists. The answer came in 1973, when Gillaspie *et al.*, Maramorosch *et al.*, and Teakle *et al.*, independently demonstrated the association of a xylem limited bacterium with this disease. Electron microscopic studies of infected cane tissues by Worley and Gillaspie (1975) revealed that the bacterium is confined to xylem elements of diseased plant. Collapsed bacterial cells, in different stages, can be observed within the matrix material. Scanning electron microscopic studies revealed the presence of the bacterium in tracheids, parenchyma and in the lumen of the xylem vessels (Kao and Damann, 1980).

Davis *et al.* (1980) succeeded in isolating the bacterium in axenic culture and proved the Koch's postulate that the associated non-motile Gram (+) coryneform bacterium (0.25-0.5 μm x 1-4 μm, length may occasionally go up to 10 μm) is the actual etiological agent of ratoon stunting disease. In 1984, the Genus *Clavibacter* was erected and the RSD bacterium was identified as *Clavibacter xyli* subsp. *xyli* (Davis *et al.*, 1984) and its close relative causing Bermuda grass stunt as *Clavibacter xyli* subsp. *cynodontis*. Later, a new genus *Leifsonia* was erected to place the bacteria having 2,4 –diaminobutyric acid in the cell wall and the RSD bacterium was transferred to it and rechristened as *Leifsonia xyli* subsp. *xyli* (Suzuki *et al.*, 1999; Evtushenko *et al.*, 2000). Presence of mesosomes in the bacterial cell aids in quick identification of the bacterium. The bacterium is slow growing and after 2 weeks attains a colony diameter of 0.1 to 0.3 mm. Colonies are hyaline, convex and circular with entire margin. The bacterium is aerobic, non-motile, Gram positive, non-spore former, non-acid fast, catalase positive, and oxidase negative. It specifically contains 2, 4- diaminobutyric acid in the cell wall (Davis *et al.* 1984).

6.4. Host range: On artificial inoculation the RSD pathogen could be successfully transmitted to several grasses like *Cyndodon dactylon, Brachiaria milliformis, B. mutica, Echinocloa colona, Imperata cylindrica, Panicum maximum, Sorghum verticilliflorum,* and *Sporobolus capensis.* From these infected grasses, it is possible to reinfect the sugarcane crop by sett inoculation added *Sorghum halepense* and *Zea mays* to this list and found *Pennisetum purpureum* readily catches RSD infection.

6.5. Epidemiology: The pathogen mainly spreads through the harvesting and sett cutting implements during the normal cane cultivation. The juice from the infected cane remained infective even after dilution (Steindl, 1961). Reported that both the top cutter blades and the sticker chains of harvester might spread the disease from infected leaves while cutting the seed-cane. Singh (1973) reported that the disease was transmitted by dipping healthy setts in infective cane juice for 30 min. Through a cutter planter, the disease from an affected clump can spread to 50 consecutive clumps. Demonstrated that the disease can be transmitted by inoculating roots of young plants with juice from diseased plants. Spread of the disease through the mediation of rats, foxes, jackal, etc. have been implicated but needs further confirmation.

6.6. Management: In India, to date, no directed breeding has been made so far to breed resistant cane genotypes for any bacterial diseases of including RSD. As the disease is mainly transmitted through seed-pieces, maximum care should be exercised to select planting materials. While preparing setts dipping of knives should be done in 5% Lysol (neutralised cresylic acid), 1% Dettol or 10% formalin or any other surface disinfectant to avoid accidental spread of the pathogen through the cutting implements. In the past strict quarantine measures have helped in arresting the spread of RSD pathogen. As a rule seed material before shipping is treated with hot water for RSD control. The disease is still spreading its geographical area because HWT gives only 90-95% control of the disease. In the remaining 5-10%, the titre of the pathogen goes so low that it becomes difficult to detect the bacterium with the common methods presently in vogue. Like all other crops, cultural practices particularly moisture stress has a profound effect on RSD development. Paucity of water augments the disease severity. Therefore, it is necessary to irrigate the crop properly to minimise losses caused by this disease.

The disease incidence can be greatly minimised by selecting apparently healthy canes and carrying out systematic heat treatment before planting. A three-tier seed programme as outlined in red rot has to be followed for the development of healthy seed nurseries and distribution healthy seed cane. Of the various control methods available, only thermotherapy has been found effective in preventing the spread of *L. xyli* subsp. *xyli* from one location to another. Currently, four methods, viz., hot water, hot air, moist hot air and aerated steam treatments are being used (Agnihotri, 1990). On global basis, hot water treatment at 50°C for 2 hours is the most preferred one. Sugarcane genotypes also differ markedly with respect to their response to heat. Hot water treatment is good for the mature canes but not for immature ones as the germination of bud is adversely affected. The major bottleneck with heat treatment is cost of the equipment, reductions in germination. Moreover, this treatment in most cases failed to eradicate the pathogen completely from the treated cane. In recent years, methods have been developed to protect germinability by treating setts with fungicides after the treatment, selection of seed material and keeping the leaf sheath intact over the bud during the treatment. Pre treatment of seed-pieces to short treatment of heat (50°C for 10 min) has been recommended to increase bud germinability (Benda, 1972, 1978) before the final treatment at 50 °C for 2 h in the following day.

7. Leaf scald

7.1. Economic impact: Leaf scald caused by *Xanthomonas albilineans* (Ashby) Dowson, is the other major bacterial diseases causing considerable damage to the crop in different countries including India. The disease has been recorded from more than 66 sugarcane growing countries in the World (Martin and Robinson, 1961; Ricaud *et al.*, 1989). The disease was first recorded in 1915 as *'Java gum disease'* (Gomziekte) and wrongly considered identical with the Cobb's gumming disease (C.o. *Xanthomonas campestris* pv. *vasculorum*)(Groenewege, 1915). However, significant contributions in the understanding of this disease emanated from the researches of Wilbrink (1920), who developed a simple medium for isolating this slow growing xanthomonad in pure culture. The pioneering work done by North (1926) in Australia and by Wilbrink (1920) in Java (now Indonesia) presented clear-cut evidence that the new disease *'leaf scald'* is conspicuously different from both *gummosis* and *'sereh'*.

In spite of the strict quarantine measures, the pathogen is progressively spreading in new geographical locations and has established in countries where it has been hitherto unknown. Moreover, spread of the pathogen has been accentuated by the behaviour of different cultivars as *'symptomless carrier'*, which surreptitiously transfers the pathogen from one location to another and from country to country through seed materials, germplasm exchange, etc. The occurrence of this disease in India was first recorded in April 1961 by Egan during his short stay at I.A.R.I., New Delhi (Egan, 1962, 1979). However, this disease remained almost unnoticed until the report of Satyanarayana (1974) from Anakapalle, Andhra Pradesh. After 1974, the disease has been recorded from different north Indian States like Punjab, Bihar, and Uttar Pradesh (Singh and Misra, 1976; Agnihotri, 1990). Prominent among the susceptible genotypes are BO 17, BO 70, BO 90, BO 109, Co 419, Co 1158, Co 62399, Co 7301, Co 8312, Co 8315, Co 8334, Co 93016, CoS 767, CoS 90269, CoLk 7710, CoLk 7901, CoLk 8001, CoLk 8102, CoLk 8901, CoJ 64, CoJ 81, CoH 56, CoH 72, CoH 92201, CoH 94201, CoPant 84211, CoPant 84212, CoPant 84213, ISH 40 etc.

7.2. Symptoms

The pathogen induces different types of symptoms depending on the time of infection, time of the season, varietal susceptibility, growth stage of the crop, etc. There are two recognised phases of this disease, viz., *chronic phase*, and *acute phase*. The two phases are so distinct in appearance that they were regarded as two separate diseases until late 1919 when close investigation of the diseases revealed them to be the same (North, 1926; Wilbrink, 1920).

7.2.1. Chronic phase: The diagnostic symptoms of the disease appear in the chronic phase, which are characterised by several external symptoms. On the young leaves, diagnostic symptoms appear as *'white pencil-line'* usually extending almost along the entire length of the lamina until it reaches the margin. Usually a vascular strand on the upper surface of the leaf turns white and becomes distinct from the rest of the vascular strands of the leaf. This symptom is prominent in the young leaves. As the leaves become older, the thickness of the white line (vascular strand) increases sideways and as a result, loses its sharp margin and turns into a diffused white stripe. On the slightly older *pencil-line*, small reddish necrotic areas are often encountered. Depending on the weather condition, the stripe starts drying from the tip and imparts a partially burnt appearance to the leaf. Because of this diagnostic symptom, the disease is referred to as *'leaf scald'*. In this phase, in many cases, the young leaves may display different degrees of chlorosis, from total albinism to the interveinal chlorosis (especially during the summer months).

The other major symptom of the chronic phase is the activation of buds and throwing of side shoots in the standing cane in acropetal fashion and thus imparting a bushy appearance to the affected plant. The side shoots or *'lalas'* invariably bear *'white pencil-line'* on the leaves. Sometimes the side shoots may appear from the middle of the cane imparting an impression of top borer attack. At the maturity stage in the field, usually an odd stalk here and there is found with shortened internodes. The shortening of the internode is more common when the disease expresses in mild form during the month of July-August (coinciding the grand growth phase of the crop). In some cases, the top stops elongating and thus imparts a broomy appearance ; plants show gradual wilting with leaves becoming stiff and central young leaves curling inwards.

When badly diseased stalks are cut open, bright to dark red vascular strands due to the necrosis of the vascular tissues become apparent. These streaks are usually more prominent at the node and invariably found in the side shoots and help in diagnosis of the disease. The presence of the bacteria may be confirmed by the ooze test. The *'white pencil-line'* produces feeble ooze. Lysigenous cavities may also develop inside the badly affected stalks, particularly near the shoot apex.

7.2.2. Acute phase: The disease may also take the form of an acute wilt, particularly in dry weather when mature stalks suddenly wilt and die without displaying any diagnostic symptoms. Sometimes a large number of clumps in a field may suddenly wither and die without displaying any characteristic symptom of the disease. In order to ascertain the cause of the death of the plants, canes affected with chronic phase must be searched among the affected plants or in many cases one has to wait till the new tillers come out bearing the *white pencil-line* symptom (North, 1926, 1929). In India, symptoms of acute phase in commercial planting have not yet been recorded. Occasionally this phase was observed only in seedling population raised from fluff (true seed of sugarcane) (Duttamajumder, 2004).

7.2.3. Latent infections: The other most peculiar characteristic of this disease is that the pathogen is quite notorious in producing latent infection and remains symptomless; thus evading the chances of detection. The masking of symptoms is more common during the rains, particularly in tolerant varieties. The latency may be broken, due to some unknown reasons, at any time of the crop growth and the disease appears suddenly in the field. However, it is observed that the plants express symptoms due to moisture stress followed by irrigation and again followed by moisture stress. To date, the mechanism of latent infection is not clearly understood and it has formed a major blockade to check this disease from spreading.

The notoriety of *X. albilineans* for its latency has eluded many workers to detect its presence in cane stalks. Ricaud *et al.*, (1978, 1979) obtained a highly specific antiserum by intramuscular injection of bacterial cells into rabbits. Oliviera *et al.*, (1978) also reported the production of antiserum - but it was less specific (obtained the antiserum by injecting into lymphnodes of rabbits). Loville and Coleno (1976) developed a fluorescent antibody technique, which is quite effective in detecting the latent infection. By this technique, they were able to detect the presence of *X. albilineans* in 81 per cent of symptomless stalks from diseased stools and in 66 per cent of apparently healthy stools in the diseased field. Similarly, ELISA has been used by various workers with moderate success. However, according to the XMA medium provides an excellent and cheap method of detection of *Xanthomonas albilineans* comparable in efficiency with ELISA based techniques.

7.3. Causal organim: *Xanthomonas albilineans* is now regarded as one of the several taxospecies recognised in the genus *Xanthomonas*. Ashby (1929) named the leaf scald bacterium as *Bacterium albilineans*. Placed the pathogen in the genus *Phytomonas*, i.e., *Phytomonas albilineans* (Ashby) Magrou. The most accepted name was given by Dowson (1943) who erected the genus *Xanthomonas* for yellow rod shaped bacterium and placed this bacterium under the genus *Xanthomonas*. The present accepted name of the bacterium is *Xanthomonas albilineans* (Ashby) Dowson. The pathogen gets its specific epithet *albilineans* due to its characteristic *'white pencil-line'* symptom on the leaf.

The bacterium is rod shaped (0.25-0.30µm x 0.6-1.0 µm), monotrichus (bears a single polar flagellum), occurs singly or in chains, Gram negative. On Sucrose peptone Agar medium (SPA) produces buff yellow, honey yellow, naples yellow, minute, glistening, viscid slow growing colony with entire margin but not mucoid like other xanthomonads. Broth culture is turbid, doesn't liquify gelatin, doesn't change the milk, doesn't reduce nitrate, and doesn't produce H_2S, indole and ammonia. Doesn't hydrolyse starch, fair growth in sugar but without any fermentation. No growth in peptone water without carbohydrates, releases invertase and is deficient in methionine and glutamic acid.

The bacterium is slow growing and takes 5-7 days to form a visible colony. A selective medium (XAS medium), based on Wilbrink's medium supplemented with 5 g KBr, 100 mg cycloheximide, 2 mg benomyl, 25 mg cephalexin, 30 mg novobiocin, and 50 mg kasugamycin per litre, has been developed for the isolation of *X. albilineans* and authors have claimed it to be highly sensitive for the detection of the bacteria and compares fairly well with the prevalent serological methods.

7.4. Molecular characterization and toxin production: The striking *'white pencil-line'* symptom attracted the attention of various workers from North (1926) to Birch and Patil (1983). Orian (1942) postulated that formation of the white pencil line is due to the production of some toxic metabolite, which selectively acts on the chloroplasts. Later on, Orian's hypothesis was supported by researches of Patil and his students. They observed that chloroplasts were absent in the white pencil line but proplastids and etioplasts were present. They concluded that chlorosis of emerging leaves of infected plants resulted from blocked chloroplasts differentiation apparently caused by a phytotoxin. These workers also suggested that the chlorosis inducing isolates of *X. albilineans* produce various antibacterial compounds including albicidin which inhibit prokaryote DNA replication. They also established from mutation studies a correlation between the production of albicidin and chlorosis. Of a family of antibiotics produced by *X. albilineans,* albicidin is the best studied. The primary action of albicidin is the inhibition of DNA replication in the protoplastids. Besides, it also partially inhibits the protein synthesis. The bacterium present in the invaded xylem induces chlorosis by preferentially inhibiting plastid DNA replication, resulting in block chloroplasts' differentiation. In fact, the gene governing the albicidin production has been cloned and a method of detoxification of this toxin has been developed (Birch, 1999, 2001). Zhang and Birch (1997) identified gene for albidicin detoxification from *Pantoea dispersa* that encodes an esterase and attenuates pathogenicity of *Xanthomonas albilineans* to sugarcane. In fact the antipathogenesis approach of leaf scald disease control has been patented.

Variability of the pathogen is common phenomenon, especially in phytopathogenic bacteria. Studies by different workers indicate that clear-cut pathotypes of *X. albilineans* may exit in nature. Variety B 34104 is highly susceptible in British Guiana but resistant in Mauritius (Antoine and Perombelon, 1965). They suspected that a new strain of the pathogen caused the outbreak of leaf scald in resistant varieties grown before 1964. Observed that different pathogenic isolates show quite different pathogenicity reactions in sugarcane varieties. Baudin and Chatenet (1980) recorded variability in *X. albilineans* based on serological attributes. Rott (1984) studied variability in 28 isolates of this pathogen collected from 11 countries with respect to cultural and biochemical characteristics, bacteriophage typing and serological groupings *in vitro*. He found little variation in cultural and biochemical characteristics between the isolates including antibiotic resistance and could not justify their groupings into different biotypes. However, the isolates could be separated into sero- and lysogenic - groups which showed some correlation with one another. Identified eight clonal population of this bacterium in a collection of 218 strains from different parts of the world. Isolated and characterized a 25 kb size plasmid from this bacterium but plasmid cured strains retained the pathogenicity.

7.5. Host range: In addition to sugarcane, *X. albilineans* infects a few grasses and many other related plant species when inoculated artificially. Orian (1962) reported a bacterial disease closely resembling leaf scald on *Paspalum dialatatum;* the isolated organism induced leaf scald symptoms in sugarcane (Persley (1973a) and reported natural occurrence of leaf scald on *Brachiaria piligera, Imperata cylindrica* var. *major, Paspalum conjugatum.* The presene of leaf scald has been reported in *Imperata cylindrical* (Baudin, 1984), in *Panicum maximum, Paspalum* sp., *Pennisetum purpureum, Rottboellia cochinchinensis,* and *Zea mays,* and *Imperata cylindrica* and a species of *Sorghum* in India (Persley (1973b). Tested different grasses and found that only maize *(Zea mays)* and Job's tears *(Coix lacrymajobi)* showed systemic infection and concluded that *X. albilineans* has a fairly narrow host range.

7.6. Epidemiology: The expression of the disease occurs throughout the season. Major dissemination of the bacterium occurs at the time of harvesting and planting. *X. albilineans* is predominantly transmitted by harvesting and sett cutting implements and this has been demonstrated both under field and laboratory conditions. Transmission occurs more readily in young plants/suckers when cut above the growing point than a cut at the base of the cane (Koike, 1965). The spread of the pathogen in the field is very slow and quite erratic (Duttamajumder 1998a). This is perhaps due to the insidious and latent nature of the pathogen; even if the pathogen has spread, it is beyond visual detection. In spite of strict quarantine measures the pathogen is migrating from country to country through diseased seed /planting material (Wismer and Ricaud, 1969). However, the greatest danger is the existence of latent infection, particularly in tolerant / moderately resistant cultivars for which the planting materials (with the present detection technique) cannot be guaranteed as disease free.

Transmission by wind-blown rain is suggested because of the higher prevalence of this disease after cyclones (in Australia) but no experimental evidence has been put forward to support the claim. Aerial long distance spread of the pathogen is also suspected but no concrete proof in this regard has come so far to support this claim. In Australia (Persley, 1973c, 1975; Persley and Ryan, 1976) and in Mauritius observations on the spread of leaf scald indicates that it is favoured by wet seasons, especially with cyclonic conditions. It is not clear whether wind blown rain is involved in the dispersal and transmission or whether moist conditions favour symptom expression. On the other hand, in India, the disease expression is more in mature canes during winter months when the cane is at maturity. A flush of symptom is usually observed in April –May just after the germination of setts in plant crop and after ratoon initiation. Another flush usually appears just after the cane receives first pre monsoon shower in north India. Duttamajumder (1990) observed flowering in leaf scald infected canes of a moderately susceptible variety CoLk 7901 and the bacterium was detected in the fluff. Presence of bacteria in fluff was also confirmed and also provided evidences for aerial transmission and transmission through guttation water/ dew drops.

7.7. Management: The high latency of this pathogen poses several constraints for the management of this disease. Experiences of research workers in Australia and Mauritius have shown that selection of healthy planting materials from the diseased field is not at all effective due to the latent infections. To manage this disease an integrated approach is necessary starting with selection of resistant variety, healthy seed, use of three-tier seed programme for the production and distribution of disease free seed and rouging and destruction of complete clump when any symptom of the disease is observed. The most effective, practical, and economic method of managing the leaf scald is the replacement of the susceptible varieties with resistant ones. The introduction of resistant variety in cultivation was successful in countries like Australia, Mauritius, etc. So far, in India, the disease has not made any serious impact and no directed breeding has so far been attempted to breed resistant varieties against this disease. Occasional screening has been done but susceptibility is still not a deterrent in the release of sugarcane variety (Duttamajumder, 2004).

As the disease is mainly transmitted through seed-pieces, maximum care should be exercised to select planting materials. While preparing setts dipping of knives should be done in 5% Lysol (neutralised cresylic acid), 1% Dettol or 10% formalin or any surface disinfectant to avoid accidental spread of the pathogen through the cutting implements.Strict quarantine measures have greatly helped in mitigating losses caused by leaf scald. However, with its notorious latency the pathogen is still spreading surreptitiously from one location to another and from one country to another. For detecting the bacterium in the seed materials, every quarantine station should have the facility of serological and immunofluorescence diagnoses or access to PCR based diagnostic facility. The presence of the bacterium in the fluff poses problem in fluff exchange programme. Utmost care should be taken to avoid such an innocuous migration of the pathogen through the fluff and young seedlings (Duttamajumder, 1990).

Several methods of heat treatment have been tried to eradicate the pathogen from the planting materials i.e. canes. Hot air therapy (54°C for 8h) is not very effective against this disease (Bailey, 1976). Long hot water treatment (50° C for 3h) and short hot water treatment (52° C for 20 min) were tried using single bud setts but these treatments appreciably reduced the germination of buds (Steindl, 1971). Successful control of the disease has been obtained only when single bud setts of diseased cuttings were soaked in cold water for 24h before long hot water treatment (Steindl, 1971, Bailey, 1976). In India, Agnihotri (1990) also reported effective control of this disease using moist hot air treatment (54° C for 2 ½ hours). However, in several experiments with both MHAT at 54°C for 2 ½ hours and hot water treatment at 50°C for 2 hours failed to eradicate the leaf scald pathogen completely from the fully affected cane. In the treated generation, 10-15% of three-budded sett obtained from the cane showing leaf scald symptom (cv. CoLk 8001, CoLk 8102, CoS 767, Co 1158, Co 62399) showed chronic symptom of the disease. Thus, a control of 80-90% could be achieved. However, in commercial plantings these types of canes are usually get rejected and success of control goes beyond 95% (Duttamajumder, 2004).

A three-tier seed programme as advocated in red rot should be followed for the production and distribution of disease free seed cane.Sett dipping in Streptomycin (100 ppm) was also tried and was found effective against this pathogen. However, sugarcane is sensitive to streptomycin treatment as some of the plants (3 to 8%) die due to albinism (Agnihotri, 1990). As a general practice, this treatment is not recommended.

C. Viral and mycoplasmal diseases

8. Mosaic Disease

8.1. Occurrence and Distribution: Like many other sugarcane diseases mosaic was also first reported from Java (Indonesia) in 1892. In India, sugarcane mosaic was first reported in India from Pusa in 1921 on sugarcane variety D 99 (imported from USA) and Sathi 131, an indigenous cane of Bihar (Dastur, 1923). Subsequently, it was noted in other parts of the country, viz., U.P, Punjab, Central Province (Dastur, 1926); Bombay, Madras (Sundararaman, 1928). In 1925, it was found to occur as an epiphytotic form in Hemja and indigenous cane of Bihar belonging to Mungo group of *Saccharum barberi*. Dastur (1926, 1923) found symptoms produced in this country to be slightly different from those reported from other countries. Sundararaman (1928) has reported distinct types of symptoms produced by this disease.

Mosaic disease, caused by SCMV (a member of *Potyvirus* group), a seed piece transmissible disease with interveinal chlorotic specks, streaks or stripes, mosaic mottling specially on young leaves of sugarcane has been prevalent in almost all sugarcane cultivating regions of India (Agnihotri, 1990; Jain *et al.*, 1998; Rao *et al.*,1998). Incidence of this virus in commercial fields is nearly 100% in major sugarcane growing states of India. In India up to 1990's sugarcane mosaic disease was supposed to be caused by different strains of *Sugarcane mosaic virus* (SCMV) (Bhargava, 1975). Nine strains (A, B, C, D, E, F, H, I and K) of SCMV have been reported from different parts of India (Bhargava *et al.*, 1971, 1972, Khurana and Singh, 1972; Rao *et al.*, 2000a). However, Hema *et al.* (1999, 2003) reported *Sugarcane streak mosaic virus* (SCSMV) as the new casual virus of the mosaic disease in tropical India. So far, only three viruses have been identified as casual agents of mosaic disease of sugarcane in India, i.e. *Sugarcane mosaic virus* (SCMV), *Sorghum mosaic virus* (SrMV) and *Sugarcane streak mosaic virus* (SCSMV) (Rishi and Rishi, 1985; Hema *et al.*, 1999).

8.2. Symtoms: There are no authentic diagnostic symptoms to differentiate between SCMV and SCSMV. In most of the cases the symptoms for SCMV and SCSMV looks like similar. It is really hard to distinguish symptoms caused by SCMV and SCSMV due to overlapping symptoms and changing symptoms due to plant growth and environment. In many situations, both the viruses occur together. Thus, the name streak mosaic is a misnomer. However, the first symptoms of SCMV appeared in the form of diffused chlorotic spots on the newly emerging leaves.Soon after a large number of conspicuous spots of various shapes and sizes appeared all along the leaf surface. This formed a well defined mosaic pattern (Bhargava *et al.*, 1972; Jain *et al.*, 1998; Rao *et al.*, 2002). SCSMV induced most of the times systemic mosaic in the form of continuous/discontinuous streaks on sugarcane. The virus is propagated periodically by sap inoculation on *Sorghum bicolour* cv. Rio and it reacted with systemic chlorotic streaks/stripes, which was followed by necrosis (Hema *et al.*, 2001).

8.3. Causal organim: *Sugarcane mosaic virus* (SCMV) belongs to the genus *Potyvirus* and family *Potyviridae*. It is reported to easily aggregate and sediment during initial low speed centrifugation (Gough and Shukla, 1981). Much virus is lost when organic solvents are used to denature colloidal plant constituents. In some cases, there was 90% loss of infectivity when second high-speed centrifugation was given (Bond and Pirone, 1971). Many protocols for purification have been published, primarily attempting to improve extraction and clarification. Various worker attempted alone or different combination of chloroform, carbon tetra chloride, butanol followed by Triton X-100 with aim to have better clarification, minimum aggregation, less breakage and high yield in final virus preparation (Rao *et al.*, 1998a: Rishi and Rishi, 1985) .

Finally, a shorter and better protocol for purification of *Sugarcane mosaic virus* was developed. Best virus yield was achieved by extracting the viruses in borate buffer (0.5M, pH 8.0) containing 0.15% thioglycolic acid and 0.1M diethyl dithiocarbamate followed by clarification with chloroform, butanol (1: 1 v/v) + 0.5% Tween-20 and 0.5% Triton X-100. This protocol used only one cycle of differential centrifugation. Density gradient run was the final step also aided in maximum removal of host contaminants. The protocol yielded 75-100 mg virus/ kg of leaf material. The purified preparation yielded long flexuous filamentous particles of 760 x14 nm size. Rao *et al.* (1998b) developed high titred (1: 2048) polyclonal antibodies against *Sugarcane mosaic virus*, which was capable of detecting SCMV from leaf, stalk and bud tissues of sugarcane and sorghum at early stage of crop growth in serological tests. The physical properties of the virus isolate under study showed that it had dilution end point (DEP) between 10^{-3} - 10^{-4}, thermal inactivation point (TIP) between 50-55°C and longevity *in vitro* (LIV) 24 hrs at room temperature (25 + 3^0C) (Rao *et al.*, 1998 a&b).

Hema *et al.* (1999) have reported the existence of another virus causing mosaic disease of sugarcane in Andhra Pradesh, India and suggested that it is probably distinct from SCMV as it failed to react with antisera of several strains of SCMV and MDMV. Serological characterization of a virus isolate causing a mosaic disease on commercial sugarcane around Uttar Pradesh, and Andhra Pradesh revealed that it is not a strain of SCMV sub group but is a strain of a new virus named *Sugarcane streak mosaic virus* (SCSMV), which has been claimed to be a member of *Susmovirus* genus in the family of *Potyviridae* (Hema *et al.*, 1999). SCSMV-AP has DEP between 10^{-4}-10^{-5}, TIP 50 °C and LIV 1-2 days at room temperature or 8-9 days at 4°C (Hema *et al.*,1999).SCSMV-AP isolate induced laminated aggregates and pinwheels in infected sugarcane and sorghum leaves, suggesting that it is the member of the family *Potyviridae* (Hema *et al.*,2001). The electron micrograph of the purified suspension of SCSMV-AP showed flexuous filamentous particles with dimensions of ca. 890x15 nm (Hema *et al.*, 1999). Polyclonal antibodies against SCSMV-AP were produced by immunization of rabbit with purified virus preparations. The titre of the antiserum determined by DAC-ELISA was 1: 2048 with virus infected sugarcane and sorghum leaf antigens (Hema *et al.*, 1999).

8.4. Molecular characterization and diagnosis: Diagnosis of both these viruses (SCMV and SCSMV) have well been established by different serological (ELISA, DIBA, ISEM and western blots) and molecular biology (RT-PCR, NASH, Southern hybridisation) assays (Hema *et al.*, 1999; Gaur *et al.*, 2003). In DAC-ELISA and EBIA, the SCSMV strongly reacted with homologous antiserum of SCSMV-AP and antisera potyvirus of sorghum around Parbhani (Maharashtra), sugarcane virus isolate from Uttar Pradesh and weakly with Narcissus latent from UK (Hema *et al.*, 2001). The virus failed to react with antisera of potyviruses like SCMV strains A, B, D, E, H, I, MDMV-A, *Peanut mottle virus* (PMV), *Peanut stripe virus* (PStV), *Pea seed-borne mosaic* virus (PSbMV), *Soybean mosaic virus* (SMV), *Cowpea aphid borne mosaic virus* (CpABMV), *Sweet potato feathery mottle virus* (SPFMV), *Henbane mosaic virus* (HMV), *Peanut green mosaic* (PGMV) and *Potato virus Y* (PVY) and potyvirus group specific antiserum (Hema *et al.*, 2001). SCSMV was detected in sugarcane by RT-PCR assays. Additionally, nucleic acid spot hybridization (NASH) was standardized to detect SCMV and SCSMV in sugarcane and sorghum. Recently full genome of the virus has been sequenced. The single stranded positive sense RNA genome of SCSMV-IND is 9786 nucleotides in length (excluding the poly (A) tail) and it comprises a large open reading frame encoding polyprotein of 3131 amino acid residues (Viswantahan *et al.*, 2011, unpublished).

SCMV has been routinely identified serologically assays like DAC-ELISA, DAS-ELISA, EBIA, DBIA and immunosorbent electron microscopy (Rao *et al.*, 2002) in India. Further, PCR assays have also been developed to identify the SCMV infecting sugarcane and sorghum crops in India (Rao *et al.*, 2005: Gaur *et al.*, 2003). Gaur *et al.* (2003) further established serological identification of SCMV and SCSMV through ELISA in stalk juice for routine diagnosis in 6 month old affected crops.

Virus indexation was carried out by a combination of infectivity test, serological test and electron microscopy.All the tests used for virus indexation were highly sensitive. It is believed that infectivity test or sap transmission of virus is hundred times more sensitive than serological test and electron microscopy.

8.5. Host range: Though several isolates of SCMV naturally infecting sugarcane, maize (*Zea mays*), sorghum (*Sorghum bicolour*), pearl millet (*Pennisetum typhoids*) and finger millet (*Eleusine coracana*) have been reported from different parts of India, few of them have been purified or biochemically characterized (Singh, 1983; Kondaiah and Nayudu, 1984; Hema *et al.*, 2001; Rao 1998). Rao *et al.* (1990) enlisted 277 plants as natural and artificial hosts of different strains of SCMV. In India natural hosts of SCMV include *Pennisetum purpureum* (Bharagava *et al.*, 1971), *Bothriochola insulpta, Brachiaria ramosa, Chlorosis barbata, Cynodon dactylon, Cyperus rotundus, Dinebra retroflixa,* and *Echionochloa colonum.* In addition, several other graminaceous plants have been found susceptible to SCMV strains by artificial inoculations (Rao *et al.*, 1990).

SCSMV infected many sorghum differentials (Hema *et al.*, 2001) but it failed to infect other poaceous members like *Triticum aestivum, Pennisetum typhoides, Zea mays, Eleusine coracana* which were reported as hosts to certain strains of SCMV. *Sorghum caudatum, S. cernuum, S. verticilliflorum, S. controversum, S. bicolour* cv. collier were found to be good propagation hosts for the virus as almost hundred per cent infection was found after sap inoculation. Infection of *S. halepense* suggests that SCSMV-AP is not related to SCMV (Teakle *et al.*, 1990). On the cv. Atlas, SCSMV-AP induced systemic mosaic symptoms without necrosis, unlike the strain of four Potyviruses of the sugarcane mosaic virus subgroup which all induced necrosis (Tosic *et al.*, 1990; 1994).

8.6. Epidemiology: Both the viruses causing mosaic disease on sugarcane (SCMV & SCSMV) spread in nature through infected setts. In India, *R. maidis* and *Schizaphis graminum* had been reported as vector of SCMV (Chona and Seth, 1958). Bhargava *et al.* (1971) and reported several aphids, colonizing sugarcane and other plants growing in the vicinity of sugarcane plantations and reported that *Aphis gossypii, Melanaphis (Longiunguis) sacchari, Lipaphis pseudobrassicae, Myzus pericae, Rhopalosiphum maidis* and *R. rufiabdominals* could transmit SCMV from sugarcane to sugarcane, maize and sorghum, from maize to sugarcane and maize, and from sorghum to sugarcane and sorghum. Bhargava *et al.* (1972) have been shown that *Aphis nerii* is also capable of transmitting strain D of SCMV. Rao *et al.* (1994) in an epidemiological study of SCMV in relation to natural incidence and periodical increase of different available vectors of SCMV in sugarcane fields and neighboring areas observed natural transmission of SCMV up to 40 per cent in some affected sugarcane fields. Mosaic first appeared in June and the maximum increase in natural incidence occurred during the monsoon period, i.e., July to September. *Rhophalosiphum maidis* and *Melanaphis indosacchari,* which generally occurred on graminaceous hosts during monsoon period, appear to play a major role in secondary spread of SCMV in nature, while *Aphis gossipii* and *Myzus persicae* observed on malvaceous, solanaceous, cucurbitaceous and cruciferous crops and weeds may cause only chance spread due to their limitation of host plants and/ or scanty occurrence in the vicinity of sugarcane field. SCSMV is easily sap transmitted from sugarcane to *Sorghum bicolour* cv. Rio and also transmitted through sugarcane setts. However, *Aphis craccivora* and *Rhopalosiphum maidis* failed to transmit SCSMV-AP in non-persistent manner from sorghum to sorghum under experimental condition (Hema *et al.*, 2001).

8.7. Management: SCMV and SCSMV are seed cane transmitted potyviruses, use of virus free planting material is imperative for raising healthy crops. Control of mosaic disease could only be achieved through agronomical approaches, by controlling insect vectors, use of resistant varieties, heat therapy, chemotherapy and apical mestrim culture.Breeding disease resistant varieties is a long term control of SCMV and SCSMV.Meristem culture alone or in combination with heat therapy has also been found useful in generating SCMV free plants in India (Hendre *et al.*, 1975). Since last two decades, *in vitro* meristem tip culture has been playing significant role in solving the problem of viral infection in plants. This technique is being successfully applied to many sugarcane varieties affected with SCMV and SCSMV (Balamuralikrishnan *et al.*, 2002; Mishra *et al.*, 2010).

9. Sugarcane yellow leaf disease

9.1. Occurence and distribution: Yellow leaf disease (YLD) in sugarcane was first observed in 1980s, in Hawaii as yellow leaf syndrome (YLS) by Schenck (1990). It is characterized by yellowing of the midrib and lamina. In India, Yellow leaf disease (YLD) caused by *Sugarcane yellow leaf virus* (SCYLV) was observed during 1999 by Viswanathan *et al.* (1999). Rao *et al.* (2000 a, b, 2001 a, b) reported the spread of YLD in sugarcane in different regions. Subsequently Viswanathan (2002) reported the disease occurrence in detail with symptomatology. The expression of symptoms is influenced by different biotic and abiotic factors. Disease incidence may go up to 100 per cent in some susceptible varieties and severe infection significantly affects cane yield and juice quality. YLD infected sugarcane plants recorded lesser photosynthetic activity and reduced mobilization of photosynthates from the leaves to sink, thereby reducing the sucrose accumulation in the affected stalks (Rao *et al.*, 2000 a, b; Viswanathan *et al.*, 1999, 2002 a, b).

9.2. Symptoms: The symptoms appear initially on matured leaves after 5 to 6 months of crop growth. On the leaves, the symptom appears as yellowish midrib on the lower surface. The yellowing may be confined to midrib region or the yellow discolouration may spread laterally to adjoining laminar region parallel to midrib. Reddish discolouration of midrib and laminar region is also noticed in certain varieties. In most susceptible varieties, typical yellowing of midribs and laminar region is noticed on lower surface of the leaves. Finally, symptoms of necrosis of discoloured laminar region from leaf tip to bottom along the mid rib and subsequent drying of entire leaf is also noticed. Severe infection of the disease leads to shortening of internodes in the top. This effect culminates in burning appearance of leaves at the top results in premature drying of entire clump.

9.3. Causal organism: Icosahedral particles of 25-30 nm in diameter were found in a symptomatic plant which is transmitted by sugarcane aphid, *Melanaphis sacchari*. Planting of setts from virus infected canes has been found as the chief source of disease introduction in the field and secondary transmission of the disease occurs through aphid vectors to varying extent. Precise diagnostic techniques like tissue-blot, ELISA and RT-PCR have been standardized to detect the associated pathogen, SCYLV. Although association of SCYLV has been reported from many countries, there are also evidences of sugarcane yellows phytoplasmas (ScYP) causing the disease in few other countries. Efforts are being made to identify the disease-resistant sources to YLD in sugarcane germplasm and to breed YLD-resistant varieties in different countries.

9.4. Molecular characterization and Diganosis: SCYLV has been diagnosed by serological assays like ELISA, tissue blot immunoassay (TBIA) technique using polyclonal antisera to detect SCYLV (Gaur *et al.*, 2003). DAS-ELISA has also been successfully used to detect the pathogen in infected plant material (Scagliusi and Lockhart, 2000; Viswanathan, 2002; Viswanathan and Balamuralikrishnan, 2004). Reliable results were also obtained through reverse transcription-polymerase chain reaction (RT-PCR) for detecting the virus in sugarcane.

The entire genome of SCYLV has been sequenced, and the virus has been assigned to the genus *Polerovirus* of the family *Luteoviridae* (D'Arcy and Domier, 2005). The genome of SCYLV is monopartite and consists of a positive-sense single stranded RNA of 5895-5898 nucleotides. The viral genome encodes atleast six open reading frames (ORFs 0-5) and shows a genome organization typical of *Polerovirus*. In India, occurrence of SCYLV was reported in five states of India (Uttar Pradesh, Bihar, Uttaranchal, Haryana and Tamil Nadu) on the basis of symptomatology, serology and particle morphology (Rao *et al.*, 2000 a, b). Attempts have also been made on molecular characterization of virus associated with YLD in India (Viswanathan *et al.*, 2008 a, b). A multiplex reverse transcription–polymerase chain reaction (multiplex-RT-PCR) was developed for the detection of SCYLV along with other major RNA viruses widely prevailing in the sugarcane-growing regions in India viz. *Sugarcane mosaic virus* (SCMV) and *Sugarcane streak mosaic virus* (SCSMV). The newly designed primers from the coat protein genes of the respective viruses amplified fragments of ~860 bp (SCMV), ~690 bp (SCSMV) and ~615 bp (SCYLV) in multiplex-RT-PCR (Viswanathan *et al.*, 2008 a, b).

The phylogenetic analyses of the genome of SCYLV revealed the occurrence of four genotypes of SCYLV (BRA for Brazil, CUB for Cuba, PER for Peru and REU for Reunion Island) based on the geographical location where it was first detected (Abu Ahmad *et al.*, 2006a, b). Recently, Viswanathan *et al.* (2008b) reported the fifth genotype of SCYLV viz. IND from India. Additionally, variation in infection capacity and in virulence between SCYLV genotypes has recently been demonstrated. Detailed studies conducted in India based on partial sequences encoding for ORFs1 and 2 strongly established the occurrence of at least three genotypes, viz., CUB, IND and BRA in India. The genotype IND may be restricted in India (Viswanathan *et al.* 2008; Singh *et al.*, 2011).

9.5. Host range: *Saccharum* species including traditional and modern sugarcane cultivars and wild relatives are the only known natural hosts of SCYLV (Lockhart and Cronje, 2000, Schenck and Lehrer, 2000, Comstock *et al.*, 2001, Lehrer *et al.*, 2001). Apart from sugarcane, the other cereal crops viz., wheat, oats and barley were very susceptible to SCYLV infection than sorghum, which was moderately susceptible to SCYLV, and sweet corn and rice plants were also successfully inoculated with SCYLV (Schenck and Lehrer, 2000).

9.6. Epidemiology: Infected setts are the primary source for the disease in the field. Further, it was found that the disease incidence was more severe in ratoons and in poorly maintained fields.Two aphid species viz., sugarcane aphid-*Melanaphis sacchari* (Zehntner) and corn leaf aphid- *Rhopalosiphum maidis* (Fitch) transmit the virus from infected to healthy plant (Lockhart *et al.*, 1996, Scagliusi and Lochkart, 2000, Schenck and Lehrer, 2000). However, *M. sacchari* is the most important and efficient vector of SCYLV in India than *R. maidis*, which is common in sugarcane fields.

9.7. Management: Among the different management approaches, clean seed programme is found to be the most effective to control the disease. The technique of *in-vitro* culturing arises as a powerful technique for elimination of SCYLV (Mishra and Rao, 2011) .Elimination of the virus through meristem culture has been demonstrated to purify the virus from the infected planting materials and this technique needs to be adopted to supply disease-free planting materials which would sustain sugarcane production. In certain countries, it was found that spread of the viral infection to neighbouring plants in the plantation fields via aphids was relatively slow and in the range of a few metres per year. No indication of long-distance transfer could be seen.

10. Grassy Shoot disease

10.1. Occurrence and distribution: Sugarcacane grassy shoot (SCGS) disease is an important phtoplasmal disease of sugarcane in India and has an all India foot print. It was first observed in Co 419 near Belapur, Ahmadnagar and Maharashtra in 1949 (Chona 1958). Similar disease was also reported from other parts of the country and described under different names such as "new chlorotic disease", "yellowing disease", "albino disease", "bunchy disease" or "leafy tuft" (Rao *et al.* 2005). Studies by Rane and Dakshindas (1962) showed that grassy shoot, yellowing and albino symptoms are associated with the same disease and subsequently the term "grassy shoot" was accepted as common name.

10.2. Symptoms: The appearance of chlorotic leaves among the crown leaves in affected shoot is the earliest symptom in the plant crop. Sometimes straight and regular prominent chlorotic or creamy streaks of varying width are common on the whole length of the leaf lamina, which later broadens and becomes fully chlorotic. Later many thin white grassy shoot emerge from the base of affected stalks. These symptoms are more common in ratoon crops as compared to plant crop. The chlorotic tillers dry out eventually. Premature sprouting of lateral buds with chlorotic leaves and formation of aerial roots are also seen in severely affected clumps. The production of inflorescence in GSD affected plant is not observed (Vasudeva, 1956, Chona *et al.*, 1960, Bhargava *et al.*, 1971).

Each stalk that is produced from the affected stool shows shortened internodes and the development of side shoots from the bottom to the top. Affected plants do not produce millable canes. The disease is particularly pronounced in the ratoon crop where the clusters of slender tillers with reduced leaves usually growing erect give the appearance of a field full of perennial grass, and from which it has derived its popular name "grassy shoots". The symptoms of the disease on nature of infection (primary or secondary), concentration of pathogen, reaction of variety and climatic conditions. The disease is characterized by the production of numerous tillers with narrow leaves, with or without albinism.Edison and Ramakrishnan (1972) studied the symptom expression in grassy shoot with reference to the bud position. They found that disease symptom usually appear late in sprouts raised from the top buds than those raised from the basal buds, indicating gradual reduction in the titer of the pathogen from top to bottom of the cane. The anatomical studies have revealed the disorganization of conducting elements and loss of bundle sheath chloroplast in leaf (Dhumal, 1983).

GSD can cause very heavy losses particularly when planting material is obtained from infected sources, or when disease transmission occurs in the early growth stage of the crop. Yield losses in ratoon reach their maximum in crops in which primary infection appeared early in the plant crop. Up to 70 per cent or higher incidence of GSD has been recorded in some area resulting 100 per cent yield and sugar loss (Rao *et al.*, 2000a).

10.3. Causal organism: Until 1971, the causal organism of grassy shoot disease of sugarcane was believed to be a virus (Chona *et al.*, 1960, Rane and Dakshindas, 1962). But Corbett *et al.* (1971) for the first time confirmed with electron microscopy that mycoplasma like organisms (MLOs) are the causing this disease. Rishi *et al.* (1973) and Dhumal (1983) supported this finding.

The phytoplasma associated with GSD are the smallest self replicating, cell wall less, unicellular prokaryotes bounded by a single lipoprotein membrane, containing ribosomes and double stranded DNA. These are spherical and/or ovoid located in sieve tubes of phloem. Phytoplama multiply by budding and binary fission. They pass from one cell to other through sieve pores (Corbett *et al.*, 1971; Rishi *et al.*, 1973). The GSD phytoplasmas range from 300 to 400nm in size. The large cells contain DNA strands in the centre with ribosome like granules surrounding them.

10.4. Molecular characterization and diagnosis: The GSD-phytoplasma was detected in sugarcane by ELISA and immunofluorecence (Viswanathan, 2000). Universal phytoplasma-specific primer pairs P1 and P7 were used for nested PCR assays that successfully detect the SCGS phytoplasma in sugarcane and its reported leafhopper vector *Deltocephalus vulgaris* in India (Rao *et al.*, 2003; Viswanathan *et al.*, 2005; Srivastava *et al.*, 2006a) in India. Since the diseased plants exhibit various phenotypic symptoms under field conditions, diagnosis of the disease becomes difficult. Hence, a rapid method for enrichment and isolation of GSD phytoplasma from the infected plants was developed. Differential filtration approach was used to isolate and enrich the GSD phytoplasma and its genomic DNA that was detected by PCR analysis for phytoplasmal 16S rDNA. Ratio of pathogen to host plant DNA was found in the order of 10^3 and 10^5 in infected tissue and enriched fraction respectively, offering 148-fold increase in sensitivity for their detection.

Nucleotide sequence analysis of 16S rRNA genes revealed that GSD phytoplasma affecting sugarcane crops in India is very closely related to the SCWL agent and is, thus, a member of the RYD phytoplasma group. GSD and SCWL phytoplasmas shared a 16S rDNA sequence similarity which varied from of 97.5 to 98.8%. Of the phytoplasmas that cluster in other phylogenetic groups, those most closely related to GSD phytoplasma are the BGWL (='*Candidatus* Phytoplasma cynodontis') and brachiaria grass white leaf (BraWL) agents, which share 97.3 and 97.1% 16S rDNA sequence similarity, respectively (Rao *et al.*, 2008).

Nasare *et al.* (2007) have analysed 198 sugarcane plant samples exhibiting grassy shoot symptoms in India tested positive for phytoplasma through PCR amplification of 16S rRNA gene and 16S-23S rRNA SR using primers specific for phytoplasmas and concluded that the sequence homology in the present GSD - causing phytoplasma in India is more than 99%, and their homology with SCWL and BGWL is from 98 to 99%. Therefore, it can be concluded that GSD, SCWL, SGS, and BGWL belong to the same species-level taxa (Marcone *et al.*, 2004). In another study, the phylogenetic relationships of GSD phytoplasma strains to each other and to the most related phytoplasmas were examined, by sequencing both 16S rRNA gene and 16S/23S rRNA spacer region. The phylogenetic relationships of GSD phytoplasma isolates among themselves and related phytoplasmas based on 16S rRNA gene sequences showed 100% identity. No variation among 16S rRNA gene sequences of all the 18 SGGS isolates of India was observed. However, there were significant variations in phenotypic expression of GSD phytoplasmas on sugarcane, no genotypic variations could be established (Viswanathan *et al.*, 2011).

10.5. Epidemiology: Sugarcane being clonally propagated, the pathogen enjoys the advantage of both uniform genotype and rapid vegetative multiplication. The phytoplasma are well protected inside the host tissue and pass undetected to the resultant crop; this causes progressive accumulation of pathogen, which finally leads to the withdrawal of a variety from cultivation. Planting diseased setts and infected plants scattered throughout fields are important factors in disease spread.

In severely affected crop, the disease tillers may not survive for more than a month. Round to elongated smaller buds are common in phytoplasma affected canes. Premature sprouting of lateral buds and the formation of the aerial roots completely destroyed the seed sett germination capability. If disease setts are used for planting material the germination per centage is highly reduced to 30-60 per cent (Dhumal, 1983). This results in the redctuion of millable canes per hectare and finally reduction on yield. It was reported 40-90% loss in cane yeild due to GSD. While others have recorded adverse effect of GSD on commercial cane sugar (CCS) and sugar recovery. The affected canes have very poor milling quality, the juice shows in reduction in brix, pol, CCS% and purity, but increase in the invert sugar (Dhumal and Nimbalkar, 1983; Rao *et al.*, 2000a).

The primary transmission of GSD is through infected seed setts it plays a major role in the spread of the disease. Setts from infected seed sources show disease incidence wihithn 2 to 3 months (Visawnathan, 2000). Field observations in India indicated a higher disease incidence during the summer months than during the monsoon (Rao *et al.*, 2000a). Jha *et al* (1973) reported that The GSD phytoplasma can be transmitted from diseased to healthy plant through *Cuscuta campestris.* Recently secondary spread of GSD in nature has been reported in nature by a leafhopper, *Deltocephalis vulgaris* (Srivastava *et al.*, 2006).

10.6. Management: The identification of phytoplasma diseases are mostly relied on the identification of symptoms but sometimes it is too difficult to identify the diseased plant because absence of peculiar symptoms. Hence, identification and management of phytoplasma disease are very difficult. In order to prevent the spread of sugarcane phytoplasmas through seed cane, it is necessary to reinforce the inspection and quarantine facilities with molecular diagnostic tools. Treatment of cuttings with moist hot air (MHAT) at 54^0 C for 2½ hours and hot water treatment at 50^0 C for 2h are recommended (Rao, 2006). Setts after heat treatment should always be treated with fungicide (Carbendazim 0.1%) to reduce the entry of soil pathogens through cut ends. No single approach can provide effective and long lasting management of sugarcane phytoplasmas. Judicious integration of phytoplasma free seed-cane developed using a three tier seed prgroamme, appropriate cultural practices and resistant clones has to done to achieve the goal (Rao, 2006).

REFERENCES

Abbott, E. V. 1938. Red rot of sugarcane. U.S. Dept. Agr. Tech. Bull. 641. 96.

Abbott, E. V., and Hughes, C. G.1961. Red rot. In, *Sugarcane Diseases of the World,* Vol. I. (Eds. J.P. Martin, E.V. Abbott and C.G. Hughes), Elsevier, Amsterdam. 262-287.

Abu Ahmad, Y., Royer, M., and Daugrois, J. H. 2006a. Geographical distribution of four *Sugarcane yellow leaf virus* genotypes. Plant Dis. 90: 1156-60.

Abu Ahmad, Y., Rassaby, L., and Royer, M. 2006b. Yellow leaf of sugarcane is caused by at least three different genotypes of *Sugarcane yellow leaf virus*, one of which predominates on the Island of Reunion. Arch Virol. 151: 1355-71.

Agnihotri, V. P. 1990. Diseases of sugarcane and sugarbeet, Oxford and IBH publishing Co, New Delhi.

Agnihotri, V. P., and Duttamajumder, S. K. 1994. Disease management practices in sugarcane: past, present and future. In *Current Trends in Sugarcane Pathology.* (Eds.) Rao, G. P. *et al.*, Int. Books and Periodicals Supply Service, New Delhi, 333-345.

Agnihotri, V. P., Budhraja, T. R., and Singh, K. 1979. Role of diseased sett and soil and the annual recurrence of red rot in sugarcane. Inter. Sugar Jour. 82: 263-265.

Alexander, K. C., and Ramakrishnan, K. 1977. Studies on the smut disease (*Ustilago scitaminea* Syd.) of sugarcane: 4 Parasitism, germination, dikaryotisation and infection. Proceedings of the International Society of Sugarcane Technologists 16: 469–472.

Alexander, K. C., and Rao, M. M. 1976.Identification of genetic stocks possessing high resistance to red rot and smut. Sugarcane Breeding Newsl. 37: 10.

Alexander, K. C., and Rao, M. M. 1981a. Resistance to smut in the hybrid varieties of sugarcane. Sugarcane Path. Newsl. 27: 13.

Alexander, K. C., and Rao, M. M. 1981b. Sources of resistance to smut in the different species of *Saccharum*. Sugarcane Path. Newsl. 27: 7- 12.

Antonie, R., and Perombelon, M. 1965. Leaf scald. Ann. Rep 1964. Maurit. Sug. Ind. Res. Inst. 56-58.

Ashby, S. F. 1929. The bacterium which causes gumming disease of sugarcane with notes on two other bacterial diseases of the same host. Trop. Agric. (Trinidad).6: 135 –138.

Bailey, R. A. 1976. Heat treatment and leaf scald. Sugarcane Pathol. Newsl.17: 14-16.

Balamuralikrishnan, M., Doraisamy, S., Ganapathy, T., and Viswanathan, R. 2002. Combined effect of chemotherapy and meristem culture on sugarcane mosaic virus elimination in sugarcane. Sugar Tech.4: 19–25.

Barber, C. A. 1901. Sugarcane diseases in Godavari and Ganjam districts. Madras Dept. Land Records and Agri. 2. Bull. 43: 181-194.

Baudin, P., and Chatenet, M. 1980. Determination d'une souche de *Xanthomonas albilineans* (Ashby) Dowson isolee de Haute Volta. Agron. Trop. 35: 288-291.

Benda, G. T. A. 1972. Hot water treatment for mosaic and ratoon stunting disease control. Sugar Jour. 34: 32-39.

Benda, G. T. A. 1978. Increased survival of young seed cane after hot water treatment for RSD control. Sugar Bull. 56: 7-8, 13-14.

Bhargava, K. S. 1975. Sugarcane mosaic- Retorspect and prospects. Indian Phytopath. 28: 1-9.

Bhargava, K. S., Joshi, R. D., and Lal, K. M. 1972. Strain D of *Sugarcane mosaic virus* in India. Sugarcane Pathol. Newsl.8: 23.

Bhargava, K. S., Joshi, R. D., and Rizvi, S. M. A. 1971. Some observations on the insect transmission of *Sugarcane mosaic virus*. Sugarcane Pathol Neswl.6: 20-21.

Birch, R. G. 1999. Albicidin detoxification – A case study in plant genetic engineering to destroy toxins from microbial pathogens. In Dietzgen, R. G. (ed) Elimination of Aflatoxin contamination in peanut. ACIAR, Canberra.50-53.

Birch, R. G. 2001. *Xanthomonas albilineans* and the antipathogenesis approach to disease control. Mol. Pl. Pathol. 2: 1-11.

Birch, R. G., and Patil, S. S. 1983.The relation of blocked chloroplast differentiation to sugarcane leaf scald disease. Phytopathology 73: 1368-1374.

Bond, W. P., and Pirone, T. P. 1971. Purification and properties of sugarcane mosaic virus strains. Phytopath. Z. 71: 56-65.

Booth, C. 1971. The genus *Fusarium*. CMI, Kew, Surrey, England.127-129.

Burner, D. M, Grisham, M. P., and Legendre, B. L. 1993. Resistance of sugarcane relatives injected with *Ustilago scitaminea*. Plant Dis.77: 1221–1223.

Butler, E. J. 1918. Fungi and Diseases in Plants. Thacker and Spink. Calcutta, India. 547.

Butler, E. J., and Khan, A. H. 1913. Some new sugarcane diseases, Part I. Wilt. Mem. Dept. Agri. India, Bot. Ser. 6: 180-190.

Butler, E. J. 1906. Fungus diseases of sugarcane in Bengal. India Dept. Agr. Mem. Bot. Ser. 1: 2-24.

Carvajal, F., and Edgerton, C. W. 1944. The perfect stage of *Colletotrichum falcatum*. Phytopathology 34: 206-213.

Chand, J. N., Dang, J. K., and Kapoor, T. R. 1974. Systemic chemicals as sugarcane sett protectants. Sci. & Cult. 40: 69-70.

Chona, B. L. 1956. Address of the Chairman (Pathology Section). Proc. ISSCT.9: 975-986.

Chona, B. L. 1980. Red rot of sugarcane and sugar industry - a review. Indian Phytopath. 33: 191-206.

Chona, B. L. and Bajaj, B. S. 1953. Occurrence in nature of *Physalospora tucumanensis* Speg. the perfect stage of sugarcane red rot organism in India. Indian Phytopath.6: 63-65.

Chona, B. L., and Nariani, T. K. 1952. Investigations on the survival of *Colletotrichum falcatum* in soil. Indian Phytopath. 5: 152-157.

Chona, B. L., and Padwick, G. W. 1942. More light on the red rot epidemic. Indian Farming.3: 70-73.

Chona, B. L., and Srivastava, D. N. 1952. The perithecial stage of *Colletotrichum falcatum* went in India. Indian Phytopath.5: 158-160.

Chona, B. L., Capoor, S. P., Verma, P. M., and Seth, M. L. 1960. Grassy shoot disease of sugarcane. Indian Phytopath. 13: 37-47.

Chona, B. L. 1958. Some diseases of sugarcane reported from India in recent years. Indian Phytopath.11: 1-9.

Chona, B. L., and Seth, M. L. 1958. *Aphis maidis* Fitch as a vector of sugarcane mosaic in India. Ind. J Agr. Sci.28: 257-260.

Comstock, J. C., and Raid, R. N. 1994. Sugarcane common rust. In: Current Trends in Sugarcane Pathology (Prof. K.S. Bhargava Festschrift), Rao, G.P., Gillaspie, A.G., Upadhyaya, Jr. P.P., Bergamin Filho, A., Agnihotri, V.P., and Chen, C.T. (Eds).1-10. Dehli, India, International Books and Periodicals Supply Service.

Comstock, J. C., Miller, J. D., and Schnell, R. J. 2001. Incidence of *Sugarcane yellow leaf virus* in clones maintained in the world collection of sugarcane and related grasses at the United States National Repository in Miami, Florida. Sugar Tech 3: 128-133.

Corbett, M. K., Misra, S. R., and Singh, K. 1971. Grassy shoot disease of sugarcane, IV. Association of mycoplasma like bodies. Plant Sci. 3: 80-82.

Croft, B. J. 1998. Improving the germination of sugarcane and the control of pineapple disease. Proceedings of the Australian Society of Sugar Cane Technologists Conference 20: 300-306.

Croft, B. J., and Braithwaite, K. S. 2006. Management of an incursion of sugarcane smut in Australia. Australas. Pl. Pathol. 35: 113–122.

D'Arcy, C. J., and Domier, L. L. 2005. *Luteoviridae*. In: Fauquet CM, Mayo MA, Maniloff J, Desselberger U, Ball LA (eds) Virus Taxonomy. VIIIth Report of the International Committee on Taxonomy of viruses. Academic Press, NewYork, 343–352.

Damann, K. E. Jr. 1988. Alkaline induced metaxylem auto- fluorescence: A diagnostic symptom of ratoon stunting disease of sugarcane. Phytopathology. 78: 233-236.

Daniels, J., and Daniels, C. A. 1976. A note on red rot disease, *Physalospora tucumanensis* Spe, and the origin of sugarcane. Sugarcane Breeder's Newsl. 37: 17-19.

Dastur, J. F. 1923. The mosaic disease of sugarcane in India. Agric. Jour. India 18: 505-509.

Dastur, J. F. 1926. A mosaic like disease of sugarcane in Central province. Agric. Jour. India 21: 429-432.

Davis, M. J. 1985. Direct count technique for enumerating *Clavibacter xyli* subsp. *xyli* which causes ratoon stunting disease of sugarcane. Phytopathology 75: 1226-1231.

Davis, M. J., Gillaspie, A. G., Vidaver, Jr. A. K., and Harris, R. W. 1984. *Clavibacter*: a new genus containing some phytopathogenic coryneform bacteria, including *Clavibacter xyli* subsp. *xyli* sp. nov., subsp. nov. and *Clavibacter xyli* subsp. *cynodontis* subsp. nov., pathogens that cause ratoon stunting disease of sugarcane and bermuda grass stunting disease. Int. J. Syst. Bacteriol.34: 107-117.

Davis, M. J., Gillaspie, A. G., Harris, Jr. R. W., and Lawson, R. H. 1980. Ratoon stunting disease of sugarcane. Isolation of the causal organism. Science 240: 1365-1367.

Deerr, N. 1949. The History of Sugar. Chapman & Hall, London. I: 43- 258.

Dhumal, K. N., and Nimbalkar, J. D. 1983. Studies on grassy shoot disease affected sugarcane cultivars Co 419 and Co 740. Indian Phytopath. 36: 448-452.

Dhumal, K. N. 1983. Physiological studies in sugarcane (Comparative physiological studies in healthy and GSD affected sugarcane cvs Co 419 and Co. 740). PhD Thesis Shivaji University Kohlapur, Maharashtra, India 11-53.

Dowson, W. J. 1943. On the generic names *Pseudomonas, Xanthomonas* and *Bacterium* for certain bacterial plant pathogens.Trans. Brit. Mycol. Soc. 26: 4-14.

Duttamajumder, S. K. 1990. Evaluation of sugarcane genotypes against *Xanthomonas albilineans*, the incitant of leaf scald disease. Indian Phytopathol. 43: 265.

Duttamajumder, S. K. 1990. Fluff transmission of *Xanthomonas albilineans* - the incitant of leaf scald disease of sugarcane. Curr. Sci. 59: 744-745.

Duttamajumder, S. K. 1998a. Dynamics of field spread of *Xanthomonas albilineans* (Ashby) Dowson causing leaf scald disease in sugarcane. In *Plant Pathogenic Bacteria (Proceedings of the 9th International Conference).* Mahadevan, A. (Ed.). Centre for Advanced Study in Botany, Univ. of Madras, Chennai, India. 324-330.

Duttamajumder, S. K. 2001.Surreptitious spread of sugarcane ratoon stunting disease pathogen *Leifsonia xyli* subsp. *xyli* in the sub-tropical India. Indian Phytopathol. 54: 481-483.

Duttamajumder, S. K. 2004. Bacterial diseases of sugarcane in India: A bird's eye view. In Sugarcane Pathology, Vol. III. Bacterial and nematode diseases. Eds. Rao, G.P., Saumtally, A.S. and Rott, P. Oxford & IBH Publishing Co. Pvt. Ltd., New Delhi.15-50.

Duttamajumder, S. K. 2008. Redrot of Sugarcane. Indian Institute of Sugarcane Research, Lucknow.171 (inclusive of 39 colour plates).

Duttamajumder, S. K., and Agnihotri, V. P. 1992. Management of sugarcane diseases-problem and progress. In, *Farming systems and integrated pest management.*(Eds.

J.P. Verma and A.Varma) Malhotra Publishing House, New Delhi.145-158.

Edgerton, C. W. 1959. Sugarcane and its diseases. 2nd ed. Louisiana State Univ. Press, Baton Rouge. 301.

Edgerton, C. W., and Carvajal, F. 1944. Host-parasite relations in red rot of sugarcane. Phytopathology 34: 827-837.

Edison, S., and Ramakrishnan, K. 1972. Symptom appearance in grassy shoot disese of sugarcane with reference to bud position. Curr Sci. 41: 571.

Egan, B. T. 1962. The diseases of sugarcane in Ceylon, Proc. ISSCT. 11: 809-811

Egan, B. T. 1979. Leaf scalds disease in India: letters to the editor, Sugarcane Pathol. Newsl, 22: 53.

Evtushenko, L. I., Dorofeeva, L. V., Subbotin, S. A., Cole, J. R., and Tiedje, J. M. 2000. *Leifsonia poae* gen. nov., sp. nov., isolated from nematode gall on *Poa annua*, and reclassification of '*Corynebacterium aquaticum*' Leifson 1962 as *Leifsonia aquatica* (ex Leifson 1962) gen. nov., nom. rev., comb. nov. and *Clavibacter xyli* Davis *et al.* 1984 with two subspecies as *Leifsonia xyli* (Davis *et al.* 1984) gen. nov., comb. nov. Int. J. Syst. Evol. Microbiol. 50: 371-380.

Ferreira, S. A., Comstock, J. C. 1989. Smut. *In*: Diseases of Sugarcane. Major Diseases. C. Ricaud, B.T. Egan, A.G. Gillaspie Jr and C.G. Hughes (Eds). 211-229. Amsterdam, The Netherlands, Elsevier Science Publishers.

Gams, W. 1971. *Cephalosporium*-artige Schimmelpilze (Hyphomycetes). Stuttgart, Germany: Gustav Fischer Verlag.

Ganguly, A. 1964. Wilt. In, *Sugarcane Diseases of the World, Vol. II*. Eds C. G. Hughes, E. V. Abbott and Wismer, C. A. Elsevier Publishing Co., Amsterdam. 354.

Ganguly, A., and Chand, J. N. 1963. Longevity of *Cephalosporium sacchari* Butler causing wilt disease of sugarcane. Sci. & cult. 29: 347-348.

Ganguly, A., and Khanna, K. L. 1955. Annual report of the scheme for investigation and control of wilt disease of sugarcane for the year ending 31st May 1955, Sugarcane Research Institute, Pusa, Bihar.

Gaur, R. K., Singh, A. K., Singh, M., Singh, A. K., Upadhyaya, P. P., and Rao, G. P. 2003. Reliability of serological identification of sugarcane mosaic potyvirus and sugarcane yellow leaf luteovirus from cane stalk juice. Sugar Cane International, UK, Sept/Oct: 18-21.

Gillaspie, A. G., Jr., Davis, R. E., and Worley, J. F. 1973. Diagnosis of ratoon stunting disease based on the presence of a specific microorganism. Plant Dis. Rept.57: 987-990.

Gough, K. H., and Shukla, D. D. 1981. Coat protein of potyviruses.I. Comparisions of four Australian strains of *Sugarcane mosaic virus*. Virology.111: 455-462.

Groenewege, J. 1915. De gomziekte van het suikerriet veroozaakt door Bacterium vasculorum Cobb. Meded. Proefst Java- Suikerind. Arch. Suikerind. Ned. Indie, 23.

Guevarra, M. M. 1990. Evaluation of *Gliocladium* spp. in the control of pineapple disease of sugarcane caused by *Ceratocystis paradoxa* (Dade) Moreau. Philippine Sugar Quarterly 1: 20-28.

Hema, M., Savithri, H. S., and Sreenivasulu, P. 2003. Comparison of direct binding polymerase chain reaction with recombinant coat protein antibody based dot-immunobinding assay and immunocapture-reverse transcription-polymerase chain reaction for the detection of sugarcane streak mosaic disease in India. Curr Sci 85: 1774-1777.

Hema, M., Joseph, J., Gopinath, K., Sreenivasulu, P., and Savitri, H. S. 1999. Molecular characterization and interviral relationships of a flexuous filamentous virus causing mosaic disease of sugarcane in India. Arch. Virol. 144: 479-490.

Hema, M., Savithri, H. S., and Sreenivasulu, P. 2001. *Sugarcane streak mosaic virus*: Occurrence, purification, characterization and detection.37-69. In: Sugarcane Pathology Vol. II Viral and Phytoplasmal Diseases (eds; G.P. Rao, R.E. Ford, M. Tosic and D.S. Teakle), Science Publishers Inc, Enfield, NH, USA.

Hendre, R., Mascarenhas, S. F., Nadgir, A. L., Pathak, M., and Jagannathan, V. 1975. Growth of mosaic virus free sugarcane plants from apical meristems. Indian Phytopathol. 28: 175-178.

Hughes, C. G. 1955. Some recent development in the study of ratoon stunting disease. Cane Gr. Quart. Bull.19: 27-28.

Jain, R. K., Rao, G. P., and Varma, A. 1998. Present Status of management of Sugarcane mosaic virus .In: Plant Virus Disease Control (eds) A Hadidi, R.K. Khetrapal and H. Koganezawa, APS Press, St Paul, Minnesota, USA.495-506.

Jha, A., Prasad, H. C., and Misra, B. 1973. Dodder, a new vector for transmitting spike and grassy shoot virus diseases of sugarcane. Indian Sugar. 23: 515-516.

Kao, J., and Damann, K. E., Jr. 1980. *In situ* localization and morphology of the bacterium associated with ratoon stunting disease of sugarcane. Can. J. Bot. 58: 310-315.

Khanna, K. L., Sharma, S. L., Srivastava, R. C., and Sinha, J. N. 1958. Search for ratoon stunting disease of sugarcane in Bihar. Proc. Ind. Acad. Sci. 47: 1-14.

Khurana, S. M. P., and Singh, S. 1972. Sugarcane mosaic starin E and C in India and new sorghum differentials. Sugarcane Pathol. Newsl. 9: 6-8.

Kirtikar., Singh, G. P., and Shukla, R. 1972. Role of seed material in carry over of wilt disease of sugarcane. Indian Sugar. 22: 89-90.

Koike, H. 1965. The Aluminium cap method for testing sugarcane varieties against leaf scald disease. Phytopathology. 55: 317-319.

Kondaiah, E., and Nayudu, M. V. 1984. A key to identification of of Sugarcane mosaic virus strains. Sugarcane. 6: 3-8.

Kumar, A., and Satyavir. 1998. Evaluation of biological control agents against red rot (*Colletotrichum falcatum*) of sugarcane. Ann appl. Biol.132: 72-73.

Kumar, A., and Satyavir. 1999. Efficacy of bioagents in the management of red rot of sugarcane under field condition. J. Mycol. Plant Pathol. 29: 277-278.

Kumar, N., Jhang, T., Satyavir., and Sharma, T. R. 2010. Molecular and pathological characterization of *Colletotrichum falcatum* infecting subtropical Indian sugarcane. J. Phytopath. 159: 260-267.

Lee-Lovick. 1978. Smut of sugarcane – *Ustilago scitaminea*. Rev. Pl. Pathol. 57: 181-188.

Lehrer, A. T., Schenck, S., Fitch, M. M. M., Moore, P. H., and Komor, E. 2001. Distribution and transmission of *Sugarcane yellow leaf virus* (SCYLV) in Hawaii and its elimination from seed cane. In: Proc, 24th International Society of Sugar

Cane Technologists Congress, Brisbane 2001. The Australian Society of Sugar Cane Technologists, Mackay. 439–443

Leoville, F., and Coleno, A. 1976. Detection de *Xanthomonas albilineans* (Ashby) Dowson, agent de l'echadure de la canne a sucre dans des boutures contaminees. Ann. Phytopathol. 8: 233-236.

Lockhart, B. E. L., and Cronje, C. P. R. 2000. Yellow leaf syndrome, p. 291-295. In: A guide to sugarcane diseases (Eds. P. Rott, R.A. Bailey, J.C. Comstock, B.J. Croft, A.S. Saumtally, CIRAD-ISSCT, Montpellier, France.339.

Lockhart, B. E. L., Irey, M. J., and Comstock, J. C. 1996. *Sugarcane bacilliform virus, sugarcane mild mosaic, and sugarcane yellow leaf syndrome*. In: Sugarcane Germplasm Conservation and Exchange. B.J. Craft, C.M. Piggin, E.S. Wallis and D.M. Hogarth (eds.), ACIAR Proccedings No. 67, Canberra.113-115.

Madan, V. K., Mandal, B., Ansari, M. L., Srivastava, A., Soni, N., Solomon, S., and Agnihotri, V. P. 2000. RAPD-PCR analysis of molecular variability in the red rot pathogen (*Colletotrichum falcatum*) of sugarcane. Sugarcane Intern. 3: 5-8.

Maramorosch, K., Plavsic-Banjac, B., Bird, J., and Liu, L. J. 1973. Electron microscopy of ratoon stunted sugarcane: Microorganisms in xylem. Phytopath. Z. 77: 270-273.

Marcone, C., Schneider, B., and Seemuller, E. 2004. '*Candidatus* Phytoplasma cynodontis', the phytoplasma associated with Bermuda grass white leaf disease. Inter Jour Sys & Evol Microbiol. 54: 1077-1082.

Martin, J. P., and Robinson, P. E. 1961. Leaf scalds, in *Sugarcane diseases of the world*, vol. I, Martin, J. P., Abbott, E. V. and Hughes, C. G., Eds., Elsevier, Amsterdam.78-107.

Mishra, S., and Rao, G. P. 2011. Regeneration of Sugarcane yellow leaf virus free sugarcane plantlets through meristem culture. Jour Southern Agril. 42: 1-5.

Mishra, S., Singh, D., Tiwari, A. K., Lal, M., and Rao, G. P. 2010.Elimination of *Sugarcane mosaic virus* and *Sugarcane streak mosaic virus* by tissue culture. Sugar Cane International 28: 119-122.

Mundkur, B. B. 1939. Taxonomy of sugarcane smuts. Kew. Bull.10: 525-533.

Nasare, K., Yadav, A., Singh, A. K., Shivasharanappa, K. B., Nerkar, Y. S., and Reddy, V. S. 2007. Molecular and symptom analysis reveal the presence of new phytoplasamas associated with sugarcane grassy shoot disease in India. Plant Dis. 91: 141

Natarajan, S., and Muthuswamy, S. 1981. Effect of aerated steam treatment on the incidence of red rot (*Colletotrichum falcatum* Went) of sugarcane. Inter. Sugar Jour. 83: 300-301.

Nirenberg, H. I. 1976. Untersuchungen uber die morphologische und biologische Differenzierung in der *Fusarium-* Sektion *Liseola*. Mitt Biolog Bundesanstalt fur Land-u Forstw (Berlin-Dahlem). 169: 1–117.

North, D. S. 1926. Leaf scald, a bacterial disease of sugarcane, The Colonial Sugar Refining Co. Ltd., Sydney. Agric. Report No. 8. 80.

North, D. S. 1929. *Leaf scald disease of sugarcane and its control* (condensed version), The Colonial Sugar Refining Company Ltd., Sydney. Agric. Report No. 9. 48.

Oliviera, A. R., Nakamuma, T., Liu, H. P., and Sugimori, M. H. 1978. Serological tests applied to leaf scald disease of sugarcane.Proc. ISSCT, 16: 459- 468.

Orian, G. 1962. A disease of *Paspalum dialatum* in Mauritius caused by a bacterial species closely resembling *Xanthomonas albilineans* (Ashby) Dowson.Review Agric. Sucriere Ile Maurice.41: 7-20.

Persley, G. J. 1973a. Epiphytology of leaf scald in the central district of Queensland, Proc. Queensl. Soc. Sugar Cane Technol.40: 39-52.

Persley, G. J. 1973b. Naturally occurring alternative hosts of *Xanthomonas albilineans* in Queensland. Plant Dis. Reptr.57: 1040-1042.

Persley, G. J. 1973c. Pathogenic variation in *Xanthomonas albilineans* (Ashby) Dowson, the causal agent of leaf scald disease of sugarcane. Aust. J. Biol. Sci. 26: 781-786.

Persley, G. J. 1975. Leaf scalds disease in Q93 at Bundaberg, Australia Sugarcane Pathol. Newsl.13/14, 23, 1975.

Persley, G. J., and Ryan, C. C. 1976. Epidemiology of leaf scalds in the Moreton district of Queensland.Proc. Queensland. Soc. Sugar Cane Technol. 43: 79-82.

Purdy, L. H., Krupa S. V., Dean, J. L. 1985. Introduction of sugarcane rust in the Americas and its spread into Florida. Plant Dis. 69: 689-693.

Raboin, L. M., Selvi, A., Oliveira, K. M., Paulet, F., Calatayud, C., Zapater, M. F., Brottier, P., Luzaran, R., Garsmeur, O., Carlier, J., and D'Hont, A. 2007. Evidence for the dispersal of a unique lineage from Asia to America and Africa in the sugarcane fungal pathogen *Sporisorium scitaminea*. Fung. Genet. Biol. 44: 64–76.

Raid, R. N., Perdomo, R., Powell, G. 1991. Influence of seedpiece treatment and seeding density on stalk population and yield of a pineapple disease susceptible sugarcane cultivar. Jour. Amer. Soc. Sugar Cane Technol. 11: 13-17.

Rane, M. S. and Dakshindas, D. G. 1962. The sugarcane disease 'albino' or 'grassy shoot'. Indian Sugar. 12: 179-180.

Rao, G. P. 2006. Management of Phytoplasma Diseases of Sugarcane. pp, 1-16 In: Plant Protection in New Millennium Vol. II (Eds. Ashok V, Gadewar and B.P. Singh), Satish Serial Publishing House, Delhi.

Rao, G. P., Jain, R. K., and Varma, A. 1998a. Identification of sugarcane mosaic and maize dwarf mosaic potyviruses infecting poaceous crop in India. Indian Phytopathol. 51: 10-16.

Rao, G. P., Jain, R. K., and Varma, A. 1998b. Characterization and Purification of an Indian Isolate of sugarcane mosaic virus. Sugar Cane, UK, 1: 8-10.

Rao, G. P., and Ford, R. E. 2001a. Vectors of Virus and Phytoplasma Diseases of Sugarcane: An Overview. pp. 267-318 In: Sugarcane Pathology Vol.II: Virus and Phytoplasma Diseases. (eds) G.P.Rao, R.E.Ford, M.Tosic and D.S. Teakle.373.

Rao, G. P., Gaur, R. K., Singh, M., Viswanathan, R., Chandrasena, G., and Dharamwardhaanhe, N. M. N. N. 2001b. Occurrence of sugarcane yellow leaf virus in India and Srilanka. Proc. of Int. Soc. of Sugar Cane Technol. 24: 469-470.

Rao, G. P., Singh, M., Rishi, N. and Bhargava, K. S. 2002. Century status of sugarcane virus diseases research in India. pp. 223-254. In: Sugarcane Crop Management (Eds) S.B.Singh, G.P.Rao and S. Easwaramoorthy. SCI Tech Publishing LLC, Houstan, Texas, USA. Pp734.

Rao, G. P., Ford, R. E., Tosic, M., and Teakle, D. S. 2000a. Sugarcane Pathology, Vol II, Virus and Phytoplasma Disease. Science Publisher's Inc, New Hampshire, USA. 377.

Rao, G. P., Gaur, R. K., Singh, M., Srivastava, A. K., Virk, A. S., Singh, N., Patil, A. S., Viswanathan, R. and Jain, R.K.2000b. Existence of Sugarcane Yellow leaf Luteovirus in India. Sugar Tech. 2: 37-38.

Rao, G. P., Gaur, R. K., and Singh, M. 2003. Distribution and serological diagnosis of sugarcane mosaic potyvirus in India. Sugar Cane International, UK. Jan/Feb: 6-11.

Rao, G. P., Singh, A., Singh, H. B., and Sharma, S. B. 2005. Phytoplasma diseases of sugarcane: Characterization, diagnosis and management. Indian Jour. Pl. Pathol. 23: 1-21.

Rao, G. P., Singh, M., and Singh, H. N. 1990. Alternative hosts of sugarcane diseases Sugarcane, UK, Autumn Supplement. 8-26.

Rao, G. P., Srivastava, S., Gupta, P. S., Singh, A., Singh, M. and Marcone, C. 2008. Detection of Sugarcane grassy shoot phytoplasma infecting sugarcane in India and its phylogenetic relationships to closely related phytoplasmas. Sugar Tech. 10: 74-80.

Rao, G. P., Singh, M., Singh, H. N., and Shukla, K. 1994. Epidemiological studies on *Sugarcane mosaic virus* in eastern UP.In: Virology in Tropics (Eds. N. Rishi, B.P. Singh and K.L. Ahuja) Malhotra Publ. Co., New Delhi, India. 343-350.

Ricaud, C., Sullivan, S., and Autray, J. C. 1976. Systemic infection of sugarcane by the bacterium associated with symptoms of ratoon stunting disease. Rev. Agri. Sucr. Maurice. 55: 159-162.

Ricaud, C., Felix, S., and Ferre, P. 1979. A simple serological technique for the precise diagnosis of leaf scalds disease in sugarcane, Proc. IVth Int. Cong. Plant Pathogenic Bac., Angers, France. 337–340.

Ricaud, C., Sullivan, S., Felix, S., and Ferre, P. 1978. Comparison of serological and inoculation methods for detecting latent infection of leaf scald. Proc. ISSCT. 16: 439-448.

Rishi, N., and Rishi, S. 1985. Purification, electron microscopy and serology of strains A nd F of *Sugarcane mosaic virus*. Indian Jour. Virol.1: 79-86.

Rishi, N., Okuda, S., Arai, K., Doi, Y., Yora, K., and Bhargava, K. S. 1973. Mycoplasma like bodies, possibly the cause of grassy shoot disease of sugarcane in India. Ann. Phytopath Soc. Japan. 39: 429-431.

Roach, B. T., and Jackson, P. A. 1992. Screening sugarcane clones for resistance to ratoon stunting disease. Sugarcane. 2: 2-12.

Rott, P. 1984. Apport des cultures in vitro a l'etude de l'echaudure des feuilles de canne a sucre (*Saccharum* sp.) causee par *Xanthomonas albilineans* (Ashby) Dowson, Ph.D. Thesis, Univ. Paris-Sud, Centre d'Orsay.185.

Sampang, R. C. 1991. Biological control with *Trichoderma* spp. of pineapple disease of sugarcane caused by *Ceratocystis paradoxa*. Philippine Sugar Quarterly. 2: 29-36.

Sarma, M. N. 1976. Wilt disease of sugarcane. Sugarcane Pathol. Newsl. 15/16: 30-33.

Satyanarayana, Y. 1974. Leaf scald of sugarcane: a new disease in Andhra Pradesh. Indian Sugar. 24: 23-24.

Scagliusi, S. M., and Lockhart, B. E. 2000. Transmission, characterisation and serology of a luteovirus associated with yellow leaf syndrome of sugarcane. Phytopathology. 90: 120-124.

Schenck, S. 1990. Yellow Leaf Syndrome - a new disease of sugarcane. Annual Report of the Hawaiian Sugarcane Planters Association Experimental Station.38.

Schenck, S., and Lehrer, A. T. 2000. Factors affecting the transmission and spread of sugarcane yellow leaf virus. Plant Dis. 84: 1085-1088.

Schexnayder, C. A. 1960. The use of sugarcane 'Test Plants' as a means of detecting the presence of ratoon stunting disease in sugarcane. Proc. ISSCT.10: 1069-1071.

Singh, V., Srivastava, S. N., Lal, R. J., Awasthi, S. K., Joshi, B. B. 2008. Biological control of red rot disease of sugarcane by *Trichoderma harzianum* and *T. viride*. Indian Phytopath. 61: 486- 493.

Singh, D., Rao, G. P., Snehi, S. K., Raj, S. K., Karuppaiah, R., and Viswanathan, R. 2011. Molecular detection and identification of thirteen isolates of Sugarcane yellow leaf virus associated with sugarcane yellow leaf disease in nine sugarcane growing states of India. Australasian Pl. Pathol. 40: 522–528.

Singh, G. R. 1969. An indicator sugarcane variety for ratoon stunting disease. Curr. Sci. 38: 221-222.

Singh, K. 1966. Ratoon stunting disease of sugarcane in India. Indian Sugar. 16: 335-337.

Singh, K. 1973. Hot air therapy against red rot of sugarcane. Plant Dis. Reptr. 57: 220-222.

Singh, K. 1983. Seed-piece transmissible diseases of sugarcane and three tier seed programme. In, *Recent Advances in Plant Pathology*. (Eds. A. Husain, K. Singh, B.P. Singh and V.P. Agnihotri). Print House (India), Lucknow. 331-343.

Singh, K., and Alexander, K. C. 1970. Laminar infection of sugarcane leaves by red rot (*Physalospora tucumanensis*) organism in nature. Indian Phytopath. 23: 114-115.

Singh, K., and Misra, S. R. 1976. Leaf scald disease of sugarcane in BO 70. Sugarcane Pathol. Newsl.38: 83.

Singh, K., and Singh, R. P. 1989. Red rot. In, *Sugarcane Diseases*. (Eds. C. Ricaud, B.T. Egan, A.G. Gillaspie, Jr and C.G. Hughes). Elsevier, Amsterdam.169-188.

Singh, K., Budhraja, T. R., and Agnihotri, V. P. 1977. Survival of *Colletotrichum falcatum* in soil, its portals of entry and role of inoculum density in causing infection. Inter. Sug. Jour. 79: 43-44.

Singh, K., Misra, S. R., Shukla, U. S., and Singh, R. P. 1980. Moist hot air therapy of sugarcane: Control of sett-borne infections of GSD, smut and red rot. Sugar J. 43: 26-28.

Singh, K., Singh, R. P., and Agnihotri, V. P. 1975. Taxonomy and pathogenicity of fungi causing sugarcane wilt syndrome. Indian Phytopathol. 28: 86-91.

Singh, K., Singh, R. P., and Misra, S. R. 1980. Development of wilt disease in sugarcane through sett and soil inoculations. Sugarcane Pathol. Newsl. 24: 23-25.

Singh, N., and Duttamajumder, S. K. 1995. Bio-control potential of some fungi in checking the development of red rot disease in standing cane. Ind. J. Sugarcane Tech.10: 111-114.

Singh, N., Somai, B. M., and Pillay, D. 2005. Molecular profiling demonstrates limited diversity amongst geographically separate strains of *Sporisorium scitaminea*, *FEMS* Microbiology Letters. 247: 7–15.

Singh, R. P., and Agnihotri, V. P. 1987. Thermotherapy of sugarcane for disease control. In, *Review of Tropical Plant pathology*. (Eds. S.P. Raychaudhuri and J.P. Verma) Today and Tomorrow's Printers & Pub. New Delhi, vol. IV. 305-329.

Singh, V., Singh, P. N., Yadav, R. L., Awasthi, S. K., Joshi, B. B., Singh, R. K., Lal, R. J., and Duttamajumder, S. K. 2010. Increasing the efficacy of *Trichoderma harzianum* for nutrient uptake and control of red rot in sugarcane. J. Hortic. and Forestry. 2: 66-71.

Spegazzini, C. 1896. Hongos de la cana de azucar. Rev.Fac. Agron. y Vet. 2: 227-258.

Sreeramulu, T., Vittal, B. P. R. 1970. Periodicity in the uredospore content of air within and above a sugarcane field. Journal of the Indian Botany Society.50: 39-44.

Srinivasan, K. V., Chenulu, V. V. 1956. A preliminary study of the reaction of *Saccharum spontaneum* variants to red rot, smut, rust and mosaic. Proceedings International Society of Sugar Cane Technologists Congress 9: 1097-1107.

Srinivasan, K. V. 1964. Some observations on sugarcane wilt. Jour. Ind. Bot. Soc. 43: 397-408.

Srinivasan, K. V. 1965. Towards the ideal of red rot resistance - ends and means. Proc. Int. Soc. Sugar Cane Technol. 12: 1108-1117.

Srinivasan, K. V. 1971. Hot water treatment for disease control.Sugarcane Pathol. Newsl. 6: 46.

Srivastava, S., Singh, V., Gupta, P. S., Sinha, O. K., and Baitha, A. 2006. Nested PCR assay for detection of sugarcane grassy shoot phytoplasma in the leafhopper vector *Deltocephalus vulgaris*: a first report. Plant Pathol. 55: 25-28.

Steindl, D. R. L. 1950. Ratoon stunting disease. Proc. ISSCT.7: 457-465.

Steindl, D. R. L. 1949. Q 28 disease. Cane Gr. Quart. Bull.12: 191-193.

Steindl, D. R. L. 1961. Ratoon stunting disease. In, *Sugarcane Disease of the World*. vol. I. (Eds.) Martin, J.P., Abbott, E.V. and Hughes, C.G., Elsevier Publishing Co., Amsterdam. 433-459.

Subramaniam, L. S., and Chona, B. L. 1938. Notes on *Cephalosporium sacchari* (causal organism of sugarcane wilt). Indian Jour. Agri. Sci. 8: 189-190.

Suman, A., Lal, S., Shasany, A. K., Gaur, A. K. and Singh, P. 2005. Molecular assessment of diversity among pathotypes of *Colletotrichum falcatum* prevalent in sub-tropical Indian sugarcane. World J. Microbial Biotechnology. 21: 1135-1140.

Sundaraman, S. 1928. Mosaic disease of sugarcane in south India. Madras Agr. Dept. Bull.2: 5-13.

Sutton, B. C. 1980. The Coelomycetes. CMI, Kew, England. 696.

Suzuki, K., Suzuki, M., Sasaki, J., Park, Y. H., and Komagata, K. 1999. *Leifsonia* gen.

nov. a genus for 2,4-diaminobutyric acid-containing actinomycetes to accommodate *"Corynebacterium aquaticum"* Leifson 1962 and *Clavibacter xyli* subsp.*cynodontis* Davis *et al.* 1984. J. Gen. Appl. Microbiol. 45: 253-262.

Teakle, D. S., Smith, P. M., and Steindl, D. R. L. 1973. Association of a small coryneform bacterium with the ratoon stunting disease of sugarcane. Aust. J. Agri. Res. 24: 869-874.

Teakle, D. S., Shukla, D. D., and Ford, R. E. 1990. *Sugarcane mosaic virus* (revised) No. 343. In: Description of Plant Viruses CMI/Assoc. Appl. Biol., Kew England.

Tosic, M., Ford, R. E. and Rao, G. P. 1994. Differentiation of *Sugarcane mosaic virus, Maize dwarf mosaic virus, Johnsongrass mosaic virus* and *Sorghum mosaic virus* on the basis of differential host reactions. 199-220. In: Current Trends in Sugarcane Pathology (Eds G.P. Rao, A.G. Gillaspie, P.P. Upadhyay and A. Bergamin Flho, V.P. Agnihotri and C.T. Chen), International Book and Periodicals Supply Service, Delhi, India.

Tosic, M., Ford, R. E., Shukla, D. D., and Jilka, J. 1990. Differentiationof sugarcane, maize dwarf, Johnsongrass and Sorghum mosaic virus based on reactions on oat and some sorghum cultivars. Plant Dis. 74: 549-552.

Vasudeva, R. S. 1956. Some diseases of sugarcane newly found in India. F. A. O. U.N. Plant Protect. Bull.4: 129-131.

Viswanathan, R., Balamuralikrishnan, M., and Poongothai, M. 2005. Detection of phytoplasmas causing grassy shoot disease in sugarcane by PCR technique. Sugar Tech. 7: 71-73.

Viswanathan, R., Balamuralikrishnan, M., and Karuppaiah, R. 2008a. Characterization and genetic diversity of sugarcane streak mosaic virus causing mosaic in sugarcane. Virus Genes 36: 553-564.

Viswanathan, R., Balamuralikrishnan, M., and Karuppaiah, R. 2008b. Identification of three genotypes of sugarcane yellow leaf virus causing yellow leaf disease from India and their molecular characterization. Virus Genes. 37: 368-379.

Viswanathan, R, Chinnaraja, C., Karuppaiah, R., Ganesh Kumar, V., Rooba, J. J., and Malathi, P. 2011. Genetic diversity of sugarcane grassy shoot (SCGS)-phytoplasmas causing grassy shoot disease in India. Sugar Tech.13: 220-228.

Viswanathan, R. 2000. Detection of phytoplasmas causing grassy shoot disease in sugarcane by immunofluorescence technique. Indian Phytopathol.53: 475-477.

Viswanathan, R. 2002. Sugarcane yellow leaf syndrome in India: Incidence and effect on yield parameters. Sugar Cane Intern.5: 17-23.

Viswanathan, R., and Balamuralikrishnan, M. 2004. Detection of sugarcane yellow leaf virus, the causal agent of yellow leaf syndrome in sugarcane by DAS-ELISA. Arch Phytopathol Plant Protect. 37: 169-176.

Viswanathan, R., Padmanaban, P., Mohanraj, D., Ramesh Sundar, A., and Premachandran, M. N. 1999. Suspected yellow leaf syndrome in sugarcane. Sugarcane Breeding Institute Newsletter.18 (3): 2-3.

Viswanathan, R., Poongothai, M., and Malathi, P. 2011. Pathogenic and molecular confirmation of *Fusarium sacchari* causing wilt in sugarcane. Sugar Tech. 13: 68-76.

Von Arx, J. A. 1957. Die Arten der Gattung *Colletotrichum* Corda. Phytopath. Z. 29: 413-468.

Von Arx, J. A., and Muller, E. 1954. Die Gattungen der Amerosporen Pyrenomyceten. Beitrage zur Kryptogamenflora der Schweiz.11: 5-434.

Went, F. A. F. C. 1893. Het Rood Snot. Arch. Java Suikerindus.1: 265-282.

Wilbrink, G. 1920. *De gomziekte van het suikerriet, hare oorzaak en hare bestrijding.* Archief Suikerind. Ned. Indie No.33- 34: 1399-1525.

Wismer, C. A. 1961. Pineapple disease. *In*: Sugar-Cane Diseases of the World, Vol. 1. J.P. Martin, E.V. Abbott and C.G. Hughes (Eds), p. 223-245. Amsterdam, the Netherlands, Elsevier Publishing Company.

Wismer, C. A., and Bailey, R. A. 1989. Pineapple disease. *In*: Diseases of Sugarcane. Major Diseases. C. Ricaud, B.T. Egan, A.G. Gillapsie Jr and C.G. Hughes (Eds), 145-155. Amsterdam, The Netherlands, Elsevier Science Publishers

Wismer, C. A., and Ricaud, C. 1969. Methods for testing the resistance of sugarcane to disease, (6) Leaf scald. Sugarcane Pathol. Newsl. 2: 24-25.

Worley, J. F., and Gillaspie, A. G., Jr. 1975. Electron microscopy *in situ* of the bacterium associated with ratoon stunting disease in Sudan grass. Phytopathology. 65: 287-295.

Xu, L., Que, Y., and Chen, R. 2004. Genetic Diversity of *Ustilago scitaminea* in Mainland China. Sugar Tech. 6: 267– 271.

Zhang, L., and Birch, R. G. 1997a. Mechanisms of biocontrol by *Pantoea dispersa* of sugarcane leaf scald disease caused by *Xanthomonas albilineans*. J. appl. Mirobiol.82: 389-398.

Zhang, L., and Birch, R. G. 1997b. The gene for albicidin detoxification from *Pantoea dispersa* encodes an esterase and attenuates pathogenicity of *Xanthomonas albilineans* to sugarcane. Proc. Natl. Acad. Sci. (USA). 94: 9984-9989.

Diseases of Field and Horticultural Crops   *Pages* **166-195**
Editor-in-Chief: **P. Chowdappa**
Published by: **Daya Publishing House, New Delhi**

6

Diseases of Tobacco

S. Roy and S.K. Dam

Tobacco is one of the important cash crops in India earning a sizeable foreign exchange, internal excise revenue and generating employment to many people. It is grown in almost all the states in India, traditionally in black-soils though its cultivation is now extended considerably to light soils. In India tobacco is grown as important commercial crop in an area of 0.45 M ha which accounts for 0.31% of net cultivable area in the country with 750 M production. India stands second in tobacco production and exports in the world.

Cultivated tobacco coming under *Nicotiana tabacum* L. and *N. rustica* L. is susceptible to several fungal, bacterial, viral and nematode diseases. These diseases not only reduce the yield of tobacco but also impair the quality parameters of the cured leaf. In normal years average crop loss due to diseases is estimated to be 5 to 10 per cent. Since tobacco is grown as a monoculture, year after year in the same traditional belt it provides ideal conditions for gradual development and perpetuation of pathogen causing serious disease problems under favourable weather condition during its active growth period. Leaf being the economic end product in tobacco, any blemish due to pathogen results in lowering the market value. Considerable research, both fundamental and applied on some of these diseases has been carried out since last 50 years at the Central Tobacco Research Institute, Rajahmundry, with special emphasis on evolving suitable control measures and resistant varieties to combat them. Details of research work on various aspects of different diseases are compiled in this chapter.

1. Damping off

Damping off is one of the most important diseases of tobacco, especially in nurseries, showing decay of stems of young seedlings. This is responsible for poor stand of seedlings or complete loss of nursery beds. The disease is most common wherever tobacco nurseries are grown, irrespective of soil type, variety, weather parameters, nutritional status of the soil etc. damping off disease has been reported in almost all tobacco growing regions.

1.1. Symptoms: The disease appears in the nurseries any time upto 35 days after sowing. It may appear soon after seedling emergence (pre-emergence damping off) and may be confused with poor germination. The affected areas in the nursery beds appear in small patches, usually circular in shape, gradually enlarge resulting sometimes in complete failure of the nurseries. If such areas are ignored, the disease usually radiates from initial infection points causing large areas, in which nearly all the seedlings are killed, depending on the weather conditions. The disease may appear in seed beds at any stage but maximum damage (even upto 75%) is observed 4 to 6 weeks after sowing under favourable wet weather conditions. In early stage, tiny seedlings seem to disappear due to rotting causing daily reduction in seedling stand. Older seedlings show shriveling and dark brown discolouration of stem at the base and ultimately collapse and topple over. The wet rotting and collapse of seedlings start in circular patches and may extend to the entire bed, if unchecked (Fig. 6.1).

Fig. 6.1. Symptoms of Damping off

The disease attacks the root region or stem region near the soil surface. The infected portion initially becomes water soaked resulting in soft rot. The young seedlings are girdled and because of soft rot, they topple to the ground. Such seedlings may remain green for few days or rot and disappear depending on the weather conditions. Under humid conditions, white cottony growth of the pathogen appears on the infected seedlings gradually dry off and become papery white in colour. These are the important diagnostic characters of this disease. The seedlings that look healthy may collapse by the next day. Under favourable weather conditions, this disease is responsible for as much as 90% death of seedlings. Under normal conditions, losses may vary from 25-75% of seedlings, if left unchecked.

1.2. Causal organism: This disease is mainly caused by a soil-borne facultative parasite, *Pythium aphanidermatum*, belonging to the family Pythiaceae, under class Phycomycetes but other species such as *Pythium myriotylum* also be involved to cause damping off disease in tobacco nurseries (Institute Research Committee Meeting, 2012). This organism is the most common in India and other countries, though several other fungi like *Phytophthora, Sclerotinia, Sclerotium, Rhizoctonia* and other species of *Pythium* have been reported to cause damping off disease in tobacco nurseries (Lucas, 1975). The disease occurs in almost all types of soils but more severe in poorly drained and compact soils. Since *Pythium* grown well at low temperatures, high soil moisture coupled with low temperature of 20-25^{0}C, the disease development is faster. Over-crowding of seedlings and presence of excessive organic matter predispose the seedlings to damping off disease. Same pathogen causes damping off in vegetable nurseries also.

1.3. Epidemiology: *Pythium* spp. in general is universal in almost all types of soils, wherever vegetable or solanaceous nurseries are grown. This fungus is primarily a saprophyte and in the presence of suitable host, it becomes a facultative parasite on the young roots, stems or any decaying organic matter.Low soil pH, low moisture and high temperature above 25° C are normally not congenial for the fungus. The fungus normally survives in the form of oospores or chlamydospores. Exudates from roots of germination seeds or actively growing seedlings stimulate these resting structures, which germinate by germ tube or form zoospores. These zoospores or hyphae infect the young seedlings at the soil surface and the fungus proliferates fast. The secondary spread occurs during wet weather by means of surface or drainage water. The pathogen produces oospores which are released into the soil, when the infected plants decay. Snails and earthworms also help in transmission of the pathogen, though to a lesser extent.

Optimum conditions under which the fungus thrives better were reported to be temperatures below 24^0 C with a high relative humidity of > 85%, soil pH 5.5 to 7.5 and high soil moisture. Such conditions prevail as a result of high seed rate, continuous rain and cloudy weather during nursery season. Under these conditions, the fungus *Pythium* develops too rapidly resulting in sudden and quick spread of the disease (Mathrani and Murthy, 1965).

1.4. Disease management

1.4.1. Cultural and preventive measures: Since this disease is soil borne in nature, careful management of nursery beds is of immense value in management of the disease. Several means of controlling this disease have been tried and it was observed that no single method will give complete control of the disease. Careful integration some of the following measures have to be adopted to manage the disease.

(a) Selection of nursery site: The site for raising the nurseries should preferably be slightly elevated with good drainage. Light soils with good proportion of sand also facilitate good drainage. Sandy to loamy sand soils are ideal for raising the nursery beds. The organic manure should be well decomposed. The nursery site should be changed every year to avoid build up of the inoculums. Otherwise, periodical soil treatment of the nursery site is very important.

(b) Deep summer ploughing: Ploughing the nursery site 3-4 times during summer months with disk, followed by harrowing exposes the fungal propagules to high temperatures, resulting in reducing the inoculums potential. This also helps in reducing other soil-borne pathogens including nematodes, insect pests, weed seeds etc.

(c) Raising the seed beds: The nursery beds should be raised by 15 cm and cut to a convenient size (usually 10 x 1 m) with channels of 0.5 m around, to facilitate adequate drainage.

(d) Rabbing seed beds: Any slow burning agricultural waste liken paddy husk, wheat bhusa, ground nut shells etc. can be spread on the raised nursery beds @ 6-8 kg/sq. m. and burnt slowly. This is having multiple advantages like control of fungal propagules, weeds, nematodes, insect pupae etc. The ash which is rich in potash should be incorporated into the top soil that helps in promoting rapid seedling growth (Krishnamurty *et al.*, 1979).

(e) Optimum seed rate: Seed rate should be 3.0 to 3.5 kg/ha. This rate gives fairly sufficient healthy and strong seedlings and avoids crowding of seedlings. Farmers and commercial nursery growers usually use high seed rate anticipating more number of seedlings which results in over-crowding of seedlings encouraging high humidity, favourable for disease development.

(f) Regulation of watering: Number of watering per day may be regulated depending on the need and weather parameters, so as to avoid excess humidity / dampness on the nursery beds. Excess watering increased the soil moisture to more than 90% and resulted in the incidence of damping-off up to 56.5% (Devaki *et al.*, 1991).

1.4.2. Chemical control: Several fungicides have been tried since last 4 to 5 decades. Drenching nursery beds with Bordeaux mixture @ 1.0% before sowing and after sowing drenching @ 0.4% 4-6 times at 3-4 days intervals was recommended earlier (Mathrani and Murthy, 1965; Pillai and Murthy, 1967). However with the introduction of several new fungicides, they were screened against damping off disease and the results were reported from time to time. Mathrani and Murthy (1965) tried 15 different proprietary copper fungicides at 1.36 or 2.72 g / 1 litre, along with Bordeaux mixture (0.2 and 0.4%). They concluded that drenching or spraying either Bordeaux mixture or any of the readily available copper fungicides for the control of damping off disease. However, Bordeaux mixture had an edge over other fungicides.

Chandwani *et al.* (1973) studied the effect of non-copper fungicides on the control of damping off and reported that soil drenching with brestan, brassical and RH 90 reduced germination, while agallol and brestan were toxic resulting in mortality of the seedlings. Blitane, bisdithane, captan, cuman, mancozeb, zineb, RH 341, Thiram sprayed at 0.125% were as good as Bordeaux mixture in the control of the disease. Brassicol showed poor control of the disease. Metalaxyl, a systemic fungicide has been reported to be very effective against not only *Pythium*, but many other fungi like *Phytophthora, Sclerotinia, Peronospora* etc. This fungicide in different formulations was studied in detail in the last 20 years. Spraying

Ridomil 25 WP twice @ 0.1, 0.2 and 0.3% concentrations showed highly effective in controlling both damping off and leaf blight diseases. Spraying/drenching of Bordeaux mixture (0.4%), Difolatan and Fytolan (0.2%) six times during the susceptible stage of the nursery also showed promise in the control, while Bavistin (0.1%) and neem kernel suspension (1.0%) were least effective (Nagarajan and Reddy, 1980 and 1982).Shenoi and Abdul Wajid (1988) studied the chemical Metalaxyl @ 0.02% drenched before sowing and subsequently drenched at 20 to 30 days after sowing (DAS) or, subsequently drenched with Bordeaux mixture (1.0%) at weekly intervals after every rain received effectively controlled the above two soil-borne Phycomyceteous diseases and thus resulted in increased production of healthy transplants. Shenoi and Abdul Wajid (1992) again suggested schedules of Ridomil MZ 72 WP for overall management of three diseases viz., damping off, blight and black shank in tobacco nurseries, i.e., (i) pre-sowing drench of fungicide to the soil 0.1% concentration with knap-sac sprayer using flat-fan nozzle, to be followed by a foliar spray @ 0.2 % concentration at 30 days after sowing (DAS) and (ii) only foliar sprays commencing early at 20 DAS, to be followed at 35 DAS and at 50 DAS if required @ 0.2% concentration.

New molecules like Folio gold, Ridomil gold @ 0.2% concentration at 25 DAS and 35 DAS showed superiority over other fungicides but at par with Metalaxyl (0.2%) in controlling the damping off disease (Raju and Dam, 2011) and Sectin 0.3% concentration at 25 DAS and 35 DAS showed superiority over other fungicides in controlling the damping off disease (paid trial, 2008 and 2009 nursery season at CTRI, Rajahmundry).Timely application of fungicides is a prime factor in getting control of the disease.

1.4.3. Biological control: Incorporation of inoculums of *Trichoderma harzianum* (prepared in wheat bran saw dust medium) in artificially infested soil significantly reduced damping-off disease. Use of metalaxyl treated seeds coupled with application of *T. harzianum* inoculums gave excellent control of the Damping-off disease under greenhouse conditions (Mukhopadhyay *et al.*, 1986).

Sreeramulu *et al.* (1998) observed in the nursery that the dual inoculation of VA mycorrhiza *Glomus fasciculatum* and *Trichoderma harzianum* was more effective in controlling Damping off and Black shank diseases than the individual inoculation and resulted in better germination count and improved the plant growth parameters in comparison with Blitox 50 WP (0.2%) and Ridomil MZ 72 WP (0.2%).

2. Leaf blight and Black shank

Among the major diseases of tobacco in India, leaf blight and black shank are important which occur both in nursery and field planted crop. It is soil borne and wide spread in Karnataka, Andhra Pradesh, Bihar and Gujarat states. Loss in yield due to this pathogen varies from 2 to 10 per cent annually depending on weather conditions.

2.1. Symptoms: The symptoms of the disease vary with the age of the plant and weather conditions. When young tiny seedlings are affected in the nursery, they rot and die suddenly. Seedlings show blackening of roots and stem at ground level (Fig. 6.2). Under continuous wet weather conditions, large circular to irregular water soaked patches appear on the leaf surface causing leaf blight. Symptoms of black-shank on the transplanted tobacco are seen in the form of blackening of roots and stalk. Blackening of the stalk starts at the base near the soil gradually extends upwards. The leaves turn yellow and the whole plant wilt and die. When the affected stem is split in two halves length wise, the pith is found dried with plate like dark brown discs. During rainy and damp weather, the lower leaves are attacked with pale or dark brown lesions which enlarge rapidly, coalesce and result in blighting.

Fig. 6.2. Symptoms of leaf blight and Black shank

2.2. Causal organisms: This disease is caused by *Phytophthora parasitica* f. sp. *nicotianae* (Breda de Hann).

2.3. Epdemiology: Heavy rain fall after planting, continuous wet weather and high moisture lead to severe incidence. The fungus remains viable in soil for several years and possibly over winters as mycelia in dead tobacco stalks and roots. It spreads through irrigated water or through the air-borne sporangia, splashed during the rain. All types of tobacco viz., Cigarette, Cigar, Natu, Bidi, Chewing, Cheroot and snuff are susceptible to this pathogen in India.

2.4. Disease management

2.4.1. Cultural practices: Deep ploughing in summer to reduce inoculum,raising of seed beds 15 cm high with channels around to provide drainage,rabbing the seed bed before sowing with slow burning, farm waste materials like paddy husk, tobacco stubbles, waste grass,use seed rate @ 3.5 kg/ha only to avoid overcrowding of seedlings and regulating water to avoid excessive dampness on bed surface are the cultural practices to be followed to prevent the disease incidence..

2.4.2. Chemical control: Work at C. T. R. I., Rajahmundry showed that leaf-blight in nursery can be affectively controlled by periodical spraying of Bordeaux mixture 0.4% or Blitox 0.2% or Foltaf 0.2%. Timely spraying of Ridomil MZ 72 WP 0.2% was found to be highly promising Black shank disease in field can be checked by drenching the planting hole with 0.4% Bordeaux mixture or 0.2% foltaf or Blitox. Spraying 0.2% Ridomil MZ 72 WP 4 weeks after planting offers good control. Rouging the affected plants and spot drenching with any of the above fungicides checks the spread of the disease. Spot drenching of Bordeaux mixture has been found beneficial for the control of black shank of tobacco at this research station. Nagarajan *et al.* (1977) also recommended spot drenching with Bordeaux mixture (5: 5: 500) or 0.2% Blitox 50, Cuman or Dithane Z - 78 for the control of black shank of tobacco. Gupta and Patel (1979) observed that Bayer 5072, Blitox 50, Blue Copper 50, Bordeaux mixture (6: 3: 100), Burcop, Copper Sulphate, Fytolan, Macuprax, Miltox, Ziram, Daconil 2787, Difolatan, Dithane M - 45, Lonacol and Rovral exhibited *in vitro* and *in vivo* inhibition against *Phytophthora parasitica* var. *nicotianae* of tobacco.

Nagarajan and Reddy (1980) indicated that the most promising fungicide for the control of both damping off and leaf blight disease was Ridomil 25 WP @ 0.1, 0.2, & 0.3% concentrations with high transplantable & total healthy seedlings. Next best were 0.4% Bordeaux mixture, 0.2% Difolatan and 0.2% Fytolan. Elias *et al.* (1981) studies on control of black shank (*Phytophthora parasitica* var. *nicotianae*) in FCV tobacco nursery involving chemicals viz., MBC, Benlate, Bavistin, Bayleton, Ridomil, Dithane M-45, Brassicol, Thiram, Difolatan + Blitox and Bordeaux mixture and revealed that one spray of Ridomil at 10 days after sowing effectively controlled the disease. Another nursery trial with different formulations, doses and applications of Ridomil showed that pre sowing application either in the form of wettable powder (25 WP) or granules (5G) was superior to spray application of Ridomil (25 WP) 15 days after sowing in disease control.

Shenoi and Abdul Wajid (1988) showed that control of damping off and black shank diseases in tobacco nurseries with a metalaxyl formulation, Apron 35 SD, works excellently when used as drench. The chemical @ 0.02% drenched before sowing and subsequently drenched at 20 to 30 days after sowing (DAS) or, subsequently drenched with Bordeaux mixture at weekly intervals after every rain received effectively controlled the above two soil-borne Phycomyceteous diseases and thus resulted in increased production of healthy transplants. Shenoi and Abdul Wajid (1992) again showed that for the overall management of damping off, blight and black shank diseases in the tobacco nursery, it is ideal to have the combination of pre-sowing drench @ 0.1% concentration with knap-sack sprayer using flat-fan nozzle, and foliar spray schedule with Ridomil MZ 72 WP @ 0.2% concentration at 30 days after sowing.Single application of Ridomil MZ 72 WP in the planting hole along with planting water, @ 0.1 % concentration (150 ml / hole) gave 100% control of black shank upto 35 Days after Transplanting (DAT) and 70% Control upto 65 DAT, the efficacy getting reduced by 100 DAT (Abdul Wajid and Shenoi, 1992).

2.4.3. Sources of resistance: Growing resistant varieties is the best method of controlling the disease. At CTRI resistant / tolerant donors were identified in 1) Beinhart 1000-1 (Cigar); 2) *N. plumbaginifolia* (wild *Nicotiana* species)- resistant; and 30 Mc Nair-12 (FCV)- tolerant to race '0' which is predominant in East and West Godavari Districts of Andhra Pradesh. Resistance found in the above donors was transferred to commonly cultivated FCV, Barley and Chewing varieties of tobacco. These are being screened for yield and quality parameters and are in advanced stages of release.

Out of the 122 entries screened, only seven viz. Beinhart 1000-1, Coker 411, F 180, F 210, MC 1610, Reams 64 and Va 770 showed resistant reaction to all the three races of *P. parasitica* var. *nicotianae*. Georgia 1469 and Waimea exhibited moderate susceptibility to all three races. While 72 entries exhibited complete susceptibility to these races, certain cultivars showed variable reactions. Some of the entries like Coker varieties, FCH lines. NC 73. Speight G 28 etc. having resistance to the new race only (Abdul Wajid *et al.*, 1987).

2.4.4. Biological control: Sreeramulu *et al.* (1998) observed in the nursery that the dual inoculation of VA mycorrhiza *Glomus fasciculatum* and *Trichoderma harzianum* was more effective in controlling Damping off and Black shank diseases than the individual inoculation and resulted in better germination count and improved the plant growth parameters in comparison with Blitox 50 WP (0.2%) and Ridomil MZ 72 WP (0.2%). Patel and Patel (1999) also observed that *Trichoderma harzianum* was highly antagonistic to race 'o' of *P. paracitica* var. *nicotianae* under *in vitro* conditions.

3. Brown spot

Brown spot of tobacco caused by *Alternaria alternata* (Fries) Keissler has wide host range. The target pathogen, attacks all kinds of tobacco i.e. FCV and non-FCV tobacco in India, and elsewhere in the world. Brown spot disease of tobacco, in general, is a disease of senescence.

3.1. Symptoms: The first indication of infection on the older leaves of the plant is small water- soaked lesions which enlarge quickly. As the spots enlarge, the centers die and become brown, leaving a sharp line of demarcation between diseased and healthy tissue. On the lower leaves, the brown spot lesions with concentric rings are mostly circular and usually range from 1-3 cm or more in diameter (Fig. 6.3). On the upper leaves, the spots are usually smaller and range up to 1.5 cm in diameter. In warm weather (30^0 C) under high humidity, the spots enlarge, 1-3 cm in diameter, centres are necrosed and turn brown with characteristic marking giving target board appearance with definite outline. In severe infection spot enlarge, coalesce and damage large areas making leaf dark brown, aged and worthless.

Fig. 6.3. Brown spot symptoms on the leaf

3.2. Variability of the pathogen: In *Motihari* tobacco (*N. rustica*), the spots produced by isolate Aa 1 were large (1.2 to 2.7 cm), round, dark brown in colour having concentric rings with prominent yellow halo. The spot produced by Aa 2 isolates were medium (0.3 to 0.5 cm) size having concentric rings having less yellowing and halo. Maximum coverage of leaf area was observed for Aa 1 isolate under field conditions in *Motihari* tobacco (Dam *et al.*, 2010a). Three different isolates showed differences in their morphological variability and growth character (Dam *et al.*, 2010a). The isolates exhibiting higher degree of virulence had slower growth rate (10 mm/day) as compared to less virulent ones (11 mm/day). The observation is in conformity with Dong and Wang (1990) who observed that more virulent strains of *A. alternata* showed slower growth of colony, vigorous extension of aerial hyphae and less sporulation. Maximum conidial length (48) and beak length (12.07) were measured in isolate Aa 3 while in Aa 1 conidial length (26.0) and beak length (7.6) was lower. The study (Dam *et al.*, 2010a) indicated that higher the ratio of beak length to conidial length, the virulence of *A. alternata* was more on *Motihari* tobacco. The virulence pattern of three isolates (Aa 1, Aa 2 & Aa 3) of the pathogen, *Alternaria alternata* collected from brown spot affected *Motihari* tobacco in North Bengal were inoculated on 35 day old plants of var. Dharla with spore suspension (10^5 spores/ml) in atomizer. Highest number of spots was observed on isolate Aa 3 but maximum necrotic area was observed in Aa 1 in observations made after 6 days of inoculation. From the perusal of the data, it is evident that isolate Aa 1 is more virulent and aggressive when compared to other two isolates of *Motihari* tobacco under *terai* agro-ecological region of West Bengal (Dam *et al.*, 2010a). Dong and Wang (1990) obtained twenty strains from 10 different locations in China. The more virulent strains showed slow growth of colony, vigorous extension of aerial hyphe and weak sporulation, while it was inverse in case of weakly virulent strains.

3.3. Epidemiology: The studies (Dam, 2008., Pers. communication) carried out in *terai* zone of West Bengal indicated that the appearance of the disease during first week of February in the field and its intensity increased gradually up to the maturity of the crop. In the crop seasons 2005-06, the disease score was highest as per observations recorded during 11-18th March, 2006. Therefore, the weather factors like temperature (25.1^{0}C to 13.4^{0}C), relative humidity % (96 to 54), total rainfall (9.3 mm) and bright sunshine hours (7.4) prevalent during this period were favourable for rapid build-up and spread of the disease.

3.4. Disease management

3.4.1. Cultural practices: Monga (1990) reported that intensity of disease in different date of planting from northern part of West Bengal in India. Out of 4 different dates of planting, brown spot index was found to be lowest on 20th November compared to 6th November, 30th November and 15th December. Compared to other planting dates, the planting on 20th November registered higher productivity of quality leaf. Drastic yield reduction in late planting cigar wrapper (*Nicotiana tabacum*) tobacco in North Bengal has also been earlier observed by Srivastava and Subba Rao (1984)

3.4.2. Host resistance: Nagarajan *et al.* (1984) tested 740 collections of *N. tabacum* and *N. rustica* against brown spot pathogen and none were found to be resistant although a few had shown slightly less degree of susceptibility. Monga and Dobhal (1987) tested 25 cigar wrapper tobacco lines against brown spot of tobacco and reported that no line was highly resistant. However, sixteen lines were in the resistant category while three and six lines fell in the susceptible and highly susceptible categories. Monga and Dobhal (1989) found that out of 121 hookah and chewing tobacco lines, most fell in moderately resistant to moderately susceptible groups. No line was under highly resistant category. Among the various *Nicotiana* species, only *N. debneyi* has shown immunity to brown spot, whereas *N. exiqua, N. glutinosa, N. longifolia, N. nesophila* and *N. plumibaginifolia* have shown resistance (Reddy *et al.*, 1976). Nagarajan and Shenoi (1998) observed that there were no resistant varieties of *N. tabacum* while resistant donor is available in wild *Nicotiana spp.* Three entries in accordance to rating scale developed by Shenoi *et al.* (2002) viz. Vaishali Special, PT-76 and Gandak Bahar were found to be resistant out of a total of 60 germplasm accessions of *Jati* tobacco (*N. tabacum*) screened for resistance to brown spot in North Bengal (Ann. Scientific Rept. CTRI. 2007-08)

3.4.3. Chemical control: Chari and Nagarajan (1994) suggested Dithane M-45 and Foltaf for brown spot control. Monga (1991) reported two sprays, one at disease appearance and another after ten days either with Thiram at 0.2% a.i. or Mancozeb at 0.2% a.i. were promising in reducing brown spot of *Motihari* tobacco with higher monetary returns. The fungicides, Tilt, Score, Foltaf and Indofil M-45 reduced PDI and per cent leaves with severe damage besides improving the yield. Mahtabi *et al.* (2001) found Propiconazole (Tilt), Tabuconazole (Folicur) and Mancozeb fungicides to be promising of brown spot disease of tobacco in the field.

In vivo evaluation of chemical fungicides showed that Baycor, Bayleton, Beam, Score and Tilt were found promising against brown spot of FCV tobacco (Nagarajan and Shenoi, 1998). Shafik and Taha (1984) found that Dithane M-45 when sprayed @ 2.0 g / l water proved effective in controlling the disease. Dam *et al.* (2010c) indicated that pooled means of all the fungicidal treatments (screened from *in vitro* study) significantly reduced the disease index as compared to control. Results of field evaluation indicate that two sprays, one at disease appearance and another at 15 days with propiconazole @ 0.1% or mancozeb @ 0.25% were promising in the management of *Motihari* tobacco with higher monetary returns (Dam *et al.*, 2010c). In FCV tobacco, Nagarajan and Shenoi (1998) reported the effectiveness of hexaconazole, propiconazole and beleton for the management of brown spot disease of tobacco in Andhra Pradesh.

3.4.4. Biological control: Dam *et al.* (2010b) have observed that both *Trichoderma viride* and *Pseudomonas aeruginosa* restricted the growth of *Alternaria alternata in vitro* though in the former, the inhibitory effect was more. It was observed that soil application + two foliar sprays, one at disease appearance and another after 15 days with *Trichoderma + Pseudomonas + Azotobactor* (B: C ratio 1: 1.79) was at par with two foliar sprays with propiconazole (B: C ratio 1: 1.74) in reducing brown spot disease of *Motihari* tobacco. Pande (1985) reported that three isolates of *Trichoderma viride* retarded growth of *Alternaria alternata*.

4. Sore shin

Rhizoctonia diseases occur throughout the world. They cause losses on almost all vegetables and flowers, several field crops, turf grasses, and even perennial ornamentals, shrubs, and trees. Symptoms may vary somewhat on the different crops, with the stage of growth at which the plant becomes infected, with the prevailing environmental conditions. The most common symptoms on most plants are damping-off of seedlings and root rot, stem rot, or stem canker of growing and grown plant.

4.1. Symptoms: A seedling stem cancer, known as sore shin, is common and destructive in tobacco. Sore shin lesions appear as reddish-brown, sunken cankers that range from narrow to completely girdling the stem near the soil line (Fig. 6.4). Injury is more common during warm weather. As soil temperature rises later in the season, affected plants may show partial recovery due to new root growth. On transplants, the dark brown discolouration of the stalk at or near the soil line extending upward for several centimeters is a characteristic symptom of sore shin. Growth of *Rhizoctonia solani* is stimulated by exudates from germinating seedlings and actively growing roots. Once contact has been made *R. solani* can penetrate plant tissue by (i) direct penetration, (ii) forming hyphal aggregations on the roots which cause discolouration and death of root cells with subsequent penetration through this dead tissue, and direct invasion through natural cracks and wounds.

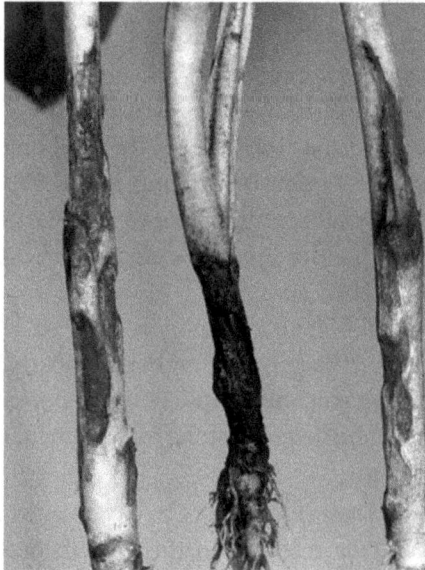

Fig. 6.4. Sore shin lesions on infected plant

4.2. Disease management: Disinfestations of seed-bed soils with steam or methyl bromide helps prevent sore shin in the plant bed. Hartill (1968) in Rhodesia tested Terraclor, Chloroneb, Captan, Folpet, Difolatan, Thiram, Plantvax and several other chemicals for control by applying them in the transplant water. Those chemicals that gave good control unfortunately were phytotoxic to the tobacco nursery different chemicals viz., propiconazole, thiophanate methyl, carbendazim, chlorothalonil and copper hydroxide were tested and observed that propiconazole (0.1%), a triazole compound showed superiority in controlling the disease over all other fungicides. The next best chemical identified was carbendazim (CTRI, 2010).

5. Bacterial Wilt

Bacterial wilt is another important disease of both *Jati* and *Motihari* tobacco in *terai* region of North Bengal. Bacterial wilt of tobacco is known to be caused by *Ralstonia solanacearum* and was coined 'Granville wilt' as it was first recognized in FCV tobacco in Granville County during 1880. Except for the prevalence of this disease in *terai* agro-ecological region of West Bengal, the disease has not been reported from FCV or non-FCV tobacco tobacco growing states of India.

5.1. Symptoms: The disease initiates right from nursery stage and in adult plants. In the field the first symptom of the disease is drooping of 1-2 leaves during day which may recover during evening (Fig. 6.5). Only half of the affected leaves become flaccid, a characteristic symptom of bacterial wilt of tobacco. On slow progression of the disease, the affected leaves turn light green and may gradually turn yellow, midribs and veins get flaccid and large leaves may droop in an umbrella like fashion.

Fig. 6.5. Drooping of tobacco leaves due to Bacterial wilt

5.2. Epidemiology: Survey of the disease was carried out for *Jati* (*N. tabacum*) and *Motihari* tobacco (*N. rustica*) in villages in and around Dinhata sub-division of Cooch Behar district for the crop season 2003-04 and 2004-05 (CTRI, Ann. Rept. 2004-05 and 2005-06). The survey of the villages revealed that the plants aged 35-60 days were found to be highly susceptible to bacterial wilt infection in the field following high temperature (25-30^{0}C) and rainfall. Few water logged plots at CTRI

Research Station farm during 2003-04 due to lower elevation were predisposed to rapid build-up and spread of the disease. Inspite of latent infection in plants, the cause attributed to non-expression of the disease in some plants was due to adult plant resistance. However, impending danger always lay ahead in survival of the pathogen in soil, weed hosts and solanaceous plants (CTRI, 2004).

The studies on survey, epidemiology and management of *Motihari* for the villages in and around tobacco revealed that i) increase in temperature from 20-30⁰C favoured wilt development, ii) at lower temp. (< 20⁰ C) disease remain latent and symptom expression did not occur, iii) high soil moisture, temp. in the range of 20-30⁰ C and high relative humidity (> 80) in atmosphere were found to be highly favourable predisposing factors for rapid build- up and spread of the disease (CTRI, 2005).

5.3. Disease management: Cultural management of the disease includes keeping the fields weed free, not to throw the uprooted diseased plants in the field but by burning outside the field and deeply burying it. The recommended cultural practice from Central Tobacco Research Institute Research Station, Dinhata (W.B.) against bacterial wilt of tobacco is application of lime @ 1 tonne/ha followed by ploughing and laddering of field and to keep it fallow for 20-30 days. Incorporation of dhaincha as green manure has also found to be beneficial (CTRI, 2011). There is every likelihood that owing to extensive cultivation of bacterial wilt susceptible crops over the last 2-3 decades in *terai* region of North Bengal like potato, brinjal, chilies, tomato, jute and mesta have created enough scope for the progressive build-up and spread of the pathogen in soil and crop debris. This is the reason why the pathogen could establish itself as serious endemic disease of economic importance in *terai* region of North Bengal (Roy *et al.*, 2008).

6. Hollow stalk

Hollow stalk of *Motihari* tobacco caused by *Erwinia carotovora* sub sp. *carotovora* (syn. *Pectobacterium carotovorum* sub sp. *carotovorum*) poses serious threat in *terai* region of North Bengal as the disease is endemic in nature and the losses on an average caused to the crop range from 0.5 - 30 %.). In case of higher latent infection in plants, the entire crop can be wiped out in the event of high rainfall and water logging of soils (CTRI, 2004 and 2005). In FCV tobacco (*N. tabacum*) hollow stalk disease has been reported from U.S.A, Canada and China affecting mainly stem (black leg) and cured leaves in stores. From India hollow stalk has been reported from *N. rustica* type of tobacco only and to a lesser extent in *Jati* tobacco of *N. tabacum* type from *terai* region of North Bengal (Ann. Rept. 2004-05; 2005- 06; Roy *et al.*, 2008).

Soft rot Erwinias are one of the most important groups of phytopathogens that are destructive to a wide variety of plants both in culture and in storage (Perombelon and Kelman, 1980). Members of this group such as *Erwinia (Pectobacterium) carotovora* sub sp. *carotovora* can infect a multitude of plant species under right environmental conditions, particularly high humidity (Lucas, 1975; Perombelon and Kelman, 1980; Kotoujansky, 1987). *Erwinia carotovora* sub sp. *carotovora* does not contain avirulence genes nor appear to cause a hypersensitive reaction in plants, nor has any genetically defined resistance to the pathogen is described (Collmer and Keen, 1986; Kotoujansky, 1987; Keen, 1990). Exogenously added salicylic acid (SA) has been found to induce resistance of the tobacco plants to *Erwinia carotovora* sub sp. *carotovora* (Palva *et al.*, 1994). As per reports in literature that susceptibility of a given tissue to maceration is not due to enzyme specificity of soft rot pathogen but depends upon such factors as the amount of calcium pectate present, type of pectic enzyme inducing maceration, association of non- pectic poloymers, concentration of pectic enzyme inhibitors (phenolic compounds) and degree of tissue hydration (Lucas, 1975; Turner and Bateman, 1968).

6.1. Symptoms: Hollow stalk in the field usually appears first at topping and suckering time. The disease may appear at any time following stem injury but it is commonly observed after 35-40 days during topping operations. Soon after infection a rapid browning of the pith develops followed by soft rot and eventual collapse of the tissue (Fig. 6.6). The top leaves wilt and the infection spreads downwards. Hollow stalk phase of the disease characterized by hollowness of the stem following soft rot which has been reported from FCV (*N. tabacum*) tobacco (Lucas, 1975) as well as from *N. rustica* tobacco (*Motihari* tobacco) and to a very limited extent in *Jati* tobacco (*N. tabacum*) from terai region of West Bengal. Black leg phase of the disease characterized by formation of blacks stripes or bands girdling the stalk and cured leaves have been reported from USA, Canada (Lucas, 1975) and China (Xia and Mo, 2007) but not from India.

Fig. 6.6. Hollow stalk symptoms

6.2. Disease management: Utmost care should be exercised during field operations not to use unsterilized sickle /knife lest the disease is spread in healthy plants. Field operations like desuckering and topping should be avoided during damp and cloudy weather. Effective prophylactic management of the disease has been recommended as slurry/paste spot application of Bordeaux mixture or Blitox at topped end and desuckered points of leaf (CTRI, 2005). After the onset of disease, the most effective management strategy lies in providing deep incision with sterilized knife at the stem base or desuckered points of leaf followed by paste application of Bordeaux mixture or copper oxy choride. Keeping into consideration the non-occurrence of hollow stalk disease in FCV tobacco (*N. tabacum*) under Indian conditions and very low incidence (0.1- 1%) of the disease in *Jati* tobacco (*N. tabacum*), it was felt worthwhile to search for source of resistance in accessions of *N. rustica* type being maintained in the germplasm bank of Central Tobacco Research Institute Research Station, Dinhata. Out of a total of 185, *N. rustica* germplasm accessions screened for resistance to hollow stalk disease, only two accessions viz. White Pathar and Bengthuli exhibited resistant disease reaction (< or upto 2 cm linear soft rot of pith on artificial inoculation) (CTRI, 2010).

7. Tobacco Mosaic Virus

Tobacco mosaic virus (TMV) is a single-stranded, positive-sense, rod shaped RNA virus that is worldwide in distribution and is found in all countries where tobacco is grown. Tobacco mosaic virus has a wide host range, including 199 species from 30 families. However, other *Solanaceous* hosts are the only important sources of inoculum for the tobacco crop (Shew and Lucas, 1991). Tobacco mosaic virus reduces cured tobacco yield, quality, and average price. Because of its adverse effects on tobacco and the high value of the crop, TMV is an economically important disease. Since then TMV has been acknowledged as preferred didactic and symbolic model to illuminate the essential features that define virus. Today, TMV is used as a tool to study host –pathogen interactions and cellular trafficking, and as technology to express valuable pharmaceutical proteins in tobacco (Scholthof, 2004).

Beijernick (1898) contended that the filterable agent of tobacco mosaic disease was neither bacterium nor any corpuscular body, but rather that was 'contagium vivum fluidum'. The path-breaking work on tobacco mosaic virus was that it could pass through a filter capable of retaining bacteria (Ivanowski, 1892; Zaitlin, 1998). TMV was the first to be chemically purified (Stanley, 1935; Bawden *et al.*, 1936), to be detected in centrifuge and in electrophoresis apparatus and to be visualized in an electron microscope (Kausche *et al.*, 1939). The TMV coat protein (CP) was the first virus protein to be sequenced (Anderer *et al.*, 1960; Tsugita *et al.*, 1960) and TMV's particle structure was the first to be elucidated in atomic detail (Bloomer *et al.*, 1978; Namba *et al.*, 1989).

7.1. Symptoms: Characteristic symptoms include an irregular pattern of dark green and light green leaf areas intermingled, stunted plant growth, leaf malformation, and mosaic burn. This mosaic pattern is the result of intermingled yellow-green mottling on the foliage of the tobacco plant. Young leaves of infected plants are often malformed and may show leaf puckering, or wrinkling of the leaf tissue (Fig. 6.7). Nearly mature leaves that are infected may show "mosaic burn". Mosaic burn is characterized by large, irregular, burned or necrotic areas on the foliage that can cause extensive damage to the tobacco crop (Lucas, 1975).

7.2. Epidemiology: Tobacco mosaic virus is sap transmissible and is one of the most infectious plant viruses. Primary sources of infection may include perennial weeds and infected crop debris in the soil. Primary infections are usually only

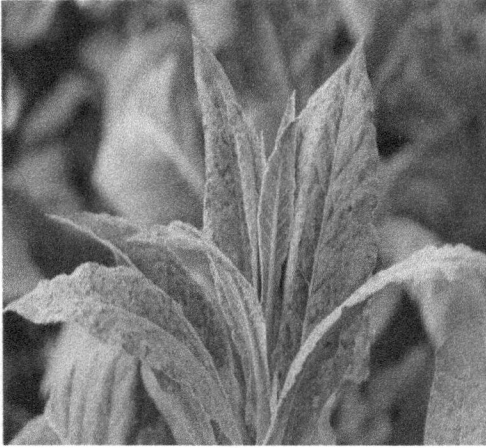

Fig. 6.7. Tobacco leaves infected with TMV

responsible for a small proportion of infected plants in a field. Gooding (1969) reported that only about 10% of TMV infection in North Carolina is the result of primary infection by crop debris in the soil. Therefore, secondary spread of the virus accounts for most TMV infections. The virus is mainly spread by contact. Secondary infections may occur when a worker's hands, clothing, or equipment, previously in contact with an infected plant, comes in contact with a healthy plant. Cultivation practices such as hoeing, topping, spraying pesticides and insecticides, and other field operations can also spread the disease (Lucas, 1975). Another source of new infection is air-dried tobacco. Workers who smoke cigarettes, or who use chewing tobacco or snuff containing air-cured tobacco, may introduce the virus to the plants. This is especially common when the worker is performing plant bed operations. Leaf symptoms do not usually have time to develop before transplanting, so mosaic is rarely seen in the plant beds. However, if TMV is present in the plant beds, then it could easily be spread at transplanting time when the infected plants come in contact with workers and equipment.

7.3. Disease management

7.3.1. Preventative and Eradicative: Attempts to control the spread of TMV have included both preventative and eradicative measures in cultural practices as well as host plant resistance. According to Lucas (1975), "The most efficient way to control mosaic is to keep the crop TMV-free". Some preventative measures include keeping hands, clothing, and equipment free of TMV through rigorous sanitation; rotating crops to prevent infection from perennial weeds and other crop debris; plant bed sanitation, which would help to reduce the primary infections; and using cultivation practices that do not call for much handling of the plants, for instance, using herbicides as weed killers in the field instead of weeding by hand, in order to reduce secondary spread of the disease. This would reduce the incidence of infection by TMV by limiting contact between an infected plant and a healthy plant through equipment and laborers, thus reducing secondary spread via worker contact. Hare and Lucas (1959) found that milk acted as an antiviral agent. When applied 24 h before transplanting, milk could be used to control the mechanical spread of TMV. Two fungicides, dodine and glyodin, have also been found to reduce the amount of damage caused by TMV when applied in the field, by acting as foliar protectants for controlling contact transmission of the virus (Chow and Rodgers, 1973). However, neither of these fungicides is used practically in controlling transmission of the virus.

Nagarajan and Murty (1975) observed that out of 25 plant species screened for potential inhibitions of Tobacco mosaic virus (TMV) infection, plant extract of *Clerodendron fragrans, Agava Americana, Bougainvillaea spectabilis, Capsicum annuum, Coccinia indica, Basella alba* and *Dianthus caryophyllus* contained potent indictors, the per centage of inhibition ranging from 92 to 99. Patel and Patel (1979) found that among 12 plant species tested, leaf extracts of *Ailenthus excellsa.* L. and *Clerodendron inerme* L. were found to possess potent inhibitors of tobacco mosaic virus (TMV) infection, the per centage of inhibition being 100 on both test plants, *Nicotiana glutinosa* L. and *N. tabacum* L. (VFC culture TMVRR. 2a). Leaf extracts of *Gynandropsis pentaphylla* DC, *Pithecolobiuln dulce* Benth and *Solanum xanthocarpum* S & W were also found to contain relatively strong TMV inhibitors. Patel (1981) again observed the presence of possible mechanism of inhibition of tobacco mosaic virus (TMV) infection in the leaf extract of *Ailenthus excelsa* L. Patel (1980) also found that the leaf extract of *Parkinsonia aculeate* possesses strong (98.5%) inhibitor of TMV infection. The leaf extracts of *Coccinia indica* and *Melia ozadirachta* were found to inhibit TMV infection to a considerable extent (82.4 to 88.3%).

7.3.2. Host resistance: Modified cultural practices are only one way to limit the spread and reduce the damage caused by TMV. Another way to limit the damage caused by TMV is to incorporate resistance to the pathogen into the host. Host resistance is an effective means of controlling most pathogens, in many cases host resistance eliminates or limits dependence on pesticides for disease control. This makes host resistance both economical and environmentally sound. Apple *et al.* (1963) found that host resistance derived from *N. glutinosa* reduced losses in yield and value to TMV more than did milk treatment of seedlings.

Control of the spread of TMV in the tobacco crop includes such measures as crop rotation and sanitation practices as well as host plant resistance. The first resistance to TMV incorporated into tobacco came from the Colombian variety 'Ambalema' (Nolla and Roque, 1933). This resistance was controlled by two single recessive genes as well as some modifying genes. Once resistance was incorporated into the tobacco genome, the plants were of poor agronomic quality. Holmes (1938) used inter-specific hybridization to incorporate resistance to TMV from *Nicotiana glutinosa* into the Turkish tobacco variety Samsoun. This resistance was governed by a single dominant gene and was designated the 'N' gene. The 'N' gene causes a necrotic reaction when the plant is inoculated with TMV.

Resistance derived from *Nicotiana glutinosa* ('N' gene) was successfully incorporated into burley tobacco. However, when the 'N' gene was incorporated into flue-cured tobacco, Chaplin *et al.* (1961) reported reductions in cured tobacco yield and value, and Chaplin and Mann (1978) concluded that the N factor may be inherently difficult to disassociate from the adverse yield and quality characteristics in flue-cured tobacco. The N gene was identified molecularly and sequenced and is believed to be a cytoplasmic protein involved in a signal transduction pathway. It is hypothesized that the N gene acts as a cytoplasmic receptor to the TMV gene product and elicits a defensive hypersensitive response in the plant that results in a necrotic or local lesion reaction. A family of related genes clustered around the N locus was identified by restriction fragment length polymorphism (RFLP) (Whitham *et al.*, 1994).

Chaplin and Gooding (1969) screened 907 tobacco introductions (TI) for reaction to TMV. They found that there are three possible reactions to infection with TMV in the *Nicotiana* genus, a necrotic or local lesion reaction, a reaction that displays no visual symptoms, and a susceptible reaction that exhibits the characteristic mosaic response. Thirty-six TIs were identified as resistant to infection with TMV. Eleven of the lines exhibited the local lesion response and 25 lines exhibited no visual symptoms upon inoculation with the virus. Resistance to TMV derived from *N. glutinosa* is characterized by a local lesion reaction. It has not been determined if the eleven TIs that exhibit the local lesion reaction possess the same N factor for resistance as that derived from *N. glutinosa*. If the gene for resistance in these 11 TIs is found to be the same as the N gene, then it is probable that the same undesirable traits that cause reduced yield and value in flue-cured tobacco will be linked to the gene in these TI lines. If that is the case, other TIs that display no visual symptoms upon infection with TMV should be assessed as possible new sources of resistance to TMV. Tobacco introductions could provide a useful and alternative source of resistance to TMV.Out of 34 mosaic resistant Tobacco (*Nicotiana tabacum* 1.) introductions (T.I.) were evaluated under flue-cured cultural conditions for each of two years. Twenty-four of the lines were symptomless carriers of the virus and ten had the local - lesion type reaction. Tobacco Introductions in the symptomless group produced higher leaf yields but lower leaf quality than the local lesion group (Gwynn, 1977).

Reddy and Nagarajan (1981) observed that among the 268 collections of *N. tabacum* and *N. rustica* screened 8 FCV, 10 Burley, 10 Air cured, 3 *rustica*, 2 each of hookah and cigar and 1 each of natu and snuff types were found resistant, the resistance being *glutinosa* type, showing typical hypersensitive local lesion reaction. Some of the important resistant varieties are N.C. 73, MRS. 1,2,3,4 Va- 770, Ky-21, Burley-21; Burley-49, Dubeck 565, Trapizond 174, SamSun 47(10 and DG-3. Among the 39 *Nicotiana* species, ten species viz. *N. benthamiana*, *N. glutinosa*, *N. goodspeedii*, *N. ingulba*, *N. nesophila*, *N. repanda*, *N. simulens*, *N. solanifolia*, *N. stocktonii* and *N. undulata* were resistant to TMV. Sastri *et al.* (1981) found that three Tobacco mosaic Virus (*Nicotiana* 1) resistant flue-cured tobacco variety viz. TMVRR-1, 2 and 3 were superior over their susceptible counter parts viz. Virginia gold, CTRI Spl. and Kanakaprabha under heavy infection condition. Shenoi *et al.* (1992) screened 216 *Nicotiana* germplasm/advanced breeding lines for resistance and found that out of which 36 entries were resistant to TMV showing hypersensitive *glutinosa* reaction, and one as tolerant to TMV.

7.3.3. Management with botanicals: Nagarajan and Murty (1977) screened 27 leaf extract of several plants and found that among them *Acacia Arabica*, *Boerhaavia* *diffusa, Lawsonia inermis, Nerium odourum, Peltophorum ferrugenium, Pithecolobium duice* and *Prosopis specigera* where the TMV inhibition was more than 80 per cent. Stem and /or twig extract of *B. diffusa, P. dulce, Telanthera ficoidea, Andrographis paniculata* and *A. echioides* also inhibited the virus infection to a greater extent. Found that application of Viroson 2% (27.7% disease incidence) followed by Bougainvillea leaf extract 5% (30.2% incidence) and neem 1500 ppm (31.8%) reduced the TMV incidence. However, higher plant height, leaf length and leaf width were recorded in Viroson, neem 1500 ppm and cow urine application indicating their role in triggering host defense and plant growth promotion.

8. Tobacco Leaf Curl Virus (TLCV)

This disease is caused by a virus belonging to 'Gemini' virus group. The virus is transmitted by insect-vector whitefly, *Bemisia tabaci* Gennadius (Lucas, 1975). It is neither seed borne nor transmitted mechanically. Intensity and spread of the disease depend on plant age at the time of infection and build-up whiteflies. Whiteflies are more abundant and active in relatively dry season and as a result, leaf curl is more prevalent during dry weather (Lucas, 1975). Whereas high rainfall, more number of rainy days and higher relative humidity, accompanied by less bright sunshine hours favoured low leaf curl incidence (Monga and Tripathi, 1988). The virus remains persistent in the whitefly vector. In severe infection, losses vary from 60-70% as infected leaves do not cure well. In mild type of infection, the yield loss is very marginal.

Valand and Muniyappa (1992) reported the incidence of disease caused by tobacco leaf curl geminivirus (TbLCV) in ten tobacco growing areas of India ranged from 1.2% to 77%. The highest incidence of disease was observed in Andhra Pradesh (77%) followed by Gujarat (59%), Karnataka (17%), Bihar (11.6%) and West Bengal (5.4%). Under field conditions, an average of 32 adult whiteflies (*Bemisia tabaci*) per plant was recorded in Andhra Pradesh followed by Gujarat (20), Karnataka (12), Bihar (8) and West Bengal (5). In sequential sowings at Bangalore, all the plants were infected within 90 days in plots planted from February to June. Infection in plots planted later was progressively less. There was a positive correlation between whitefly catches and the final incidence of leaf curl disease in plantings. TbLCV was transmitted by *Bemisia tabaci* to 35 plant species, including *Beta vulgaris, Capsicum annuum, Carica papaya, Cymopsis tetragonoloba, Lycopersicon esculentum, Sesamum indicum, Phaseolus vulgaris* and *Petunia hybrida*. Forty five TbLCV isolates from different parts of India induced four distinct types of symptoms on tobacco cultivars Samsun and Anand 119. Group 1 isolates caused severe curling and cup-shaped enations; group II isolates induced pale green leaves, pit-like depressions and thorny enations: group III isolates caused leathery leaves, narrow and tiny protruding enations between the veins, and group IV isolates induced irregular thickening and swelling of veins and green flap-like enations on veins. Nylon net covers protected tobacco seedlings in nursery beds for 45 days. *Ricinus communis* and *Helianthus annuus* sown around the tobacco nursery bed as barrier crops attracted adult whiteflies and decreased the number found on tobacco.

8.1. Symptoms: In tobacco, TLCV causes stunting; the stems are twisted and the leaves are small, curled, twisted and puckered, often with green thickenings or enations along the veins (Fig. 6.8).

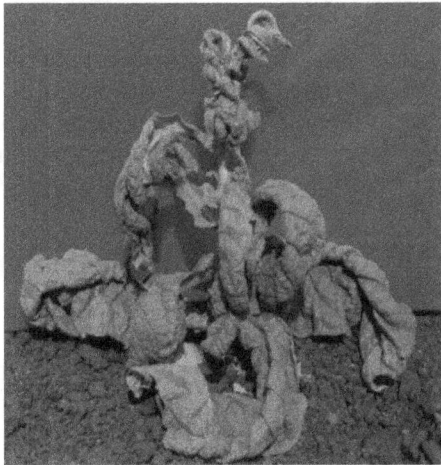

Fig. 6.8. Tobacco leaves infected with TLCV

8.2. Epidemiology: Infective whiteflies insert the virus into phloem tissues by means of the stylets while feeding on the leaves (Pollard, 1955). The incubation period varies from 12 to 33 days depending on temp., plant age and vigor. The disease is seldom seen in the plant bed, but usually appears 2 to 3 weeks after transplanting.A common vector of TLCV is the whitefly, known generally as *Bemisia tabaci* Gennadius, a member of the family Aleyrodidae. The virus persisted in the vector for at least 6 days. Single viruliferous whiteflies were found to be efficient vectors.

8.3. Disease management: As TLCV is a virus, no direct method for control is as yet available. The incidence of TLCV can be decreased by vector control. The application of chemicals has been reported to be effective in the control of *Bemisia tabaci*.

1. Discard leaf curl infected seedlings while planting.
2. Remove and destroy the diseased tobacco plants within one month after planting if the infected plants are less than 2%.
3. Alternate weed hosts for whitefly should be removed and destroyed, in and around tobacco fields.
4. Soon after harvesting the tobacco crop should be destroyed to prevent overwintering.
5. Do not grow crops like brinjal and sunflower in the vicinity of tobacco fields.
6. Install 12 yellow sticky traps (castor oil coated) per hectare to monitor the whitefly population. If 100 whiteflies stick to the trap the following insecticide, spray schedule has to be adopted; Imidacloprid 200 SL 2.5 ml / 10 l water - 1st spray, Thiamethoxam 25 WG 2.0 g / 10 l water- 15 days after 1st spray.

9. Cucumber mosaic virus

The cucumber mosaic virus has one of the broadest host ranges. The virus is distributed worldwide and the symptoms it causes are easily mistaken for tobacco mosaic (Phillips, 1942).

9.1. Symptoms: Typical mottling and mosaic patterns appear, sometimes accompanied by stunting and narrowing and distortion of the leaves (Fig. 6.9). Severe strains may cause intervainal discolouration and the oak-leaf pattern of necrosis on the lower leaves. 'Mosaic-burn' or sun-scald frequently appears on the upper leaves of infected plants. Mild strains cause only a faint mottling of the leaves (Hidaka, 1960).

Fig. 6.9. Tobacco leaves infected with
Cucumber Mosaic Virus

9.2. Epidemiology: The cucumber mosaic virus (CMV) overwinters in perennial weeds and may be transmitted to healthy plants by aphid vectors or by mechanical means. The cucumber mosaic virus cannot withstand drying or persist in the soil. It also is more difficult than tobacco mosaic to transmit mechanically. Thus, cucumber mosaic tends to progress more slowly than tobacco mosaic in a field. CMV is introduced in to the leaf through wounds, principally those made by aphid. Bradley (1968) showed that the mid veins and secondary veins were excellent sources of CMV and that aphids probing infected tissue for less than a min became highly viruliferous.

9.3. Disease management: The virus is readily transferred by aphids and survives on a wide variety of plants. Varietal resistance is the primary management tool, and eliminating weeds and infected perennial ornamentals that may harbor the virus is critical. Virus diseases cannot be controlled once the plant is infected. Therefore, every effort should be made to prevent introduction of virus diseases into the field. Sanitation is the primary means of controlling virus diseases. Infected plants should be removed immediately to prevent spread of the pathogens. Perennial weeds, which may serve as alternate hosts, should be controlled in and adjacent to the field. Disease incidence can be reduced through management of vectors by timely spraying of imidacloprid (200 SL 2.5 ml / 10 l water - 1st spray) and thiamethoxam (25 WG 2.0 g / 10 l water- 15 days after 1st spray)

Couzzo *et al.* (1998) demonstrated coat protein (CP) mediated protection in CMV. Quemada *et al.* (1991) engineered the coat protein (CP) gene from cucumber mosaic virus (CNV) strain C and transferred them into genome of tobacco (*N. tabacum* 'Xanthi'). They infected transgenic tobacco plants with CMV strain C of chi of sub group I and strain WL of subgroup II, transmitted mechanically or by aphids. They found significant degree of protection when challenged with CMV strains of either sub-group. Rizos *et al.* (1996) reported that transgenic plants expressing coat protein (CP) of an Australian isolate of cucumber mosaic virus (CMV) were resistant to infection with the homologous and two heterologous CMV strains. The level of resistance observed in CP - expressing plants was related to the virulence of challenging CMV isolate and not to the similarity between the CP expressed by transgenic plant and CP of the challenging virus. A mutant of the cucumber mosaic virus subgroup I A strain Fny (Fny-CMV) lacking the gene encoding the 2b protein (Fny- CMV delta2b) induced symptomless systemic infection in tobacco. Both the accumulation of Fny-CMV delta2b in inoculated tissue and systemic movement of the virus appear to proceed compared to wild type Fny-CMV (Soards *et al.*, 2002).

10. Orobanche (Broomrape)

This is popularly known as 'bodu' or 'malle' in Andhra Pradesh and broomrape is the common name. *Orobanche* is a member of plant family Orobanchaceae in Scrophulariales. This is a complete root parasite on many solanaceous plants, including tobacco. The shoots emerge in clusters and their basal portion is attached to tobacco roots through which it draws nourishment. It is prevalent in all the tobacco growing tracts in our country and predominant in A.P. Affected plants become stunted. Leaves turn pale and wilt. Initially leaf tips droop and as the attack intensify, later all the leaves wilt with characteristic ribbing of midribs (Fig. 6.10). High soil moisture due to irrigation or rain after planting, low soil temperature during winter months encourage heavy incidence of *Orobanche*. It is more severe in black cotton

Fig. 6.10 Symptoms of Orobanche

soils than in light soils. Yield losses due to early infection were estimated to be around 30%. Leaf quality is also greatly affected.

10.1. Life cycle: The parasite perennates through seed. Seeds of *Orobanche* remain viable in the soil more than 10 years in absence of the host. Once tobacco is planted, within 2 weeks the root exudates induce germination of *Orobanche* seed and produces a shallow disc or cup-like appressorium which surrounds the host root, penetrates it with a mass of undifferentiated, polymorphic cells that extend into the xylem of the host root and absorb nutrients and water from it.

10.2. Management

Sustainable management of the parasite can only be achieved by reducing the soil seed bank.

10.2.1. Cultural and preventive measures

(a) Deep summer ploughing of the sick field with disk plough 2-3 times exposes the *Orobanche* seed to the hot sun and results in desiccation. The seed on the upper layers of the soil are likely to be buried deep in the soil, beyond the reach of tobacco roots.

(b) Thick sowing of any of the trap crops like Jower, Gingelly, Greengram, Black gram, Pillipesara (*Phaseolus trilobus*) etc. in Kharif induce germination of *Orobanche* seed upto 30%. The germinated seeds die as they cannot infest the trap crops.

(c) Avoid growing of brinjal, tomato, bhendi and other Solanaceous crops in the sick fields.

(d) Application of any of the copper fungicides @ 0.2% at the time of planting delays emergence of *Orobanche* by few days and also protects the seedlings from Black shank.

(e) Periodical removal of emerged *Orobanche* spikes either manually or by using 'spear' an instrument to cut the spikes at soil level, before flowering or seed setting helps in reducing the inoculums in the soil. Following this method meticulously for 3 to 4 years reduces *Orobanche* incidence greatly (Pal and Gopalachari, 1957; Krishnamurty and Krishnan, 1967; Krishnamurty *et al.*, 1991).

(f) Late planting is better to reduce *Orobanche* infestation in Tobacco (Okazova-AG, 1975).

10.2.2. Chemical management: Krishnamurty *et al.* (1976) have observed that swabbing *Orobanche* shoots with kerosene oil gave maximum mortality (84.8%) and quick knock-down effect. Spray/drenching with 0.2% allyl alcohol gave moderate kill (47.6%) followed by 0.125% of I.C. 21 (18.8%) and 0.125%, DNOC (18.0%). Krishnamurty *et al.* (1991) were tested Eucalyptus, Pongamia, Rice Bran, Soybean and Tobacco seed oils for the control of broomrape (*Orobanche cernua*) in tobacco. The oil was applied to young, unflowered broomrape shoots at 1-5 drops/shoot with a dropper. All the oils effectively killed the parasitic shoot, their optimum doses being 1 drop/shoot for Eucalyptus, Pongamia, Soybean and Tobacco seed oils and 2 drops for Rice bran oil. In Italy, Sandri *et al.* (1998) observed that Glyphosate @ 1 lit/ha gave good control of *Orobanche ramosa* in Burley tobacco. In

India, Dhanapal *et al.* (1998) have reported that Glyphosate at 500 g ai/ ha at 60 days after transplanting (DAT) and Imazaquin at 10 g ai/ha at 30 DAT reduced the spikes by 75 to 80%, respectively.

10.2.3. Biological management: *Orobanche cernua* is a serious parasite in most of the tobacco growing regions in India causing quantitative and qualitative losses. *Sclerotium rolfsii* was reported to cause a serious disease in *Orobanche* parasitizing tobacco, effecting the seed production of the parasite. However, since this is a polyphagous pathogen and under high moisture conditions, it is likely to attack the host crop tobacco, the pathogen is still under investigation to be considered as biocontrol agent for *Orobanche* (Raju *et al.*, 1995).

Among other fungal pathogens infecting *Orobanche* spp., *Fusarium* spp. seems to be highly widespread. Many scientists reported its association with the parasite in different crops (Ramaiah, 1987; Mazheri *et al.*, 1991; Bedi and Donchev, 1991 and 1995; Linke *et al.*, 1992 and Parker and Riches, 1993). Though other fungi like *Alternaria, Urocladium* etc. have also been observed to be associated with diseased *Orobanche* spikes, *Fusarium* spp., especially *oxysporum* f. sp. *orthoceros* and *laterticum* appear to be highly potential and also widespread. It is interesting to note that *Fusarium* controlled 50 to 90 per cent of *Orobanche* in different countries. The fungus applied as conidiophores was the most virulent and *Orobanche* was most susceptible between germination and tubercle formation (Bozoukov and Kouzmanova, 1994).

Other fungi like *Alternaria, Trichoderma, Aspergillus, Penicillium, Mucor, Rhizopus, Cladosporium, Trichothecium* etc. have also been reported to be associated with the diseased *Orobanche* spikes but their role in the biological control has not been established (Abdel-Kader *et al.*, 1996 and Al-Menoufi, 1994). Okazova-AG (1973), Klein and Kroschel (2002) have reported that the larvae of *Phytomyza orobanchia* mine in *Orobanche* shoots and capsules. As a consequence, a natural reduction of *Orobanche* seed production by 30 to almost 80% has been reported from different countries. To strengthen the natural population and its impact, inundative releases of *P. orobanchia* adults at the beginning of *Orobanche* emergence have to be undertaken.

REFERENCES

Abdul Wajid, S. M., Shenoi, M. M., and Moses, J. S. L. 1987. Screening *Nicotiana* germplasm for resistance to Black shank disease. Tob. Res. 13: 111-15.

Abdul Wazid, S. M., and Shenai, M. M. 1992. Chemical control of Black shank disease in field planted Tobacco. Tob. Res.18: 97-103.

Anderer, F. A., Uhlg Weber, E., and Schramm, G. 1960. Primary structure of protein of tobacco mosaic virus. Virology. 40: 344 - 56.

Ann. Sci. Rept. 2010. Central Tobacco Research Institute, Rajahmundry. Andhra Pradesh (India).

Anonymous. 1951-1976. Annual reports of All India Coordinated research Project on Tobacco, ICAR and Cooperating agencies.

Bawden, F. C., Pirie, N. W., Bernal, J. D., and Fankuchen, I. 1936. Liquid crystalline substances from virus- infected plants. Nature. 138: 1051 - 55.

Bedi, J. S., and Donchev, N. 1991. Results on mycoherbicide control of sunflower broomrape under glasshouse and field conditions. In: Proceedings of the Fifth

International Symposium on Parasitic Weeds, Nairobi, Kenya.76-82.

Bedi, J. S., and Donchev, N. 1995. A technique for macro conidia production by *F. oxysporum* f. sp. *orthoceras*, a biocontrol agent for *Orobanche*. In: Plant Disease Research. 10: 62-63.

Beijernick, M. W. 1898. Ube rein contagium vivum fluidum als Ursache der Flekenkrankheit der Tabaksblatter. Verh. K. Akad. Wet. Amsterdam. 65: 3-21. [Translation published in English as Phytopathological Classics No.7 (1942)], American Phytopathological Society, St. Paul, MN.

Bloomer, A. C., Champness, J. N. Bricogne, G. Staden, R., and Klug, A. 1978. Protein disk of tobacco mosaic virus at 2.8^0 A resolution showing the interaction within and between sub- units. Nature. 276: 362 -68.

Bradley, R. H. E. 1968. Virology. 34: 172-73.

Chandwani, G. H., Reddy, T. S. N., and Pillai, S. N. 1973. Effect of fungicides on damping off tobacco seedlings. Pesticides. 7: 20-23.

Chaplin, J. F., and Mann, T. J. 1978. Evaluation of tobacco mosaic resistance factor transferred from burley to flue-cured tobacco. J. Hered. 69: 175-78.

Chaplin, J. F., Mann, T. J., and Apple, J. L. 1961. Some effects of the *Nicotiana glutinosa* type of mosaic resistance on agronomic characters of flue-cured tobacco. Tob. Sci. 5: 80-83.

Chaplin, J. F., and Gooding, G. V. 1969. Reaction of diverse *Nicotiana tabacum* germplasm to tobacco mosaic virus. Tob. Sci. 13: 130-33.

Chari, M. S., and Nagarajan, K. 1994. Management of pests and diseases of tobacco. CTRI. Rajahmundry, India.

Chow, H. S., and Rodgers, E. M. 1973. Efficacy of dodine and glyodin as foliar protectants for control of tobacco mosaic virus. Phytopathology. 63: 14-28.

Collmer, A., and Keen, N. T. 1986. The role of pectic enzymes in plant pathogenesis. Annu. Rev. Phytopathol. 24: 383-409.

Couzzo, M. K., O'Connel, W., Kaniewski, R. X., Fang Chua, N. H., and Turner, N. E. 1988. Viral protection in transgenic plants expressing the cucumber mosaic virus coat protein or its antisense RNA. Biotechnology. 6: 549 -57.

CTRI. 2004. Annual Scientific Rept.2004-05. Central Tobacco Research Institute, Rajahmundry, Andhra Pradesh, India. 88- 92.

CTRI. 2005. Annual Scientific Rept. 2005-06. Central Tobacco Research Institute, Rajahmundry, Andhra Pradesh, India.54-61.

CTRI. 2010. Annual Scientific Report 2010-11. Central Tobacco Research Institute, Rajahmundry, Andhra Pradesh, India.31.

CTRI. 2011. Proceedings of the IRC Meetings- 2011. Central Tobacco Research Institute, Rajahmundry, Andhra Pradesh, India.23.

Dam, S. K., Dutta, S., Laha, S. K., Roy, S., and Bhattacharya, P. M. 2010c. Effect of fungicides for management of brown spot disease of Motihari tobacco. J. of Mycopathol. Res. 48: 221-25.

Dam, S. K. 2008. Variability and management of brown spot in *Motihari* tobacco caused by *Alternaria alternata* (Fries) Keissler. Thesis in partial fulfillment for the degree of Ph.D. from Uttar Banga Krishi University.95.

Dam, S. K., Dutta, S. Laha, S. K and Roy, S. 2010a. Variability of *Alternaria alternata*, incitant of brown spot disease of Motihari tobacco. J of Mycol and Plant Pathol. 40: 59-62.

Dam, S. K., Roy, S., and Dutta, S. 2010b. Biological control of brown spot caused by *Alternaria alternata* in *Motihari* tobacco (*Nicotiana rustica*) in West Bengal, India. J Mycol and Plant Pathol. 40: 356-59.

Devaki, N. S., Shankar Bhat, S., Shenoi, M. M., and Abdul Wajid, S. M. 1991. Influence of soil moisture on the incidence of damping-off disease in tobacco nurseries. Tob. Res. 17: 45-46.

Dhanapal, G. N., Brog ter, S. J., Struik, P. C., and Borg ter. S. Y. 1998. Post emergence chemical control of nodding broomrape (*Orobanche cernua*) in bidi tobacco (*Nicotiana tabacum*) in India. Weed -Technology, 12: 652-59.

Dong, H. S., and Wang, Z. F. 1990. Pathogenicity differentiation of *Alternaria alternata* and induced resistance of tobacco brown spot by weak virulence strains. Bul. Spec. Coresta, Symposium Kallithea.162.

Elias, N. A., Bhaktavatsalam, G., Moses, J. S. L., and Abdul wajid, S. M. 1981. Chemical control of tobacco black shank caused by *Phytophthora parasitica* var. *nicotianae*. Tob. Res. 7: 46-51.

Gooding, G.V. Jr. 1969. Epidemiology of tobacco mosaic virus on tobacco in North Carolina. North Carolina Agric. Exp. Stn. Tech. Bull. 195: 24.

Gupta, B. M., and Patel, R. C. 1979. Preliminary screening of fungicides against *Phytophthora parasitica* dastur var. *nicotianae* (Breda De Hann) Tucker of Tobacco. Tob. Res. 5: 50-53.

Gwynn, G. R. 1977. Evaluation of tobacco mosaic virus resistant germplasm 1. Tob. Res. 3: 89-95.

Hare, W. W., and Lucas, G. B. 1959. Control of contact transmission of tobacco mosaic virus with milk. Plant Dis. Report. 43: 152-54.

Hartill, W. F. T. 1968. Rhodesia J. Agr. Res. 6: 13-18.

Hidaka, Z. 1960. Hatano Exp. Sta. Bull. 48. 149.

Holmes, F. O. 1938. Inheritance of resistance to tobacco mosaic disease in tobacco. Phytopathology. 28: 553-61.

Institute Research Committee Meeting. 2012. Crop Improvement Session, CTRI, Rajahmundry, July 25-27. 62- 63.

Ivanowski, D. 1892. Uber die Mosaikkrankheit der Tabakspflanze. Bull. Acad. Imp. Sci. St. Petersb., Nauv. Ser. 35: 67-70. [Translation published as Phytopathological Classics No. 7 91942)], American Phytopathological Soc, St. Paul, MN.

Kausche, G. A., Pfankuch, E., and Ruska, H. 1939. Die Sichtbarmachung von pflanzlichem Virus im Ubermikroskop. Naturwissenschaften.27: 292-99.

Keen, N. T. 1990. Gene for gene complimentarily in plant-pathogen interactions. Annu. Rev. of Genet. 24: 447-63.

Klein, O., and Kroschel, J. 2002. Biological control of *Orobanche* spp. with *Phytomyza orobanchia*, a review. Biocontrol. 47: 245-77.

Kotoujansky, A. 1987. Molecular genetics of pathogenesis by soft rot Erwinias. Annu. Rev. Phytopathol. 25: 405-30.

Krishnamurty, G. V. G., Chari, M. S., Ransom, J. K. Musselman, L. J., Worsham, A. D., and Parker, C. 1991. Proceedings of the 5th international symposium of parasitic weeds, Nairobi, Kenya, 108-10.

Krishnamurty, G. V. G., Nagarajan, K., and Lal, R. 1979. Multiple advantages of rabbing with paddy husk on tobacco seed-beds. Tob. News 2: 17-20.

Krishnamurty, T., and Krishnan, A. S. 1967. Control of *Orobanche* under natural and artificially infested conditions. Indian J. Agron. 12: 277-82.

Linke, K. H., Scheibel, C., Saxena, M. C., and Sauberborn, J. 1992. Fungi occurring on *Orobanche* spp. and their preliminary evaluation for *Orobanche* control. Trop. Pest Management 38: 127-30.

Lucas, G. B. 1975. Disease of tobacco. 3rd ed. Biological Consulting Associates, Raleigh, NC. 161.

Mahtabi, R. A., Zamanizadh, H. R., and Javid, K. 2001. Chemical control of brown spot disease of tobacco caused by *Alternaria alternata*. Coresta Meet Agro-Phyto Groups, Capetown, abstr. PPOST 8.

Mathrani, D. I., and Murthy, N. S. 1965. Control of damping off of tobacco seedlings by fungicides, Indian Tob. 15: 33.

Mazheri, A., N. Moazami, M. V., and Moayed-Zadeh, N. 1991. Investigations on *F. oxysporum*, a possible biological control of broomrape (*Orobanche* spp.). In: Proceedings of the Fifth International Symposium on Parasitic Weeds, Nairobi, Kenya. pp. 93-95.

Monga, D. 1990. Brown spot disease incidence of tobacco as influenced by different varieties and their planting dates. Tob. Res.16: 83-88.

Monga, D. 1991. Chemical control of brown spot (*Alternaria alternata*) on *Motihari* tobacco. Tob. Res. 17: 129-33.

Monga, D., and Tripathi, S. N. 1988. Factors affecting the incidence of Leaf Curl in Jati Tobacco (*Nicotiana tabacum* L). Tob. Res. 14: 21-24.

Monga, D., and Dobhal, V. K. 1987. Note on reaction of cigar wrapper tobacco (*Nicotiana tabacum* L.) germplasm to brown spot disease. Tob. Res. 13: 69-70.

Monga, D., and Dobhal, V. K. 1989. Hookah and chewing tobacco germplasm evaluation. II. Screening for resistance to brown spot. Tob. Res. 15: 150-52.

Mukhopadhyaya, N., Brahmbhatt, A., and Patel, G. J. 1986. *Trichoderma harzianum* - a potential bio-control agent for tobacco Damping - off. Tob. Res. 12: 26-35.

Murthy., Karunakara, K., and Shenoi, M. M. 2001. Evaluation of various chemical fungicides against brown spot disease of FCV tobacco. Tob. Res. 27: 1-6.

Nagarajan, K., and Shenoi, M. M. 1998. Chemical control of brown spot disease of tobacco caused by *Alternaria alternata*. Bull. Spec. Coresta, Congress Brighton, p. 119.

Nagarajan, K., and Murty, N. S. 1975. Effect of certain plant extracts and plant latex on the inhibition of Tobacco mosaic virus infection. Tob. Res. 1: 122-25.

Nagarajan, K., and Murty, N. S. 1977. Further screening of plants possessing anti -viral properties against tobacco mosaic virus. Tob. Res. 3: 108-10.

Nagarajan, K., and Reddy, T. S. N. 1980.Preliminary evaluation of Ridomil against Damping-off and Leaf blight diseases of Tobacco in seed beds. Tob. Res. 6: 127-31.

Nagarajan, K., and Reddy, T. S. N. 1982. Evaluation of recent fungicides on the control of seed bed diseases of tobacco. Bul. Spéc. CORESTA , Symposium Winston-Salem, p. 65

Nagarajan, K., Reddy, T. S. N., and Chandwani, G. H. 1977. Control of black shank disease of tobacco. Indian Farming. 27: 10-1.

Nagarajan, K., Reddy, T. S. N., Lal, R., Krishnamurthy, G. V. G., and Murthy, N. S. 1984. Salient research findings on tobacco diseases. Bulletin. 1: 23.

Namba, K., Pattanayek, R., and Stubbs, G. 1989. Visualization of protein –nucleic acid interactions in a virus. Refined structure of intact tobacco mosaic virus at 2.9^0 A resolution by x-ray fiber diffraction. J. Mol. Biol. 208: 307-25.

Nolla, J. S. B., and Roque, A. 1933. A variety of tobacco resistant to ordinary tobacco mosaic. J. Puerto Rico Dept. Agric. 17: 301-03.

Okazova, A. G. 1973. *Phytomyza* in the control of broomrape in the tobacco fields in the Crimea. Zashchita-Rastenii 18: 21-22.

Okazova, A. G. 1975. Planting times and infestation of tobacco by broomrape. Zashchita-Rastenii 20: 52.

Pal, N. L., and Gopalachari, N. C. 1957. A note on the root parasite *Orobanche* on tobacco and its control by weeding. First Conf. Tob. Res. Workers, Bangalore.

Palva, T. K., Hurtig, M., Saindrenan, P., and Palva, E. T. 1994. Salicylic acid induced resistance to *Erwinia carotovora subsp. carotovora* in tobacco. Molecular Plant Microbe Interactions. 7: 356-63.

Pande, A. 1985. Biocontrol characteristics of some moulds. Biovigyanam. 11: 14-18.

Parker, C., and Riches, C. R. 1993. Parasitic Weeds of the World: Biology and Control. CAB International, U. K. p. 332.

Patel, B. N. 1980. Inhibition of tobacco mosaic virus infection by some more plant extracts. *Tob. Res.* 6: 57-59.

Patel, B. N. 1981. Leaf extract of *Ailenthus excelsa*: an inhibitor of Tobacco mosaic virus infection. Tob. Res. 7: 192-95.

Patel, B. N., and Patel, G. J. 1979. Inhibition of tobacco mosaic virus infection by some plant extracts. Tob. Res. 5: 33-36.

Patel, D. N., and Patel, B. N. 1999. Evaluation of plant extracts and *Trichoderma harzianum* Rifai against *Phytophthora paracitica* var. *nicotianae*. Tob. Res. 25: 4-8.

Perombelon, M. C., and Kelman, A. 1980. Ecology of the soft rot *Erwinia*, Annu. Rev. Phytopathol. 18: 361-87.

Phillips, J. H. H. 1942. Can. J. Res. 20: 329-35.

Pillai, S. N., and Murthy, N. S. 1967. Relative efficacy of some copper fungicides for the control of damping off in tobacco nurseries. Indian Phytopath. 20: 381-83.

Pollard, D. G. 1955. Ann. Appl. Biol. 43: 664-71.

Quemada, H. D., Gonsalves, D., and Slightom, J. L. 1991. Expression of coat protein gene from cucumber mosaic virus strain C in tobacco: Protection against infection by CMV strains transmitted mechanically or by aphids. Phytopathology 81: 794-802.

Raju, C. A., Krishna Murty, G. V. G., Nagarajan, K., and Chari, M. S. 1995. A new disease of *Orobanche cernua* parasitizing tobacco, caused by *Sclerotium rolfsii*. Phytoparasitica 23: 307-313.

Raju, C. A., and Dam, S. K. 2011.Bio-efficacy of new fungicides against damping-off disease in FCV tobacco nursery. Paper presented in XIV National symposium on Tobacco - 2011 New Frontiers in Tobacco Science held at CTRI, Rajahmundry, December 20-22,

Ramaiah, K. V. 1987. Control of *Striga* and *Orobanche* – A review. Proc. of the Fourth International Society for Plant Parasitic Weeds, pp. 637-64.

Reddy, T. S. N., Nagarajan, K., and Chandwani, G. H. 1976. Sources of resistance in *Nicotiana* species to important tobacco diseases in India. Tob. Res. 2: 63-8.

Reddy, T. S. N., and Nagarajan, K. 1981. Studies on tobacco mosaic virus on tobacco (*Nicotiana tabacum* L.) I. Evaluation of germplasm and loss in yield and Quality of FCV tobacco. Tob. Res. 7: 75-80.

Rizos, H., Gillings, M. R., Pares, R. D., Gunn, L. V., Frankham, R., and Daggard, G. 1996. Protection of tobacco plants transgenic for cucumber mosaic cucumovirus (CMV) coat protein is related to the virulence of the challenging CMV isolate. Australasian Plant Pathology. 25: 179-85.

Roy, S., Amarnath, S., Arya, R. L., Krishnamurthy, V., and Dam, S. K. 2008. Symptomatology and integrated management of bacterial diseases of Jati (*N. tabacum*) and Motihari (*N. rustica*) tobacco in *terai* agro – ecological zone of West Bengal. Environment & Ecology. 26: 528-32.

Sandri, G., Sandri, A., and Martini, G. 1998. Protection of tobacco against *Orobanche*. Informatore-Agrario 54: 74-75.

Sastri, A. B., Krishnamurty, K. V., Nagarajan, K., Krishna Murty, A. S., Prasada Rao, P. V., Reddy, T. S. N., and Mohan, M. 1981. Performance of tobacco Mosaic Virus resistant Flue- Cured Tobacco varieties under field conditions. Tob. Res. 7: 145-49.

Scholthof, K. B. G. 2004. Tobacco Mosaic Virus: A model system for Plant Biology. Ann. Rev. Phytopath. 42: 13-34.

Shafik, J., and Taha, K. H. 1984. Chemical control of brown spot of tobacco in northern Iraq. Indian Phytopathology. 37: 669-672.

Shenoi, M. M., Moses, J. S. L., Rao, S. V., Abdul Wajid, S. M., and Subrahmanya, K. N. 1992. Screening *Nicotiana* germplasm against tobacco mosaic Virus and studies on TMV tolerant FCV special mutant line FCH 6248. Tob. Res. 18: 93-6.

Shenoi, M. M., and Abdul Wajid, S. M. 1988. Control of Damping off and Black shank diseases by Metalaxyl (Apron 35 SD) in tobacco nurseries. Tob. Res. 14: 97-101.

Shenoi, M. M., Srinivas, S. S., Karunakaramurthy, K., and Abdul Wajid, S. M. 2002. A new rating scale for assessment of brown spot disease in tobacco. Tob. Res. 28: 142-45.

Shenoi, M. M., and Abdul Wazid, S. M. 1992. Management of Damping off, blight and black shank diseases with Ridomil MZ 72 WP in FCV tobacco nurseries of Karnataka. Tob. Res. 18: 53-58.

Shew, H. D., and Lucas, G. B. 1991. Compendium of Tobacco Diseases. APS Press, USA.

Soards, A. J., Murphy, M., Palukaitis, P., and Carr, J. P. 2002. Virulence and differential local and systemic spread of cucumber mosaic virus in tobacco are affected by the CMV 2b protein. Mol. Plant Microbe Interactions. 15: 647 – 53.

Sreeramulu, K. R., Onkarappa, T., and Narayana Swamy, H. 1998. Biocontrol of Damping off and Black shank disease in tobacco nursery. Tob. Res. 24: 1-4.

Srivastava, R. P., and Subba Rao, D. 1984. Effect of nitrogen fertilization and plant densities under varying dates of planting on certain chemical quality characteristics of cigar wrapper tobacco in the northern sandy soil of West Bengal. Tob. Res.10: 1-6.

Stanley, W. M. 1935. Isolation of a crystalline protein possessing the properties of tobacco mosaic virus. Science. 81: 644-45.

Tsugita, A., Gish, D. T., Young, J., Frankel – Conrat, H., Knight, C. A., and Stanley, W. M. 1960.The complete amino acid sequence of the protein of tobacco mosaic virus. Proc. NaH. Acad. Sci. USA. 46: 1463 -69.

Turner, M. T., and Bateman, D. F. 1968. Maceration of plant tissues susceptible and resistance to soft rot pathogens by enzymes from compatible and host pathogen combinations. Phytopathology. 58: 1509-15.

Valand, G. B., and Muniyappa, V. 1992. Epidemiology of tobacco leaf curl virus in India. 120: 257-67.

Whitham, S., Dinesh-Kumar, S. P., Choi, D., Hehl, R., Corr, C., and Baker, B. 1994. The product of the tobacco mosaic virus resistance gene N: similarity to toll and the interleukin-1 receptor. Cell 78: 1101-1115.

Xia, Z. Y., and Mo, X. H. 2007. Occurrence of black leg disease of tobacco caused by *Pectobacterium carotovorum* sub sp. *carotovorum* in China. Plant Pathol. 56: 348.

Zaitlin, M. 1998. The discovery of the causal agent of the tobacco mosaic virus disease. In discoveries in Plant Biology, Kung, S.D; Yang, S.F. eds. (Hong Kong World Scientific Publishing Co. Ltd.) , 105-10.

Diseases of Field and Horticultural Crops   *Pages* **196-222**
Editor-in-Chief: **P. Chowdappa**
Published by: **Daya Publishing House, New Delhi**

7

Diseases of Bast Fibre Crops

Chinmay Biswas and S.K.Sarkar

Bast fibres are basically stem fibres composed of sclerenchyma cells united together into small or large strands or bundles, arranged in the form of layer(s). In India the most important bast fibre crops are jute (*Corchorus olitorius* and *C. capsularis*), mesta (*Hibiscus sabdariffa* and *H. cannabinus*), sunnhemp (*Crotalaria juncea*) and ramie (*Boehmeria nivea*).

DISEASES OF JUTE

Jute (*Corchorus olitorius*: *tossa* jute and *C. capsularis*: whitejute) is the most important bast fibre crop of the world, which is mainly grown in the South East Asian countries like India, Bangladesh, Nepal, China, Indonesia, Thailand, Myanmar and few South American countries. Jute fibre has wide applications in making ropes, sacks, bags, carpets, shoes, geo-textiles, jewellery and home decorations. Tender jute leaves are also used as a leafy vegetable and in soups which are rich in antioxidants. Despite tough competition from synthetics, use of jute fibre is increasing because of its eco-friendly, biodegradable and recyclable nature.Most important disease of jute is stem rot caused by *Macrophomina phaseolina*.Other important diseases are anthracnose (*Colletotrichum corchorum and C. gloeosporioides*), black band(*Botryodiplodia theobromae*), soft rot (*Sclerotium rolfsii*), jute mosaic (a begomo virus) and Hoogly wilt (*Pseudomonas solanacearum* and *Macrophomina phaseolina*).

1. Stem rot (*Macrophomina phaseolina* TassiGoid)

Stem rot is the most important disease of jute, equally affecting both the species viz., *Corchorusolitorius* (*tossa* jute) and *C. capsularis* (white jute). It is prevalent in all the jute growing areas of the world. In India this disease is severe particularly in eastern parts covering Assam, West Bengal, Bihar and Orissa. It causes significant reduction in yield and quality of the fibre. Average yield loss due to this disease is about 10%, but it can go up to 35-40% in case of severe infection under favourable climatic conditions (Roy *et al.*, 2008).

1.1. Symptoms: The pathogen attacks any part of the plant at any stage of growth. It can cause damping off, seedling blight, leaf blight, collar rot, stem rot and root rot (Ghosh and Mukherjee, 1970). Sometimes just after germination deep brown spots are found on the cotyledonary leavesgiving a blighted appearance and it is known as seedling blight. If dry weather prevails at that time browning reaches to the roots and the seedling dies. This is called damping off. After the seedling grows up browning of the collar region may be noticed which is called collar rot. In the months of June and July when the plants are about more than seventy days oldseveral small brown spots are found on the leaves. Those small spots gradually increase in size and coalesce with each other to form a bigger brown rotted area on the leaves. The disease generally spreads from infected leaves in two ways. The pathogen may enter into the stem through the petioles of the infected leaves or the infected leaves fall and adhere to stem surface where the infection may occur. The most characteristic symptom of the disease is formation of blackish brown lesions or depressions on the stem which increase in size and several such lesions may coalesce and finally girdle the stem. Sometimes longitudinal streaks are formed along the length of the stem and the cortex becomes shredded exposing the fibres. The pathogen may also attack the roots causing wilting and death of the plant and on uprooting blackish brown discolouration of the roots are seen. All the leaves fall from the infected plant and the stem looks black or dark brown.In case of seed crop late infection may cause spotting on the capsules and formation of pycnidia and sclerotia on the capsule and seed (De and Kaiser, 1991). The infected seeds are

Fig. 7.1. Symptoms of stem rot

smaller,shrivelled and light in colour. If these seeds are sown germination is found to be very less and the small seedlings get infected.

1.2. Scoring of the disease: As the pathogen produces symptoms both on leaves as well as stems and the stem rot pattern varies significantly depending on the severity of the disease scoring of this disease is not easy. Simple incidence or per cent infected plants does not reflect the actual disease scenario. De *et al* (2008) reported a scoring method based on stem lesions. The stem rot was scored by observing lesion size (1-4 score), position of lesion on the stem (1-4 score) and lesion type (1-8) with maximum value of 16(4+4+8). Lesion size of less than 0.5 cm^2 was scored with 1.0, lesion size 0.6- 1cm^2 was rated as 2.0, likewise lesion size of 1.1-2 cm^2 with 3.0 and lesion size of more than 2.0 cm^2 with 4.0. Position of lesion on the first quarter of the stem from the top of the plant carried 1.0 score, similarly on the second quarter it would be 2.0, on the third quarter 3.0 and on the last quarter it would be 4.0. Lesion type covering less than 10% of the stem diameter is given 1.0 score, lesion covering 10.1-25% of the stem diameter scored 2.0, 26.1-40% with 4.0 and more than 40% is scored 8.0. The per cent disease index (PDI) is calculated by summation of all numerical ratings multiplied by 100 and dividing it with number of plants observed multiplied by highest value.

1.3. Etiology: *Macrophomina phaseolina* TassiGoid.isa dreaded pathogenand it can infect more than 500 plant species in about 72 families. *M. phaseolina* (Tassi) Goid. (syns. *M. phaseolina* (Maubl) Ashby is the pycnidial stage of the pathogen and *Rhizoctonia bataticola* (Taub.) is the sclerotial stage. The pycnidial and sclerotial stages are responsible for the disease and the perfect stage, *Orbiliaobscura* is very rarely seen (Ashby, 1927; Mandal, 1990). It is a soilborne as well as seed borne pathogen belonging to the phylum Deuteromycetes and class Coelomycetes. It is highly variable, with isolates differing in microsclerotial size and presence or absence of pycnidia. The pycnidial stage is common on jute and peanut, but not on soybean. The pycnida are initially immersed in host tissues, then erumpent at maturity. They are about 100-200 μm in diameter; dark to greyish, becoming black with age; globose or flattened globose; membranous to subcarbonaceous with an inconspicuous or definite truncate ostiole. The pycnida bear simple, rod-shaped conidiophores, 10-15 μm long. Conidia (14-33 x 6-12 μm) are single celled, hyaline, and elliptic or oval. Microsclerotia of *M. phaseolina* are black in colour and appear smooth and round to oblong or irregular. Across isolates, microsclerotia vary in size and shape and on different substrates. Microsclerotia are formed from aggregates of hyphal cells joined by a melanin material with 50 to 200 individual cells composing an individual microsclerotium. Colonies in culture range in colour from white to brown or gray and darken with age. Hyphal branches generally form at right angles to parent hyphae, but branching is also common at acute angles.

1.4. Epidemiology: Alluvial and lateritic soils with low pH (5.6-6.5), high level of nitrogen, high rainfall, moderate temperature and high humidity favour infection of *Macrophomina phaseolina*. Higher soil temperature and low soil moisture predispose the older plants to stem rot infection. The severity of the disease depends on the date of sowing. Generally, the early sown crops (sown in March) suffer heavily from this disease. The pathogen survives as sclerotia in the soil and on infected crop debris. The sclerotia serve as the primary source of inoculum and have been found to persist within the soil up to three years (Dhingra and Sinclair, 1977). The sclerotia are black, spherical to oblong structures that are produced in the host tissue and released into the soil as the infected plant decays. These multi-celled structures allow the persistence of the fungus under adverse conditions such as low soil nutrient levels and temperature above 30^0 C.Germination of the sclerotia occurs throughout the growing season when temperatures are between 28^0C and 35^0 C. Sclerotia germinate on the root surface, germ tubes form appresoria that penetrate the host epidermal cell walls by mechanical pressure and enzymatic digestion or through natural openings (Bowers and Russin, 1999). The rate of infection increaseswith higher soil temperatures and low soil moisture will further enhance disease severity. *M. phaseolina* can grow and produce large amounts of sclerotia under relatively low water potentials allowing this disease to be severe under water stress. Population of *M. phaseolina* in soil will increase when susceptible hosts are cropped in successive years. The seed borne inoculum also causes infection of *M. phaseolina* in jute leading to damping off or seedling blight. Under favourable conditions viz., cloudy weather, high rainfall and temperature of about 30-35⁰C secondary spread of the disease takes place through airborne conidia by rain splash or air.

1.5. Management : As the source of inoculum may be soil, seeds or infected plants the disease management practices should be adopted well in advance, otherwise the disease may cause havoc. Some disease management options are discussed here.

1.5.1. Cultural Practices

(a) Crop establishment

(i) *Field preparation*: For growing this crop heavy soils which lack proper aeration should be avoided. The crop debris and weeds should be removed and burnt because the pathogen may harbour therein. If clean cultivation is ensured through proper sanitation possibility of the disease attack can be reduced considerably.

(ii) *Soil pH*: Continuous cultivation of jute crop in the same field may cause depletion of calcium, potassium and other basic substances from soil and make the soil acidic which favours the incidence of stem rot disease. Therefore, once in 3-4 years soil pH has to be tested for liming as per requirement. Generally, lime or dolomite @ 2-4 t/h is applied about one month before sowing for correction of soil pH.

(iii) *Spacing*: The incidence of stem rot disease increases under close spacing. Therefore, optimum spacing has to be maintained. In case of line sowing row to row distance should be 25-30 cm and plant to plant distance should be 5-6cm. While sowing by broadcasting care should be taken so that proper distance is maintained. In both the cases plant to plant distance can be corrected by thinning.

(b) Nutrient management: In organic matter deficient soil compost or FYM has to be applied @ 7-8 t/h. Chemical fertilizer has to be applied in a judicious manner because higher dose of nitrogen (beyond 80 kg/h) promotes stem rot disease. However, application of potash (K_2O @ 50-100 kg/h) reduces the disease severity.

(c) Crop rotation: Jute cultivationin the same field year after year should not be allowed. Moreover, non-host crops i.e, the crops which are not attacked by *M. phaseolina* or *Rhizoctonia bataticola* like rice, wheat, mustard etc must be brought in the cropping sequence to follow jute.

(d) Drainage: Although white jute can withstand water-logging to some extent, *tossa* jute can not. Water-logging in general increases disease intensity. Therefore, proper drainage has to be made to avoid water stagnation.

1.5.2. Chemical control

(a) **Seed treatment:** As infected seed is a major source of the disease seeds must be treated before sowing. Seed treatment not only eliminates seed borne inoculum but also gives protection against soil borne pathogen for few weeks. Carbendazim 50 WP (Bavistin, Derosaletc) @ 2g /kg or Dithane M 45 (mancozeb) @ 5g/kg are very effective fungicides for seed treatment.

(b) **Spraying:** When the disease incidence is 2% or more spraying of Carbendazim 50WP @ 2g/l or Dithane M 45 @ 5g/l or Blitox 50 WP (copper oxychloride) @ 5-7 g/l is recommended. In case of severe infection 3-4 sprays are recommended at 15-20 days interval.

(c) **Induced resistance:** Induction of resistance refers to heightened resistance in a plant towards pathogens as a result of a previous treatment with a pathogen, an attenuated pathogen or a chemical that is not itself a pesticide. Pre-treatment with certain chemical elicitors have been reported to induce resistance against many crop diseases (Biswas *et al.* 2009). We tried to induce resistance against stem rot pathogen by pre-treatment with some chemical elicitors. Jute seeds of cultivar JRO 8432 were treated with chemicals of different concentrations *viz.* Chitosan @ 1.0%, 5.0% and 10.0%; Salicylic acid @ 0.1mM, 1.0mM and 10.0mM; Indole acetic acid (IAA) @ 0.01mM, 0.1mM and 1.0mM; Beta amino butyric acid (BABA) @ 1mM, 5mM and 10mM; Dipotassium hydrogen phosphate (K_2HPO_4) @ 10mM , 25mM and 50mM; Calcium chloride ($CaCl_2$) @ 15mM, 20mM and 25mM; Carbendazim @ 0.05%,0.1% and 0.2%. Four days old seedlings were then sprayed with cultural suspension of *M. phaseolina* and per cent infection was recorded after three days.

No infection of *M.phaseolina* was observed in treatments *viz.* Carbendazim @ 0.1% and 0.2%; $CaCl_2$ @ 20mM and 25mM; BABA @ 5mM and 10mM; K_2HPO_4 @ 50mM and Chitosan @ 5% and 10%. But the per cent infection in case of untreated seedlings was as high as 86.5%. All the chemical elicitors reduced the infection considerably except IAA. However, these results are based on preliminary observations and further investigation is required to study these effects. It is most important to find out how long the protection is provided by the inducers.

1.5.3. Biological control:
Although stem rot of jute is mainly controlled by applying fungicides some bio-agents viz., *Trichoderma viride, Aspergillus niger* AN27, some species of fluorescent *Pseudomonas* etc. have been reported to check the disease. Seed treatment with *T. viride* and its soil application reduced the disease incidence at various locations (Anonymous, 1999). Seed inoculation with PGPR, *viz.* fluorescent *Pseudomonas, Azotobacter, Azospirillum*etcreduced stem rot incidence and increased the fibre yield considerably (Bandopadhyay and Bandopadhyay, 2004). Bandopadhyay *et al.* (2008) reported that *T. viride, Gliocladium virens, Aspergillus niger* and *A. fumigatus* were very effective against *Macrophomina phaseolina* causing stem rot and root rot disease in jute. Nonvolatile components isolated from *T. viride* JPT1, *G. virens* JPG1, *Gliocladium* sp. JPG4 and *A. niger* AN15 and *Aspergillus* sp. A26 isolates inhibited mycelial and sclerotial growth of *M. phaseolina*.

1.5.4. Varietal resistance:
The best way of disease management is use of resistant varieties. But no jute variety is truly resistant to the stem rot pathogen and the degree of resistance /susceptibility varies with climatic conditions and

concerned pathotypes present in a particular area. *Tossa* jute varieties namely JRO 524 and JRO 632 and white jute cultivars JRC 212 and JRC 321 showed differential reactions at different geographical locations. Six *C.capsularis* accessions viz., CIM 036, CIM 064, CIN 109, CIN 362, CIN 360 and CIN 386 out of total196 accession tested showed resistant reaction even at the hot spot of the disease in Sorbhog, Assam (Mandal *et al.*, 2000). De *et al* (2008) reported four *C. olitorius* accessions namely OIN 125, OIN 154, OIN 651 and OIN 853 as moderately resistant to stem rot pathogen. De and Kaiser (1991) evaluated the diallel crosses among seven varieties excluding the reciprocals to analyse the resistance to the stem rot disease and it was found that additive gene effect was predominant over non-additive gene effect. General combining ability (gca) effect showed that the parents Russian red, Sudan green, Tanganyka and JRO 878 were the best general combiners towards desirable direction of resistance to the disease, while the specific combining ability (sca) effect was recorded to be highest in the cross Peking x JRO 7835 where both parents had undesirable gca effect. Some of the good general combiners also produced desirable sca effect in some of the cross combinations such as Russian red x JRO 878, Sudan green x JRO 878 and Tanganyka x Russian red. Average disease index of the parents and their F1 progenies further showed that resistance to stem rot pathogen *M. phaseolina* was dominant over susceptibility.

2. Anthracnose

This is also a very important disease mainly of *C. capsularis*. Two different species of *Colletotrichum* are responsible for anthracnose in the two different species of *Corchorus*. In *C. capsularis* it is *C. corchorum* and in *C. olitorius* it is *C. gloeosporioides*. However, in *C. olitorius* infection takes place at a very later stage than *capsularis* jute.

2.1. Economic importance: It is more serious in *C. Capsularis* jute. *Olitorius* jute is rarely affected. In Assam, severity of anthracnose in *olitorius* jute was also noticed. At Barrackpore the disease has been noticed recently in seed crops of *olitorius* jute. In *capsularis* jute, numerous spots appear on the stem and plant die in many cases. In case where the plant survives, the fibre is specky or knotty. It causes considerable damage to the fibre quality as well as loss of yield.

2.2. Geographical distribution: The disease is of regular occurrence in *capsularis* belt of India viz. Assam, North Bengal, Bihar, and Uttar Pradesh. The disease is prevalent in Bangladesh also (Ghosh, 1957; Mandal, 1990). In all probability anthracnose caused by *Colletotrichum corchorum* entered India during thirties along with jute germplasm from Southeast Asia, particularly Taiwan unknowingly. There was no mentionof this disease in India prior to 1945 when it was first observed on "Jap-Red', a capsularis introduction from Formosa (Taiwan) at Dacca Farm, now in Bangladesh (Ghosh 1999). Then from Dacca it spread to other parts of Bangladesh. Later it entered India through Assam. Continuous rain, high humidity and temperature around 35^{0}C are congenial for this disease. Epidemic of anthracnose was noticed in exotic variety, Japanese Red at Chinchura, West Bengal during 1950-51 (Ghosh, 1957). By preventing release of hybrids involving South Asian susceptible or selections thereof, India was enjoying comparative freedom from anthracnose while in Bangladesh, anthracnose established itself firmly. In recent years the disease is appearing in very severe form in *capsularis* jute belt at Bahraich areas in Uttar Pradesh and also in some places of North Bengal, Bihar and Assam (Anonymous, 2004).

2.3. Symptoms: In *C. capsularis* at seedling stage the disease appears on leaves and stems as brownish spots and streaks followed by drying up of the entire stem. In mature plants initially light yellowish patches can be seen on stem which turn to brownish depressed spots. The spots are irregular in shape and size. Several spots may coalesce causing deep necrosis showing cracks on the stem and exposing the fibre tissues. The coalescing spots may often girdle the stem and in such cases the plants break at that point and die. Affected plants when survive show necrotic wounds all over the stem. Fibres extracted from such affected plants are specky and knotty and fall under very low grade. Capsules of diseased plants are also affected showing depressed spots and seeds collected from such fruits are also infected. The infected seeds are lighter in colour, shrunken and germination is poor. In *C. olitorius* incidence occurs at the later stage of crop growth. Generally it starts during mid July and continues up to September. Brown to black sodden spots appears on the stems.

2.4. Etiology: The pathogen is *Colletotrichum corchorum*incapsularis jute and *Colletotrichum gloeosporioides* in*olitorius* jute.Acervuli: black, superficial (errumpant) scattered, stroma: subglobose to hemispherical, sometimes lenticular, Setae: 10-30 in number, mostly 20-30 in each acervulum; deep dark brown scattered all over the stroma, septa 2-5; 52-138 µm mostly 80 µm, 3.9-5.2 µm at the base. Conidiophores: simple straight hyaline bearing conidia singly, 25-25-30µm. conidia: acrogenous, falcate, hyaline to subhyaline, guttulate, 17-24x2.6-3.25µm mostly 20x3.5µm, germinate forming appresoria, in mass dirty flesh coloured to warm buff. The fungus grows very slowly on PDA. Growth starts with pale grayish-white aerial mycelium. Innumerable acervuli develop on the surface. Colour of the mycelial mat is masked by formation of abundant acervuli which are blackish – brown to black in maturity. In culture the acervuli are arranged in concentric zones. In about a week's time at a temperature of 32^{0}C a pale dirty whitish cream coloured mass covers the acervuli. This mass is the conidial accumulation in a semi viscid matrix. In older cultures the colour of the conidial mass changes to warm buff to dirty flesh tint.

2.5. Epidemiology: The mycelium enters through the epidermis and attacks the parenchymatous tissues between the wedges of bast fibre bundles. The entire parenchymatous tissue of cortex gradually disintegrates. Under favourable conditions the mycelia attack the thin walled phloem tissueswithin the phloem bundles. The phloem parenchyma disintegrates and the bundles are exposed. The mycelia often reach the cambial layer but seldom attack the wood. During humid days acervuli appear on the surface of the affected tissues and are visible with naked eyes; under a magnifying lens they appear bristled hemispherical or slightly lenticular eruption. The pathogen is also seed borne. The disease starts in the hot and very humid months of July when the crop is about 8-10 weeks old. The damage is most severe near about the harvest time. Continuous rain, high relative humidity and temperature of around 35⁰C are congenial for the faster development of this disease. Whitejute is more susceptible than*tossa* jute.

2.6. Management: Removal of affected plants from the field and clean cultivation.Seed treatment with carbendazim @ 2 g kg⁻¹ of seed or captan @ 5 g kg-1 of seed to eliminate primary source of infection.Spraying of carbendazim @ 2 g l⁻¹ or Dithane M 45 @ 5 g l⁻¹ can check the disease. Seeds having 15% or more infection should not be used even after treatment.

3. Hooghly wilt

3.1. Geographic distribution: This disease was mostly prevalent in the areas where jute is followed by potato or other solanaceaous crops. The disease was observed in the districts of Hooghly, parts of Howrah, North 24 Parganas, Burdwan and Nadia of West Bengal. This was reported from Bangladesh also. *Olitorius* jute generally suffers from it. There is no report of *capsularis* jute being attacked by this disease. During late seventies to end eighties, more than 40 % infection was recorded in Hooghly district of India. But presently, this disease is not a concern.

3.2. Economic importance: During late forties and early fifties, a typical wilt disease in jute (*Corchorus olitorius*) was observed in Tarakeswar areas of Hooghly district of West Bengal. The malady was very widespread and the damage to crop was so severe that a new Pest and Disease Control Centre had to be established at Tarakeswar in the heart of Hooghly district (Annual Report JARI, 1949-56). The disease later widely spread to other areas of the district and also to the adjoining districts where jute used to be followed by potato in winter. The benchmark survey estimated 30 - 34% loss of jute crop each year during 1950-1954 (Annual Report JARI, 1949-56). During late eighties and early nineties, 5-37% disease was recorded in Kamarkundu area of Hooghly district and 2-20% in some areas of Nadia and North 24-Parganas districts (Mandal and Mishra, 2001). The disease was recorded in *olitorius* jute in North Bengal (Coochbehar) although *capsularis* jute was disease free.

3.3. Etiology: Ghosh (1999) coined the name Hooghly wilt and isolated seven fungi and a bacterium from the diseased or dead plant. The term 'Hooghly wilt' was well accepted by the working pathologists. Later on by repeated experimentation it was established that *Ralstonia solanacearum* (= *Pseudomonas solanacearum*) is primary pathogen and *R. bataticola* and *M. incognita* are associated pathogens. Presence of these root rot pathogens (*R. bataticola*/*M phaseolina*) and root knot nematode (*M. incognita*) increases the disease since they create wounds in the root, and thus facilitate the entry of the primary bacterial pathogen, *R. solanacearum* (Mandal and Mishra, 2001).

3.4. Management: Potato or other solanaceaous crops in the rotation are to be avoided. Jute: Paddy-Paddy or Jute-Paddy-Wheat could be the most effective rotation. In case where solanaceaous crop is the main crop of the area in the *rabi* season, it should be replaced by paddy or wheat at least for two years. Removing wilt affected plants from the field, burning of the dead plants of solanaceaous plants and burning rejected and rotten potato tubers (Ghosh and Mukherjee 1970, Sarkar *et al.*, 1998) are important cultural practices to keep the disease under control. By adopting cultural practices particularly the appropriate crop rotation in Hooghly district the disease came down to 1-2% compared to above 40% in the late eighties. Seed treatment with Carbendazim @ 2 g kg-1 of seed and spraying the same fungicide @ 2 gl^{-1} of water. This fungicide helps to reduce root rot and restricts the entry of the bacteria.

4. Jute mosaic

The diseased was first observed in 1917 in undivided Bengal province of India by Finlow, 1917. Since then it is variously known as jute yellow mosaic, jute leaf mosaic and jute golden mosaic etc. (Ahmed, 1978, Ahmed *et al.*, 1980). Recently the incidence of the disease has increased from 20-40 per cent (Ghosh *et al.*, 2007a).

4.1. Geographic distribution: The disease was reported in *capsularis* jute from different jute growing belts of India and Bangladesh. Recently, from Vietnam some sequence of Jute golden mosaic virus has been deposited in GenBank, but no details is available. In India, the disease was reported in *capsularis* jute from West Bengal (Roy *et al.*, 2006, Ghosh *et al*, 2007b) and Assam).

4.2. Symptoms: The disease is characterized by appearance of small yellow flecks on leaf lamina in the initial stage, which gradually increased intermingled with green patches and produced a yellow mosaic appearance. Leaves in some cases produce small enation along the mid vein. In extreme cases the infected plant gets stunted and leads to reduced plant height to the extent of 20% (Das *et al.*, 2001). The incidence was about 50% on some leading cultivars such as JRC 7447 and JRC 212 (Ghosh *et al.*, 2007b).

4.3. Particle morphology: As jute leaves are rich in mucilaginous substances, observation of virus particles is very difficult. So far no report on the purification of this virus is available and possibly that become a great hindrance to identification of the particle from purified virus material. Presence of mucilage also became a great factor in viewing the particle morphology through dip method.

4.4. Detection, identification and phylogenetic relationship: Based on sympotomatology and transmission, the virus was found to be a member of Begomovirus under family Geminiviridae. PCR based detection with Begomovirus group specific primer amplified a 1.2 kb DNA A fragment of the virus (Ghosh *et al.*, 2007a) Cloning and sequencing of this amplicon revealed that it consisted of 1263 nucleotides (Accession No EU 047706) and shared the highest nucleotide sequence identity (91.2%) with Corchorus golden mosaic Vietnam virus (DQ 641688). The nucleotide sequence at the origin of replication was found to be CATTATTAC instead of conventional TAATATTAC. Such unique feature was also noticed in case of Corchorus golden mosaic Vietnum virus. Phylogenetic analysis with other begomovirus revealed that the begomovirus from jute grouped with other begomovirus reported to be associated with Corchorus species and clustered with new world begomoviruses. Two other primers have been designed to amply the complete DNA A and DNA B component of the viral genome which gave expected 2.7 kb amplicon in each case.

4.5. Transmission: The causal virus was reported to be transmitted by white fly (*Bemisia tabaci*) (Ahmed, 1978, Roy *et al.*, 2006). Some workers also reported that the virus was transmitted by seeds (Das *et al.*, 2001). Adult whitefly can acquire the virus within 30 minutes of its access to a diseased plant and acquisition become maximum after 8 hours of access. A viruliferous whitefly can inoculate the virus to the test plant within 30 minutes of its inoculation. A viruliferous vector can retain the virus upto 10 days. A typical symptom of the disease appeared on the plants of cv. JRC 7447 and JRC 212 after 10 days of whitefly transmission with 60 per cent transmission efficiency when acquisition and inoculation access period were 24h and 12 h respectively.

4.6. Management: No detailed work has been carried out. In general, rouging of diseased plants and spraying of imidacloprid, thiamethoxan and acetamedrid could prevent the spread of the disease. Studies in Bangladesh revealed that a combination of collection and use of seeds from healthy plants, one insecticidal spray around 30 days after emergence (DAE), field sanitation with rouging several times during growth period and application of an extra booster dose of nitrogen at around 45DAE reduced the spread.

5. Black band (*Botryodiplodia theobromae*)

Earlier, it was a minor disease but gradually the occurrence of this disease is increasing.

5.1. Symptoms: The pathogen affects both the species of jute and causes serious damage to the jute crop from July onwards. The disease first appears as small blackish brown lesions which gradually enlarge and encircle the stem resulting in withering of epical and side branches. Stems infected at the lower portion often break at that point. The affected plants shed leaves, turn brown to black and remain standing as dry sticks in case of severe infection. On rubbing the stem surface, unlike stem rot profuse black shooty mass of spores adhere to the fingers. Crops raised from infected seeds show seedling blight symptoms also.

5.2. Predisposition: The disease is soil and seed borne. Hot and humid conditions favour the disease development.

5.3. Management: Clean cultivation, seed treatment with carbendazim 50 WP @ 2g/kg and foliar application of carbendazim 50 WP@ 2g/l of water or Cu-oxychloride @ 5-7 g/l of water or mancozeb @ 4-5 g/l of water provides effective management

6. Soft rot (*Sclerotium rolfsii*)

This was also a minor disease earlier but like black band it is also gradually spreading affecting both the species of jute.

6.1. Symptoms: On stems soft brown wet patches are observed at the point of infection. The skin peels off and the exposed fibre layers turn rusty brown and the affected plants wilt. White mycelial growth and brown globose to sub-globose mustard like sclerotia are also observed at the site of infection.

6.2. Predisposition: The pathogen is soil-borne and has a large number of hosts. High soil moisture with high temperature favours the disease.

6.3. Management: Clean and weed free cultivation, summer ploughing, and spraying of Cu-oxychloride @ 5-7 g/l of water at basal region reduces the disease.

DISEASES OF MESTA

Mesta (*Hibiscuss abdariffa*: roselle and *H. cannabinaus*: kenaf) is the second most important bast fibre crop after jute. The principal producing countries are India, China, Thailand, Egypt, Sudan, Brazil and Australia.In India it is mostly concentrated in Andhra Pradesh, Orissa and West Bengal.Besides fibres it has other uses as well such as production of paper and pulp and some people prepare pickle from mesta. Although mesta is more adaptive to adverse soil and climatic conditions, it suffers from a number of diseases which are discussed below.

1. Foot and stem rot

1.1. Economic importance: Foot and stem rot is the most important disease of mesta in India (Ghosh, 1983) causing loss to the extent of 10 – 25%. The disease can attack the crop from very early stage to maturity affecting both quality and quantity of fibre. It is more serious in roselle (*Hibiscus saffdariffa*) than in kenaf (*H. cannabinus*). In case of severe infection, more than 40% crop loss was observed in roselle.

1.2. Symptoms: It is prevalent in all the mesta growing areas of India. In foot and stem rot the symptom appears on the stem generally a few inches above the ground level but the spots may be seen at higher or lower level also. The spots are deep brown to blackish in colour with variable size. Larger spots very often girdle the whole stem and as a result the plants break at the point of infection. No fibre is obtained from such plants.

1.3. Pathogen: The disease is caused by *Phytophthora parasitica* var. *sabdariffae*.

1.4. Predisposition: It is favoured by high temperature and high humidity. Water stagnation which is common during mesta growing season in West Bengal predispose the plants to infection. Continuous drizzling, high rainfall, cloudy condition from May to September may be responsible for epidemic. The mean monthly temperature and relative humidity during the period were 33⁰C and 93-70 per cent.

1.5. Management

1.5.1. Varietal resistance: None of the cultivated varieties are resistant to the disease, but AMV 1, Roselle Type 1 and AP 481 were observed to be moderately resistant. Red bristled *H. saffdariffa* lines are more resistant than others.

1.5.2. Chemical control: Seed treatment with Dithane M 45 (mancozeb) @ 5.0 g kg-1 followed by soil drenching (0.2%) twice and spraying of Blitox 50 WP or Fytolan 50 WP (copper oxychloride) @ 5.0 – 7.0 g l-1 or Bavistin 50 WP (carbendazim) @ 2.0 g l-1 of water (Mukherjee and Basak, 1973). De *et al.*, (2008) reported that pre-sowing seed treatment with copper oxychloride was more effective than carbendazim, mancozeb, metalxyl, ediphenphos, carboxim, hexaconazole and thiophenate methyl. Copper oxychloride reduces the disease incidence by 50.3-45.5% at 30 days after sowing and at maturity, respectively and increased 20.6 %

fibre yield of HS 4288.

1.5.3. Biological control: Disease control was reported by the application of *Trichoderma viride* + *Azotobacter* and also by *Pseudomonas* + *Azotobacter* (Bandopadhyay and Bandopadhyay, 2004).

2. Yellow vein mosaic

Yellow vein mosaic of mesta was found in a devastating form in all mesta growing areas in India. It has been reported from eastern (Chatterjee *et al.*, 2005) and northern part of India (Ghosh *et al.*, 2007b).

2.1. Symptoms: The disease was first observed in few plants from Andhra Pradesh and then it was reported from west Bengal and Uttar Pradesh. The characteristic symptom of the disease is yellowing of veins and veinlets followed by complete chlorosis of the leaves of affected plants at an advanced stage of infection. The disease is developed one or both halves of the leaves or with the appearance of erratic chlorotic flecks in the vein and veinlets (Chatterjee *et al.*, 2006). In severe cases, those flecks gradually increase in size and yellow netting. Along with the vein yellowing the leaf lamina started showing foliar discolouration. This discolouration is green yellow at first stage which in advance forms yellow mosaic, chlorotic yellowing and complete chlorosis of foliage in sequential chronology. Occasionally the stem of diseased plant becomes yellow, partially or totally. The flowers and fruits are malformed causing low seed yield. It the plants are infected by the virus at their early stage of growth they do not flower even. Infected plants in general show stunted growth with reduced leaf size. Survey revealed that 90 per cent incidence of the disease is yield loss due to this disease alone was found to be 12.78- 17.45 per cent with respect to fibre yield and 18.91- 23.83 per cent with respect to seed yield (Roy *et al.*, 2007)

2.2. Transmission: Viruliferous whitefly could effectively transmit the virus with 78-85 per cent transmission efficiency (Chatterjee *et al.*, 2005). Three whiteflies per plant were found to be effective for disease transmission. Typical symptom appeared after a minimum incubation period of nine days under glass house condition. Minimum acquisition and inoculation access period for the vector was observed to be 12h and 4 h respectively.

2.3. Host range: The virus has a very narrow host range. Besides the two species of mesta, it could be experimentally transmitted to *Vigna unguiculata* and *Vigna umbellate* (Chatterjee *et al.*, 2007).

2.4. Particle morphology: Transmission Electron Microscopy with 2 per cent uranyl acetate stains form typical symptomatic leaves of *H. cannabinus* revealed that the disease is associated with geminivius. The size of the geminate particle is 20nm x 30 nm (Chatterjee *et al.*, 2005).

2.5. Detection, identification and phylogenetic relationship: Positive hybridization signal in Southern blot using radio labeled probe of cotton leaf curl Rajasthan virus DNA A and a-DNA confirmed the involvement of a Begomovirus with the disease (Chatterjee *et al.*, 2005). From the northern India a strain of äDNA containing tobacco leaf curl New Delhi virus has been reported to be associated with this disease. Characterization of coat protein gene of associated Begomovirus and the satellite DNA Beta molecule from six different geographical isolates of eastern and northern India revealed isolates obtained from eastern and northern India formed two distinct groups indication the existence of two distinct Begomovirus complexes associated with the yellow vein mosaic disease of mesta in two different geographical location in India (Roy *et al.*, 2007). Nucleic acid based diagnostic have been developed for identification of these complexes (Chatterjee *et al.*, 2007). Characterization of DNA molecule of an east India isolate revealed its close similarity with beta DNA molecule associated with cotton leaf curl disease (Chatterjee and Ghosh, 2007a). The Full length DNA A homologue (-2.7 kb) of begomovirus have been characterized from two east Indian isolates which revealed that these two isolates consisted of 2728 and 2752 nucleotide respectively (Chatterjee and Ghosh, 2007b, Roy *et al.*, 2007) and shared 84.3% sequence identity with cotton leaf curl Bangalore virus followed by Malvestrum yellow vein virus from china. It has been noticed that this DNA A molecule shared 89 % sequence identity with any known begomovirus sequence and hence it is proposed as a new species of begomovirus with a tentative name mesta yellow vein mosaic virus. Individual ORF analysis revealed that this DNA A evolved as a recombinant molecule from at least three distinct begomovirus from north, south and China.

2.6. Management: Initial record on vector control revealed that imidachlorprid and Thiamethoxam were effective against reducing the population of whitefly. The vector of yellow vein mosaic of mesta (*B. tabaci*) can be managed effectively by imidaclorprid 17.8 SC (3ml/10lit. of water) and thiamethoxam 25 WG (5g/20 lit. of water).

DISEASES OF SUNNHEMP

Sunnhemp (*Crotalaria juncea* L) commonly known as Indian hemp, Madras hemp, brown hemp is a multipurpose Fabacious crop grown widely in India. The fibre obtained from sunnhemp is slightly lignified, light in colour somewhat course, strong and long lasting. It is used for various purpose like making ropes, strings, twines, floor mat, fishing nets, handmade paper etc in cottage industries and paper pulp in paper industry. Apart from these industrial values, the plant being a legume is advantageous to grow on poor, fallow or freshly reclaimed soil playing major role as soil builder or renovator as well as deterrent to nematode. One of the major constraints of sunnhemp cultivation is the incidence of diseases especially in monsoon-sown crop.

1. Sunnhemp wilt

This disease has been recorded from several countries. Vincens (1921) reported the occurrence of sunnhemp wilt from Tonkin which was described as a collar disease similar to pigeon pea and cotton wilt. He isolated several fungi from the affected plants and found constant presence of *Fusarium* closely allied to *F.udum* Bult. *(F.vasinfectum* Atk*)* but failed to prove the pathogenecity. Wilt of *C. usaromoensis* was reported from Dutch East India. It was recorded from Trinidad and found that the pathogen was in resemblance with *F. vasinfectum* Atk. infecting pigeon pea.

Hean (1947) reported this disease from South Africa. In India this disease was reported by Uppal and Kulkarni (1937) from the then Bombay Province. Mitra (1934) observed this disease and found that the pathogen similar to *F.vasinfectum* Atk. at Pusa, Bihar. Desai *et al.*, (1984) reported premature wilting of sunnhemp in *in vitro* conditions. Studies conducted at Sunnhemp Research Station, Pratapgarh, U.P. also showed heavy incidence of vascular wilt in late sown crop (Sarkar *et al.*, 1998). Thakur (1971) reported the occurrence of wilt (*F.udum* Bult.) in *C.verrucosa*. Armstrong and Armstrong (1950, 1951) isolated *Fusarium* from seven *Crotalaria* species from India and USA. He found three races (1,2,3) based on selective pathogenecity and in contrary to other workers, they found that the pathogen did not attack pigeon pea. In spite of appreciable difference in cultural characteristics between Indian and USA isolates, they belong to same pathovar, *F. udum* var *crotalariae*. In India, it is reported from sunnhemp growing areas of Uttar Pradesh, Bihar, Maharashtra and Madhya Pradesh.

1.1. Economic importance: Generally the incidence was 10-12 % (Mitra, 1934) but under favourable conditions the incidence may go up to 60-80%. Uppal and Kulkarni (1937) also reported 88% incidence under green house conditions. More than 60 % loss was observed in seedling condition. In seed crop the incidence is much more because of favourable conditions.

1.2. Symptoms: The description of symptoms was reported by Mitra, (1934). The affected plants gradually whither, droop, hang down and later on turn brown, and ultimately die within one or two days. Usually the whole plant wilts but partial wilting is also noticed. In grown up plants the wilting parts droop at the tips and defoliation starts which consequently die. The sporodochium of the fungus with pinkish tinge are produced on the dead stem or the dead portion of the stem where the infection is confined to one side.The discolouration of the tissues could be traced to the main tap root or lateral roots. In the early stage, the fungus is confined to the lateral roots especially in the tip portion and subsequently attacks the vascular bundle of meristem (Fig. 7.2)

Fig. 7.2. Sunhemp wilt

1.3. Causal organism: The disease is caused by *Fusarium udum* (Bult) f.sp *crotalariae* (Kulkarni) earlier named as *F.vasinfectum* Atk. v. *crotalariae*, *Fusarium lateritium* f.sp. *crotalariae* (Padwick). As the pathogen is a facultative parasite, it survives in the crop stubbles. The pathogen produces pink coloured sporodochia on which enormous micro and macro conidia are produced. Fungal hyphae and spores plug the xylem vessels of the infected part causing the death of the plant. Microconidia 1- celled, hyaline, mostly curved and scattered. Macroconidia are subulate, falcate, narrowed towards either end, 1-3 rarely 4-7 septate and pedicellate. Conidia produced from the tips of phialides are borne on more or less verticillate branced conidiophores forming slimy groups at the tips of the phialides, later scattered on the mycelium to form pale pinkish masses sporodochia on the host usually formed in sporodochia of similar colour. Chlamydospores are 4-10µ in diameter, usually intercalary, ochre yellow, stroma mostly immersed, more or less spread out, plectenchymous, at first pale to pinkish, then salmon orange to cinnabarinous when dry.

1.4. Epidemiology: The fungus survives in the soil as well as in crop residues as facultative parasite (Saxena, 1989). It enters into the plants through the thinner roots, rootlets and even through the cracking in the basal portion of the stem. The pathogen produces enormous spores (both macro and micro condia) in the pink coloured sporodochium that are also capable of infecting the growing crop. The fungus was also noticed on the pod and in many cases in the seeds of diseased pod (Mitra, 1934, Sarkar and Tripathi, 2003). The infected discoloured seed can also initiate the infection in the field. Temperature plays an important role in the development of this disease. Uppal and Kulkarni (1937) reported that the incidence of this disease declined with high temperature at Bombay. Mitra (1934) and Mundkur (1935) recorded similar observation at Pusa. At Sunnhemp Research Station, Pratapgarh, U.P., Sarkar *et al.,* (2000) studied the effect of weather parameters (temperature, relative humidity and rain fall) on the crop vis-a-vis incidence of this disease during the months of July to October and found that wilt started at about 62 days after sowing and increased significantly with the age of the crop. Temperature was found to be negatively correlated with the incidence. Relative humidity and maximum temperature were found to be non-significant in disease development. Rainfall during last fortnight of October was found to be conducive for the spread of this disease. The incidence of disease is less in early (mid-April) sown crop. But on the contrary, Saxena (1989) reported that the disease incidence increased with high temperature and high moisture. In low-lying areas wilt incidence was also found to be more.

1.5. Management

1.5.1. Host resistance: K-12 yellow – a selection from K-12 (black) was found to be largely resistant against wilt (Ghosh *et al.,* 1977, Sarkar *et al.,* 1998). An improved self-compatible strain (Bidhan Shan), developed at Bidhan Chandra Krishi Viswavidyalaya, West Bengal, through bulk selection, was found to be highly resistant against diseases (Anonymous, 2003). Uppal and Kulkarni (1937) found that D-IX was highly resistant in decan condition. Considerable work has been done on screening of germplasm against wilt by Sarkar *et al.*(1998). But as the crop is highly cross-pollinated in nature, whatever promising germplasm were found to possess resistance became ultimately susceptible. Sarkar *et al.* (1998) tested a number of wild *Crotalaria* species and found that *C. brevidence, C. mucronata, C. verucosa and C. striata* are highly resistant against wilt under natural conditions. This character may be transferred to the cultivated species but so far attempts on interspecific cross between different *Crotalaria* spp. remained unsuccessful (Kundu, 1964).

1.5.2. Seed treatment: Seed treatment with Benlate @ 3 g ha^{-1} or spraying of Bavistin @ 3 g kg^{-1} was found to reduce the disease considerably. Considerable work on soil application of neem cake along with seed treatment and application of ZnSO$_4$ has been done at Sunnhemp Research Station, Pratapgarh, U.P and CRIJAF. It was finally recommended that sowing of treated seeds with Bavistin @3 g kg^{-1} and *Rhizobium* and soil application of neem cake @1 q ha^{-1} + 25 kg ZnSO$_4$ ha $^{-1}$ is the best option. In contrast to this, Saxena (1995) tested nine fungicides including Bavistin but the results were not encouraging.

1.5.3. Nutrient management: Management of *Fusarium* wilt was also attempted with application of NPK but no encouraging result was observed (Saxena, 1989, Sarkar, *et al.,* 1998). Sarkar *et al.,* (2000) studied the effect of boron, zinc and iron but no significant decrease in wilt was observed.

1.5.4. Sowing dates: Sunnhemp can be successfully grown between April to August in North Indian conditions. But it was observed that delay in sowing time increased this disease (Mitra, 1934). In April sowing, the crop almost escapes the disease (Sarkar *et al.,* 1998, Sarkar and Tripathi, 2003).

1.5.5. Intercropping: Intercropping with sesame and sorghum reduced the incidence of wilt (Saxena, 1995 and Sarkar *et al.,* 2000).

2. Anthracnose

Three anthracnose diseases of sunnhemp have been reported. Petch (1917) described this disease from Peradeniya, Srilanka on *Crotalaria striata* and named the pathogen as *Colletotrichum crotalariae* Petch. In southern USA the same fungus was reported on *C. striata* and *C spectabilis* (McKee and Enlow, 1931). Mitra (1937) reported severe incidence of this disease in seedling stage at Pusa.

2.1. Symptoms: The seedlings are attacked first on the cotyledon and subsequently infection spreads to the stem and growing point. The disease makes an appearance in the form of soft discoloured areas on the cotyledon. Later, brownish spots are formed on all parts of host except underground parts. The affected seedlings droop from the point below cotyledon. The affected cotyledons themselves drop from the petiole. The infection spreads downward and acervuli are formed with copious spores on the infected areas within two days. The young seedling when infected generally dies. When older plants are infected the disease is restricted on leaf and stem and the heavily infected leaves fall off. The spots on older leaves appear on one side of the leaf but gradually enlarge and extend to the opposite side. These spots are grayish brown to dark brown, roundish or irregular. Several spots coalesce and cover the entire leaves or large portion of the leaf. They may also be formed on the midrib.

2.2. Causal organism: The disease is caused by *Colletotrichum curvatum* Briant and Martyn. The mycelium is at first localized in the tissues of a lesion either on leaf or stem of a young seedling and does not extend internally to other parts. In severe cases, however, it not only penetrates in the cortical tissues but also invades the vascular bundles. The acervulli are formed in abundance in the epidermis of the diseased area. They are white or faint pinkish in colour and consist of simple erect closely packed conidiophores and setae. From the tip of the conidiophores the conidia are budded off epically one at a time and pile up in a light pinkish heap on the top of acervullus. A mucilaginous secretion holds the conidia together. Conidia are one celled, hyaline, falcate and acute at each end measuring around

15-24 x 3-4 µm. The conidia germinate by producing germ tube, which forms an appresorium to infect the host. The setae are brown to dark brown in colour, septate, tapering towards the tip and swollen at the base. These are found between the conidiophores, measuring 66-140µ x 4-6µ.

2.3. Epidemiology: Weather conditions and age of the plant influence the disease greatly. The cloudy weather accompanied by continuous rain favour the rapid spread of disease in the thickly populated crop. The infection is found to be severe in the seedling stage. Rain splashing helps the spores to spread in adjacent plants (Mitra, 1937).

2.4. Management

2.4.1. Seed treatrment: Seed borne infection can be checked by disinfecting the seed with suitable fungicides (Mitra, 1937). Sarkar and Tripathi (2003) tested a number of fungicides and found that Thiram or Bavistin are highly effective in reducing this disease.

2.4.2. Spraying: Spraying with Bordeaux mixture (0.5%) at seedling stage also reduces the disease incidence.

2.4.3. Sowing time: As the disease is favoured by rain and high humidity early sowing in dry season i.e. mid-April to mid-May helps to escape this disease.

2.4.4. Resistance: Amongst the cultivated varieties, K-12 yellow was found to be largely resistant to the disease (Ghosh *et al.*, 1977). But due to cross-pollinating nature of the crop resistance might have depleted. Dey *et al.*, (1990) tested 66 germplasm and found 11 germplasm to be resistant. Recently, a variety named SH-4 has been released, which is resistant against wilt under natural conditions.

3. Leaf blight

Sarkar and Pradhan (1999) reported heavy incidence of leaf blight (54%) from Pratapgarh, Uttar Pradesh. The blight started from the margin of the leaf and proceeds inwards. Under moist and warm conditions with intermittent rains, the whole leaf may be blighted. But with the fall in temperature along with receding rains severity was restricted. Often black pinhead like fungal structure is noticed on blighted site. In the early morning, the blighted leaves look greyish and water soaked, which ultimately become brownish with broad yellow margin. Subsequently, the infected leaf becomes weak and droops down from the plant, which gives a sickly appearance of the whole

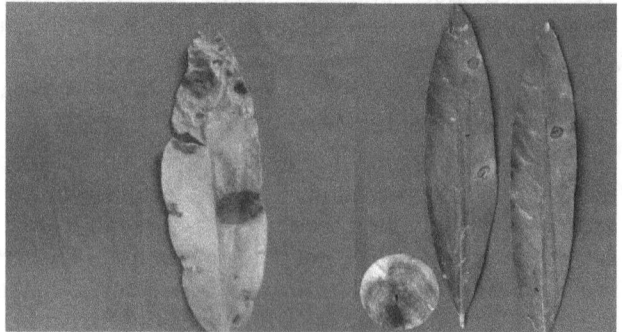

Fig. 7.3. Leaf Blight

field. This disease is caused by *Macrophomina phaseolina* (Tassi) Goid. It is a soil borne disease (Fig. 7.3).

4. Choanephora leaf blight and tip rot

Sporadic incidence of this disease was reported from Varanasi, U.P. The authors also observed the disease at Sunnhemp Research Station, Pratapgarh, U.P in highly humid conditions. About 20 % incidence of this disease was recorded in the month of August. This disease is characterized by the rotting of terminals. Brown discolouration occurred just below the infection point. Affected portions decay, break and droop. White mycelial growth along with black coloured sexual bodies of the fungus is seen on the affected parts from the tip towards petioles. A characteristic whitish brown discolouration in leaves precedes the disease development. Infected leaves lose their chlorophyll and droop from the stem. The pathogen affects the epidermal and outer cortical layers, and cells in these regions get disintegrated. The disease has also been reported from the wild species like *C. spectabilis, C.sericea* and *C. retusa.*

5. Leaf spot

There are three types of leaf spots reported. These are as follows:

5.1. Pleospora leaf spot: Pathak and Chauhan (1976) reported this disease from Agra, U.P., occurred in the month of September-October. It is characterized by the development of pale yellow spot measuring 7-15 mm in diameter. The spots turn into black later on.

5.2. Phoma leaf spot: It was also reported by Pathak and Chauhan (1976). Here the spots appeared as yellow, which turn to brown. Size of the spots is 3-5mm.

5.3. Pringsheimia leaf spot: Mishra (1976) reported this disease. It is characterized by the development of brown spot (bound by vein and veinlets) on the mature green leaves during the month of September. Microscopic observations showed the presence of minute globose black erumpent perithecia scattered all over the necrotic spots. The perithecia contain a number of pyriform, hyaline asci with a thin but tough wall. The ascus containaing 8 ascospores are hyaline, somewhat thick walled, 4-7 celled, muriform with longitudinal septum appearing in one or two cells only.

6. Sunnhemp mosaic

Raychaudhury (1947) first reported this disease from Delhi and described the symptoms. The first visible symptom appeared within 10 to 12 days after inoculation when mottling was seen on the youngest leaves. As the disease progressed, patches of light and dark green areas became more prominent (Fig. 7.4.). Diseased leaves were smaller than the normal. In severe cases of infection, growth of the lamina generally became abnormal. Frequently dark green raised areas on the upper surface were seen with corresponding depression on the lower surface. The infected plant remained shorter in height

Fig. 7.4. Sunhemp mosaic

and therefore, fibre and seed yields were greatly reduced.

6.1. Causal organism: The disease is caused by virus.The structure of the virus was determined under electron microscope (Dasgupta *et al.*, 1951). The virus consists of spherical particles measuring 26-40 µm. Anand and Sahambi (1965) studied serological relationship of sunnhemp mosaic virus with other mosaic virus and they found that the sunnhemp mosaic virus produces a flocculent precipitate resembling with bottle gourd mosaic virus, tobacco mosaic virus and cowpea mosaic virus.

6.2. Transmission and host range: The disease was reported to be transmitted mechanically. Nariani and Chadrashekher (1963) showed the successful transmission through root inoculation. It was found that the hosts of the disease are *C. mucronata, Cyamopsis tetragonoloba, Pisum sativum, Nicotiana tabacum, Datura strumonium* and *Lycopersicon esculenta.* Jen sen (1950) reported that sunnhemp mosaic virus of Trinidad can easily be transmitted to cowpea. The virus movement was studied by Sastry and Vasudeva (1963) and they observed the systemic transmission of the virus. They also noticed that optimum dose of nitrogen and increased dose of phosphorus increased the spread of the virus within the plant but the reverse was true in case of potassium. The concentration of the virus was found to be more in optimum nitrogen application whereas, potash application was found to reduce the concentration of the virus (Sastry and Vasudeva, 1963).

6.3. Epidemiology: Epidemiological studies on the virus have not been studied so far. But at Sunnhemp Research Station, Pratapgarh, U.P. it was observed that the severity of the virus increases with the onset of monsoon. As the virus is reported to be sap transmitted and having wide host range, the disease may be initiated from the other wild or weed host(s). The same observation was also noticed in the farmers' field in and around Pratapgarh. Higher dose of nitrogen increases while potassium application reduces the disease (Sastry and Vasudeva, 1963).

6.4. Management

6.4.1. Sowing time: Time of sowing is very important to avoid the disease. Drastic reduction of this disease was noticed in the early sown (mid- April) crop (Sarkar and Tripathi, 2003) than the monsoon crop.

6.4.2. Resistance: Ghosh *et al.*, (1977) reported K-12 yellow as a largely resistant variety. Being a cross-pollinated crop resistance might not be observed at present. Recently released SH-4 variety was found to be resistant to sunnhemp mosaic. Wide variation in disease reaction amongst the different germplasm was noticed at Sunnhemp Research Station, Pratapgarh, U.P. as well as at Gobind Ballabh Pant University of Agriculture and Technology, Pantnagar.

7. Southern sunnhemp mosaic

Sunnhemp is extensively grown for fibre and green manure (in South India) in rotation with rice, wheat and sugarcane. Severe incidence of mosaic disease, which caused appreciable loss in fibre yield, seed and green matter, was reported by Capoor (1950, 1962). This mosaic disease is widespread in nature.

7.1. Symptoms: Initial symptom of the disease appeared within 9-20 days depending on weather conditions. It is characterized by faint discoloured patches appearing first on young leaves. Distinct mosaic with puckering and blistering of leaves developed. Subsequently thin elongated enations running more or less parallel to each other developed on the under surface of the leaves. The characteristics mosaic mottle and varying degree of leaf distortion occurred in advanced stage of the disease. Plants became dwarf with reduced leaves and bear scanty flush of flower resulting in poor pod setting and consequently, reduction in quantity of seed (Capoor, 1962; Chandra *et al.*, 1975). The incidence of the disease is more in monsoon months when temperature ranges between 22-41⁰C and relative humidity 80%.

7.2. Host range: The host range of southern sunnhemp mosaic virus was confined mainly to the families, Leguminosae and Solanaceae. Systemic mosaic mottling appeared at about 10-20 days after inoculation in *C. retusa, C. mucronata, C. laburnifolia, C usaramoensis, C.spectabilis, C. lanceolata, Pisum sativum, Vigna sinensis, V. unguiculata, V. cylindricum, V sesquipedalis, Phaseolus vulguris, P latanus, Cassia tora, Cajanus cajan, Solanum nigrum, Datura inoxia* and *Nicandra physaloides.* Local lesion develops on *Nicotiana tabacum, N glutinosa, N. sylvestris, D strumonium, D. aegyptiana.* and *Gompherena globosa.*

7.3. Transmission: The virus was readily transmitted by leaf rubbing without the help of any abbrasive or by pinpricking. It was also transmitted by wedge or patch grafting of diseased tissues to healthy sunnhemp plants. However, the virus is not transmitted through seed and soil (Capoor, 1962).

7.4. Virus particles and physical properties: The electron monograph revealed that the particles are rod shaped with 300 µm x 18 µm (Anand, 1968 and Nariani *et al.*, 1970). On the basis of cross-inoculation studies, Capoor (1962) concluded that it is a strain of tobacco mosaic virus. Serological studies also proved that it is related with the type strain of TMV. Physical properties of the virus was studied by Capoor (1962) and it was found highly stable with TIP-95⁰C, DEP 10⁻⁷ and aging *in vitro* for 6 years. The virus can withstand complete desiccation.

8. Sunnhemp rosette

This disease is characterized by mosaic mottling, rosetting of leaves, stunted growth and sparse flowering, poor fruiting and seed setting (Verma and Awasthi, 1976). They reported around 40-60 % incidence of this disease in eastern U.P.

8.1. Causal organism: The disease is caused by a rod shaped RNA virus measuring 400 µm x 17 µm. The particle shape although conforms to TMV the length is more as compared to TMV. The physical properties like thermal inactivation point, dilution end point and *in vitro* longevity are 93⁰C, 5x10⁻⁵ and 54-56 days at 20-26 ⁰C, respectively.

8.2. Transmision: The virus is transmitted mechanically through sap, but it is also seed transmitted (5-10%) in nature if the plants were infected before flowering (Verma and Awasthi, 1976).

8.3. Host range: Unlike other virus diseases of sunnhemp, rosette virus did not produce systemic symptom as in other member of family Leguminosae, but it produces hypersensitive response in *Cyamopsis tetragonoloba, Vigna sinensis* (Leguminosae) *and Chenopodium amaranticolour* (Chenopodiaceae).

8.4. Management: Practically no information is available to control the disease at the field level. But early sown (mid- April) crop generally escapes the disease in eastern U.P. Amongst different varieties, K-12 yellow showed lowest disease incidence (Ghosh *et al.*, 1977). At present systemic induced resistance is gathering momentum in the field of disease management. In this context, a number of plant extracts like *Cleodendrum aculeatum, Chenopodium murale, Boerhaavia diffusa* (root extract), *Cuscuta reflexa, Solanum melongena* and *Pseuderanthemum bicolour* were found to reduce the disease under laboratory conditions (Verma and Awasthi, 1978, Verma and Khan, 1984; Srivastava and Verma, 1995, Verma and Varsha, 1995).

Diseases of Ramie (*Boehmeria nivea*)

Ramie (*Boehmeria nivea*) produces the strongest and finest fibre which can be blended with any other natural fibres. China is the leading producer in the world. World production of ramie is about 180 thousand tonnes from a cultivated area of 109 thousand hectares. It is grown in tropical, sub-tropical and temperate regions and the main countries where it is grown are China, Brazil and the Philippines. In India, ramie cultivation is concentrated in North Eastern parts of the country. The diseases of ramie are discussed in relation to their occurrence, symptoms and control measures. The major and most widespread diseases are leaf spot caused by *Cercospora* spp., seedling rot caused by *Rhizoctonia solani*, eye rot caused by *Myrothecium roridum*, anthracnose caused by *Colletotrichum boehmeriae* Swada and *Colletotrichum gloeosporioides*, cane rot by *Rhizoctonia bataticola* and stem rot of ramie caused by *Corticium rolfsi*.

1. Cercospora leaf spot (*Cercospora boehmeriae* and *C. crugiana*)

It is a common and widespread disease in all the ramie growing tracts of the state of Assam. Affected plants are not killed outrightly. When the spotting is very severe and if the plant is very young the growth is retarded. The disease appears as circular to angular spotson the upper surface of the leaves. Sometimes the spots which are 1-8 mm in diameter and dark brown to nearly black in colour are limited by the leaf veins. The centre of older spots, however, turn paler and become greyish brown. Adjacent spots may coalesce. No symptom has been seen on stem. The petiole however shows brown elongated spots. The disease has been observed through the year but the severest infection occurs from the later part of September to early January. The development of this disease is favoured by moist and cool weather. Infection is usually greater on the lower leaves than on the upper leaves. In case of severe infection more than 60% of leaves are affected and 30-60% of the leaf area on an average gets affected by spotting. Severely spotted leaves turn yellow and fall prematurely. The fungus grows well in a number of culture media. The disease can be controlled by spraying with copper fungicides.

2. Damping off (*Rhizoctonia solani*)

Damping off of ramie is caused by *Rhizoctonia solani*. The fungus attacks foot of the stem or crown of the roots under certain conditions rendering tissues at the region weak and ultimately results in seedlings collapse. In seed beds (when ramie is raised through seed for breeding purpose etc.) the seedlings may be attacked by damping off in humid and moist conditions. Affected seedlings are pale green and show a girdle of brown decaying cortex. The seedlings then collapse together get decomposed. *R. solani* has been found to kill many young plants causing breakage of young plants at the point of infection. Moisture content of soil and humidity are important factors for seedling disease. Sterilization of soil of seed bed by using formaldehyde (1: 50) is the best remedy. Treating planting stalks with captan is also effective in controlling *Rhizoctonia* seedling disease in ramie.

3. Eye rot (*Myrothecium roridum*)

The disease attacks leaves as well as stems. Infection of leaves results in both qualitative and quantitative loss of fibre. The first symptom appears as irregular, small, round tan coloured spots of about 1 mm diameter on the upper surface of the lamina. As the disease advances the spots become circular and elongated to irregular, 1-16 mm in diameter and brown to dark brown in colour. In case of severe infection, the incidence of disease is reported to be 7-10%. The disease is controlled by spraying copper fungicides.

4. Cane rot (*Rhizoctonia bataticola*)

The symptom of cane rot disease appears as necrotic lesions on the leaves which gradually cover the entire leaf blade. The leaves then crinkle, rot, adhere to the canes and ultimately shed off. Root system become weak and turns brown. The disease is prevalent during rainy season on mature clumps. Brown sunken circular or elongated lesions are common on the stalks of ramie specially inthe basal regions. They increase in size and several such lesions coalesce and girdle the stalk. When the lesions streaks along the length of the stalk it shrivels resulting in complete drying. A number of stalks may be infected in a single clump. The organism causing the symptoms exists as brown mycelia on the affected region of the stalk and form black *sclerotia*.

5. Anthracnose (*Colletotrichum boehmeriae* Swada and *Colletotrichum gloeosporioides*)

In China, anthracnose is regarded as one of the most widespread and devastating diseases of ramie. Seriously infected seedlings always lodge and even die due to this disease. Adult plants have perforations on infected leaves, yellow and caducous leaves, red fibers which are difficult to bleach and decrease the fiber fineness, fiber strength and filament elongation ratio, which in all negatively affect the production and quality of ramie fiber. Lesions are initially small, scattered, round, and gray with brown margin on leaves. As the disease progresses, irregular spots develop and expand until the leaves are withered. Initial lesions on stems are fusiform and expanded, causing the stem to break. Finally, the fibers rupture. Conidia are single celled, colourless, straight, oval, obtuse at both ends, and 11 to 18 × 3 to 6 μm with an average of 14.89 × 4.32 μm. Conidiophores are dense and 11 to 22 × 4 to 5 μm with an average of 15.82 × 4.43 μm. Setae were few, dark brown, one to two septa, and 62 to 71 × 4 to 5 μm with an average of 65.13 × 4.46 μm.

6. Stem rot

A new disease of ramie, named stem rot, was observed and reported for the first time in the Philippines. The disease is caused by a fungus, *Corticium rolfsii* Cursi. Symptoms of the disease were wilting and water soaking of the basal portion of the plant. Severely infected plants turn brown, get defoliated and ultimately die. Profuse white mycelia of the fungus are found covering the infected stem. The fungus has a wide host range.

REFERENCES

Ahmed, M. 1978. A whitefly-vectored yellow mosaic of jute. FAO, Plant Protection Bulletin 26: 169-171.

Ahmed, Q. A., Biswas, A. C, Farukuzzaman, A. K. M., Kabir, M. Q., and Ahmed, N. 1980. Leaf mosaic disease of *jute.* Jute and Jute Fabrics, Pakistan. 6: 9-13.

Anand, G. P. S., and Sahambi, H. S. 1965. Serological relationship of sunhemp mosaic virus with some other mosaic viruses commonly occuring in Delhi. Indian Phytopath. 18: 204-205.

Anand, G. P. S. 1968. Particle shape of southern sunnhemp mosaic virus. Curr.Sci. 37.706.

Anonymous. Annual Report, AINPJAF, 2001-2004.

Anonymous. 2003. ICAR News 9: 3.

Anonymous. Annual Report JARI. 1949-1956

Anonymous. 1999. Fifty years research in jute and allied fibres agriculture (1948—1947). Central Research Institute for Jute and Allied Fibres (CRIJAF), Barrackpore.

Armstrong, J. K., and Armstrong, G. M. 1951. Physiological races of the Crotalaria wllt fusarium. Phytopathology 41: 714.

Armstrong, J. K., and Armstrong, C. M. 1950. Biological races of the fusarium causing wilt of cowpeas and soybeans. Phytopathology 40: 785.

Ashby, S. F. 1927. *Macrophomina phaseoli* (Maubl.) Comb. Nor.: the pycnidial stage of *Rhizoctonia bataticola* (Taub.) Butl. Transactions of the British Mycological Society.12: 141-147

Bandopadhyay, A., and Bandopadhyay, A. K. 2004. Beneficial traits of plant growth promoting rhizobacteria and fungal antagonist consortium for biological disease management in bast fibre crops. Indian Phytopath. 57: 356-357.

Bandopadhyay, A., Bandopadhyay, A. K., and Samajpati, N. 2008. Invitro antifungal activity of some biocontrol fungi against jute pathogen *Macrophomina phaseolina.* Indian Phytopath.61 (2): 204-211.

Biswas, C., Biswas, S. K., and Srivastava, S. S. L. 2009. In: Recent advances in biopesticides: biotechnological applications. (Ed. Johri, J.K.), New India

Publishing agency, New Delhi, pp 105-116.

Bowers, G. R., and Russin, J. S. 1999. In: Soybean production in the mid-south. L. G. Heatherly and H. F. Hodges. CRC Press.

Capoor, S. P. 1950. A Mosaic Disease of Sunn Hemp in Bombay. Curr. Sci. 19: 22.

Capoor, S. P. 1962. Southern sunn-hemp mosaic virus: a strain of tobacco mosaic virus. Phytopathology 52: 393-397.

Chandra, S., Singh, B. P., Nigam, S. K., and Srivastava, K. M. 1975. Effect of some Naturally Occuring Plant Products on Southern Sunnhemp *Mosaic* Virus (SSMV). Current-Science 44: 511-512

Chatterjee, A., and Ghosh, S. K. 2007a. Association of a satellite DNA â molecule with mesta yellow vein mosaic disease. Virus Genes 35:835-844

Chatterjee, A., and Ghosh, S. K. 2007b. A new monopartite begomovirus isolated from *Hibiscus cannabinus* L.in India. Arch. virol. 152:2113-2118

Chatterjee, A., Roy, A., and Ghosh, S. K. 2006. In: Characterization, diagnosis and management of plant viruses (Eds. G. P. Rao, S.M. Paul Khurana and S.L. Lenardon) Vol. 1 Industrial crops Studium Press, Texas, USA. Pp 497-505.

Chatterjee, A., Roy, A., Padmalatha, K. V., Malathi, V. G and Ghosh, S. K. 2005. Occurrence of a Begomovirus with yellow vein mosaic disease of mesta (*Hibiscus cannabinus* and *Hibiscus sabdariffa*). Australasian Plant Pathol.34: 609-610

Chatterjee, A., Sinha, S. K., Roy, A., Sengupta, D. N., and Ghosh, S. K. 2007. Development of diagnostics for DNA A and DNA â of a *Begomovirus* associated with mesta yellow vein mosaic disease and detection of geminiviruses in mesta (*Hibiscus cannabinus* L. and *H. sabdariffa* L.) and some other plants. J. Phyto. Pathol. 155: 683-689.

Dasgupta, N. N., De, M. L., and Raychoudhury, S. P. 1951. Structure of Sannhemp (Crotalaria juncea Linn.) Mosaic Virus with the Electron Microscope. Nature.168: 114.

Das, S., Khokon, A. R., Haque, M. L., and Ashrafuzzaman, M. 2001. Detection of the causal agent of leaf mosaic of jute. Pakistan J.Biol. Sci. 4: 1500-1502.

De, D. K., and Kaiser, S. A. K. M. 1991. Genetic analysis of resistance to stem rot pathogen - Alice. Pesq.agropec.bras., Brasilia, 26: 1017-1022.

De, R. K., Mandal, R. K. and Mahapatra, A. K. 2008. Sources of resistance in jute (*Corchorus olitorius* L) against stem rot pathogen. Jaf News, Central Research Institute for Jute and Allied Fibres, Barrackpore 6: 12-14.

Desai, S. A., Siddaramaiah, A. L., Hegde, R. K., and Kulkarni, S. 1984. Efficacy of some fungicides against Fusarium lateritium f. crotalariae causing pre-mature wilting of sunnhemp in-vitro. Current-Research, University-of-Agricultural-Sciences, Bangalore 13: 10/12, 75-76.

Dey, D. K., Banerjee, K., Singh, R. D. N., and Kaiser, S. A. K. M. 1990. Effect of 2,4-D preplant soil spraying in the establishment of Centrosema pubescens. Environment and Ecology: 1217-1219

Dhingra, O. D., and Sinclair, J. B. 1977. An annotated bibliography of Macrophomina phaseolina. 1905-1975. Universidade Federal de Vicosa, Minas Gerais, Brazil.

Finlow, R. S. 1917. Historical notes on experiments with jute in Bengal. Agric. J. India. 12: 3-29.

Ghosh, T., and Mukherjee, N. 1970. *Macrophomina phaseoli* (Maubl) Ashby on jute; plant disease problems. In International Symposium on Plant Pathology, 1., Proceedings. New Delhi: IARI, p.369-370.

Ghosh, R., Paul, S., Das, S., Palit, P., Acharyya, S., Mir, J. I., Roy, A., and Ghosh, S. K. 2007a. Begomovirus and satellite DNA beta associated with yellow vein mosaic disease of malachra. In: Proceedings of Xth Plant Virus Epidemiology Symposium. October 15-19, ICRISAT, India.pp 115.

Ghosh, R., Paul, S., Roy, A. Mir, J. I., Ghosh, S. K., Srivastava, R. K. and Yadav, U. S. 2007b. Occurrence of Begomovirus Associated with Yellow Vein Mosaic Disease of Kenaf (Hibiscus cannabinus) in Northern India. Online Plant Health Progress DOI 10. 1094/PHP-2007-0508-01-RS

Ghosh, T. 1999. In: Jute and Allied Fibres – Agriculture and Processing (Eds. P. Palit, S. Pathak and D.P.Singh), Central Research Institute for Jute and Allied Fibres, Barrackpore, India.

Ghosh, T. 1957. Anthracnose of jute. Indian Phytopath. 10: 63-70.

Ghosh, T. 1983. Handbook on jute, FAO plant production and protection. Paper no 51. Rome, Italy.

Ghosh, T., Mohan, K. V. J., and Prakash, G. 1977. Sunnhemp Variety. Jute Agricultural Research (ICAR), Barrackpore, West Bengal, India.

Hean, A. F. 1947. A South African Virus Disease of Crucifers. Science Bull. Department of Agric. South Africa.15: 255

Jensen, D. D. 1950. A Crotalaria mosaic and its transmission by aphids. Phytopath. 40: 512-515

Kundu, B. C. 1964. Sunnhemp in India. Soil Crop Sci. Soc. Florida Proc. 24: 396-404.

Mandal, R. K., Sarkar, S., and Saha, M. N. 2000. Field evaluation of white jute Corchorus capsularis L. germplasms. Environ. Ecol. 18: 814-818.

Mandal, R. K. 1990. Jute diseases and their control. In: Proc. National Workshop cum Training on jute, mesta, sunnhemp and ramie held at CRIJAF, Barrackpore.

Mandal, R. K., and Mishra, C. D. 2001. Role of different organisms in inducing Hoogly wilt symptom in jute. Environment & Ecology 19: 969-972.

Mckee, R., and Enlow, C. R. 1931. Crotalaria, a new legume for the south. U.S. Dept. Agric. Cir. Bull.137, 30pp.

Mishra, C. B. P. 1976. A new host record of Pringsheimia sp. on sannhemp *Crotalaria juncea* L. Curr. Sci. 45: 468.

Mitra, M. 1934. Wilt disease of*Crotalaria juncea* Linn. (sann-hemp). Indian J. Agric. Sci., 4: 701-714.

Mitra, M. 1937. An anthracnose disease of Sann-Hemp. Indian J. Agric. Sci. 7: 443-449.

Mukherjee, N., and Basak, S. L. 1973. Chemotherapeutic control of foot and stem-rot of roselle *Hibiscus sabdariffa* var. altissima caused by *Phytophthora parasitica* var. sabdariffae. Indian Phytopath 26: 567-577

Mundkur, B. B. 1935. Parasitism of *Sclerotium oryzae* Calt. Indian J. Agric. Sci., 5: 609-618.

Nariani, T. K., and Chandrashekher, B. K. 1963. Transmission of Sunnhemp mosaic virus by mechanical root inoculation. Indian Phyto Path. 16: 171-73.

Nariani, T. K., Kartha, K. K., and Prakash, N. 1970. Purification of the Southern Sannhemp mosaic virus using butanol and differential centrifugation. Curr. Sci. 23: 539-541.

Pathak, P. D., and Chauhan, R. K. S. 1976. Two new leaf spot disease of *Crotalaria juncea* L. caused by *Pleospora infectoria* Fuckel and *Phoma glomerata* Corda. Curr. Sci. 45:206

Petch, T. 1917. Additions to Ceylon fungi. Ann Royal Botanic Gardens Perademya 6: 239

Raychaudhury, S. P. 1947. Note on Mosaic Virus of Sann-Hemp (Crotalaria Juncea Linn.) and its Crystallisation. Curr. Sci. 16: 26-28.

Roy, A., De, R. K., and Ghosh, S. K. 2008. Diseases of bast fibre crops and their management in jute and allied fibres, pp.327. In: Karmakar, P. G., Hazara, S. K., Subramanian T. R., Mandal, R. K., Sinha, M. K. and Sen, H. S (Eds.). Updates Production Technology, Central Research Institute for Jute and Allied Fibres, Barrckpore, West Bengal, India.

Roy, A., Ghosh, R., Paul, S., Das, S., Palit, P., Acharyya, S., Mir, J. I., and Ghosh, S. K. 2007. In Proceedings of Xth plant virus epidemiology symposium. October 15-19, ICRISAT, India.Pp 85.

Roy A., A. Bandyopadhyay, A.K. Mahapatra, S.K. Ghosh, N.K. Singh, K.C. Bansal, K.R. Koundal and T. Mohapatra. 2006. Evaluation of genetic diversity in jute (*Corchorus* species) using STMS, ISSR and RAPD markers. Plant Breed. **125**: 292-297.

Sarkar, S. K., and Pradhan, S. K. 1999. Incidence of leaf *blight* of Sunnhemp caused by *Macrophomina phaseolina*. J. Mycol. Pl. Pathol. 29: 143.

Sarkar, S. K., and Tripathi, M. K. 2003. Summer crop of sunhemp escape major pests and diseases. ICAR News 9: 16.

Sarkar, S.K., Pradhan, S. K., and Tripathi, S. N. 2000. Influence of boron, zinc and iron on the incidence of sunhemp wilt (Fusarium udum f. sp. crotalariae). J. Mycol. Pl. Pathol. 30: 116-118.

Sarkar, S. K., Pradhan, S. K., and Prakash, S. 1998. Influence of dates of sowing, fertilizer levels and varieties on the incidence of wilt (*Fusarium udum f.sp. crotalaria*) in sunhemp. Legume Res. 21: 225-228

Sastry, K. S. M., and Vasudeva, R. S. 1963. Effect of host plant nutrition on the movement

of sunnhemp mosaic virus. Indian Phytopath. 16: 143-150.

Saxena, P. 1989. Ann. Report, Central Res. Instt. For Jute and Allied Fibres, Barrackpore, West Bengal pp.135.

Saxena, P. 1995. Ann. Report, Central Res. Instt. For Jute and Allied Fibres, Barrackpore, West Bengal pp.113

Srivastava, N., and Verma, H. N. 1995. Prevention of plant virus disease by Boerhaavia diffusa inhibitor. Indian Phytopath. 48: 177-179.

Thakur, R.N. 1971. Occurrence of stem rot of Grotalaria juncea in Jammu and Kashmir. Indian J. Mycol. Plant Pathol.1: 147-48.

Uppal, B. N., and Kulkarni, N. T. 1937. Studies in Fusarium wilt of sann-hemp. I. The physiology and biology of Fusarium vasinf actum Atlj. Indian J. Agric Sci., 7: 413-442.

Verma, H. N., and Awasthi, L. P. 1976. Sunnhemp rosette. A new virus disease of sunnhemp. Curr. Sci. 45: 642-43.

Verma, H. N., and Awasthi, L. P. 1978. Further studies on a rosette virus of *Crotalaria juncea*. Phytopathology, 92: 83-87.

Verma, H. N., and Khan, M. M. A. A. 1984. Management of plant virus diseases by *Pseuderanthemum bicolour* leaf extract. Zeitschrift-fur-Pflanzenkrankheiten-und-Pflanzenschutz. 91: 266-272

Verma, H. N. and Varsha. 1995. Induction of systemic resistance by leaf extract of *Clerodendrum aculeatum* in sunnhemp against sunnhemp rosette virus. Indian-Phytopath. 48: 218-221

Vincens, M. F. 1921. summary report of the plant pathology laboratory, Scientific institute of Indochina, 1 Jan 1915- 1 July 1921. Bull. Agric. Inst. Sci. Sudu, Saigon, 3: 381-384.

8

Diseases of Groundnut

P.P. Thirumalaisamy

Groundnut (*Arachis hypogaea* L.) or peanut is an important oilseed and ancillary food crop of the world. It is the world's fourth most important source of edible oil and the third most important source of vegetable protein. A native of South America, groundnut is cultivated in tropical, sub-tropical, and warm temperature regions of the world. Groundnut requires long and warm growing season in a climate having well distributed rainfall in the range of 500-1000 mm. It grows best in temperature range of 23-30⁰C in sandy loam soils which permit easy growth of peg in the soil and harvest of mature pods. The optimum soil pH for groundnut is 6.0-6.5. China, India, Nigeria, USA and Myanmar are the top five producers (Table 8.1). Though India ranks first in area, it ranks second (after China) in production. India accounts for 25.4% of global area and contributed 19.7% to the world production with an average productivity of 1220 kg/ha which is nearly one-third of those of USA (3731 kg/ha) and China (3339 kg/ha) and even lower than that of the world (1579 kg/ha).

Among the seven commercial edible oilseed crops of the India, *viz.*, groundnut, rapeseed-mustard, soybean, sunflower, sesame, safflower and niger, currently groundnut is the third most important oilseed crop after rapeseed-mustard and soybean. Currently, six states *viz.*, Gujarat, Andhra Pradesh, Karnataka, Tamil Nadu, Maharashtra and Rajasthan account for more than 90% of the groundnut area. Madhya Pradesh, Uttar Pradesh, Orissa and West Bengal are the other states having substantial areas under this crop. In addition, Chhattisgarh, Jharkhand, Punjab, Goa and Kerala also contribute to national groundnut production to some extent. In the last decade, the crop has been successfully introduced in the north eastern states of India (Misra and Rathnakumar, 2011).

Table 8.1. Global Scenario of groundnut production (Misra and Rathnakumar, 2011)

Country	Ares(lakh ha)	Production(lakh tones)	Yield(kg/ha)
China	42.1	140.6	3339
India	59.7	72.9	1220
Nigeria	23.9	29.0	1211
USA	5.1	19.0	3731
Myanmar	8.0	13.0	1613
Indonesia	6.4	9.8	1530
Sudan	8.3	7.4	890
Senegal	8.3	7.0	837
Tanzania	5.5	4.0	724
Niger	5.5	2.4	431
Others	61.7	65.4	-
World	234.6	370.3	1579

Constraints in Production

The major factors which hamper the improvement in productivity of groundnut in India are i) Cultivation of old and inferior varieties on by and large marginal lands under low-input and rain dependent situations (~5.0 to 6.0 million hactares). ii) Flawed application of fertilizers, and inefficient use of water resources. iii) The biotic and abiotic stresses. Among the biotic stresses, the foliar fungal diseases (early leaf spot, late leaf spot and rust), viral diseases (peanut bud necrosis diseases and peanut stem necrosis disease) soil borne diseases (stem rot, collar rot and pot rot complexes), and the insect pests like defoliators (red hairy caterpillar, tobacco caterpillar, gram pod borer and leaf miner) and sucking pests (jassids, aphids and thrips) are the major ones. Problems of nematodes and white grubs are also encountered in certain areas. The major abiotic stresses are soil-moisture deficit stress at one or the other stage of crop during rainy season, and low-temperatures prevailing during germination as well as vegetative stages followed by high-temperatures during the pod-filling and maturation stages during summer season. In addition, build-up of salinity and acidity and deficiencies of micronutrients in certain areas, lower the productivity. The pre- and post-harvest invasions of groundnut by the fungi of *Aspergillus* group, which produce aflatoxin, lower the quality. The aflatoxin contamination is now regarded as the major hurdle in export of groundnuts and its products from India. In case of summer crop, sometimes due to early rains, the soil-moisture becomes high at the time of harvest and the problem of *in situ* germination is often encountered. Another peculiar problem associated with both *rabi* and summer produce is that of rapid loss of seed viability during post-harvest storage which renders the seed unfit for raising the new crop in the next *rabi* or summer season. There is need for cold-tolerant varieties for post-rainy season, for varieties with ability to germinate at low-temperatures to allow early sowing for assured-irrigation for summer season, and short-duration varieties possessing fresh-seed dormancy for spring season (Mayee, 1987; Ghewande *et al.*, 2002; Basu and Singh, 2004; Misra and Rathnakumar, 2011)

More than 70 diseases have been reported from this crop caused by fungal, bacterial, viral, phytoplasma, nematode, and also nutrient disorders. While some diseases are major and widely distributed, appearing in epidemic form causing heavy economic losses, the others are minor and are of less important as per the severity and yield losses. The major diseases and its casual agents are listed in Table 8.2. Diseases *viz.*, stem rot, collar rot, dry root rot, afla-root, leaf spots (early and late), rust and bud necrosis which affects the groundnut production at *kharif*, *rabi* and summer. Diseases reduce the pod yield of groundnut and also the quality of haulm. Among foliar fungal diseases, leaf spots (early and late) and rust are economically important which are widely distributed and can cause yield losses to the extent of 70%. In the past several years, *Alternaria* leaf blight occurs severely on summer groundnut at Saurashtra region of Gujarat. Of the soil borne diseases, collar rot, stem rot, afla root and dry root rot are major diseases. These can cause severe seedling mortality resulting in 'patchy' crop stand in sandy loam soils and reduce pod yields from 25 to 40%. Among viral diseases, peanut bud necrosis, peanut mottle, peanut clump and peanut stem necrosis are economically important which can cause yield losses to the extent of 60%. Besides this, few nematode diseases *viz.*, root knot, root lesion and brownish yellow lesion locally known as 'Kalahasti Malady' have also been reported from India. Aflatoxin contamination in groundnut and its processed product is another major challenge to groundnut growers, processors, and exporters (Rangaswami, 1996; Ghewande *et al.*, 2002; Basu and Singh, 2004; Singh, 2005; Sreenivasulu, 2005).

SOIL BORNE DISEASES

1. Collar rot (*Aspergillus niger* van Tieghem)

1.1. Distribution and Economic Importance: Collar rot was first reported from India by Jain and Nenra (1952). It is prevalent in almost all groundnut-growing states principally in Punjab, Andhra Pradesh, Tamil Nadu, Uttar Pradesh, Gujarat, Maharashtra, Rajasthan, Karnataka and Orissa, especially in sandy loam and medium soils. This disease is extensive in *kharif* than in *rabi* (Ghewande, 1983). In India, under Punjab conditions, the losses may amount to 40 to 50% in terms of mortality of plants (Aulakh and Sandhu, 1970, Chohan, 1972). The loss due to this disease was reported 28 to 47% (Bakhetia, 1983).

1.2. Symptoms: The characteristic diagnostic symptoms are pre-emergence rotting of seeds (Fig. 8.1), rotting of cotyledons, rotting in the collar region (Fig. 8.2), post emergence seedling blight and occasional wilting of plants at later stage. The affected tissue is covered with abundant black spores. At this stage, the hypocotyl and cotyledon are particularly rotted. As the infection spreads, the whole collar region becomes shredded and covered with black spores. In case of infected mature plants, lesions develop on the stem below the soil and spread upwards. The dead and dried branches easily can detach from the disintegrating collar region and has abundant black spores inside (Fig. 8.3).

Fig. 8.1 Seed rot-Aspergillus niger

Fig. 8.2 Collar rot - sporulation of *Aspergillus*

niger

Fig. 8.3 Collar rot

1.3. Disease Development: The pathogen survives in the soil, in the decayed plant debris and seed. It may be carried on externally as seed borne but mainly soil borne inoculum serves as the primary source of infection. The fungus develops best at temperature between 31 and 35^0C. During heavy and incessant rains very low incidence of pre and post emergence seedling blight are observed. High temperature and humidity conditions prevailing in the first fortnight of July immediately after the rainfall enhance the seedling mortality. Generally the disease appears within a month after sowing, occasionally in later stage plants also get wilted. Late sowing, insect feeding, high soil temperature and drought stress in the first few weeks after sowing have been associated with the incidence of this disease.

2. Afla-root / yellow mold (*Aspergillus flavus* (Link) Fries.)

2.1. Distribution and Economic Importance: This disease is prevalent in almost all groundnut-growing states. It is very common during *kharif*, but is sporadic in most of the groundnut grown areas. Though the plant mortality is very less, infected isolated plants are seen in most of the fields. However, compared to collar rot the disease severity is very low.

2.2. Symptoms: yellow mold fungus, *Aspergillus flavus*, is commonly found in the seed of both rotten and apparently healthy pods of groundnut. It first appears on cotyledons after the emergence of seedlings. Infected plants are stunted and often chlorotic (Fig. 8.4). The leaflets are reduced in size with pointed tips, widely varied in shape and sometimes with vein clearing. Such seedlings lack a secondary root system, a condition known as "aflaroot" (Fig. 8.5). *A. flavus* is recognized by its yellowish green colour and its colonies develop on over mature and damaged seeds and pods. During their metabolism these fungi not only produce aflatoxins but also deteriorate the nutritive value of the associated food products.

Fig. 8.4. Afla root

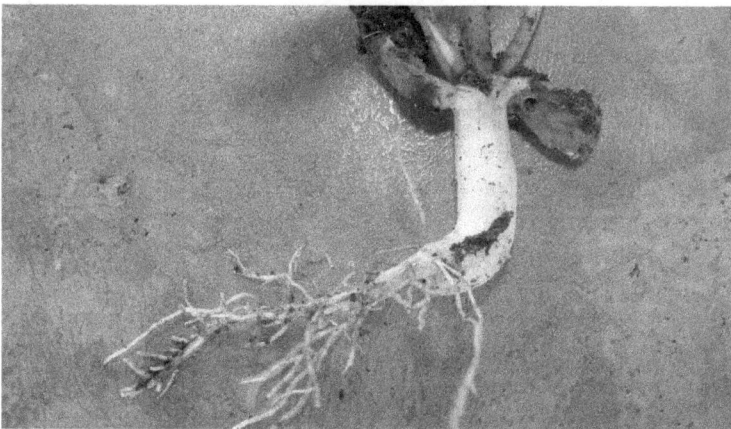

Fig. 8.5. Damage of secondary roots due to Afla root

2.3. Disease Development: This disease is seed and soil borne. It is usually associated with drought stress or high moisture conditions. The pathogen can tolerate low soil moisture and the fungus develops best at temperature between 25 and 35⁰C.

3. Stem rot (*Sclerotium rolfsii* Sacc. teleomorph: *Athelia rolfsii* (Curzi) Tu & Kimbrough)

3.1. Distribution and Economic Importance: The stem rot pathogen, *Sclerotium, rolfsii* is polyphagous plant pathogen. In India, stem rot occurs in all groundnuts growing states, particularly severe in Maharashtra, Gujarat, Madhya Pradesh, Karnataka, Andhra Pradesh Orissa and Tamil Nadu where it is estimated that over 50,000 ha of groundnut fields are infected with *S. rolfssi* (Mehan *et. al.*, 1995). In India, losses in yield of groundnut about 27% or more have been reported (Singh and Mathur, 1953, Chohan, 1974). Zaved *et al.*, (1983) reported that *S. rolfsii* also causes indirect losses such as reduction in both dry weight and oil content of groundnut kernels. This disease is most severe in Maharashtra, particularly where irrigated groundnut cultivation is expanding in Marathwada region of Maharashtra and also in Saurashtra region of Gujarat where mostly monocropping and set furrow systems are practiced. The disease incidence on farmers' fields ranges from zero to 60%

Fig. 8.6. Stem rot-white Mycelial growth

particularly in Spanish bunch varieties in light soil and yield losses of over 25% have been reported (Mayee and Datar, 1988). Pod rots incited by *S. rolfsii* are also economically important in central and southern Maharashtra, and Raichur area of Karnataka. Of late, this disease is assuming serious proportion in south Saurashtra zone of Gujarat particularly in Junagadh district both in medium black and light calcareous soils. The survey conducted by the author during *Kharif* 1996-99 in farmers' fields of Junagadh district under IVLP programme

Fig. 8.7. Stem rot-brown Sclerotia

reveled that the average incidence of stem rot ranged from 14.3 to 24.0%. The post harvest observations of four years (1996-1999) indicated that the average pod infection, seed infection and seed colonization was 23.4%, 9.5% and 3.6% respectively (unpublished). Latur in Maharashtra, Raichur and Dharwad in Karnataka and Hanumangarh in Rajasthan have been identified as hot spot locations under All India Coordinated Research Project on Groundnut.

3.2. Symptoms: The first symptom is yellowing and partial or complete wilting of the stem or one or more branches. In advance stage of the disease, white mycelium growth at the junction of stem, branches or sometimes on leaves, roots, pods, pegs and soil can be observed (Fig. 8.6). Abundant sclerotia which is white initially, present on the infected areas, soil surface and other crop debris turns brown at later stages (Fig. 8.7). Pods and damaged pegs are usually shed before harvest. At times, a severe pod rot may occur just before harvest without any apparent damage to the rest of the plant (Fig. 8.9). Scattered area or hot spots of dead or drying plants may be seen in those fields where disease activity is light to moderate (Fig. 8.8). As the soil surface dries or weather becomes cool, the white fungal mat disappears but the sclerotia remains. Fallen leaves,

Fig. 8.8. Stem rot infected plants

pods and other plant debris on the soil surface, which stimulate sclerotial germination, are quickly colonized by the fungus using them as a food base by it to attack the nearby healthy plant.

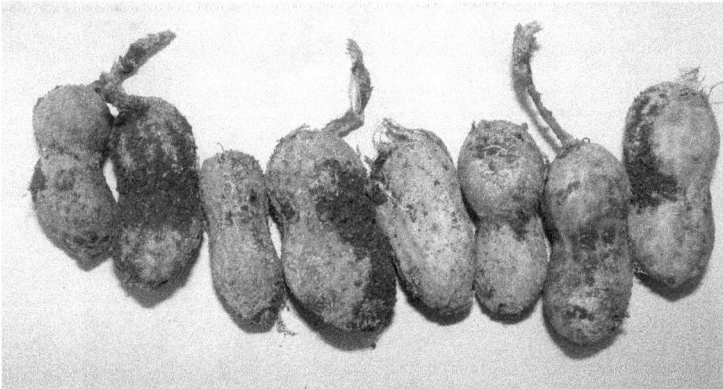

Fig. 8.9. Pods infected by S. *rolfsii*

3.3. Disease Development: The disease spread through sclerotia and infected plant debris in soil. Sclerotia survive in the soil for more than 4 years. Besides, the pathogen has wide host range. It colonizes either living plant tissues or on plant debris. Sudden wilting or drooping of one or more branches or stems starting in mid-late July is favoured by humid and hot weather. Normally it appears during hot weather in July and August at any time from seedling to the matured plants. Disease development is favoured by several days of hot, dry weather followed by a few showers. Often, two or more separate outbreaks may occur in disease prone fields. Crop residues, soil moisture to the extent of 40 to 50% of water holding capacity, temperature 29 to 32⁰C during day and 25⁰C during nights favours the pathogen infection and disease development. Lengthy periods of wet weather due to incessant and heavy rains or extreme drought conditions appear to suppress the sclerotium rot.

4. Dry root rot [*Macrophomina phaseolina* (Tassi.) Goid.]

4.1. Distribution and Economic Importance: Dry root rot is caused by *Macrophomina phaseolina*, the sclerotial stage of which is known as *Rhizoctonia bataticola* (Taub) Butl. Dry root rot or charcoal rot is sporadic in nature and occurs in Gujarat, Rajasthan, Uttar Pradesh, Tamil Nadu, Andhra Pradesh and Maharashtra. The pathogen causes severe seedling mortality results in patchy crop stand, thus reduce optimum plant population. Root rot is higher in bunch varieties as compared to the spreading type.

4.2. Symptoms: The initial symptoms starts as necrosis at hypocotyl and the stem of infected plants become straw to brown in colour. Water soaked necrotic spots appear on the stem just above the ground level. The lesions darken as the infection spread upwards to the aerial part and down to the roots resulting in sudden wilting. The infected stem portion is shredded and with the development of sclerotia which becomes black and sooty in appearance (Fig. 8.10). Occasionally when only the roots are attacked, the taproot turns black and later it rots and shreds. The kernels also turn black with abundant sclerotia on its inner wall.

Fig. 8.10. Dry root rot

4.3. Disease Development: Dry root rot has wide host range. The pathogen is facultative saprophyte and a soil dweller. Infected soil, plant debris and pods serve as the sources of inoculum. The disease increases with the increase in level of inoculum. The disease may appear at any stage of the crop. But generally it appears around 50 to 60 old crop when there is stress condition. The optimum temperature for seedling infection is 29 to 35^{0}C, for pods invasion is between 26 and 32^{0}C. The sclerotia are disseminated through plant debris, soil, infected pods, shell, and kernel. Sudden wilting appears in patches at dry weather conditions. A mild irrigation is enough to recover from the wilt under such conditions.

Foliar diseases

5. Early and late leaf spot disease (*Tikka* disease)

Early leaf spot caused by *Cercospora arachidicola* S. Hori (teleomorph; *Mycospharella arachidis* Deighton) and late leaf spot caused by *Phaeoisariopsis personata* (Berk. & M.A. Curtis) Arx = *Cercosporadium personatum* (Berk. & M.A. Curtis) Deighton (teleomorph; *Mycospharela berkelevi* Jenk.) are commonly called as *tikka* disease.

5.1. Distribution and Economic Importance: *Tikka* disease is prevalent in all groundnut grown areas. However, it is more severe and widespread in southern, central and north-eastern states of India during *kharif*. The loss in yield has been estimated to be 60%. Besides the loss in pods and kernel yield, the value of fodder is also adversely affected. The incidence is sporadic and the intensity is very less in summer groundnut. Both early and late leaf spots are commonly present wherever groundnut is grown. However, the incidence and severity of each disease varies with localities and seasons and there can be both short and long-term fluctuations in their relative proportions (Kolte, 1984, McDonald, *et al.*, 1985). In India, early leaf spot is more prevalent in all groundnut growing states. Recently, it started assuming serious proportion in southern and central states of India also. Late leaf spot is more severe in southern and central India. In India, late leaf spot is however, more severe than early leaf spot. Both early and late leaf spots cause damage to the plant by reducing the photosynthetic area, by intense lesion formation, and by stimulation leaflet abscission.

5.2. Symptoms: Early leaf spot consists of sub-circular, light to dark brown spots on the upper leaflet surface. A yellow halo is often seen around the brown spots (Fig. 8.11). The spots are of lighter shade of brown on the lower side of leaflets. Oval to elongate spots are also seen on stems and petioles. Leaflets become chlorotic and are shed prematurely during severe infection. The disease normally occurs 30 days after sowing. Late leaf spot symptoms are dark brown to black, circular to sub-circular lesions on the upper surface of the leaves having a rough appearance (Fig. 8.12). It exhibits circular

Fig. 8.11. Early leaf spot

rings of the fruiting structures of the fungus on the lower leaf let surface. Oval to elongate spots similar to early leaf spots are also formed on stems and petioles. Under severe conditions the disease leads to shedding of leaflets resulting in premature ageing of the crop. Late leaf spot attack is usually seen along with rust. The disease normally occurs on 60 days old crop to till harvesting. Presence of early and late leaf spots simultaneously on the standing crops is the most common phenomena in field (Fig. 8.13).

5.3. Disease Development: Early and late leaf spots diseases are spread through both soil and air born inoculum. Conidia survive on infeced plant debris in soil or on the infected groundnut shell. The late leaf spot is more severe and at times it assumes epidemic form at favourable weather conditions, having high humidity of more than 80 % and temperature between 25 and 30°C. Prolonged wetness on leaf favours infection and disease development. The pathogen also survives from one season to another on volunteer groundnut plants.

Fig. 8.12. Late leaf spot

Fig. 8.13. Tikka Leaf spot

6. Rust *Puccinia arachidis* Speg.

6.1. Distribution and Economic Importance: After the initial report of groundnut rust from Punjab in India (Chahal and Chohan 1971) there have been records of its incidence from different parts of the country (Bhama, 1972, Misra and Misra 1975, Ramakrishanan and Subbayya, 1973, Sinde and More, 1975, Tripathi and Kaushik, 1978, Yadav *et al.,* 1975). Rust was severe particularly in the southern states of India (Subrahmanyam *et al.,* 1979). Surveys conducted by the National Research Centre for Groundnut (NRCG), Junagadh during the rainy seasons of 1980-81, 1981-82 and 1982-83 revealed that the rust with moderate to heavy severity was distributed in all groundnut-growing districts of Saurashtra region of Gujarat (Ghewande and Misra, 1983). Rust was present in almost all rabi/summer groundnut cultivation areas in India (Subrahmanyam and McDonald, 1987). It is now evident from the literature that the rust disease of groundnut is prevalent throughout India. In India, losses in yield due to rust alone have been reported in the range of 10-52% depending upon the variety (Ghuge *et al.,* 1981, Ghewande *et al..* 1983, Siddaramaihah *et. al.,* 1981, Subrahmanyam and McDonald, 1983). The loss in oil content due to rust infection has been estimated to be about 7 to 10%. Considering the magnitude of losses, and extent of prevalence, rust is now economic importance in almost all groundnut growing areas of the country.

6.2. Symptoms: Rust first appears as yellowish-green fleck on the upper leaf surface. Almost simultaneously pustules appear as small raised bumps on the lower leaf surface. Rust can be readily recognized as orange coloured pustules (uredinia) that appear on the lower surface and rupture to expose masses of reddish brown urediniospores (Figs. 8.14 and 8.15). The pustules range from 0.5 to 1.4 mm in diameter. In highly susceptible cultivars the original pustules may be surrounded by colonies of secondary pustules. They may also appear on upper surface of leaflets. Severely infected leaves turn necrotic and desiccate but remain attached to the plant. The rust may be formed in all aerial parts except the flower (Fig. 8.16). Leaf spots and rust occurs simultaneously on the crop (Fig. 8.17).

Fig. 8.14. Rust-Orange colour pustule

Fig. 8.15. Rust pustules in brown colour

Fig. 8.16. Rust disease on petiole and bracts

Fig. 8.17. Rust disease in field

6.3. Disease Development: Rust is known to perpetuate and produce severe disease outbreaks by means of urediniospores. Volunteer groundnut plants enable survival of the uredinal stage. An optimum temperature of 25°C, prolonged leaf wetness and high humidity favour infection and disease development. Epidemiological factors coupled with virulence of the pathogen play an important role in disease development. The pustules release reddish brown spores which are blown by the wind, spread the disease from plant to plant and also carry it to other groundnut fields.

7. *Alternaria* leaf spot and blight

Three species of *Alternaria* causes diseases in groundnut.

7.1. Distribution and Economic Importance: In the past few years, *Alternaria* leaf blight occurs severely on summer groundnut. The disease reduced pod (up-to 22%) and fodder (up-to 63%) yield depending on severity of the disease (Vinod Kumar *et al.*, 2012). It reduces the quality of the fodder and kernels from the diseased plants were shriveled. Species of *Alternaria viz.*, *A. tenuissima*, *A. arachidis*, *A. alternata* and *A. longipes* were reported from diseased groundgnut plants (Balasubramanian, 1979; Ghewande *et al.*, 1982; Giri and Murugesan, 1996; Patil and Hiremath, 1989; Subramaniyam *et al.*, 1981).

7.2. Symptoms: Initially blighting starts from apical portions of leaflets, which turn light to dark brown colour 'v' shaped spots (Fig. 8.18). Later the disease extends to midrib and the entire leaf shows blighted appearance. In the later stages of infection, blighted leaves curl inward and become brittle. Adjacent lesions join together, giving the leaf a ragged and blighted appearance. Sometimes the disease progression was so fast that the full canopy was blighted in a week covering the entire field (Fig. 8.19).

Fig. 8.18. Alternaria leaf blight

Fig. 8.19. Alternaria blight in field

7.3. Disease Development: The epidemiology of the disease is not fully studied. It is speculated that changes in the climate may be reason for major influenced the pathogen to cause severe disease in the last few years.

Viral diseases

One of the major production constraints of groundnut is the reduction in yield caused by the various diseases including those caused by viruses. Although natural infection of more than 30 plant viruses representing 14 virus groups have been recorded on groundnut from different countries, *Groundnut bud necrosis virus, Peanut stem necrosis virus, Tobacco streak virus, Peanut mottle virus* and *Indian peanut clump virus* are economically important virus diseases in India (Table 8.3). The magnitude of losses caused by these virus diseases may vary from 1.2 to 100% depending upon the extent of incidence and stage of crop growth. Currently, PBND and PSND are potential threat to groundnut cultivation.

Table 8.3. Groundnut viruses reported from India, their group and vectors

Virus	Group	Vector	References
Groundnut bud necrosis virus	Tospovirus	*Thrips palmi* Karny (Thysanoptera)	Vijayalakshmi, 1994; Reddy *et al.*, 1998
Peanut stem necrosis virus	Ilarvirus	*Frankliniella schultzei*	Prasad Rao *et al.*, 2003
Indian peanut clump virus	Pecluvirus	*Polymyxa graminis* Ledingham	Reddy *et al.*, 1983; Nolt *et.al.*, 1988
Peanut chlorotic leaf streak	Caulimovirus	*A.craccivora, Bemisia tabaci M.persicae* (Homoptera)	Reddy *et al.*, 1993
Peanut stripe virus	Potyvirus	*A. craccivora, M.persicae, A.gossypii, Rhopalosiphum maidis, A. glycines, A. citricola, Hysteroneura setariae* (Homoptera)	Prasada Rao *et al.*, 1988

1. Peanut Bud Necrosis Disease (*Groundnut Bud Necrosis Virus*)

1.1. Distribution and Economic Importance: Until the mid 1960s, peanut bud necrosis disease (PBND) was a minor disease in groundnut (Reddy, 1988) which later occurred in epidemic proportions Vijayalakshmi, 1994). The incidence of PBND was reported to range from 1.2 to 100% (Chohan, 1974; Ghanekar *et al.*, 1979; Mayee 1987; Singh and Gupta, 1989; Singh and Srivastava, 1995; Dharamraj *et al.*, 1995). The hot spot locations for PBND are Jagtiyal, Kadiri and Hyderabad in Andhra Pradesh; Latur in Maharashtra, Tikamgarh in Madhya Pradesh; Raichur in Karnataka, Mainpuri in Uttar Pradesh and Saurashtra in Gujarat. GBNV has been considered as one of the major virus diseases of groundnut in Andhra Pradesh, Uttar Pradesh, Madhya Pradesh, Tamil Nadu, Karnataka, Gujarat and Maharashtra on *kharif* groundnut. It has also been occurring on post rainy season crop of groundnut in Saurashtra region of Gujarat, Nizamabad, Nalgonda and Mahaboob Nagar districts of Andhra Pradesh and, Northern and Vidarbh regions of Maharashtra and north eastern parts of Karnataka. In India, PBND may cause 30-90% yield losses depending upon the time of infection on plant growth stage (Ghanekar, 1979; Basu, 1995).

1.2. Symptoms: Initially, faint chlorotic spots and mottling occurs on young leaflets later, developed into chlorotic and necrotic rings and streaks. Petiole bearing fully expanded leaflets become flaccid and droops. Necrosis of the terminal buds is a characteristic symptom of *Groundnut bud necrosis virus* (GBNV) infected groundnut plant (Fig. 8.20). If the infection is on one month old plant, entire plant die and pod setting will be absent. Mild ring spots, necrosis in the petiole and portion of stem were recorded if it infects at the later growth stages of crop. Stunting and proliferation of axillary shoots, leaf lets formed on the axillary shoots shows a wide range of symptoms *viz.*, decrease in size of leaf lamina, distortion of laming, mosaic mottling, chlorosis, rarely shoe-string appearance of leaf lets, bushy appearance of plant are characteristic secondary symptoms observed on GBNV infected groundnut. Infected plants produce small and shriveled seeds with testa in red or brown or purple mottling. Late infected plants may produce seed of normal size but the testa on such seed are often mottled and discoloured. Stains of tomato spotted wilt virus also causes symptoms similar to bud necrosis disease of peanut and transmitted by *Scirtothrips dorsalis* and *Frankliniella schultzei* (Amin *et al.*, 1981)

Fig. 8.20. Groundnut Bud Necrosis Disease

1.3. Disease Development: The virus causing PBND is transmitted by *Thrips palmi*. The majority of individual adult thrips transmitted the virus for more than half of their life period, indicating the degree of erratic transmission. Groundnut bud necrosis virus and *T. palmi* have extremely wide host range and survives on ornamentals (zinnia, cosmos and sunflower), weeds (*Ageratum conyzoides, Cassia tora, A. hispidum, D. triflorum*) and crop plants (tomato, brinjal, green gram, black gram, beans and pea). Temperature 30^0C and a wind speed of 10km/h favour migration of thrips. The thrips population increases rapidly in late August and September. The population builds up again during January and February and hence *rabi* crop also suffers very seriously.

2. Peanut Stem Necrosis Disease (*Peanut Stem Necrosis Virus*)

2.1. Distribution and Economic Importance: A disease characterized with the necrosis of the stem and terminal leaflets followed by death of plant was epidemic resulted in severe crop losses (Rupees ~ 300 crores) in about 225,000 ha in Anantpur district of Andhra Pradesh, India during rainy season of 2000. It was presumed to be caused by GBNV, because of the characteristics necrosis of terminal buds. In subsequent studies however, *Tobacco streak virus* (TSV) was found associated with the disease. The disease was named as 'Peanut Stem Necrosis Disease' (PSND). This was the first report of occurrence of TSV in groundnut in India. TSV was also shown to cause sunflower necrosis disease in sunflower (Ravi *et al.*, 2001; Ramaiah *et al.*, 2001; Bhat *et al.*, 2002). Although PSND and GBND are caused by two district viruses belonging to the *Ilarvirus* and *Tospovirus* groups respectively, the symptoms produced by them on groundnut are very similar making it difficult to distinguish between these two diseases in the later stages based on symptoms alone.

2.2. Distribution: PSND is prevalent in the district of Anantapur, Kurnool, Cuddapah and Chittor in Andhra Pradesh, and Raichur in Karnataka. Limited surveys carried out in Gujarat (Porbandar, Rajkot and Junagadh) and Maharashtra (Jalgaon and Dhulia) did not show significant levels of incidence of this disease. However, more extensive surveys and monitoring are needed in Karnataka, Tamil Nadu and Maharashtra, where sunflower necrosis disease caused by TSV is prevalent in sunflower crop.

2.3. Symptoms: Initial symptoms appear as large necrosis lesion on young quadrifoliate leaves, which coalesce and cover the entire leaflet followed by the necrosis of the stem below the necrotic leaves (Fig. 8.21). If young plants (less than one month old) are affected, the entire plant is often necroses whereas in case of older plants, one or more branches will have necrosis. These plants are stunted and do not show any axillary shoot proliferation. Majority of pods also will have necrotic spots and in severe infection the pod size will be reduced and kernels will become not marketable. The symptoms, however, have been shown to very among varieties of groundnut. Infection by GBND and PSND both the viruses can be distinguished precisely by ELISA tests (Table 8.4).

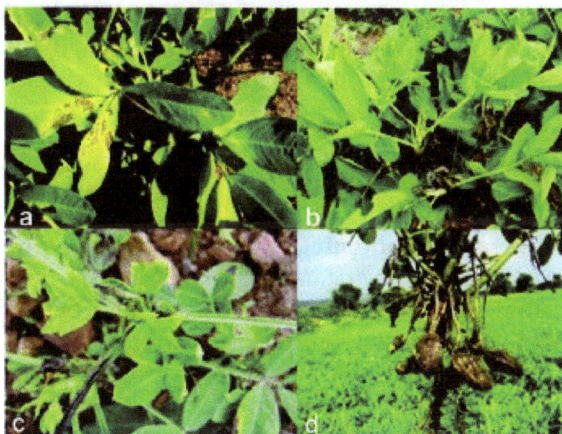

Fig. 8.21. Peanut stem necrosis disease caused by *Tobacco streak virus* (a) Necrotic symptoms on young leaves (b) Death of to growing buds on main stem and primaries (c) Infected plant showing axillary shoot proliferation (d) Infected plant showing Pod necrosis with reduced size

Table 8.4. Distinguishing features of PSND and PBND (Modified from Prasad Rao et al., 2003)

Characteristics	PSND	PBND
Causal Virus	Peanut stem necrosis virus	Groundnut bud necrosis virus
Group	Ilarvirus	Tospovirus
Groundnut Plant Showing Characteristic Symptoms of the Disease	Necrotic lesions on terminal leaflets, complete stem necrosis and often total necrosis of entire plant. Axillary shoot proliferation, restricted to apical portion may occur. Necrotic spots on pods. Testae are not discolored or mottled.	Chlorotic lesions on terminal leaflets, ring spots and often necrosis of terminal bud. Axillary shoot proliferation with small and deformed leaflets. Infected plants remain stunted and seldom die. No Necrotic spots on pods and testa are discolored and mottled.
Serological Cross Reaction	Ilarvirus in Bromoviridae reacts with many tobacco streak virus antisera.	Distinct Tospovirus and reacts only with peanut bud necrosis virus antiserum.
Transmission Vectors	Seed-transmitted in many hosts	Not seed-transmitted in any of the hosts
	Transmission by Frankliniella Schulte in a persistent manner	Transmitted by several species of Thrips. Relationship is passive
Primary Spread	Mostly weed hosts	Weed and crop plants
Secondary Spread	Selected weeds and crop plants	Selected weeds and crops plants

The PSND spreads mainly through the weed of crop species which are alternate host to the *Peanut stem necrosis virus*. The most commonly occurring weeds where the disease was epidemic were *Parthenium hysterophorus, Abutilon indicum, Ageratum conyzoides, Croton sparsiflorous, Commelina benghalensis, Cleome viscosa, Euphelina hirta, Lagasca mollis* and *Tridax procumbense,* all of which were infected. Although the three thrips species (*F. schultzei, M. Ustatus* and *S. doralis*) were experimentally shown to transmit TSV in the presence of infective pollen, the flower inhabiting *F. schultzei* played a major role in the field spread of the virus.

Nematode diseases

1. Root knot (*Meloidogyne arenaria, M. hapla* and *M. javanica*)

1.1. Distribution and Economic Importance: Three different species of *Meloidogyne* viz., *Meloidogyne arenaria, M. hapla and M. javanica* cause root knot in groundnut that commonly produces enlarged roots and pegs. Because of their wide spread distribution and high frequency of occurrence, the root knot nematodes are considered to be very important pathogen of groundnut. *M. arenaria* (Neal) Chitwood is the most important pathogen and the most damaging species as compared to *M. hapla* and *M. javanica*. The economic losses caused by *M. javanica* and *M. arenaria* together are about 21.6% in India (Khan *et al.*, 2010). The avoidable losses may reach up to 59% in individual field. *M. arenaria* is an economically important plant pathogen that parasitizes thousands of plant species worldwide.

It is commonly encountered in warmer regions. It has been reported to cause damage to groundnut crop in various parts of the country *viz.*, Andhra Pradesh, Tamil Nadu, Punjab and Saurashtra region in Gujarat (Walia and Bajaj, 2003).

1.2. Symptoms: The groundnut plants grown in soil infested with root knot nematode generally show prominent above and below ground symptoms. The soil contains numerous root knot nematodes, eggs and infective juveniles. Juveniles infect groundnut roots soon after planting. Conspicuous galling and egg masses appear on the roots at 60 to 90 days after sowing. The disease appears initially in patches in groundnut fields. The patches widen and spread along the direction of plough and flow of irrigation water. The dwarfing/stunting, chlorosis and yellowing of the leaves, loss of vigour etc are the common symptoms (Fig. 8.22). Above-ground symptoms are similar to those caused by nutrient deficiency or water stress. In many infected fields, plants remain extremely stunted with chlorotic dwarf leaves and profusely galled roots with few lateral roots and may bear very few pods or no pods. The primary knots/galls are small, which coalesce to form large knots and the females become conspicuous in these knots (Walia and Bajaj, 2003).

Fig. 8.22 Root knot infected plant

2. Kalahasty malady (*Tylenchorhynchus brevelineatus*)

Serious disease in Andhra Pradesh, particularly in Tirupathi areas.Infected plants appear in patches in the field, and are stunted with greener than normal foliage. Small brownish yellow lesions appear on the pegs, and on young developing pods (Fig. 8.23). Peg length is reduced. In advanced stages of the disease the entire pod surface becomes blackened and kernels become small (Walia and Bajaj, 2003).

Fig. 8.23 Kalahasti maladi

3. Root lesion (*Pratylenchus brachyurus*)

Roots are restricted in length and total volume, and tend to be discoloured as the nematodes feed on the tissues of roots, pegs and pods. The pod lesions begin as tiny, tan to brown coloured, pin point areas on the shell surface. As the nematode feed and reproduce, the affected areas on pods become larger and darker. Older lesions are characterized by a blotchy appearance and indistinct margins. Severely attacked plants are stunted and chlorotic with reduced root systems (Walia and Bajaj, 2003).

Post harvest problems

Aflatoxin

Aflatoxins are potent toxic, carcinogenic, mutagenic, immunosuppressive agents, produced as secondary metabolites by few species of *Aspergillus* on variety of food products. *A. flavus* and *A. parasiticus* are major aflatoxigenic fungus grow on variety of food products and feeds. The aflatoxin problem was first recognized in 1960, when there was severe outbreak of a disease referred as "Turkey 'X' Disease" in UK, in which over 100,000 turkey poults were died. The cause of the disease was shown due to toxins in peanut meal infected with *A. flavus* and the toxins were named as aflatoxins. Among 18 different types of aflatoxins identified, major members are aflatoxin B1, B2, G1 and G2. Presently, aflatoxin contamination in groundnut and its processed product is a major challenge to groundnut growers, processors, and exporters.

Aflatoxin contamination is influenced by weather conditions such as rainfall and temperature. Surveillance and monitoring of farm as well as confectionery and feed products are yet to be conducted in a systematic way from many countries in Asia. Nevertheless, it is fortunate that several important groundnut producing countries have recognized the problem and are supporting research and monitoring/control activities aimed at mitigating the problem (Vinod Kumar *et al.,* 2005).

Factors influencing Aflatoxin contamination

The optimum limits for growth of *A. flavus* and *A. parasiticus* are 82-85% relative humidity and temperature of 30-32°C. Fungal growth is optimum when moisture levels of the substrate range from 10 to 30%. The optimum conditions for aflatoxin production are between 25°C and 30°C at 85% relative humidity. When the atmospheric relative humidity is near 70%, the seed moisture content will equilibrate between 7 and 9%, which is a level unfavourable for the growth of the *A. flavus* group (Figs. 824 and 8.25). The nature of the strain of the fungus, substrate, pH, temperature relative humidity, moisture content of the substrate and aeration have been found to influence the quality and quantity of aflatoxins produced. The soil population of *A. flavus* varies from farm to farm depending on soil types and crop rotations.

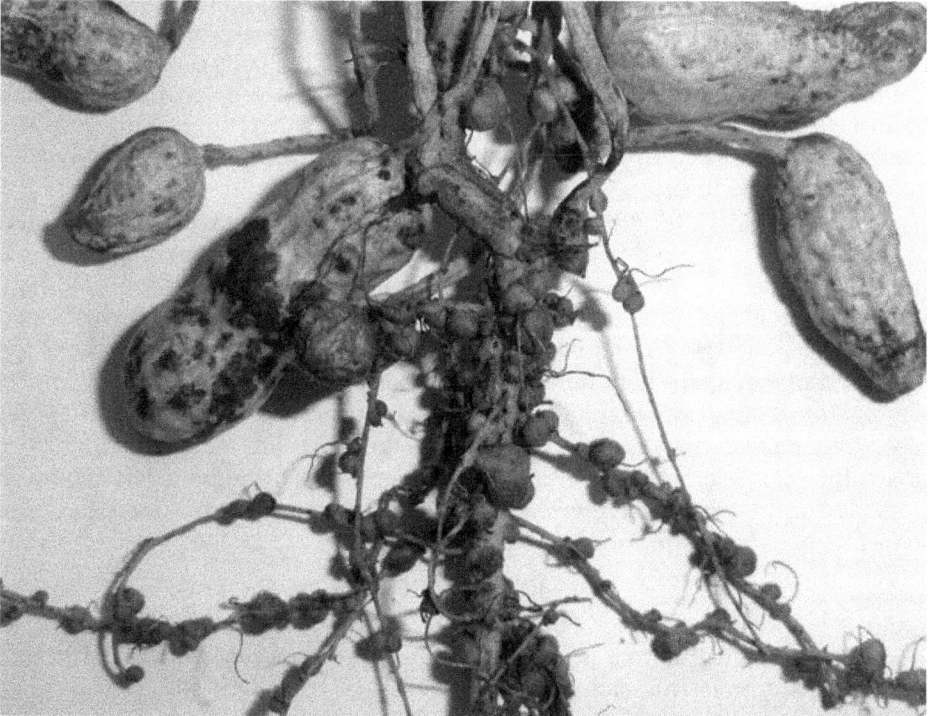

Fig. 8.24 Lesiond on roots and pods

Fig. 8.25. (a) Pods and (b) Kernel infected by *A. flavus*

Reasons for increasing *A. flavus* infection and subsequent contamination of aflatoxin during pre-harvest, harvesting and post-harvest groundnuts are disturbances in soil-water-nutrient balance during the crop growth period activate these fungi leading to infection and subsequent aflatoxin production in the kernels, development of cracks during pods growth (growth cracks), mechanical injury to pods during intercultural operations, infestation of insect-pests (termites, pod borers and wire worm) causing damage to pods, death of plants caused by diseases (stem, root and pod rots) at pod maturity stages, nematode damage to the pod, high atmospheric temperature (30-40°C) in conjunction with reduced soil moisture availability, protracted dry spell (more than 20 days) before harvest (sandy soils become hot spots in such conditions), harvesting of premature pods,

The important post harvest factors leads to increasing aflatoxin contamination in groundnut are delayed harvest (over-mature crop), mechanical damage to the pods during harvest, harvesting the crop immediately after irrigation with high initial pod moisture at the time of processing and storage, stacking the harvested plants before bringing the pod moisture level to less than 10%, stacking the harvested plants under high humidity conditions, damage to the pods by insects during storage, presence of mites in stored pods are capable of carrying the spores of *A. flavus* increased the chances of infection, storing haulms along with immature or small pods, gleaning pods from the soil after harvest (the pods left behind in the soils during harvest get infected very easily) and rewetting of stored pods due to factors like ground-moisture or leakage of roof.

REFERENCES

Amin, P. W., Reddy, D. V. R., Ghanekar, A. M., and Reddy, M. S. 1981. Transmission of tomato spotted wilt virus, the causal agent of bud necrosis disease of peanut by *Scirtothrips dorsalis* and *Frankliniella schultzei*. Plant Dis. 65: 663-665.

Aulakh, K. S., and Sandhu, R. S. 1970. Reactions of groundnut varieties against *Aspergillus niger*, Plant Dis. Rep., 54: 337.

Bakhetia, D. R. C. 1983. Control of white grub (*Holotrichia consenguinea*) and collar rot (*Aspergillus niger*) of groundnut sown in different dates in Punjab. Indian J. Agric. Sci., 53: 846-850.

Balasubramanian, R. 1979. A new type of alternariosis in *Arachis hypogea* L. Curr. Sci., 48: 76-77.

Basu, M. S. 1995. Peanut Bud Necrotic Disease: Activities in the Indian national program. In: Recent studies on peanut bud necrotic disease: Proceedings of a meeting, 20 March, 1995, ICRISAT Asia Center, India (Buiel A A M Parleviliet J E and Lenne J M eds.) Patancheru 502 0324, Andhra Predesh, India.

Basu, M. S. and Singh, N. B. 2004. Groundnut research in India. National Research Centre for Groundnut, Junagadh, Gujarat, India. 488pp.

Bhama, K. S. 1972. A rust on groundnut leaves near Madras. Curr. Sci., 41: 188-189.

Bhat, A. I., Jain, R. K., Chaudhary, V., Krishna Reddy, M., Ramaiah, K., Chattannavar, S.N., and Varma, A. 2002. Sequence conservation in the coat protein gene of Tobacco streak virus isolates causing necrosis disease in cotton, mungbean, sunflower and sun hemp in India. Indian J Biotech., 1: 350-356.

Chahal, A.S., and Chohan, J. S. 1971. Puccinia rust on groundnut. FAO Plant Prot. Bull., 19: 90.

Chohan, J.S. 1972. Final Progress Report, ICAR Scheme for research on important disease of groundnut in the Punjab for the period 1957-1667. Department of Plant Pathology, Punjab Agricultural University, Ludhiana, India 117 pp.

Chohan, J. S. 1974. Recent advances in diseases of groundnut in India. In Current Trends in Plant Pathology (Eds, S. P. Raychanduri and J. P. Verma), Lucknow University, India. 171-184 pp.

Dharmaraj, P. S., Naragund, V. B. and Somasekhar. 1995. Peanut Bud Necrotic Disease in Karnataka. In: Recent studies on peanut bud necrotic disease: Proceedings of a meeting, 20 March, 1995, ICRISAT Asia Center, India (Buiel, A. A. M., Parleviliet, J. E. and Lenne, J. M. eds.) Patancheru 502 324, Andhra Predesh, India. 69-72 pp.

Ghanekar, A. M., Reddy, D. V. R., Uzuka, N., Amin, P. W. and Gibbons, R. W. 1979. Bud Necrosis of Groundnut (*Arachis hypogaea* L.) in India Caused by tomato spotted wilt virus. Ann. Appl. Biol., 93: 173-179.

Ghewande, M. P. 1983. Effect of cultural practices on the disease (bud necrosis, collar rot, and stem rot) incidence and yield of groundnut. Indian Bot. Rep., 3: 98.

Ghewande, M. P. and Misra, D. P. 1983. Groundnut Rust – A challenge to meet. Seeds & Farms. 9: 12-15.

Ghewande, M. P., Desai, S. and Basu, M. S. 2002. Diagnosis and management of major diseases of groundnut. NRCG, Junagadh, India. 36pp.

Ghewande, M. P., Pandey, R. N., Shukla, A. K., and Misra, D. P. 1982. A new leaf blight diseases of groundnut caused by *Alternaria tenuissima* (Kunze ex Pers.) wilts. Curr. Sci., 51: 845-846.

Ghewande, M. P., Shukla, A. K., Pande, R. N., and Misra, D. P. 1983. Losses in groundnut yields due to leaf spots and rust at different intensity levels. Indian J. Mycol. & Pl. Pathol., 13: 125-127.

Ghuge, S. S., Mayee, C. D., and Godbole, G. M. 1981. Assessment of losses in groundnut due to rust and tikka leaf spot. Indian Phytopath. 34: 179-182.

Giri, G. S., and Murugesan, K. 1996). A first report of Alternaria longipes on groundnut from Tamil Nadu, India. International Arachis Newsletter., 6: 35.

Jain, A. C. and Nenra, K. G. 1952. *Aspergillus* blight of groundnut seedlings. Sci. Cult., 17: 348.

Khan, M. R., Jain, R. K. Sily, R. V. and Pramanik, A. 2010. Economically important plant parasitic nematodes distribution: ATLAS. Directorate of Information and Publication of Agriculture, Krishi Anusandhan Bhavan I, New Delhi.

Kolte, S. J. 1984. Diseases of annual edible oilseed crops. *Peanut Diseases*, Vol. I, Florida, CRC Press Inc. Usa, 29 pp.

Mayee, C. D. 1987. Disease of groundnut and their management. In *"Plant Protection in field crops"* (eds. Rao, M. A. B. and Sithanathan, S.) Plant Protection assistance of India, Hyderabad, pp. 235-243.

Mayee, C. D., and Dattar. 1988. Diseases of groundnut in the Tropics. Rev. Trop. Plant Path. 5: 85-118.

McDonald, D., Subrahmanyam, P., Gibbons, R.W., and Smith, D. H. 1985. Early and late leaf spots of groundnut, Information Bulletin No. 21. International Crops Research Institute for the Semi-Arid Tropics. Patancheru, A.P. 502324, India.

Mehan, V. K., Mayee, C. D., Brenneman, T. B., and McDonald, D. 1995. stem and pod rots of groundnut. Information Bulletin no.44 Patancheru, Andhra Pradesh, India. International Crops Research Institute for the Semi-Arid Tropics. 19pp.

Misra, A. K. and Misra, A. P. 1975. Groundnut rust in Bihar-Varietal reaction. Indian Phytopath. 28: 557-559.

Misra, J. B. and Rathnakumar, A. L. 2011. Vision 2030. Directorate of Groundnut Research,Junagadh, India. 36pp.

Nolt, B., Rajeswari, L. R., Reddy, D. V. R., Bharathan, N., and Manohar, S. L. 1988. Indian peanut clump virus isolates: host range, symptomatology, serological relationships, and some physical properties. Phytopathol. 78: 310-313.

Patil, P. V., and Hiremath, P. C. 1989. A new leaf blight disease of groundnut caused by *Alternaria tenuissima* (Kunze. Fr) Wiltshire in Karnataka. Curr. Sci. 58: 151.

Prasada Rao, R. D. V. J., Reddy, D. V. R., Nigam, S. N., Reddy, A. S., Waliyar, F., Reddy, T. Y., Subramannian, K., Sudhir, M. J., Naik, S. S., Bandhyopadyaya, A., Desai, S., Ghewande, M. P., Basu, M. S. 2003. Peanut Stem Necrosis: A New Disease of Groundnut in India. Information Bulletin no. 67. Patancheru, 502 324, Andhra Pradesh, India: International Crops Research Instityute for Semi-Arid Tropics. 16 pp.

Prasada Rao., Reddy, R. D. V., and Chakrabarty, A. S. 1988. Survey for peanut stripe virus in India. Indian J. Plant Pro. 16: 99-102.

Ramaiah, M., Bhat, A. I., Jain, R. K., Pant, R. P., Ahlawat, Y. S., Prabhakar, K., and Varma, A. 2001. Isolation of an isometric virus causing sunflower necrosis disease in India. Plant Dis. 85: 443

Ramakrishan, V., and Subbayya, J. 1973. Occurrence of groundnut rust in India. Indian Phytopath. 26: 574-575.

Rangaswami, G. 1996. Diseases of crop plants in India. Pentice-Hall of India private limited, New Delhi. 498pp.

Ravi, K. S., Buttgereitt, A., Kitkaru, A. S., Deshmukh, S., Lesemann, D. E., and Winder, S. 2001. Sunflower necrosis disease from India is caused by an ilar virus related to Tobacco streak virus. Plant Pathol. 5: 800

Reddy, D. D. R., Ranga Rao, G. V., and Sanower, T. G. 1993. Bio-ecology and management of groundnut leaf miner ,*Aproaerema modicella*. Group Discussion on Integrated Pest management Strategies in Oilseeds in India, held at Panjab Agricultural University, Ludhiana from 23-24 December, 1993.

Reddy, D. V. R. 1988. Virus diseases. In Groundnut Monograph. (Ed. P. S. Reddy). Indian Council of Agricultural Research, New Delhi. pp 508-525.

Reddy, D. V. R., Amin, P. W., Mcdonald, D., and Ghanekar, A. M. 1983. Epidemiology and control of groundnut bud necrosis and other diseases of legume crops in India caused by tomato spotted wilt virus. In: Plant Disease Epidemiology (eds, Plumb R T and Tresh J M) Oxford, U. K., Blackwell, pp 93-102.

Shinde, P. A., and More, W. D. 1975. Outbreak of groundnut rust at Rahuri, The Res. Jour. M.P.K.V. 6: 75-76.

Siddaramaiah, A. L., Desai, S. A., and Jayaramaiah, R. 1981. Occurrence and Severity of a mycoparsite on 'tikka' leaf spots of groundnut. Curr. Res. 10: 14-15.

Singh, A. B., and Srivastava, S. K. 1995. Status and control strategy of peanut bud necrosis disease in Uttar Predesh, In Recent studies on peanut bud necrotic disease: Proceedings of a meeting, 20 March, 1995, ICRISAT Asia Center, India (Buiel, A.A.M., Parleviliet, J.E. and Lenne, J.M. eds.) Patancheru 502 0324, Andhra Predesh, India: International Crop Research Institute for Semi- Arid Tropics; and P. O. Box 386, 6700 A. J. Wageningen, The Netherlands: Department of Plant Breeding, Agriculture University of Wageningen. pp 65-68.

Singh, B. and Mathur, S. C. 1953. Sclerotial root rot disease of groundnut in Uttar Pradesh. Curr. Sci. 22: 214-215.

Singh, B. R. and Gupta, S. P. 1989. Prevalence and losses in groundnut due to bud necrosis disease and peanut mottle disease in Uttar Predesh., Proc. 10th Conven. Indian Virology Society and National Symposium on Epidemiology of viral diseases, CPRI, Simla, 1989. 36pp.

Singh, R. S. 2005. Plant diseases. Oxford & IBH publishing Co. Pvt. Ltd., New Delhi. 686 pp.

Sreenivasulu, P. 2005. Groundnut viruses, distribution, biology, diagnosis and management. Indian. J. Virol. 16: 1-2.

Subrahmanayam, P., McDonald, D., Siddaramaiah, A. L. and Hegde, R. K. 1981. Leaf spot and veinal necrosis disease of groundnut in India caused by *Alternaria alternaria*. FAO Plant Prot. Bull. 29: 74-76.

Subramanayam, P. and McDonald, D. 1983. Rust disease of groundnut. Information Bulletin No. 13. International Crops Research Institute for the Semi-Arid Tropics, Patancheru, A.P., India.

Subramanyam, P. and McDonald, D. 1987. Groundnut rust disease: Epidemiology and control. ICRISAT (International Crops Research Institute for the Semi-Arid

Tropics) 1987. Groundnut rust diseases. Procedding of a Discussion Group Metting, 24-28 sep. 1984, ICRISAT Centre, Patancheru, A.P., India.

Subramanyam, P., Reddy, D. V. R., Gibbons, R. W., Rao, V. R. and Garren, K. H. 1979. Current distribution of groundnut rust in India, PANS 25: 25-29.

Tripathi, N. N. and Kaushik, C. D. 1978. A new record of groundnut rust from Haryana, Indian Jour. Agric. Res. 12: 273.

Vijayalakshmi, K. 1994. Transmission and ecology of *Thrips palmi* Karny, the vector of peanut bud necrosis virus. Ph. D. thesis. Andra Pradesh Agric. University, Rajendra Nagar, Hyderabad, India. pp 99,

Vinod-Kumar., Ghewande, M. P., and Basu, M. S. 2005. Safeguard groundnut from aflatoxin contamination. NRCG, Junagadh, India. 13pp.

Vinod-Kumar., Lukose, C., Bagwan, N. B., Koradia, V. G. and Padavi, R. D. 2012. Occurrence of Alternaria blight of groundnut in Gujarat and reaction of some genotypes against the disease. Indian J. Phytopath. 65: 25-30.

Walia, R. K., and Bajaj, H. K. 2003. Textbook on introductory plant nematology. ICAR, New Delhi, India. 215pp.

Yadav, H. L., Swarup, J., and Saksena, H. K. 1975. Occurrence of groundnut rust (*Puccinia arachidis Speg.*). A new record for Uttar Pradesh. Indian J. Farm Sci. 3: 109.

Zaved, M. A., Statour, M. M., Aly, A. Z., and El-Wakil, A. A. 1983. Importance of *Sclerotium* spp. On peanuts in Egypt. Egyptian J. Plant Pathol. 15: 7-15.

Diseases of Field and Horticultural Crops
Editor-in-Chief: **P. Chowdappa**
Published by: **Daya Publishing House, New Delhi**

Pages **247-269**

9

Diseases of Rapeseed-Mustard

P. D. Meena, C. Chattopadhyay, Lijo Thomas, Pankaj Sharma and Vinod Kumar

Oilseed Brassica also referred as rapeseed mustard, is the second most important edible oilseed crop in India after groundnut and accounts for nearly 30% of the total oilseeds produced in the country. The oleiferous Brassicas including *Brassica juncea* (L) Czern & Coss., *B. rapa* (syn. *B. campestris* L.) and *B. napus* L. are the important sources of edible oil in India. When compared to other edible oils, the rapeseed-mustard oil has the lowest amount of harmful saturated fatty acids. It also contains adequate amounts of the two essential fatty acids, linoleic and linolenic, which are not present in many of the other edible oils. Canada is the largest exporter of rapeseed-mustard, Australia follows Canada. The major importing countries are US, Japan and Mexico. Countries like China, India, Canada, Japan, Mexico, US and European Union, which consume rapeseed-mustard oil cake extensively that is mainly used as cattle feed. Rapeseed-mustard is cultivated all over the world mainly under rainfed conditions with low input management during *rabi* (post-rainy) season. Rapeseed-mustard group of crops are basically temperate crops. They perform well on loamy soils having pH around 7. The temperature range of 0.5 to 3°C, 20 to 35°C and 35 to 40°C can be considered to be minimum, optimum and maximum, respectively for these crops. The productivity of rapeseed-mustard was 1145 kilogram/ hectare (kg/ha) as compared to 1135 kg/ha of total oilseeds which was 854 kg/h during 2002-03. There are varieties like NRCDR-601 has been developed having the highest potential yield 3723 kg/ ha so far (Chauhan *et al.*, 2012). Why there exists a gap between production potential and actual realization in India? In India research is in its infancy as compared to developed countries. The major issues could be

- Lack of proper crop cultivation practices and harvesting methods.
- Lack of quality seed.
- Reduced soil fertility due to repeated sowing.

- Unpredictable rains and frost at the time of harvest.
- Seed quality affected by erucic acid and glucosinolates.
- Biotic and abiotic stresses.

The present productivity level is achieved mainly through varieties developed by pure line selection from indigenous germplasm. Therefore inter-varietal hybridization in Brassicas is needed to broaden the germplasm. Hybrids like *B. rapa* x *B. oleracea* (re-synthesized *B. napus*), *B. rapa* x *B. nigra* (re-synthesized *B. juncea*), *B. nigra* x *S. alba*, *S. alba* x *B. napus* are being developed. Reduction in maturity duration of the genotypes provided greater scope for expansion of this crop in non-traditional areas, especially in *B. juncea* from 150-160 days to 120-110 days and *B. napus* and *B. carinata* from 180-210 days to 150-160 days. Resistance to abiotic and biotic stresses needs to be developed. Salinity tolerant cultivars like CS 52 have been developed. High losses in yield occur due to aphid (*Lipaphis eysimi*) (90%), painted bug (*Bagrada cruciferarum*) and mustard saw fly (*Athalia proxima*), Alternaria blight (*Alternaria brassicae*), white rust (*Albugo candida*), downy mildew (*Hyaloperonospora parasitica*) and Sclerotinia rot (*Sclerotinia sclerotiorum*). For consistent yield performance over environments, cultivars with medium plant height and better response to added input are preferred. Effective CMS - fertility restorer system provide way for exploitation of the hybrid vigour thus increasing the production potential of rapeseed-mustard (*Ogura, oxyrrhina, polima, tournefortii, carinata, trachystoma, moricandia, lyratus*) and for improving oil and meal quality by developing canola cultivars is also considered to be a priority area for research.

1. Area and production

Indian mustard alone accounted for about 80 per cent of the total acreage under these crops in India during 2010-11. The estimated area, production and productivity of rapeseed-mustard in the world was 33.11 million hectares (m ha), 60.66 million tonnes (mt) and 1832 kg/ha, respectively, during 2011-12 (Fig. 9.1). Globally, India account for 20.24 per cent and 10.72 per cent of the total acreage

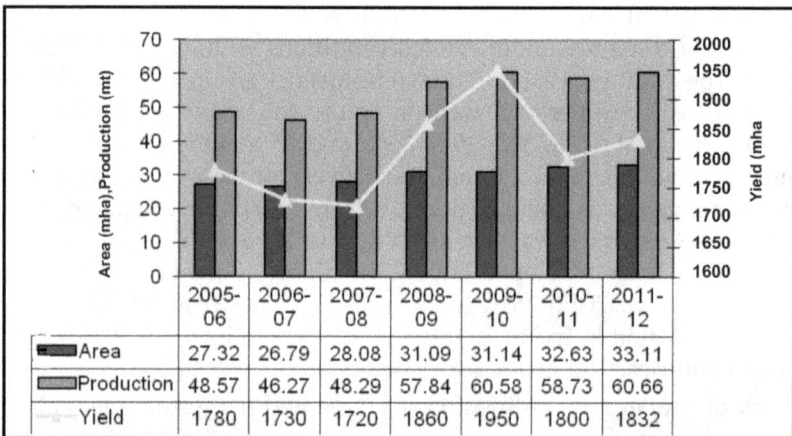

	2005-06	2006-07	2007-08	2008-09	2009-10	2010-11	2011-12
Area	27.32	26.79	28.08	31.09	31.14	32.63	33.11
Production	48.57	46.27	48.29	57.84	60.58	58.73	60.66
Yield	1780	1730	1720	1860	1950	1800	1832

Fig. 9.1. Rapeseed-mustard production trends in the world

and production (Agricultural Statistics Division, 2012). India ranks third in the world area, production and productivity after Canada and China. The rapeseed-mustard production trends represent fluctuating scenario with an all-time high production of 8.17 million tonnes from 6.69 million hectares during 2010-11 (Fig. 9.2). The productivity of rapeseed-mustard was 1145 kg/ha as compared to 1135 kg/ha of total oilseeds (Agricultural Statistics Division, 2012). Rapeseed-mustard contributed around 23 per cent and 25 per cent of the total oilseed acreage and production in India.

In India, the crop is predominantly cultivated in Rajasthan, Uttar Pradesh, Haryana, Gujarat, Madhya Pradesh and West Bengal, which together contribute 87 per cent to the total national production of the crop. A wide gap exists between the potential yield and that realized at the farmers' field, which is largely because of number of biotic and abiotic stresses to which the rapeseed-mustard crop is exposed. The emphasis should be more on higher productivity per unit area and per unit time, rather than on expansion in area especially since the scope for area expansion is limited for oilseed crops.

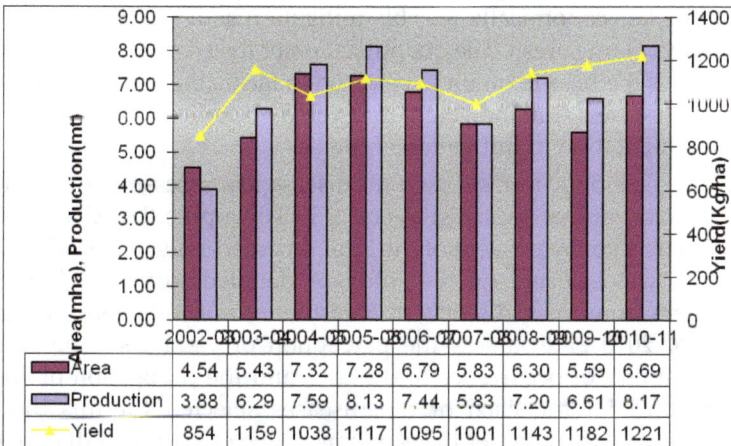

	2002-03	2003-04	2004-05	2005-06	2006-07	2007-08	2008-09	2009-10	2010-11
Area	4.54	5.43	7.32	7.28	6.79	5.83	6.30	5.59	6.69
Production	3.88	6.29	7.59	8.13	7.44	5.83	7.20	6.61	8.17
Yield	854	1159	1038	1117	1095	1001	1143	1182	1221

Fig. 9.2. Trends of rapeseed-mustard area, production and yield in India

Despite all these developments, the country is still importing 40% of the domestic oil needs. Considering the population growth rate and increased per capita edible oil consumption from the present 13.4 kg/ annum to 23.1 kg/ annum by 2030 due to improvement in living standard, about 102.3 million tons of oilseed will be required. Keeping in view the future demand, the current contribution of about 7 million tons rapeseed-mustard need to be increased to about 16.4-20.5 m t. Expression of full inherent genetic potential of a genotype is governed by inputs that go in to the production system. Production has to be increased vertically taking in to account the exploitable yield reservoir. The losses in oilseed crops due to biotic stresses is about 19.9%, out of which diseases cause severe yield reduction at different growth stages. However, the estimates of yield losses due to diseases vary between 35 and 70% as seen in different species of oilseed brassicas grown in different parts of the world. Moreover, oil yield losses due to infected seeds have been reported to be in the range of 15-36% (Ansari *et al.*, 1988). Plant protection technological component as a whole increased the productivity of rapeseed-mustard by 7-24 per cent. Various plant pathogens are reported to affect the crop. Among them, 18 are considered to be economically important in different parts of the globe. To overcome such losses, it is essential to know the causal agents, their ecosystem and means to attack the vulnerable phase. A retrospective of researches on major disease problems in rapeseed-mustard crops with special reference to Indian conditions and their management is discussed here along with brief perspectives based on emerging trends for the future.

2. Physical and Financial Loss Arising From Diseases

The plant diseases cause severe loss, both in physical and financial terms. The physical losses arise from the loss of production stemming from the incidence of disease. The financial loss is on account of both reduced production and lower value of the produce due to reduction in quality of the produce. There are different methods to assess the economic losses due to plant diseases. A proper prioritization of plant diseases in relation to the magnitude of the economic loss due to them is necessary to make economically feasible mitigation activities like disease control programmes over large areas. The quantification of the economic loss is not an end in itself. It shows the prioritization to be given to tackle the production constraints arising from diseases and indicate the level of benefits that could be gained through implementation of mitigation strategies.

Most of the economic models used for disease loss assessment use knowledge inputs from subject matter specialist as one of its key inputs. The state of Rajasthan from where more than 50 per cent of the total rapeseed-mustard production in the country comes was selected for the study. A detailed survey of 180 randomly selected farmers from three districts of Rajasthan was conducted during the *rabi* crop season of 2010-11 to assess the perception of yield loss due to different diseases, probability of occurrence of diseases and the proportion of area under rapeseed-mustard having each of the diseases selected for this disease loss assessment study. In this chapter the economic value of yield loss due to the major plant disease in rapeseed-mustard was calculated using the formula given below.

$$D_i = \{(YL_i * A_i * p_i) * TMA\} P_{rm}$$

Where,

D_i = Total expected damage in INR

YL_i = Yield loss in mustard due to the ith disease

A_i = Proportion of the area under rapeseed-mustard having i^{th} disease

p_i = Probability of occurrence of the ith disease

TMA = Total area under rapeseed-mustard in the state (Rajasthan)

P_{rm} = Market price of rapeseed- mustard during the assessment year

The value of economic loss due to major diseases in rapeseed-mustard was compared with production loss due to other biotic and abiotic stress factors. The data is presented in Table 9.1 Though yield loss due to aphid was the most serious in rapeseed-mustard, the diseases also cause significant loss in yield. The total value of loss arising from the three most commonly reported diseases was to the tune of 504 million rupees for a single crop season. Looking at this from another angle, this is the amount of money foregone by the rapeseed-mustard farmers due to disease infestation. The loss arising from pest and some other abiotic stress factors may be much more than the losses caused by diseases, but as a significant production constraint, we have to develop suitable mitigation strategies for preventing the losses from diseases in rapeseed-mustard.

Table 9.1: Economic Loses due Diseases in Rapeseed-Mustard: A Comparison

Constraint	Yield loss due to constraint (kg/ha)	Area affected by constraint (%)	Probability of) of occurrence (pi)	Value of yield loss (Rs Crore)
Aphid infestation	460.2	0.29	0.43	194.1
Frost damage	359.2	0.22	0.43	114.9
Orobanche	139.7	0.06	0.37	10.5
High temperature at seedling stage	188.4	0.15	0.43	41.1
Sclerotinia rot	76.6	0.04	0.52	5.4
White rust	232.7	0.1	0.45	35.4
Alternaria Blight	137.8	0.1	0.52	9.6

3. Impact of diseases in rapeseed-mustard cultivation

Apart from the economic impact of value of output lost due to the disease, there are other aspects which need to be taken into consideration to gain a holistic understanding on the impact of diseases in rapeseed-mustard farming. The fact that significant share of the rapeseed-mustard farmers are small holder producers make it necessary to have such an assessment so that policy makers can objectively analyze the alternate options available to them. The infestation by diseases lower prices fetched by the produce in the market since the visual observation of the produce is used to judge the quality in many local markets in fixing the price. Studies on the cost of management of these diseases have not been done in a comprehensive manner at the national level. This is especially true in case of rapeseed-mustard crops, which are usually grown with low level of inputs on marginal lands. Since large tracts of rapeseed-mustard area are rain-fed lands, the risk associated with the production is also high, preventing the farmers from investing in technology adoption. This has led to a situation where farmers seldom adopt measures for plant protection. The lack of adoption of plant protection measures makes it difficult to predict the cost of mitigation strategies against diseases. Fertilizer and plant protection chemical accounted for only 13.2 per cent of the total operational cost in rapeseed-mustard cultivation in India during the triennium ending 2009-10. From the data available on fertilizer use in this crop during the same period, it can be concluded that much of the expenditure under this head might be incurred for fertilizers and not for plant protection chemicals. India is a net importer of edible oils and mustard oil is not generally exported.

However, the oil meal, which is left after the extraction of edible oil, is exported. The plant diseases are not reported to be involved in loss of quality of oil seed meal destined for export. By improving the disease management capabilities of rapeseed-mustard producers, the avoidable losses can be prevented and their livelihood security can be strengthened. This can make a significant difference to small holder production systems in rain-fed farming tracts of India.

4. Occurrence and distribution of diseases in India

More than twenty diseases are known to affect the rapeseed-mustard group of crops in India, but diseases like Alternaria blight, white rust, downy mildew, powdery mildew, Sclerotinia rot and club root are more significant because of their wide distribution and high yield losses. The diseases affect number of *Brassica* plants of economic importance but its incidence and damage is more in mustard-rapeseed. The incidence and severity of diseases in rapeseed - mustard fields is influenced by weather and soil conditions at the time of planting and by inoculum density available in the field soil.

Table 9.2. Distribution of rapeseed-mustard diseases in India

Disease	Cause	Distribution	Yield losses (%)
Alternaria blight (Gray Leaf Spot)	*Alternaria brassicae, A. brassicicola, A. raphani, A. alternata*	All states	47
Sclerotinia Rot	*Sclerotinia sclerotiorum*	JK, Punjab, Haryana, UP, Rajasthan, MP, Bihar, WB, Assam, Jharkhand, Chhattisgarh	39.9
White Rust (Staghead)	*Albugo candida*	All rapeseed-mustard growing states	17-60
Downy Mildew	*Hyloperonospora parasitica*	All states	42.6 - 66.7
Powdery mildew	*Erysiphe cruciferarum*	All states	17
Blackleg	*Leptosphaeria maculans* (imperfect state *Phoma lingam*)	HP	-
Clubroot	*Plasmodiophora brassicae* Woronin	WB, Orissa, TN	50
Fusarium Wilt	*Fusarium oxysporum* f. sp. *conglutinans*	UP, Bihar, Haryana, Rajasthan	-
Root Rot Complex and Foot Rot	*Rhizoctonia solani, Fusarium* spp., *Pythium* spp.	Rajasthan, Haryana, MP, UP, Punjab	-
Seedling Blight (Damping Off)	*Rhizoctonia solani, Fusarium* spp., *Pythium* spp.	-	-
White Leaf Spot (Grey Stem)	*Pseudocercosporella capsellae* (*Mycosphaerella capsellae*)	-	-
Aster Yellows	Aster yellows phytoplasma	HP	-
Black Rot	*Xanthomonas campestris*	-	-
Mosaic	Sarson mosaic virus (SMV)	HP, Assam, UP, Haryana, Rajasthan	-
Phyllody	Mycoplasma-like organism	All states	-

Table 9.3. Occurrence of disease in different temperature and moisture conditions

Disease	Temperature (°C)	% RH
Alternaria blight	15-30	65
White rust	12–24	> 97
Downy mildew	15-25	95
Powdery mildew	24–30	< 90
Clubroot	18-23	70
BlackLeg	14-18	100
Damping off	16-25	90-95

5. Alternaria blight

Alternaria blight disease caused by *Alternaria brassicae* (Berk.) Sacc. infecting all above-ground parts of the plant has been reported from all the continents of the world and is considered an important constraint in husbandry of oilseed Brassicas in India. Pathogens of the disease, *A. brassicicola* and *A. raphani* are also encountered but rarely. Though total destruction of the crop due to the disease is rare and usually yield losses at harvest are 5-15%, they can reach up to 47% (Kolte, 1985) accompanied by reduction in seed quality viz., seed size, viability, lusture, etc. Alternaria blight reduced 1000 seed weight causing loss of 35.4 per cent (Kolte *et al.*, 1987). Severity of Alternaria blight on oilseed Brassicas differs among seasons, regions and individual crops within a region in India (Meena *et al.*, 2010). Pathogen is aggressive in nature (Meena *et al.*, 2012), which drastically reduces the photosynthetic area of plant parts responsible for the metabolic process is the main region of yield loss in rapeseed-mustard crops.

Alternaria blight of Brassicae caused by *A. brassicae* is favoured by low temperature, high humidity and splashing rain (Dey, 1948; Humpherson-Jones and Phelps, 1989, Meena *et al.*, 2002). In India, a temperature range of 15-25°C, relative humidity of 70-90%, intermittent winter rains, and wind velocity around 2-5 km per ha has been reported to be most conducive to Alternaria blight development in mustard (Chahal and Kang, 1979; Saharan *et al.*, 1981; Saharan and Kadian, 1984; Ansari *et al.*, 1988; Awasthi and Kolte, 1989; Saharan, 1991). Awasthi and Kolte (1989) reported the best development of Alternaria blight during rosette to flowering stages; relative humidity from 67-73%, rainfall > 70mm, 5-7 h of sunshine/day, and a minimum temperature range of 7-10°C concomitant with the maximum temperature range of 20-23°C have been found to be positively correlated (r=0.511-0.805) with the severity of disease. A minimum period of 4-hour leaf wetness is essential but the longer periods of leaf wetness at 25°C increases the infection frequency on the leaves (Kadian and Saharan, 1984; Saharan, 1991).

The optimum temperature for sporulation within 12-14 hr is 18-24°C for *A. brassicae*, and 20-30°C for *A. brassicicola* (Humpherson-Jones and Phelps, 1989). *A. raphani* infection occurs rapidly at 22-26°C under field conditions. Under high soil moisture content, infection is less at 18°C (Atkinson, 1950). High doses of nitrogen (> 80 kg N/ha), closer spacing (30 x 15 cm), and frequent irrigations are known to rapidly increase severity of Alternaria blight disease of mustard (Kadian and Saharan, 1988; Saharan, 1991; 1992). Similarly, top dressing with nitrogen in spring significantly increases the Alternaria blight intensity on rape pods (Stankova, 1972). Planting time has a major influence on the incidence of disease of mustard crops (Saharan, 1984).

Infections increased with age of leaf, the interaction between temperature and leaf age was highly significant. On older leaves infection was optimal at 25°C. On pods most infections were observed at 20°C, the highest temperature studied. Infection at each temperature increased progressively with duration of surface wetness. The minimum wet periods for infection of leaves were 3 h at 20–25°C, 4 h at 15°C, 6–9 h at 10°C and 12–24 h at 5°C and for infection of pods, between 6 h and 9 h at 10°C and 6 h (or less) at 15°C and 20°C. On leaves, dry periods limited lesion development on pods. Further infections developed when pods were re-wetted.

Meena *et al.* (2004) reported 45 days after sowing (d.a.s.) as critical stages for Alternaria blight development. Chattopadhyay *et al.* (2005) developed epidemiology and forecasting models for Alternaria blight of oilseed *Brassica* in India. They reported the first appearance of Alternaria blight disease on leaves of mustard between 42 and 139 days after sowing (DAS), 44–72, 42–61, 69–83, 45–60, 67–84 DAS having higher frequencies in 1999–2000, 2000–01, 2001–02, 2002–03 and 2003–04, respectively. It was highest in respective years at 45, 46, 75, 45, 76 DAS. The disease first appeared on pods between 67 and 142 DAS, being highest at 99 DAS Severity of Alternaria blight disease on leaves was positively correlated to a maximum daily temperature of 18–27°C, minimum daily temperature of 8–12°C, daily mean temperature > 10 °C, > 92 % morning relative humidity (RH), > 40 % afternoon RH and mean RH of > 70 % in the preceding week. Disease severity on pods was favoured by a maximum daily temperature of 20–30°C, daily mean temperature of > 14 °C, morning RH of > 90 %, daily mean RH of > 70 %, > 9 h of sunshine and > 10 h of leaf wetness. Regional and cultivar-specific models could predict the crop age at which Alternaria blight first appeared on leaves and pods, the highest blight severity on leaves and pods and the crop age when blight severity was highest on leaves and pods at least one week ahead of first appearance of the disease on the crop.

6. White rust

White rust caused by *Albugo candida* (Pers. Ex Fr.) Kuntz. can result in yield loss up to 47% (Kolte, 1985) with each per cent of disease severity and staghead formation causing reduction in seed yield of about 82 kg/ ha and 22 kg/ ha, respectively (Meena *et al.*, 2002). The disease has been reported from Brazil, Canada, France, Germany, India, Japan, Pakistan, Palestine, Romania and Turkey (Kolte, 1985). Depending on the severity of infection, the yield losses caused by white rust or a mixture of white rust and downy mildew, range between 17% and 60% (Kolte, 1996). White rust and downy mildew together produced 37-47% fewer pods and 17-32% less seed in mustard (Bains and Jhooty, 1980). The severity of white rust on leaves and the number of stagheads (diseased pods) were higher in later sown crops. First appearance of white rust disease on leaves and pods (staghead formation) of Indian mustard occurred between 36 and 131 days after sowing DAS, 60 and 123 DAS, respectively. Severity of white rust disease on leaves was favoured by > 40% afternoon RH, > 97% morning RH and 16–24°C maximum daily temperature. Staghead formation was significantly and positively influenced by 20–29°C maximum daily temperature and further aided by > 12°C minimum daily temperature and > 97% morning RH (Chattopadhyay *et al.*, 2011a).

7. Downy mildew

The disease caused by *Hyloperonospora parasitica* (Pers.) Constant. is found to appear more frequently in varying proportions, particularly in India, where rapeseed-mustard cultivation has been intensified. The disease has been reported first from India (Gaumann, 1923) and reports of its occurrence either alone or in association with white rust on leaves or inflorescence have been made from all over the world by various workers predominantly from Canada, China, France, Germany, Hong Kong, India, Japan, Pakistan, Philippines, Russia, Sweden and UK (Kolte, 1985). Reports of its occurrence either alone or in association with white rust on leaves or inflorescence (Bains and Jhooty, 1979; Kolte and Tewari, 1979) have been made. Yield losses due to downy mildew infection alone is very difficult to estimate, since in most cases it is associated with white rust. Although association of these two diseases was reported long back but their importance has been realized only recently with the assessment of losses in mustard crop from 17-54% due to mixed infection mainly on inflorescence resulting into stagheads. The downy mildew disease severity upto 70 per cent alone was observed, which caused yield reduction varying from 42.6 per cent in cultivar Rohini to 66.7 per cent in cultivar NRCDR-2 over healthy plants in *Brassica juncea*. Nevertheless, the per cent oil content reduced from 2.1 per cent in cv. Varuna to 10.6 per cent in cv. NRCDR-2 (Meena *et al.*, 2012).

Hyloperonospora produces both sexual (oospores) and asexual (conidial/ sporangia) spores, which are useful in survival and dissemination of the pathogen. Floral infection increases in the delayed sown crops. The disease is favoured by humid and cool environmental surroundings. Higher infection occurs at low temperature (8-16°C), moist weather and low light intensity. Penetration of the host occurs rapidly at 16°C. Optimum temperature for development of haustoria is 20-25°C. The infection frequency is reduced at 25°C while no infection is observed at 30°C. Leaf wetness duration of 4-6 h at 20°C and for 6-8 h at 15°C is essential for severe infection and disease development on mustard. Sporangia exposed to air are infectious for six weeks but direct sunlight may kill them within 5-6 h. Relative humidity >70 per cent helps in rapid development of the disease. The infection frequency and disease development increases significantly with the increase in duration of leaf wetness (Mehta *et al.*, 1995). In India, infection of mustard foliage starts by the end of October (cotyledon stage) and progresses up to November. The crop planted after mid-November may not contract downy mildew. However, downy mildew growth as a mixed infection with white rust on floral parts can be seen upto March (Mehta and Saharan, 1998).

8. Sclerotinia rot

Rot of mustard caused by *Sclerotinia sclerotiorum* (Lib.) de Bary has become a major disease since 2000 onward in India and elsewhere with high (up to 66%) disease incidence and severe yield losses up to 39.9% leading to discouragement of farmers growing the crop (Chattopadhyay *et al.*, 2003). There may be a great variation in losses in yield in the same area from year to year. Yield losses vary with the per centage of plants infected and the stage of growth of the crop at the time of infection. Plants infected at the early flowering stage produce little or no seeds, and those infected at the late flowering stage will set seed and may suffer little yield reduction. For predicting yield (Y), linear equation (R^2: 0.89) was fitted on disease incidence or DI (Y = 310.25 – 2.04**DI) using Sclerotinia susceptible cv. Rohini. Reduction in diseased plant over healthy for plant height was 12.3%, 36.1% for number of primary and 68.8% for number of secondary branches, 25.4% for main shoot length, 39.5% for number of siliqua per main shoot, 17% for siliqua length on main shoot and 22.8% for grains per siliqua on main shoot. The disease caused a reduction of 6.4% in test weight of seed with a 2% reduction in their oil content. Reduction in seed-yield could be due to the effect of the disease on different yield-attributes. For estimation of yield loss due to Sclerotinia rot, diseased plants were tagged at 70, 80, 90, 100 and 110 days after sowing. Seed yield losses varied from 37.18 to 92.32 per cent depending on the time of disease appearance. The highest yield loss (92.32%) resulted when disease started appearing at 70 days of crop and the least (37.18%) with the occurrence of disease at 110 days (Shivpuri *et al.*, 1999).

Table 9.4. Estimation of yield loss due to Sclerotinia rot appeared at different growth stages of the crop

Time of appearance of disease	Disease severity (%)	Seed yield (g/25 plants)	(%) Yield loss
70	100	284	92.32
80	100	1255	66.08
90	75	1627	56.02
100	62	2250	39.18
110	50	2324	37.18
125	Healthy plants	3700	-

The first record of its occurrence on rapeseed-mustard appears to have been made from Bihar, India (Shaw and Ajrekar, 1915). Since then frequent occurrences of the disease in severe form have been reported from Brazil, Canada (loss up to 28%), China, Denmark, Finland, France, Germany, India, Sweden and United Kingdom. The disease has been found to be causing severe losses in Sriganganagar, Bharatpur, Alwar, Jaipur, Hanumangarh, Dausa, Bhilwara, Ajmer and other districts in Rajasthan state as also in the states of Bihar, Uttar Pradesh, Uttaranchal and Haryana. In one of the surveys conducted, Sclerotinia rot has been rated as the most important of eight problems being faced by the farmers of Bharatpur district in mustard culture (Meena *et al.*, 2006).

The first week of Feb 2005 witnessed cold (max temp: 19-23.1°C; min temp: 1.78°C) and humid (morning r.h. 97-100%; afternoon r.h. 43-79%) with sunny (8.0-10.2 h) days accompanied by night rainfall (2.7 mm) at Bharatpur, which resulted in high soil moisture and conditions favourable for the disease. Discrimination of Sclerotinia rot infested and uninfested patches in Indian mustard was done during post rainy 2004-05 season using digital hyper-spectral data from Earth Observation-1 hyperion centred over Bharatpur (27°12′ N; 77°2′ E). Cloud free hyper ion photograph having 7.5 km swath and 142 m one track coverage were acquired on 24 Feb 2005. Sclerotinia rot incidence, ranging 10-60% was recorded from 20 mustard patches along with Global Positioning System coordinates within a study area of 20 x 5 km (Chattopadhyay *et al.*, 2005b); Dutta *et al.*, 2006). The pathogen is reported to have a wide host range, known to infect about 408 plant species from 75 families (Kolte, 1985; Boland and Hall, 1994) with no proven source of resistance against the disease reported till date in any of the hosts. The five-year (2006-11) average yield whole package was 680 kg/h and 2349 kg/h with a maximum yield gap of 29.2 % and ANMR of Indian Rupees 6423/- in Indian mustard under irrigated timely sown conditions. However, it was 508 kg/h and 1069 kg/h with a mean yield gap of 52.5% and ANMR of Indian Rupees 3986/- in Indian mustard under rainfed conditions (AICRP-RM, 2006-12).

9. Powdery mildew

Occurrence of powdery mildew (*Erysiphe cruciferarum*) on rapeseed-mustard is reported from France, Germany, India, Japan, Sweden, Turkey, the United Kingdom and the U.S. (Kolte, 1985). It is generally believed that the disease does not cause much damage to mustard and rapeseed crops except in occasional severe outbreaks, when all the leaves and siliquae get covered with the powdery growth of the pathogen, particularly at vegetative stages. In certain states of India such as Gujarat, Maharashtra, Karnataka, Haryana and Rajasthan, the disease has been found to occur quite severely, resulting in considerable loss in yield. But the exact data on yield losses are not available. Considering the differences in disease intensity from year to year, it appears the loss is proportional to the disease intensity, which varies considerably depending on the stage at which it occurs. Though total destruction of the crop due to the powdery mildew disease is rare and usually yield losses at harvest are not staggering, they can reach up to 17% (Dange *et al.* 2002).

Severity of powdery mildew on Indian mustard was higher in later sown crops. First appearance of disease on leaves of Indian mustard occurred during 50–120 d. a. s., with higher frequencies in later part of the crop period. Severity of the disease was favoured by > 5 days of 9.1 h of sunshine, > 2 days of morning (maximum) relative humidity (RH) of < 90%, afternoon (minimum) RH 24–50%, minimum temperature > 5°C and a maximum temperature of 24–30°C (Desai *et al.*, 2004). A delayed sowing results in coincidence of the vulnerable growth stage of plants as indicated earlier with warm (maximum temperature: 24–30°C; minimum temperature: > 5°C; > 9.1 h sunshine) and lower (morning: < 90 %; afternoon: 24–50 %) RH conditions. The sustenance of such favourable conditions decides the longevity of the period of mildew attack and further build-up on the crop, which consequently affects yield. Thus, the damage caused to a crop by powdery mildew is likely to be related to sowing date, i. e., late sowing results in higher mildew severity (Saharan and Kaushik, 1981).

10. Club root

Incidence and severity is greater in regions with severe winters than in regions with winter climates. It occurs more frequently in soils, which are acidic and poorly drained. More damage due to the disease results on vegetable crops such as cabbage *(Brassica oleracea* L.) and turnip *(B. campestris* var. *rapifera)* than on oilseed rape *(B. campestris* var. *oleracea* and mustard *(B. juncea).* Woronin (1878) was the first to study the disease in a systematic manner. On rapeseed-mustard, the disease is reported to occur in East Germany, Malaya, New Zealand, Poland, Sweden, the United Kingdom and the U.S. (Kolte, 1985). The disease has been reported from hills of Darjeeling (Chattopadhyay and Sengupta, 1952) and Nilgiri (Rajappan *et al.,* 1999) on vegetable Brassicas. On *B. campestris* var. *yellow sarson* (Laha *et al.,* 1985) and var. toria (Das *et al.,* 1987), the disease has been reported from West Bengal and Orissa, respectively with losses in yield being up to 50% (Chattopadhyay, 1991). In southern districts of New Zealand, losses due to club root on rape are reported high, and this factor has been the major cause of decline of popularity of that crop in that country (Lobb, 1951). However, exact information on the losses caused by the disease on rapeseed-mustard is not known.The disease can occur in both acid and alkaline soils. In acid soils, favourable conditions for incidence of the disease are a moisture content equal to about 70% of the maximum water-holding capacity and a mean air temperature of 18-23°C. Nonetheless, the level of spore load (within the limits of 103 to 2.5×10^7 spores per g. of soil) does not influence the number of diseased plants in the acid soils despite the fact that favourable conditions for attack and good growth of the host are provided. Therefore, under less favourable conditions, spore load does exercise an effect (Colhouna, 1953).

11. Fusarium wilt

Mustard is affected by *Fusarium* wilt caused by *Fusarium oxysporum* f. sp. *conglutinans* (Wr.) Snyder and Hansen. The first authentic report of *F. oxysporum* f. sp. *conglutinans* as the cause of the disease in *B. juncea* was made from India. Later it was reported to occur quite severely on *B. nigra* in India.

12. Bacterial Rot

The black rot symptoms were first observed in 1949 in *B. juncea* in India under natural conditions. The disease is now reported to occur in a severe form in the state of Haryana and West Bengal. Occurrence of the disease has also been reported in Brazil, Canada, Germany, Sweden and the U.S. (Kolte, 1985). In certain years the disease has been reported to take a heavy toll of the crop in Haryana with records up to 60% incidence in certain varieties of mustard (Vir *et al.,* 1973).

13. Bacterial Stalk Rot

The first report about occurrence of stalk rot caused by *Erwinia carotovora* (Jones) Holland appears to have been made by Bhowmik and Trivedi (1980). According to them, the disease appeared in epiphytotic proportion on the commonly cultivated *B. juncea* variety Varuna in 1979 in the State of Rajasthan, India. Presence of the disease in fodder varieties of *Brassica* spp. is also observed. The disease has been reported to cause severe losses to Indian mustard crops on farmers' fields (Meena *et al.*, 2010b). Vigorously growing succulent plants, due to an extra dose of nitrogen, as well as those growing in poorly drained soil are affected more severely.

14. Damping-off and seedling blight

Several species of fungi are involved in causing seed rot and seedling blight around the world. Among them, *Rhizopus stolonifer* is reported to be a more important cause (Petrie, 1973). Post-emergence mortality is not frequent with *Pythium aphanidermatum* (Mahmud, 1950), *P. butleri* (Aulakh, 1971), but is quite considerable in respect of infection caused by *Rhizoctonia solani* (Srivastava, 1968; Khan and Kolte, 2002), *Sclerotium rolfsii* (Upadhyay and Pavgi, 1967; Khan and Kolte, 2002), *Macrophomina phaseolina* (Srivastava and Dhawan, 1979) and *Fusarium* spp. (Khan and Kolte, 2002) are also reported to be causes of seedling blight in India.

15. Mosaics

Symptoms of the disease on rapeseed-mustard appear as vein clearing, green vein banding, mottling, and severe puckering of the leaves. The affected plants remain stunted and do not produce flowers, or very few flowers are produced on such plants. When siliquae are formed, they remain poorly filled and show shrivelling (Azad and Sehgal, 1959). The symptoms commence at or near the base of the leaf and gradually spread over the entire leaf. During the later stages of infection, numerous raised or non-raised dark-green islands of irregular outline appear in the chlorotic area between the veins, giving rise to a mottled appearance. Curvature of the midrib and distortion of the leaf blade on affected leaves can also be a prominent symptom. Plants infected early are usually stunted and are killed, but those infected late show reduced growth only slightly.

16. Phyllody

Under natural conditions, the disease has been reported to occur in India on *toria* (*B. rapa* var. *toria*), and *sarson* (*B. rapa* var. yellow *sarson*) in the states of Punjab (Vasudeva and Sahambi, 1955), Haryana (Sandhu *et al.*, 1969) and Uttar Pradesh. Similar symptoms on *B. juncea* varieties have been observed in Haryana, UP and Rajasthan.

17. Quality concern

The presence of high glucosinolate in seed-meal is the major limiting factor in export promotion, and it is likely to become a strong non-tariff barrier in the coming years. In the international market, rapeseed-mustard varieties having erucic acid and glucosinolate content as per the internationally accepted norms fetch premium price. Mustard cultivars grown in India have very high glucosinolate content. In fact, one of the glucosinolates, alyl-isothiocyante, is actually responsible for the pungency in the oil, which is actually liked by the Indian consumer. Thus, there is an urgent need to make concerted efforts for breeding varieties with levels of total saturated fatty acid less than four per cent and erucic acid less than two per cent so as to bring the Indian cultivars on par with international quality norms and also to meet the consumer's preferences and health concerns. If India is to look for earning foreign exchange through export of seed-meal and oil and fetch remunerative market price, it has to elevate nutritional value of oil and seed-meal and needs to develop single and double low ['0' and "00"] varieties for cultivation in India.

18. High yielding varieties / hybrids

Rapeseed-mustard is a *rabi* season and location-bound crop. A number of varieties and hybrids have been released for various locations in the country. During the last three decades, more than 150 improved cultivars/ hybrids have been developed in rapeseed-mustard. Some important hybrids include NRCHB 506, DMH 1, PAC 432, PAC 437 and 44 S44. This increasing trend in cultivar development in rapeseed-mustard is a promising sign.

19. Exploitable yield and technology time gap

The national average annual yield of rapeseed-mustard was 1.0 t/ha during 2000-04, while the realizable yield with improved technology was 1.5 t/ha. Thus, the realizable yield gap of 0.5 t/ha accounts for nearly one-third of the potential yield. Time to achieve the realizable yield (i.e. yield of improved varieties / hybrids in farmers' fields) was forecasted by fitting data to the growth model using the formula (Martino, 1993).

$$y = L / (1+a \ e^{-bt})$$

where, L is upper limit to growth of the variable y, t is time and a and b are the coefficients evaluated by fitting the data. Even though rapeseed-mustard is a major oilseed crop in the country and has been performing well during last few years, the time lag to achieve the realizable yield is still large. Major technological advances like that in castor (Rama Rao, 2000) and better dissemination techniques for the improved methods can reduce this gap. The technology transfer through public system for rapeseed-mustard is forecasted as to be same as at present. However, some positive changes are expected through the aggressive entry of private sector in seed marketing.

Retrospective and perspectives

In this climate change era, pathogens in rapeseed-mustard have been recognized as major forces causing economic losses with identification of the most important ones depending on their loss-causing potential based on agro-ecological zones.

Cultural practices for managing some diseases have been pinpointed. Critical stages for growth of some foliar diseases viz., Alternaria blight and white rust have been identified together with weather conditions that favour disease development (Chattopadhyay *et al.*, 2011b). Thus, now it is possible to recommend need-based sprays of botanical and chemical fungicides for effective, economic and environment-friendly disease management. Nevertheless, the area of rapeseed-mustard disease management seems to have been largely neglected by policy makers. Yield loss causing capacity of diseases in rapeseed-mustard crops has been under-estimated. Breeding for disease resistance in rapeseed-mustard has not received enough attention. Cultivars bred for agronomic benefits have been screened later for disease resistance. Rarely a systematic and conscious breeding effort was made for disease resistance. Genetics of inheritance of disease resistance and its biochemical basis have been pinpointed in some cases. Use of modern biotechnological tools has also yielded success. But, sources of resistance to diseases have remained grossly under-utilized. Though research has provided some significant tools for managing the diseases, the adaptability of the same in the farmers' field has been lacking. An oilseed farmer is resource poor, farming under harsh conditions where crop protection rarely figures in his priorities. The need of the mustard grower in the country is low monetary input production technology while research has devised mostly high input management strategies emulating cereals. There has been a total lack of extension branch of rapeseed-mustard pathology. While assessing the effects of TMO, it is clearly reflected that transfer of technology (TOT) has been a weak link in *Brassica* pathology. Even the simple practice of seed dressing is not followed by oilseed *Brassica* farmer regularly for the control of some seedling diseases and downy mildew at cotyledonary stage.

Perhaps a resource poor mustard grower would accept an ideal plant type resistant to many rather than to only one disease. Thus, the multiple disease resistance concept need to be advocated by overcoming the one of single disease resistance. The multiline cultivar concept, so successfully propounded in cereals, with resistance to major diseases, needs to be adopted to mustard crop at least. Identification of sources of resistance to disease caused by *Sclerotinia sclerotiorum* is required. Horizontal resistance to disease may be preferable for sustainability. There is also need to transfer resistance, available in wild species viz., resistance to rapeseed-mustard Alternaria blight from wild allies. Studies on biochemical and physiological basis of disease resistance or pathogenic stress physiology need to be taken up. Generation of adequate information on bio-ecology, population dynamics, mass multiplication and artificial maintenance techniques, race/biotype pattern of pathogens, functional genomics of host-pathogen interaction, rapid reliable screening and evaluation techniques against specific diseases as per priority needs to be addressed. For this purpose, there is need for creation of national facility for reference (type) cultures to speed up work on genetic enhancement. Immediate confirmation and utilization of sources of resistance/ tolerance to diseases to design well thought out crossing programmes targeted at resistance to diseases is required. Study of mechanisms and genetics of resistance to key diseases, wherever unknown, needs attention. Exploitation of modern tools for screening of large number of germplasm lines and for transfer of resistance in background of cultivated and popular varieties should be taken up. Immunization of plants through induced resistance and investigations on molecular and biochemical basis of resistance could be another approach. National crop laboratories may be given specific responsibilities and be commensurately equipped to cater to the assigned task.

From the list of diseases that are known to be important in rapeseed-mustard, it is clear how much of biotic stress the crops face under on-farm situation. Many-a-time more than one disease occurs on the same crop simultaneously. Sometimes the plant may be harbouring the pathogen(s) but may not be infected. However, that may also lead to poor yields due to latent stress caused by such deleterious microorganisms. A systems approach needs to be considered for a better crop health management. Through adaptive research at respective AICRP-RM centres and in frontline demonstrations, component-wise and in-whole package, the following measures may be combined for designing location-specific packages depending on demand of the locale:

- Choosing the sowing time is very important as they affect the disease incidence significantly.
- Deep summer ploughing.
- Use of disease resistant or tolerant or early maturing disease escaping cultivars.
- Use of high quality seeds.
- Seed treatment with biocontrol agents viz., *T. viride*, G. *virens* or botanicals like *Allium sativum* bulb extract (1% w/v) or carbendazim @ 0.1% a.i. or mixture of carbendazim with Apron 35 SD (6 g/ kg) and 250 ppm streptomycin should be made mandatory. There is a need for mixture of fungicides for avoiding resistance development in pathogens to fungicides. Use of biocontrol agents is advantageous as they are often effective against a wide range of soil-borne pathogens. Moreover, they are ecofriendly, cost effective and their use avoids the risk of development of resistance in the pathogen towards the control agent.
- Application of recommended balanced doses of N, P and K fertilizers with split application of nitrogen,
- Have only optimum plant population with recommended spacing.
- Well-drained plots are very essential. If water accumulates on one side of a plot due to slope in the land, a channel may be dug at the lower end of the slope to drain out the accumulated water.
- Clean cultivation, elimination of weeds and stubbles. Roguing out diseased plant to avoid secondary spread of disease to reduce inoculum.
- Identify the right crop for rotation and also for non-crop season as sequence crop in order to discourage the pathogen build-up. Choose a suitable intercrop as per the pathogen to be countered.
- Judicious use of irrigation, preferably only 2, depending on the stage of crop growth, soil type and rainfall availability.

- Prophylactic spraying of botanical/ bioagent/ fungicide/ antibiotic when conditions favourable for disease occurrence prevail. Timing the application of such prophylactic sprays based on databank, expert systems and epidemiological research findings. Disease forewarning/ forecasting models ensures a cost-effective management of rapeseed-mustard crops. There is need to ensure that the pesticidal sprays are avoided or minimized during peak activity period of the biocontrol agents.

To make disease management easier and successful, there is a need for integrated seed treatment with well-formulated biocontrol agent(s) having tested delivery systems. Thus, *Trichoderma* spp. tolerant to fungicide(s) and effective fungicide(s) like carbendazim and/or metalaxyl will be good for an effective biocontrol-based disease management system. Further evaluation of more effective biocontrol agents and botanicals is needed. An effective ecofriendly and cost effective tool for soil-borne disease management like soil solarisation needs to be made cost-effective. In view of the new liberal seed policy, quarantine should be strengthened to exclude introduction of pathogens or some of their biotype/races not existing in India. Holistic understanding of the different agro-ecosystems and bioecology of the pathogens is essential. Biodiversity, preservation and search for resistance in wild types will have to be focussed to pool available genes in host and pathogen. Tools for gene mapping and pooling has to be augmented.

Judicious use of biotechnology and genetic strategies are now recognized as very rewarding. Standardization of methods for rapid identification of pathogens for diagnosis of and resistance to diseases will enable screening of large germplasm collection in a short time. The modern techniques may be very useful to develop novel biocontrol agents and even disease resistant transgenics. These will necessitate rapeseed-mustard pathologists of this era to be geared for taking up such challenges to build patents.

Risks related to incidence of a disease should be accounted for while programming models for whole-farm planning. Advances in crop production strategies have resulted in a spurt of high yields in major crops like cereals. To revolutionize rapeseed-mustard production, skilful integration of various managerial tactics including disease risk analysis as related to its incidence and risk management should take into consideration the experience, beliefs and preferences of the farmers' need as a participatory mode of technology development.

Today the threats to our environment are much greater in scope and number. There is need to monitor the race pattern of different pathogens and not just diseases. Expert systems and database for information retrieval need to be prepared for developing effective disease forecasting models possibly through simulation of epidemics for all major rapeseed-mustard diseases to enable the growers to take up need-based prophylactic measures. Keeping in view the options available for designing an effective and environment-compatible disease management strategy for the crops, conceptual models, primary and secondary decision-making tools need to be used effectively. Large gaps in information in areas like biocontrol, use of locally available cheap botanicals, etc. need to be bridged. There is also need to reduce gaps between lab and land or dissemination of information by converting integrated disease management into a movement through programmes like National Network Projects and development of a community approach in recognition of pathogens. Use of resistant cultivars, seed treatment, need-based spray of pesticides have exhibited the potential of plant protection in increasing yield of oilseed Brassica crops in frontline demonstration trials. However, there is need to analyse the reasons for failure of disease management in these crops not taking root in farmers' fields. Technologies developed have been more a result of curiosity-driven research with lack of effective delivery systems. Pathologists have to be more pragmatic by rationalizing the targets for disease management with its implementation being remarkably visible. There is need to view the disease problems from the farmers' point of view by interaction of the policy maker, scientist, extension worker and the grower in an organized and harmonious way. Hence, there is an urgent need to design need-based disease management strategies with low input for sustainable agriculture (LISA) and transfer them to farmers through proper adaptive research. Plant pathologists need to advocate eco-sustainable technology for disease management, the need of the hour to save our Planet. Biomanagement of plant diseases are known to be economical and avoid ground-water pollution, death of non-target beneficial microflora and risk of new resistant pathogen strains. This could be in tune with the provision of job-led growth that may be economically, environmentally and socially sustainable with livelihood opportunities linked with agricultural competitiveness. In many European countries, organic agriculture has rapidly been transformed from a farmers' movement to an institutionalized part of agricultural policy. In certification, compliance with published organic standards is verified through annual inspections on farms (Laura Seppänen and Juha Helenius, 2004). The Indian Council of Agricultural Research has brought out a compilation of technologies available for bioformulation. These technologies if effectively utilized and industrialized, than merely planning and talking on the subject in various fora of seminars and symposia, can lead to early substitution of use of chemical fungicides for an ecofriendly and sustainable disease management program in oilseed crops. The problem of shelf life of biofungicides could be overcome by setting up small-scale biofungicide preparation systems at district level in biovillage frame with trained people and strict monitoring of quality norms, which may also provide employment while taking care of the availability aspect. Package of practices on different oilseed crops need to emphasize more on biofungicides. Proper extension coupled with appropriate trickle-down of information and feedback, timely availability, procurement of bioformulations conforming to quality norms, conviction of farmers through frontline demonstration regarding cost-effectiveness and benefits of biofungicides through proper and timely application, intervention at dealers level to effect shift from chemicals to biofungicides, etc. are also essential for better coverage by biofungicides. There is also need for a mass movement for promotion of biofungicides and secure trust of end-users about their efficacy to enable them replace chemical fungicides to the extent possible. For the success of rapeseed-mustard disease management, educating the Brassica planter with updated technology or oilseed grower awareness about diseases and their management through trained extension personnel is very essential. Personnel involved in growing, handling and storing of oilseeds need to be equipped to diagnose and manage diseases in the field and store based on available options.

In developing countries, agriculture is being intensified to produce more. The agricultural development strategies are prioritised on increasing yield, crop protection, human health, environmental hazards and social aspects. This sequential rather than integrated approach contributes to imbalance problems related to sustainability in agriculture. To improve this situation there is need to determine the potential of a zone, the biophysical environment, etc. These factors should be analysed keeping in view the Indian scenario to set priorities for oilseed research and development. In case of a negative trend in sustainability, necessary adjustments should be made in the strategy. Rapeseed-mustard crops need to be and can be made more productive by providing the required relief to the plant from stress provided by pathogens and latently associated deleterious microorganisms, through meticulous integration of the disease management options. Viewing disease management in isolation has been a major folly and needs to be looked in totality. From a myth, interdisciplinary research has to be made a reality in rapeseed-mustard scenario through intensive teamwork with a mission mode approach. Close collaboration of different disciplines on the same plane shedding "big brotherly attitude" of any discipline by broadening of horizons is another necessity. Novel pathogen management options have to be designed through reorientation of strategies may be through a shift from disease management to Total Pest Management (TPM), on ecosystem basis for rapeseed-mustard crop health. Thus, combining increase in rapeseed-mustard productivity with its improved ecofriendly protection is needed for total oil Brassica crop management programme.

REFERENCES

Agricultural Statistics Division, 2012. Agriculture Statistics at a Glance 2003. Agriculture Statistics Division, Dept. Agril. & Coop., Ministry Agriculture, GOI, New Delhi, 223.

AICRP-RM (All India Coordinated Research Project on Rapeseed-Mustard). 2006-2012. Annual Reports. Directorate of Rapeseed-Mustard Research (ICAR), Bharatpur 321303, Rajasthan.

Ansari, A. N., Khan, M. W., and Muheet, A. 1988. Effect of Alternaria blight on oil content of rapeseed-mustard. Curr Sci. 57: 1023-1024.

Atkinson, R. G. 1950. Studies on the parasitism and variation of *Alternaria raphani.* Can. J Res Sect C. 28: 288-317.

Aulakh, K. S. 1971. Damping-off of toria seedlings due to *Pythium butleri. Indian Phytopath.* 24: 611.

Awasthi, R. P., and Kolte, S. J. 1989. Effect of some epidemiological factors on occurrence and severity of Alternaria blight of rapeseed and mustard. Proc. IDRC (Canada) Oil Crops: meeting held at Pantnagar and Hyderabad, India, Jan. 4-17: 49-55.

Azad, R. N., and Sehgal, O. P. 1959. A mosaic disease of Chinese sarson *(Brassica juncea* (L.) Coss. var. *rugosa* Roxb. Indian Phytopath. 12: 45.

Bains, S. S., and Jhooty, J. S. 1980. Mixed infections by *Albugo candida* and *Peronospora parasitica* on *Brassica juncea* inflorescence and their control. Indian Phytopathol.

32: 268-271.

Bains, S. S., and Jhooty, J. S. 1979. Mixed infections by *A. candida* and *P. parasitica* on *B. juncea* inflorescence and their control. Indian Phytopathol. 32: 268-271.

Bhowmik, T. P., and Trivedi, B. M. 1980. A new bacterial stalk rot of *Brassica.* Current Sci. 49: 674.

Boland, G. J., and Hall, R. 1994. Index of plant hosts of *Sclerotinia sclerotiorum.* Canadian J. Plant Pathology. 16: 93-100.

Chahal, A. S., and Kang, M. S. 1979. Influence of the metrological factors on the development of Alternaria blight of rape and mustard in Punjab. Indian Phytopathology. 32: 171.

Chattopadhyay, A. K. 1991. Studies on the control of club root disease of rape seed mustard In West Bengal. Indian Phytopathology. 44: 397-398.

Chattopadhyay, C., Agrawal, R., Kumar, A., Meena, R. L., Faujdar, K., Chakravarthy, N. V. K., Kumar, A., Goyal, P., Meena, P. D., and Shekhar, C. 2011a. Epidemiology and development of forecasting models for white rust of *Brassica juncea* in India. Archives of Phytopath Plant Protect. 44: 751-763.

Chattopadhyay, C., Bhattacharya, B. K., Vinod-Kumar., Amrender-Kumar., and Meena, P. D. 2011b. Impact of climate change on pests and diseases of oilseeds Brassica - the scenario unfolding in India. Journal of Oilseed Brassica. 2: 1-8.

Chattopadhyay, C., Agrawal, R., Kumar, A., Bhar, L. M., Meena, P. D., Meena, R. L., Khan, S. A., Chattopadhyay, A. K., Awasthi, R. P., Singh, S. N., Chakravarthy, N. V. K., Kumar, A., Singh, R. B., and Bhunia, C. K. 2005a. Epidemiology and forecasting of Alternaria blight of oilseed *Brassica* in India – a case study. J Pl Dis Protn. 112: 351–365.

Chattopadhyay, C., Bhattacharya, B. K., Dutta, S., Meena, P. D., Kumar, V., Parihar, J. S. and Patel, N. K. 2005b. Assessing Sclerotinia rot incidence in mustard using spectral bands of satellite sensed observations. J. Mycol Pl Pathol. 35: 574.

Chattopadhyay, C., Meena, P. D., Sastry, R., Kalpana., and Meena, R. L. 2003. Relationship among pathological and agronomic attributes for soilborne diseases of three oilseed crops. Indian J Plant Prot. 31: 127-128.

Chattopadhyay, S. B., and Sengupta, S. K. 1952. Bull. Bot. Soc. Bengal 6: 57-61.

Chauhan, J. S., Singh, K. H. and Kumar, V. 2012. Compendium of rapeseed-mustard varieties notified and breeder seed production scenario in India (2006-2012). AICRP-RM, DRMR, Bharatpur 72.

Colhouna, J. 1953. A Study of the epidemiology of clubroot disease of Brassiceae. Ann Appl Biol. 40: 262-283.

Dange, S. R. S., Patel, R. L., Patel, S. I. and Patel, K. K. 2002. Assessment of losses in yield due to powdery mildew disease in mustard under North Gujarat conditions. Journal of Mycology and Plant Pathology. 32: 249–250.

Das, S. N., Mishra, S. K. and Swain, P. K. 1987. Reaction of some toria varieties to *Plasmodiophora brassicae.* Indian Phytopathology. 40: 120.

Desai, A. G., Chattopadhyay, C., Ranjana A., Kumar, A., Meena, R. L., Meena, P. D., Sharma, K. C., Srinivasa Rao, M., Prasad, Y. G., and Ramakrishna, Y. S. 2004. *Brassica juncea* powdery mildew epidemiology and weather based forecasting models for India- a case study. *Zeitschrift für P.flanzenkrankheiten und*

P.flanzenschutz. Journal of Plant Diseases and Protection. 111: 429-438.

Dey, P. K. 1948. Plant Pathology. Adm. Rep. Agric. Dept. U.P. 47: 39-42.

Dutta, S., Bhattacharya, B. K., Rajak, D. R., Chattopadhyay, C., Patel, N. K. and Parihar, J. S. 2006. Disease detection in mustard crop using EO-1 hyperion satellite data. J. Indian Soc. Remote Sensing (Photonirvachak).34: 325-329

Gaumann, E. 1923. Beiträge zu einer Monographie der Gattung *Peronospora* Corda. – Beiträge zur Kryptogamenflora der Schweiz 5: 1-360.

Humpherson-Jones, F. M., and Phelps, K. 1989. Climatic factors influencing spore production in *Alternaria brassicae* and *A. brassicicola.* Annals of Applied Biology. 114: 449-458.

Kadian, A. K., and Saharan, G. S. 1984. Studies on spore germination and infection of *Alternaria brassicae* of rapeseed and mustard. Journal of Oilseeds Research.1: 183-188.

Kadian, A. K. and Saharan, G. S. 1988. Progress of Alternaria blight of mustard in relation to cultural practices. Oil Crops Newsletter. 5: 32-33.

Khan, R. U. and Kolte, S. J. 2002. Some seedling diseases of rapeseed-mustard and their control. Indian Phytopathology, 55: 102-103.

Kolte, S. J. and Tewari, A. N. 1979. Note on the susceptibility of certain oleiferous *Brassicae* to downy mildew and white blister diseases. Indian J Mycol Plant Pathol. 10: 191.

Kolte, S. J. 1985a. Diseases of annual edible oilseed crops. Vol II. CRC Press, Inc.

Kolte, S. J., Awasthi, R. P., and Vishwanath. 1987. Assessment of yield losses due to Alternaria blight in rapeseed and mustard. Indian Phytopathol. 40: 209-211.

Kolte, S. J. 1996. Diseases. In: Chopra, V. L., and Prakash, S. (eds.) Oilseeds and Vegetable Brassicas. An Indian Pespective. Oxford and IBH Publishing Co. Pvt. Ltd. New Delhi.187-207.

Laha, J. N., Naskar, I., and Sharma, B. D. 1985. A new record of club root disease on mustard. Current Sci. 54: 1247.

Laura Seppänen and Juha Helenius. 2004. Do inspection practices in organic agriculture serve organic values? A case study from Finland. *Agriculture and Human Values.* 21 (1):1-13.

Lobb, W. R. 1951. Resistant type of rape for areas with club root. N.Z. J. Agric. 82: 65.

Mahmud, K. A. 1950. Damping-off of *Brassica juncea* caused by *Pythium aphanidermatum* Fitz. Sci. Cult. 16: 208.

Martino, JP. 1993. *Technological Forecasting for Decision Making* (3rd ed), New York, McGraw-Hill.

Meena, P. D., Chattopadhyay, C., Singh, F., Singh, B., and Gupta, A. 2002. Yield loss in Indian mustard due to White rust and effect of some cultural practices on Alternaria blight and White rust severity. Brassica. 4: 18-24.

Meena, P. D., Rani, A., Meena, R., Sharma, P., Gupta, R., and Chowdappa, P. 2012. Aggressiveness, diversity and distribution of *Alternaria brassicae* isolates infecting oilseed *Brassica* in India. African Journal of Microbiology Research. 6: 5249-5258.

Meena, P. D., Awasthi, R. P., Chattopadhyay, C., Kolte, S. J., and Kumar, A. 2010a.

Alternaria blight: a chronic disease in rapeseed-mustard. J Oilseed Brassica. 1: 1-11.

Meena, P. D., Meena, R. L., Chattopadhyay, C., and Kumar, A. 2004. Identification of critical stage for disease development and biocontrol of Alternaria blight of Indian mustard (*Brassica juncea*). J. Phytopathology.152: 204-209.

Meena, P. D., Mondal, K., Jha, S. K., Chattopadhyay, C., and Kumar, A. 2010b. Bacterial rot: a new threat for rapeseed-mustard production system in India. J Oilseed Brassica. 1: 39-41.

Meena, P. D., Sharma, A. K., Jha, S. K., and Chattopadhyay, C. 2006. Impact of fungal diseases on mustard yield- farmers' perception and garlic clove extract in management of Sclerotinia rot. Indian J Pl Prot .34: 229-232.

Meena, P. D., Sharma, P., and Asha Rani. 2012. Estimation of yield losses in oilseed brassica due downy mildew. *IN Abstracts* of *1st National Brassica Conference* on Production Barriers & Technological Options in Oilseed Brassica of Society for Rapeseed-Mustard Research during 2-4 March 2012 held at CCSHAU, Hisar 148-149.

Mehta, N., and Saharan, G. S. 1998: Effect of planting dates on infection and development of white rust and downy mildew disease complex in mustard. J Mycol Pl Pathol. 28: 259-265.

Mehta, N., Saharan, G. S., and Shrama, O. P. 1995. Influence of temperature and free moisture on the infection and development of downy mildew on mustard. Pl Dis Res. 10: 114-121.

Petrie, G. A. 1973. Diseases of *Brassica species* in Saskatchewan, 1970-72. I. Staghead and aster yellows. Can. Plant Dis. Surv. 53: 19-25.

Rajappan, K., Ramaraj, B. and Natarajan S. 1999. Knol-khol - a new host for club-root disease in the Nilgiris. *Indian Phytopath* 52 (3): 328.

Rama Rao, C. A. 2000. Growth and Efficiency in Crop Production in Andhra Pradesh. Ph.D. Thesis submitted to ANGR University, Hyderabad.

Saharan, G. S. 1984. A review of research on rapeseed and mustard pathology in India. Paper presented in the Annual Rabi Oilseed Workshop held at Jaipur, August 6-10, 1984.

Saharan, G. S. and Kadian, A. K. 1984. Epidemiology of Alternaria blight of rapeseed and mustard. *Cruciferae NewsLetter* 9: 84-86.

Saharan, G. S. 1991. Assessment of losses, epidemiology and management of black spot disease of rapeseed-mustard. *GCIRC 8th* Intl. Rapeseed Congr, July 9-11, Saskatoon, Canada: 84 (Abstr); Proc. 2: 465-470.

Saharan, G. S. 1992. Management of rapeseed and mustard disease. IN: *Advances in oilseed research.* D. Kumar and M. Rai (eds). Sci Pub., Jodhpur, India, Chapter 7: 152-188.

Saharan, G. S., and Kaushik, J. C. 1981.Occurrence and epidemiology of powdery mildew of *Brassica.* Indian Phytopathol. 34: 54–57.

Saharan, G. S., Kaushik, J. C. and Kaushik, C. D. 1981. Progress of Alternaria blight on rye cultivars in relation to environmental conditions. 3rd Internl.Symp. Pl. Pathol. IARI, New Delhi, Dec. 14-18: 236.

Sandhu, R. S., Singh, G., and Bhatia, N. L. 1969. Studies on the effect of sowing dates and spacing on incidence of phyllody in Indian rape *(Brassica campestris* L. var. *toria* Duth). Indian J. Agric. Sci. 39: 959.

Shaw, F. J. F., and Ajrekar, S. L. 1915. The genus *Rhizoctonia* in India. Mem. Dep. Agric. Lndia Bot. 7: 177.

Shivpuri, A., Sharma, K. B., and Chhipa, H. P. 1999. Some studies on the stem rot disease *(Sclerotinia sclerotiorum)* of rapeseed-mustard in Rajasthan, India. 10th Internationa Rapeseed Congress Canberra, Australia.

Stankova, J. 1972. Varietal variability of winter rape with regard to its inclination to dark leaf spot and the factors influencing the development of this disease. Rostl. Vyr. 18: 625–630.

Srivastava, O. P. 1968. Dry root and bottom rot of mustard caused by *Rhizoctonia solani.* Indian J. Microbiol. 8: 277.

Srivastava, SK and Dhawan, S. 1979. Epidemiology of *Macrophomina* stem and root rot of *Brassica juncea* (L.) Czern & Coss. In Northern India. *Proc. Indian Natn. Sci. Acad.* B 45 (6): 617-622.

Upadhyay, R., and Pavgi, M. S. 1967. Some new hosts for *Pellicularia rolfsii* (Sacc.) West. from *India.* Sci. Cult. 33: 71.

Vasudeva, R. S., and Sahambi, H. S. 1955. Phyllody of Sesamum *(Sesamum orientale* L.). *Indlan Phytopath,* 8: 124.Vir, S.; Kaushik, C. D. and Chand, J. N. 1973. The occurrence of bacterial rot of raya *(Brassica juncea* coss) in Haryana. PANS. 19: 46.

Vir, S., kaushik, C.D. and Chand J.N. 1973. The Occurrence of Bacterial rot of raya*(Brassica juncea)* in Haryana. International J. Pest Management, 19:46-47.

Woronin, M. 1878. *Plasmodiophora brassicae.* Urheberder der Kohlpflanzen-Hemie. Jahrb. Wiss Bot. 11: 548.

Diseases of Field and Horticultural Crops   *Pages* **270-297**
Editor-in-Chief: **P. Chowdappa**
Published by: **Daya Publishing House, New Delhi**

10

Diseases of Sunflower

M. Santha Lakshmi Prasad and S. Chander Rao

Sunflower (*Helianthus annus* L.) is an important oilseed crop of the country, popularly known as "Surajmukhi.". It is one of the fast growing oilseed crops in India. It occupies fourth place among oilseed crops in terms of acreage and production. The oil is used for culinary purposes, in the preparation of vanaspati and in the manufacture of soaps and cosmetics. Since its oil is considered as premium compared to other vegetable oil because of its light colour, bland flavour, high smoke point and high poly unsaturated fatty acid content. Sunflower oil is a rich source of linoleic acid (64%) which helps in washing out cholesterol deposition in the coronary arteries of the heart and thus it is good for heart patients. The oil cake contains 40-44 per cent high quality protein. It is ideally suited for poultry and livestock rations (Anon., 2006). The sunflower kernels can be eaten raw or roasted as one of the confectionary item.

Sunflower crop was introduced into India during 1969. In the early 1970s, only around 0.1 million hectares area was under sunflower cultivation. By 1993-1994, 2.7 million hectares area was under the crop. Production has multiplied 15-fold in about two decades although yield/ha has barely improved. During 2006, India has the fourth largest area under sunflower in the world. Its share in total world production is about 4% and accounts for 9% of the world acreage. However, the yield at 566 kg/ha was the lowest among the major sunflower producing countries in the world. During 2012-13, sunflower is cultivated in an area of 8.30 lakh h with a total production of 5.44 lakh tones and productivity of 655 kg/ha (Anon., 2012). The important sunflower growing states are Karnataka, former Andhra Pradesh, Maharashtra, and Tamil Nadu. Karnataka is the largest sunflower producing state in the country. It accounts for half of the total area under the crop followed by former Andhra Pradesh and Maharashtra. The losses due to diseases account up to 30% during *kharif* and 5-10% during *rabi* seasons. The economically important diseases are sunflower necrosis disease, *Alternariaster* leaf blight, powdery mildew, downy mildew, rust and *Sclerotinia* wilt (Fig. 10.1.). The diseases like *Sclerotium* wilt, charcoal rot and *Rhizopus* head rot appears in sporadic manner and damaging the crop in patches in some areas. Minor diseases are septoria leaf spot, sunflower mosaic disease and angular leaf spot. A new viral disease, sunflower leaf curl started appearing since two years.

Fig 10.1. Map showing major sunflower diseases

Table 10.1. Sunflower diseases and their causal organism

Disease	Causal organism
Sunflower necrosis disease	Tobacco streak virus
Alternaria leaf spot/*blight*	*Alternaria helianthi* (Hansf.)Tubaki and Nishihara
Powdery mildew	*Golovinomyces cichoracearum* f.sp. *helianthi* DC ex Meret; *Sphaerotheca fuligenea (Schlecht. ex.Fr.) Pollaci*
Downy mildew	*Plasmopara halstedii* (Farl.) Berl. and de Toni
Rust	*Puccinia helianthi* Schwein
Sclerotium collar rot	*Sclerotium rolfsii* Sacc
Sclerotinia wilt	*Sclerotinia sclerotiorum* (Lib.) de Bary
Head rot	*Rhizopus arrhizus* Fischer
Charcoal rot	*Macrophomina phaseolina* (Tassi) Goid
Seed rot and Seedling blight	*Aspergillus niger, Rhizopus stolonifer, R. arrhizus, Rhizoctonia sp, Sclerotium rolfsii; Pythium aphanidermatum*
Septoria Leaf spot	*Septoria helianthi* Ellis and Kellerman
White rust	*Albugo tragopogi* (Pers.) Schroct
Verticillium wilt	*Verticillium dahliae* Klebahn
Fusarium wilt	*Fusarium moniliformae* Scheld *F. oxysporum* Schlect
Phomopsis black stem canker	*Phomopsis helianthi* Muntanola-Cvetkovic *et al.*
Phoma black stem	*Phoma macdonaldii* Boerma
Botrytis head rot	*Botrytis cinerea* Pers.
Cladosporium head rot	*Cladosporium herbarum* (Pers.) Link ex S.F.Gray
Angular bacterial leaf spot	*Psuedomonas syringae* pv *helianthi* (Kaw.)Young, Dye and Wilkie
Bacterial soft rot	*Erwinia carotovora* (Jones) Holland Var. *Carotovora* Dye
Apical chlorosis	*Pseudomonas syringae* pv. *tagetis* (Hellmers) Young, Dye and Walkie
Sunflower mosaic, Yellow ring mosaic, Yellow blotch,	Sunflower mosaic virus Yellow ring mosaic virus
leaf crinkle	Virus
Phyllody	Mycoplasma like organism (MLO)
Phanerogamic parasite-Broom rape	*Orabanche cumana* waller
Storage Micro flora	*Aspergillus glaucus* group, *A. restrictus, Rhizopus* spp., *Alternaria* spp., *Aspergillus flavus, A. parasiticus*

Major disease of sunflower in India

1. Sunflower necrosis disease (Tobacco streak virus)

Sunflower necrosis disease (SND) was first reported at Bagepally region of Kolar district and around Bangalore in India during 1997 (Singh *et.al.*, 1997). Since then the disease became increasingly important in sunflower-growing areas of former Andhra Pradesh, Karnataka, Maharashtra and Tamil Nadu. This disease became a major threat to the successful cultivation of all the sunflower hybrids and varieties and devastating the crop since 1998. Significant reductions in terms of total crop loss up to 90% were reported due to early infection in the farmers' fields (Bhat *et al.*, 2001, Ramaiah *et al.*, 2001; Lavanya *et al.*, 2005). This has resulted in the substantial loss of sunflower production to 0.733 million tonnes in 2000–2001 in comparison to 2.0 million tonnes in 1998–1999 (Bhat *et al.*, 2002 ; Jain *et al.*, 2003). This has caused severe economic losses, thus forcing the farmers to switch over to other crops. Sunflower necrosis disorder has had a significant impact on sunflower production in Central Queensland (CQ) since 2004, and in 2005 caused a 20% loss ($4.5 million) across the industry. Losses have been both direct, from crop damage caused by the disorder and indirect from reduced grower confidence to plant, due to expected crop losses (Sharman *et al.*, 2009).

Severity of necrosis disease ranges from 2 to 100%. All the growth and yield parameters were significantly affected due to sunflower necrosis disease resulting in yield loss of 89% under severe conditions of 50-75% disease severity, 63% and 20% at 11-50% and 5-10% disease severity levels respectively. In general, the disease occurrence has been erratic and its incidence varies from season to season and place to place (Chander Rao *et al.*, 2003). The disease greatly reduces plant growth parameters and seed yield. Occurrence of disease incidence varies with space and time.

Sunflower crop is susceptible to necrosis disease at any stage of growth (Fig. 10.2 - a,b,c,d). The disease is observed from seedling to maturity stage (Chander Rao *et. al.*, 2002). Initially, small, irregular necrotic patches observed near to the mid rib of leaf lamina. When the disease increases, twisting of leaf occurs. Later, disease spreads in the form of black streak from one side of the lamina to the petiole and stem and finally reaches to shoot of the plant leading to paralytic symptoms. Mosaic, mottling and necrosis were observed

Fig 10.2. Sunflower necrosis disease a. Infected at seedling stage, b and c. Infected at maturity stage, d. Isometric virus particles, e. Infected flower heads.

in leaf lamina tissues of infected plants and finally spread to all other aerial parts of the plant. In some genotypes the leaves develop typical mosaic, rings and line pattern. Early infected plants remain stunted and develop malformed heads with poor or no seed setting, resulting in complete loss of the crop. The disease has been noticed in severe form both on traditional genotypes as well as hybrids. When infection occurs in the early-stages, plants become stunted, weak and die before flowering. The capitulum bends and twists due to the infection occurrence at bud formation stage. Sudden necrosis, twisting and systemic infection of floral parts like calyx, bracts and capitulum are the most destructive symptoms of necrosis disease. In necrosis infected plant, black or brown streaks are usually noticed on petiole, stem and underneath of flower head which results in twisting and splitting. When plants affected at flowering stage, flower heads without seed formation and hull was formed.

According to Ravi *et al.* (2001) SNV belongs to the ilarvirus sub group I and is related to tobacco streak virus (TSV) as the former shared 90% amino acid sequence identity with the latter. It has been reported that the SNV is a single stranded circular RNA virus with isometric virions; the sunflower ilarvirus was related to TSV on the basis of coat protein gene sequence (Prasada Rao *et al.*, 2000; Bhat *et al.*, 2002). Initially, Jain *et al.* (2000) reported that the SNV was associated with tospovirus, but later it was confirmed that the Ilarvirus, antigenically related to TSV was associated with SNVD (Jain *et al.*, 2003). The virus particles are non-enveloped, isometric particles (Fig. 10.2 e) measuring 25-30 nm in diameter (Ramaiah *et al.*, 2001). In viral genome, the virions consist of 14 % nucleic acid, 80% protein, 0% lipid and three segments (RNA 1-3) of linear positive sense single stranded RNA with coat protein of 30 KDa. Virions are found in all parts of host plant / cytoplasm.

Thrips mediated Sunflower necrosis virus transmission has already been reported in sunflower (Jain *et al.*, 2003; Lokesh *et al.*, 2005). Notably, a groundnut (*Arachis hypogaea* L.) isolate of TSV was transmitted by thrips, *Frankliniella schultzei* Trybom (Reddy *et al.*, 2002). Although ilarvirus is transmitted through seeds (Van regenmortel *et al.*, 2000), there is no report confirming the transmission of SNV through seeds in sunflower. The SND disease is not transmitted through seed and grafting but it is transmitted by sap inoculation method from sunflower to sunflower, sunflower to cowpea and back from cowpea to sunflower.

In the infected flower heads and immature seeds, the virus may be present till the plant is alive but it cannot carry virus after harvest (Chander Rao *et.al.*, 2000). Thrips species like *Scirtothrips dorsalis* Hood, *Frankliniella schultzei* (Trybom), *Megalurothrips usitatus* (Bagnall), *Karyothrips flavipes* (Jones), *Thrips hawaiiensis* (Morgan) and *Microcephalothrips abdominalis* (Crawford) acts as vector for transmission of virus on sunflower (Sonali *et.al.*,2005; Singh, 2005). In groundnut, cowpea, sunflower plants, three thrips species (*S. dorsalis*, *F. schultzei* and *M. usitatus*) plays an important role in transmission of TSV (Sonali *et.al.*, 2005) by carrying the infected pollen not by acquiring the virus in their body. The pollen grains from infected crop plants or weeds are main source of inoculum. Thrips helping in spreading the inoculum and causing the disease by carry 20-200 pollen grains over their body externally. The pollen grains present on the thrips bodies get dislodged during infection on sunflower plants and deposited on the leaves. The thrips feeding causes injury to both leaf tissue, through these injury pollen with virus enters into the plants, which causes infection.

Tobacco streak virus is able to infect other crops like groundnut, cotton, marigold, chrysanthemum, safflower, cowpea, urdbean, mungbean and sunhemp. Tobacco streak virus infection is also reported in different weed spieces like *Azeratum, Achyranthus aspera, Commelina, Euphorbia hirta, Tridax* sp. and *Parthenium hysterophorus* (Anonymous, 2001). *Parthenium* weed act as a symptom -less carrier of the virus and infective pollen of *Parthenium* play an important role in spreading the disease throughout the year. It produces several flushes during its life cycle.In *kharif* and summer season, the necrosis disease was higher and low in *rabi* season. During July-September and January-March sowings, necrosis occurrence was more whereas it was low during October- December sown crops. Favourable conditions for thrips incidence are dry weather (July-August) with moderate temperature of 30-32 ^{0}C and 55-75 % relative humidity (Singh, 2005). If the thrips population was high as a result necrosis disease was more. It indicates thrips population was directly correlated with necrosis incidence. During the prolonged dry spells followed by heavy rains resulted in highest disease incidence (Basappa and Santha Lakshmi Prasad, 2005).

July and August sowings had higher disease incidence compared with the rest of the sowing months. This higher incidence might be attributed to an increased activity of the vector. During July and August, dry spells are normally observed in the region and this might have flared the vector population. Similarly, slight increases in disease incidence was observed in January and February. These months also generally experience dry weather conditions in the region, which may enhance the vector population and thus the disease incidence. If sunflower is sown in the post rainy season, i.e., from September onwards, the necrosis disease incidence could be minimized as compared with the normal sowing period i.e. July and August. Thus, by merely adjusting the sowing period, the farmers can combat the disease without involving any extra cost on plant protection. (Shirshikar, 2003).

Management: Number of genotypes comprising of CMS, R lines, germplasm accessions and derivatives of wild sunflower were evaluated and none of them are resistant. The disease produces different types of symptoms and sometimes it becomes difficult to identify the disease based on these symptoms alone.

Intercropping of sunflower with groundnut, urdbean or mungbean has no effect on disease incidence. As by the time sunflower is in flowering stage, the other crops complete the critical periods of infection and hence even if sunflower is infected with necrosis disease, other intercrops would remain free from the disease (Chander Rao *et.al*, 2002).

Clean cultivation does not allow the vector to breed on collateral hosts in off-season and on alternate hosts in the main season. Removal of weeds particularly *Parthenium* from the field, growing wild in fallow lands, roadsides, field bunds and adjoining areas of crop is helpful in reducing the necrosis incidence as it serves as source of inoculum throughout the year. Growing chrysanthemum and marigold close to sunflower has to be avoided. Rouging of infected plants before flowering helps to destroy the virus source and spread of the disease. Border crop like sorghum / pearl millet /maize in 5-7 rows (30cm apart) has to be grown around sunflower crop which attracts the thrips population, also obstructs the wind borne thrips and inoculum carrying pollen grains of *Parthenium* from landing on sunflower plants.

Seed treatment with imidacloprid at 5g / kg of seed found to protect the crop from insect vectors in the initial stages of crop growth. Spraying of imidacloprid 0.01% or oxydemeton methyl 0.025%, three times at 15 days interval starting from 15 days after sowing was effective in control of insect vectors (Anonymous, 2003). Use of insecticides for the control of insect vectors proves to be less effective in management of necrosis disease; hence efforts on use of biotechnological tools for development of transgenic sunflower plants against necrosis disease are in progress.

2. *Alternariaster* leaf spot / blight (*Alternariaster helianthi* (Hansf.) Tubaki and Nishihara)

Alternariaster leaf blight disease was reported in countries like Europe, Australia, India, Japan, Canada, South America, North America and African countries. In Karnataka state, Alternariaster leaf blight was first observed and later it was reported throughout India wherever sunflower grows. The disease causes considerable effect on plant height, stem girth, head diameter, seed yield, seed weight and hull per centage (Mathur *et al.*, 1978). The disease has been reported to reduce the seed yield by 27 to 80 per cent and oil yield by 17 to 33 per cent (Allen *et al.*, 1981; Reddy and Gupta, 1977). The correlation between increase in disease intensity (25-96%) and reduction in yield components and oil content is negative (Balasubrahmanyam and Kolte, 1980 b). The disease also affects the seed germination and vigor of seedlings and the loss in germination varies from 23 to 32 per cent (Kolte *et.al.*, 1979). Generally the disease causes a reduction of 7 kg/ha in yield with every 1 per cent disease incidence. Allen *et al.* (1981) reported that nature of yield reduction due to leaf blight was determined by the stage of plant growth when the disease epidemic develops. Williamson (1979) revealed that severely affected crop produces only 0.1t/ha yield, otherwise the estimated yield potential of the crop is 1.25 t/ha. The disease affected 280 ha of sunflower in 1980 in Minnesota (Shane *et al.*, 1981). Hiremath *et al.*, (1990) reported 95 to 100% disease incidence in Northern Karnataka. Borkar and Patil (1995) reported severity of this disease to the extent of 52 to 85 per cent in three years of study. Simulation of epidemics of Alternaria blight at different plant growth stages resulted in yield losses as high as 61.9% in 'Morden' and 49.7% in 'APSH 11' cultivars of sunflower (Chattopadhyay, 1999). The number of seeds per head and their test weight were also significantly reduced. Infection at the late vegetative to budding stage caused greatest losses in yield.

Development of dark brown to black coloured spots of circular to oval in shape with 0.2 to 5.0 mm diameter in size. Gradually, the size of spot increases into round or irregular with concentric rings and coalesce causing blighting and withering of leaves (Fig. 10.3 a,d). Chlorotic zone is present surrounding the spots with grey white nectrotic center. Under high humidity conditions blighting of leaves and sometimes rotting of flower head also occurs. As the plant grows, spots were first observed on the lower leaves, later spreads to the middle and upper leaves. In severe disease conditions, initially lesions appear as dark flecks on stems (Fig. 10.3b). Simultaneously size of the lesions increases to form long, narrow lesions which may also coalesce to a larger blackened area resulting in stem breakage. The lesions also appear on petioles, ray florets, head as brown round spots about 1cm diameter with a slight depression in the centre and sometimes rotting of flower head in severe cases (Fig. 10.3c).

In earlier, the pathogen was classified under *Helminthosporium* spieces. Later it is confirmed as *Alternaria helianthi* due to formation of few longitudinal septa. The fungal genus *Alternariaster* was established by Simmons (2007) to accommodate *Alternaria helianthi*, a species known to cause leaf spots on *Helianthus annuus* (sunflower) worldwide (Alves *et al.*, 2013).

The fungal mycelium is septate, rarely branched, brown and 2.5 to 5μm in width. The fungus produces conidiophores of cylindrical, scattered and gregarious, pale gray, yellow, straight or curved, up to 5 septa which may be simple or branched, geniculate and in the range of 25-80 x 8-11μm. The cylindrical to long ellipsoidal conidia with rounded ends on both sides, without beaks (Fig. 10.3 e,f),

Fig 10.3. Alternariaster leaf blight a. Infected leaf b. Infected stem, c and d. Conidia of A. *helianthi*, e. Infected flower head, f. Infected plants in the field

slightly curved, pale gray-yellow to pale brown, usually with 11 transverse septa. The conidia measure about 40-110 x 13-28 µm and constricted at the septa. Two to three conidia are arranged in the form of short chains but they are not produced in chains.

The fungus over winters as mycelium in infected plant residues (Islam and Maric, 1980) and in dry conditions survives for 20 weeks in soil (Nagaraju *et.al.,* 1994). Disease is also transferred by seed upto 22.9%. The infection levels on outer and inner portions of the sunflower seeds to vary. Endosperm of the infected seed consists about 65% of fungus and 25-30% fungus was found in the embryo of infected seed. Alternaria infects seedlings in rainy season when sunflower sown in Alternaria infested land. Seedlings are more susceptible than adult plants (Sahu *et.al.,* 1991). There was a significant difference in infection levels of different components of leaf blight affected sunflower seeds. In case of completely infected seeds, infection ranged from 32- 91% in embryos, cotyledons and seed coats respectively (Santha Lakshmi Prasad *et. al.,* 2010).

From *A. helianthi* culture filtrate, crude pathotoxin was extracted and tested for its host specificity to sunflower and produced typical leaf symptoms at concentrations as low as 20 mg/kg (Sharma *et.al.,* 1993). Inhibition of sunflower seed germination as well as root elongation was affected by crude toxin (Bhaskaran and Kandaswamy, 1978). The spectral characteristics of the toxin showed strong absorbance at 220, 269, 281 and 342 nm. Favourable conditions for disease spread are 25-27°C temperature and 12 hrs of wet foliage.The spots will become much larger and coalesce with each other due to extended leaf wetness for three to four days (Mayee, 1992). The disease was reported in *rabi* season crop also (Jasbir Kaur *et.al.,* 1991). In *kharif* season, disease severity was high when compared with *rabi* season crop. The milk and waxy stages of plant development with hot weather and frequent rains favours infection. Previous sunflower crop stalks incorporated in soil play an important role in causing higher disease incidence of *Alternaria* leaf spot is reported (Nagaraju *et al.,* 1994).

Management: Different genotypes evaluated for sources of resistance under both field and artificial conditions revealed none of the genotypes were immune to the disease. Screening of thirty two *Helianthus* species against *A. helianthi* revealed resistant reaction in nine species *viz., H. maximiliani, H. mollis, H. divaricatus, H. simulans, H. occidentalis, H. pauciflorus, H. decapetalus, H. resinosus* and *H. tuberosus,* while the diploid annuals were susceptible (Sujatha *et al.* 1997). Efforts are being made to develop interspecific *Alternaria* resistant cultivars from these wild *Helianthus* sources.

Removal of previous season sunflower debris fallen on the ground and burning of the debris minimizes the inoculum levels in the soil and reduces the carryover of disease from season to season. Deep ploughing during summer months exposes the inoculum to high temperature, thereby reduces the soil inoculum levels. Crop rotation helps in minimizing the inoculum build up in the soil as continuous cultivation of sunflower may help to multiply the pathogen in soil. Healthy and clean seed should be used for sowing. As the fungus is of seed borne in nature, the initial inoculum load can be avoided.

Occurrence and severity of the disease depends on the season and planting dates. The spring crop sown in late February will have higher severity than the crop sown in first week of January. Mid September planting of sunflower in Nainital and Tarai region of Uttaranchal keeps the crop free from major diseases particularly with low leaf spot infection (Kolte and Tewari, 1977). Early planting of sunflower during *kharif* season results in low occurrence of leaf spot. Intercropping sunflower with groundnut in the ratio of 6:2 reduces the disease severity (Mayee, 1992). Closer spacing induces more disease build up. Spacing of 60 x 30cm or 45 x 30cm is optimum in reducing the build up of *Alternaria* leaf blight. Application of optimum dose of Nitrogen, P_2O_5 and K_2O fertilizers as basal resulted in low disease incidence. Organic nutrition of sunflower was found to reduce the severity of disease as compared to inorganic nutrition of the crop. Farm Yard Manure application recorded low disease incidence.

Seed treatment with thiram or captan at 2.5g or carbendazim at 1.0g /kg protects the seed from seed borne infection. Spraying mancozeb (0.3%) four times at an interval of 7-10 days was found effective in controlling the disease with an increase in yield by about 43 to 65%. Spraying of carbendazim + mancozeb (1:1) three times at 15 days interval after disease appearance improves seed yield and reduces the leaf spot disease. Fungicides *viz.*, zineb, ziram and captafol were also effective. Iprodione and propiconazole 0.1% were very effective and economical in controlling the disease (Chander Rao and Ranganath, 2003). Spraying of propiconazole 0.1% three times at 30, 45 and 60 days after sowing effectively reduced the leaf blight severity (Anonymous, 2006). Seed treatment with Quintal (iprodione ı carbendazim) 2g/kg seed followed by spraying of Quintal 0.2% or Propiconazole 0.1% two times at 15 days interval gives effective control of the disease. Seed treatment with carbendazim 12% + Mancozeb 63% (SAAF) 3g/kg seed followed by spraying of propiconazole 0.1% at 30 and 45 days after sowing effectively control leaf blight of sunflower (Venkataramanamma *et al.*, 2014)

3. Powdery mildew

Powdery mildew of sunflower is of worldwide in distribution and appears in serious form in tropical areas. In the temperate region the disease does not appear until flowering and even afterwards the intensity is quite low. The disease was first reported in 1928 from USA and now its occurrence has been reported from Argentina, Australia, Bulgaria, Canada, Chile, China, India, Italy, Romania, Sudan, USSR and USA. Seed yield loses upto 13 per cent have been reported in the north of Tamaulipas due to the disease.

Powdery mildew of sunflower is caused by *Golovinomyces cichoracearum* DC. f. sp. *helianthi* Jacz. Infection of sunflower by powdery mildew causes early senescence during the flowering stage and causes significant reduction in sunflower production in tropical areas. The loss due to powdery mildew is proportionate to the disease intensity and varies considerably depending on the stage of the plant growth at which disease occurs. Powdery mildew causes upto 15 % stunting and 81 % reduction in yield in the greenhouse.

In India, powdery mildew is seldom a problem in the Northern India but can be serious in the sunflower producing areas of the southern parts of India. The disease has not been observed to be of any economic importance but if the disease occurs in early stage of the crop it causes severe losses. Powdery mildew was rarely observed in Karnataka before 2006, however during 2006, high severity of foliar infections of Powdery mildew ranging from 70 to 100 % was observed in different parts of Karnataka. From then the disease is regularly seen in different parts of Karnataka in a moderate to severe form. Sudden outbreak of this disease is one among the various reasons for significant reduction in area under sunflower cultivation. Maximum mean per cent disease severity (PDI) was observed in Koppal district (74.11%) followed by Haveri district (66.61%). Whereas, minimum PDI was noticed in Bagalkot district (30.94%) followed by Belgaum district (34.11%) (Dinesh *et al.*, 2010).

The first sign of powdery mildew usually appears primarily as white grey mildew spots on the upper surface of infected leaves; which increase in size, coalesce and develop to cover much of the plant area with white powdery growth (Fig. 10.4 a,b). The mildew becomes severe on all aerial parts of the plant under heavy infection during the blooming stage of the plant. During late stages of the infection, superficial mycelia may enlarge and merge until most of the plant surface is covered. As plants mature black pin head sized are visible in white mildew areas. Affected leaves lose lustre, curl, turn chlorotic brown and finally collapse. The disease is prevalent more under dry conditions especially at the end of the winter months.

G. cichoracearum has a described host range of >300 plant species and is a biotroph fungus. The fungus is ectoparasite with superficial mycelium, which is hyaline, branched and septate with haustoria in epidermal host cells. The conidia are produced on long or short chain (Fig. 10.4 c). Conidia are oval, single celled, hyaline and measure 28-60 x 11-28µm in size.

Fig 10.4. Powdery mildew disease, a. Infected leaf, b. Infected plant at maturity stage, c. *Conidia of G. cichoracearum*

In unseason, pathogen remains as cleistothecia containing ascospores on perennial sunflower plants and these ascospores germinate to cause fresh infection and when suitable host plants are grown. Secondary infection is caused by conidial germination. The fungus produces conidia capable of germinating even under dry conditions with low humidity and hence the secondary infection spreads very rapidly by wind. The disease is favoured by presence of infected tissue on the plant, vigorous, succulent plant growth, warm temperatures and dry days followed by nights with high humidity.

The plants of 50 days old were highly susceptible to the disease. The per cent reduction of chlorophyll a, chlorophyll b and total chlorophyll was more in susceptible variety than in resistant cultivar. In susceptible variety higher amount of total sugar, reducing sugar and non-reducing sugar were recorded than in resistant cultivar. Whereas, non reducing and total sugar were synthesized at faster rate in susceptible cultivar than in resistant cultivar (Dinesh *et al.*, 2010). Powdery mildew generally occurs late enough in the crop season, hence control measures are not needed. Wild species of sunflower i.e., *Helianthus debilis, H. californicus* DC, *H.ciliaris* DC, *H. decapetalus* L., *H. lacinatus* Gray and *H. rigidus* (Cass.) Desf. show resistance to powdery mildew. Application of wettable sulphur 0.2% or karathane 0.2% or propiconazole 0.1% or difenoconazole 0.05% three times at 15 days interval effectively controls the disease.

4. Downy mildew [(*Plasmopara halstedii* (Farl.) Berl. and de Toni]

The downy mildew is first reported in North America. Seed is main source for the transfer of disease to many countries. Under Indian conditions, the sunflower crop was free from downy mildew until 1985. However, Ramnath *et al.* (1981) detected the presence of downy mildew oospores on sunflower seeds imported from Bulgaria. The first appearance of downy mildew was reported from Latur, in Marathwada region of Maharashtra, India, where the crop was extensively grown (Mayee and Patil, 1987). Long time, the disease was restricted to Marathwada region, afterwards the disease was spread into neighbouring districts like Andhra Pradesh, Karnataka and Tamilnadu. Later, 38% of sporadic disease incidence was observed on morden and EC-68414 in Northern Karnataka, Bangalore, Hyderbad etc. (Appaji *et.al.*, 1996). The disease was also reported from Madhya Pradesh and Punjab. Shirshikar (1997) conducted a major survey covering six districts of Marathwada region to establish the status of downy mildew in the farmers' fields. The results revealed that 36.67% of sunflower fields were infected by downy mildew, with disease intensity ranging from 1 to 30%. Downy mildew disease causes severe yield losses in temperate regions compared to tropical and sub tropical regions where sunflower is grown (Gulya *et al.*, 1997).

In cases of severe disease, 95% of plants destruction was noticed under favourable weather conditions. The disease leads to damping off and loss of total seedling resulting in yield losses. Seed filling doesn't occur in infected plants even though if they produce seeds, they are unequal in size and partial filled leads to low oil content.

Downy mildew symptoms are of different types. They are damping off seedlings, systemic infection, local foliar lesions and root galls. In cool and moist conditions, damping off of seedlings occurs. These seedlings are killed before or soon after emergence under field conditions in order to reduce the spread of disease. Finally infected plants become dry and wind blown. The first pair of leaves become yellowing. Sporangiospores and sporangia with whitish growth present on the cotyledonary leaves of young seedlings in cool and high humidity conditions (Fig. 10.5 c). The fungus spreads from lower to upper leaves, the chlorotic area expands and chlorosis appears on leaves and stem also.

Generally, height of the healthy sunflower plant ranges from 1.5 to 1.8 m at flowering stage. The down mildew infected plants become stunted and chlorophyll content of the upper leaves is lost. Severely infected plant height may vary from 0.1 to 1.0 m. It is a systemic infection in which the fungal mycelium was observed in all plant tissues except meristems. Initially, white downy growth of the fungus was observed on lower surface of the leaves and covers large areas. Later infected plant becomes abnormal thick with curled leaves showing prominent yellow and green epiphyllous mottling. Finally infection results in the brittelness of stem and sterility, stiff and upwards facing of flower heads. Seeds are not produced or improper seeds are formed on such heads. Reduced stem elongation, prevention of photosynthetic process and negative responses were observed in systemically infected plant. Increased synthesis of scopoletin was found in affected sunflower tissues, which is a fluorescent compound present naturally in small amounts (Lehman and Rice, 1972). Water retention capacity and water permeability of the surface layers were severely reduced in infected plants. In infected sunflower plant leaves, accumulation of potassium and phosphorus was reported and also high rate of transpiration and respiration resulting in the decrease of carbohydrate concentration (Ventslavovich *et al.*, 1971).

Small spots with angular shape, greenish yellow in colour appear on leaves. Large area of leaf was infected by spreading the spots and becomes coalesce. In humid conditions, on the lower surface of the diseased area fungal growth was noticed (Fig. 10.5 a,b). Root gall is not a systemic infection because symptoms of this infection occur independently. Base of the primary roots looks like discoloured, scurfy and hypertrophied as a result of galls formation and also reduced growth of secondary roots leading to lodging and less vigour of the plants (Patil and Mayee, 1988).

Fig 10.5. Downy mildew disease A and B . Infected plantc C. White downy growth on lower side of leaf

P. halstedii is an obligate parasite. The sporangiophores consisting of zoosporangia present individually at the tips of branches are slender, monopodially branched at nearly right angles. Roots and leaves infecting sporangia are different from each other. Leaves are infected by entering the sporangiophores through stomata which measure about 150-750μm with 1-6 monopodial terminal branches, short subulate, straight and 6-9μm long; zoosporangia present in the leaves infecting sporangiophores are elliptic with an apical papilla and flattened wall, 17-30 x 15-21μm. The sporangiophores infecting roots are 375-1200μm, monopodial or sympodial, with 2-3 terminal branches, long subulate, curved and 21-23μm long; root infecting sporangiophores consisting zoosporangia are limoniform, pyriform, oval 36-66 x 39-40μm with 1-3 papillae. Germination of zoosporangia occurs by formation of biflagellate zoospores or by germtubes. In interecelluar spaces of roots, stem and seeds, fungus produces oogonia and antheridia. The oospores are particularly formed in the cortex of the fine adventitious roots which are brown with a slightly paler wall and measures about 27-32μm in diameter.

Previous year sunflower crop plant debris present in the soil serves as residue for the preceding sunflower crop (Shopov, 1979). Pathogen survives on the pericarp and testa of seeds collected from systemically infected plants in the form of oospores for next year (Zimmer, 1975; Zad, 1978). Some oospores remain dormant up to 14 years. In all infected roots, oospores of the fungus were observed. Under wet conditions, oospores germinate to produce zoospores which cause fresh infection. Low per centage of seed borne infection occurs from systemically infected plants. Soil borne inoculum from a previous sunflower crop or wind- borne inoculum from systemically infected seedlings in surrounding fields are main sources for severe disease.

Favourable conditions for infection spread are rain in the early phases of growth and age of seedlings. Systemic infection more frequently observed in seedlings of 3 days old (Wehtje *et.al.*, 1979). As the age of seedlings increases occurrence of disease decreases. Favourable conditions for disease spread are cool weather with 16-18°C temperature, cloudy weather with winds and light drizzle or high relative humidity. Disease develops by heavy soils with poor drainage. Intensity of disease increases upto 58% due to heavy showers immediately after sowings (Raoof, 1999). Important factors for disease spread are tillage, running water and plants suffering from boron deficiency (Virupakshappa *et.al.*,1999).

P. halstedii as an oomycete fungus produces motile zoospores that, exclusively depend on the presence of free water to complete its life cycle. In this context, the most susceptible stage of host development is between germination and emergence (Meliala *et al.* 2000). Severe downy mildew outbreaks were observed in 2005 in some North Dakota fields where unusually high amounts of precipitation occurred during June, immediately following planting (Gulya 2006). Similarly, Tourvieille *et al.* (2008) investigated the incidence of spring rainfall on the severity of primary downy mildew attack of sunflower in field trials with staggered sowing.

Disease risk appeared greatest if there was heavy rainfall when sunflower seedlings were at their most susceptible stage, whereas heavy rainfall before sowing or after emergence had no effect on the per centage of diseased plants. Furthermore,

Göre (2009) recorded a serious outbreak of sunflower downy mildew in Turkey during the spring of 2007 and 2008. Low temperature and extensive spring rains encouraged the disease, resulting in approximately 85% yield loss and caused reduced sunflower production in the Marmara region of Trace. Interestingly, Baldini *et al.* (2006), aiming at developing a prediction model, studied the main climatic factors affecting development and spread of downy mildew in high-oleic sunflower. By calculating the accumulated rainfall and average temperatures for different sowing date conditions, they did not find water availability as a limiting factor between accumulated rainfall and per centage of infection. In contrast, air temperature clearly had an effect on the per centage of disease, the most favourable mean air temperatures being between 10–15°C during the 5 days after sowing.

Besides environmental conditions, disease intensity may also be influenced by the aggressiveness of the pathogen population. Based on the latent period and sporulation density, Sakr *et al.* (2009) successfully differentiated between the two pathogen strains of *P. halstedii ie* 100 and 710 in terms of their aggressiveness on two sunflower lines differing in their level of quantitative resistance

Apart from the soilborne nature of *P. halstedii*, since oospores of this pathogen give rise to systemically infected plants (Spring and Zipper, 2000), local lesions on leaves as a result of infection by wind-borne sporangia may occasionally occur in some areas. From such leaves the pathogen tended to spread along the veins reaching the petiole. Subsequent microscopic observations gave evidence of a close correlation between the presence of hyphae in leaf and corresponding petiole. In fact, such secondary systemic spread of the pathogen has become more common in different sunflower cultivars under extremely favourable (wet and cool) weather conditions. A high potential for transition from local to systemic infection was found by Spring (2008). Though localized lesions are usually not considered to be of economic importance, in the case of seed production, seeds from such plants may carry-over the pathogen to the next season and to various locations (Spring, 2008).

Management: Downy mildew disease was noticed in India in 1984 and presently race 1 is present. North American race of *P. halstedii* is highly virulent than Indian race. Quarantine measures should be strictly followed to prevent the entry of other races of the pathogen into India. Some sunflower growing areas in India are still free from the disease, so it is necessary to adopt local quarantine measures strictly to restrict the spread of the disease to disease free areas.

Growing resistant cultivars is the most effective and economically practical for control of downy mildew. Four different genes PL 1, Pl 2, PL 3 and PL4 are reported to condition resistance to downy mildew in sunflower. At present PL 2 gene is being used to give mildew resistance in breeding programmes in many parts of the world. Two downy mildew resistant hybrids LSH-1 and LSH-3 for endemic pockets of Maharashtra were released. High yielding hybrids such as BSH-1, KBSH-1, ICI-302, PKVSH- 27, DSH-1, NSH-22, ITC-601, MSFH-17, SPIC-105, Jwalamukhi, NARF-114 and varieties *viz.*, PKVSF-9, Sidheswar are resistant to downy mildew. Most of the recently released hybrids are resistant to downy mildew, while many varieties are susceptible.

Healthy seed obtained   from disease-free areas should be sown in well-drained fields. Early planting during rainy season generally escapes from disease. Late rainy season planting particularly July-August should be avoided. Shallow sowing (3cm) is desirable to minimize the disease incidence. Optimum spacing has to be followed for sowing. Spacing of 60 x 30cm or 45 x 30cm recorded least disease severity compared to 60 x 15cm or 45 x 15cm (Patil *et.al.*, 1992). Pre sowing irrigation followed by one irrigation at 10 or more days after sowing reduces disease considerably in endemic areas. Frequent irrigation up to 25 days after sowing or water stagnation due to unseasonal rains induces higher disease incidence in susceptible cultivars /hybrids. Water stagnation should be avoided as it favours quick development of pathogen (Raoof, 1999).

Rouging of mildew infected seedlings during thinning, removal and destruction of infected plants as and when they appear before flowering reduces spread of the disease. It also reduces oospore build up for the following season and the inoculum is destroyed before the seed is infected. Elimination of volunteer downy mildew infected plants at the onset of the monsoon is advisable to avoid the spread of the disease. Infected plant debris of the previous crop season should be removed and destroyed from the field. This practice helps in reducing the spread of the disease. To minimize inoculum level in the soil, crop rotation with non host crops in the same field is necessary. Six-year crop rotation in sunflower with groundnut and pigeonpea is advisable in endemic areas. Cropping sequence of sunflower followed by groundnut reduces the disease severity compared to sunflower after sunflower.

Metalaxyl fungicide ie APRON 35 ES was found to be highly effective in reducing the severity of downy mildew disease, even in sick plot conditions, when seeds were treated with the chemical at 105g a.i./100 Kg (3ml / kg of seed). Seed treatment with Apron 35 SD 6g/ kg followed by foliar sprays of ridomyl MZ 72 at 0.2% two times at 15days interval provides effective control.

5. Rust (*Puccinia helianthi* Schwein)

The disease was reported in all sunflower growing areas throughout the world-wide such as Australia, Canada, North America, Southern Europe, Isreal, Japan and India. This disease is also distributed in India where the sunflower is cultivated. It appears severely in the early stages of crop during *rabi* season and late in the *kharif* season which causes a considerable yield reduction. About 11-33% of seed yield loss was reported in severe disease conditions because it is unpredictable if it occurs in later stages (Fick and Zimmer, 1975). Head diameter, seed size, seed quality, test weight, oil content and formation of more hull seed are the effects of rust disease in sunflower.

Initially disease appears on lower leaves as small reddish brown circular uredial pustules, covered with rusty coloured (Fig. 10.6). The pustules are produced on younger leaves and spread over the entire vegetative surface covering stems, petioles, floral bracts, when conditions are favourable and stem with petals having pustules present linearly rather than round. Large areas of effected plant parts are covered by uredia often coalesce. At maturity stage of crop, the uredosori present on leaves is replaced by telia and appears as black rust results in drying of leaves.

Fig 10.6. Rust disease

Pustules are often surrounded by chlorotic halos on susceptible cultivars and adjacent pustules may merge whereas in highly resistant varieties, uredia are not formed at the site of infection and only small chlorotic or necrotic flecks developed.

Puccinia helianthi Schwein produces all stages of spore forms on sunflower and it is a macrocyclic, heterothallic and autoecious fungus. Dark cinnamon brown coloured uredosori and uredospores of brown coloured, subglobose to obvate shaped and measuring about 25-32 x 10-25µm are produced by fungus. The spore wall is cinnamon brown in colour and 1-2µm in thick. At time of maturity, uredosori of the infected plant is replaced by teliosori, which containing bicelled, smooth, brown, oblong, elliptical teliospores that are strictly constricted at the septum. The teliospores are pedicellate and pedicel may be colour less, rounded above or attenuated below and measures about 40-60 µm x18-30µm with chestnut brown wall, 1.5 –3 µm thick at the sides, 8 to 12 µm thick above with an apical pore. On young seedlings or the cotyledons or on true leaves, Pycnia of flask shaped, amber coloured, 1mm diameter with small, oval hyaline pycniospores are formed. Aecia are hypophyllous with white laciniate margins. It is orange and red in colour appears close to pycnia. Aeciospores are echinulated with four germ pores and orange coloured, typically ellipsoidal and measure 21-28 x 18-20µm in size.

Infected soil surface and infected leaves left in the field serves as inoculum for the spread of disease. The fungus mainly survives through teliospores, which are thick walled, resting spores. Sunflower seedlings, wild sunflower or new young plants in nearby fields are infected by sporidia produced by germination of spore in the spring season. In the early stages of infection pycnia are formed and later developed into aecial pustules on upper surface of leaves or cotyledons. The aecia occur individually or in small groups, which are small, orange coloured cup-shaped pustules. The aeciospores are transferred from volunteer to the other wild sunflower host plants by wind. Later aeciospores produce cinnamon-brown uredial pustules.

The teliospores present in soil, seed and plant debris exhibit dormancy for a period of variable length and viable for longer time (Miah and Sackston, 1970). The fungus survives on previous year's crop, plant debris of volunteer seedlings which carries uredia, sporidia, pycnia and aecia causes primary infection at high altitudes and it is transmitted by wind (Sackston, 1960). Uredospores are produced repeatedly in a crop season, those spores acts as inoculum for secondary infection. Spore production leads to infection continuity which ultimately results in the spread of disease from one field to other field by wind. Favourable conditions for disease spread are 25.5 to 30.5°C temperature with relative humidity of 86-92%. Warm temperature with frequent rain or dew favours rapid increase of rust disease. High relative humidity is developed by excess nitrogen source and abnormally high seeding rates promoted excessive foliage, which in turn, increases rust severity. It indicates that rust disease is positively correlated with relative humidity. Occurrence of disease also depends on the plant age. The maximum incidence of rust disease was observed on 75 days old plant.

Management: Growing of resistant varieties offers the best means of control. Tall and late ones are more resistant to rust while dwarf and early ones are significantly more susceptible (Zimmer *et.al.,* 1973). Majority of the hybrids released for cultivation are tolerant or highly resistant to rust. Wild species of sunflower viz., *Helianthus tuberosus, H.praecox* ssp. *praecox, H. praecox* ssp.*runvonii, H.petiolaris* ssp. *petiolaris* show resistance to this disease.

Removal and burning of infected crop residues minimizes initial inoculum. Clean cultivation and removal of volunteer sunflower plants carrying infection reduces the primary inoculum. Deep ploughing in summer exposes the inoculum present in the soil to high summer temperatures and kills the inoculum. Crop rotation with non host crops for 3 years reduces the inoculum load in the soil. Avoid high nitrogen rates and high plant populations.

Spraying of mancozeb or zineb at 0.25% two to three times at 10 days interval is very effective in control of the disease. Application of sulphur fungicides, sulphur dust (15 kg/ha) or wettable sulphur 0.2% or mixture of sulphur + zineb also gives good control of the disease. Spraying of oxycarboxin 20 EC and benodanil 50WP at 0.4% two times at 30 days interval also effectively control the disease (Sivaprakasam *et.al.,* 1977). Application of boron to soil reduces the rust incidence.

6. Sclerotium collar rot (*Sclerotium rolfsii* Sacc)

Collar rot of sunflower incited by *Sclerotium rolfsii* is limited in distribution to sunflower grown in tropical and subtropical areas. It has been reported on sunflower in the warmer regions of Asia, South America and countries bordering the Mediterranean specifically India, Pakisthan, Portugal, Spain, Egypt, Iran, Isreal, Uruguay, Argentina, Australia and South Africa. In most countries, the disease on sunflower is considered of minor importance except in India and Pakisthan (Gulya *et.al.,* 1997). The disease was reported in Uttar Pradesh, Maharashtra, Punjab, Haryana, Andhra Pradesh and Karnataka. The disease causes 10-11% yield loss with 10-11% disease incidence in sunflower crop planted in July or August or in February or March in Nainital and Tarai region of Uttaranchal (Kolte and Tewari, 1977). The major reduction in yield observed in plants infected by the disease is due to rapid wilting and loss of leaf tissue similar to drought or defoliation.

Initial symptoms of the disease are noticed 40 days after sowing. Stem is usually infected at or near the soil line (Fig. 10.7 a). A brownish lesion develops at the base of the stem and eventually girdles the plant. A white fan-like mycelial growth forms over the infected tissues and often radiates over the soil surface (Fig. 10.7 b). Sickly appearance of plants can be spotted from a distance and a row effect can be observed in heavily infected soil. Later the entire plant withers and dies. The lesion grows up the stem, destroying the cortical tissue and leaving the fibrous vascular strands as the tissues dry out. White cottony mycelium and mustard seed type

Fig 10.7. Sclerotium collar rot, a. Infected plant, b. white mycelium growth near stem base, c. Sclerotia

sclerotial bodies are conspicuous on the affected stem near soil level. This also causes collar rot in the seedling stage (Datar and Bindu, 1974) (Fig. 10.7 c).

The fungus produces inter and intra cellular mycelium, which invades the host tissues and weakens them. The sclerotia are tan to dark brown, uniformly round 100 200µm in diameter, formed on the host and also in culture media.The fungus survives as sclerotia or mycelium in infected plant residue and soil (Punja, 1985). Aerobic conditions, high temperatures, and high humidity favour the germination of sclerotia and mycelial growth. The mycelium colonizes organic debris and produces the characteristic fan-like mats of hyphae that infect the plant. The fungus can move from plant to plant by this mycelial growth. Large amounts of organic debris in combination with high humidity and temperature in fields, with a high population of the fungus are factors favourable for disease development. During rainy season and also when the plant is weakened due to physiological conditions or due to insect attack, the fungus becomes active. The incidental or other types of wounds in the roots, including those caused by ectoparasitic nematodes favour the entry of the fungus. Once inside the host, the organism becomes virulent, causing severe damage to the plant.

Management: Crop rotation of 3-4 years helps to reduce the disease. Removal of weeds and elimination of infected plant residue minimizes the disease severity. Deep ploughing of soil exposes the sclerotia and infested crop debris to sunlight and lowers the amount of the pathogen. Mid-September sowing of sunflower escapes the disease in Uttaranchal (Kolte and Tewari, 1977).

Addition of amendments like oat straw and finely grounded castor and neem oilcakes in the infested soil reduce the inoculum in the soil and disease incidence (Gautam and Kolte, 1979). Such reduction in incidence has been attributed to inhibition of the growth of *S. rolfsii* due to water or ether- soluble toxic substances as released in the decomposition process and also due to enhanced antagonistic activity of increased microflora. The growth of plants also enhances in amended soil, so the plants become less susceptible to the disease.

Biocontrol agent *Trichoderma harzianum* Rifai grown on finger millet seeds or on straw pieces incorporated into soil reduces the soil pathogen (Anil Kumar and

Gowda, 1983). Seed dressing with thiram + carboxin (2:1) at 3-6 g/ kg of seed is found to be effective control of the seedling phase of the plant.

7. Sclerotinia wilt and rot (*Sclerotinia sclerotiorum* (Lib.) de Bary)

Wilt caused by *Sclerotinia sclerotiorum* (Lib.) de Bary reduced yield and quality of sunflower seeds. The fungus causes root rot and wilt, stem or stalk rot and head rot (Fig. 10.8). The amount of reduction depended upon the stage of plant development when wilt occurred. Less than 0.5% of the plants were wilted at the bud stage and 7.2% by the start of anthesis. Thereafter approximately 6% of the plants wilted each week until 8 weeks after flowering when 60% of the plants had been killed. Seed yields were reduced more than 70% when wilting occurred within 4 weeks of flowering. This reduction was primarily due to lower seed weight. Oil content increased from 32.7% for plants wilted in the first 2 weeks, to 46.4% when wilting was delayed until 8 weeks after flowering. Protein content of oil-free meal was fairly stable and averaged 53.1% during the

Fig 10.8. Sclerotinia wilt

first 5 weeks, but increased to 57.7% thereafter. Fatty acid composition was relatively unaffected by wilting as linoleic acid content varied from 74.4 to 76.8%. Thus, the oil was considered to be of excellent quality regardless of when wilting occurred.

Yield loss depends on the number of infected plants and stage of the crop the greater the number of infected plants and the earlier the infection, the greater the yield loss. On the average, an infected plant yields less than 50 per cent of a healthy plant. In addition, the seed oil content is reduced on infected plants. Equally important, however, infection leads to increased levels of *Sclerotinia* in the soil. This can result in the removal of fields from sunflower production for many years. This has occurred in numerous infested fields in the Red River Valley. *Sclerotinia*, therefore, not only reduces yield, but it affects future production and economic gain from sunflowers. Head rot and wilt are important diseases of sunflower in Manitoba and are limiting factors in production. Isolates from sunflower with Sclerotinia wilt in Manitoba , Montana, and Wisconsin all belonged to the same species. Sunflower production in Manitoba dropped from about 77,000 hectares in 1972 to 8,500 hectares in 1974. This decline was due in part to severe yield losses caused by Sclerotinia head rot and wilt.

The fungus survives as sclerotia and mycelia strands in soil and seed. Wet soil conditions with < 27°C and dry soil with 5°C are favourable for sclerotial survivability and viability for 2-3 years. During winter season, the mycelium grows saprophytically on sunflower stalks. During anthesis and seed development stage, *Sclerotinia* wilt generally appears. Root-rot, stem-rot, leaf blight, and stalk rot caused by myceliogenic phase of sclerotia whereas head rot is caused by carpogenic phase of sclerotia. Indian environmental conditions are favourable for myceliogenic phase of sclerotia and carpogenic phase is active under European climatic conditions. Generally, high soil moisture and long periods of rainfall favours the carpogenic germination of sclerotia. Sclerotia buried at 2-5cm in the soil are capable of producing apothecia, with highest frequency of apothecial formation close to the soil surface (Singh and Singh, 1983).

Fungus invades roots at any site when it comes in contact with host plant and spreads from one plant to adjacent plant through root contact. Along from hypocotyls to the base of the stem, water soaked lesions spreads continuously and also present on some fibrous roots. The root systems of affected plants are completely destroyed. Leaves of the plant become yellowing, which results in rapid death and desiccation of the entire plant. White dense growth of the fungus was observed on lesions present on the stem and stem below the soil level. The diseased plants are present individually in a row and afterwards they infected in groups. Stems of severely infected plants shred in to vascular strands, when the stems dried becomes straw coloured, weak and plants lodged easily. The flower heads are infected by head rot disease at any stage from budding to seed maturity stage and symptoms are visible in any part of receptacle. Head rot are two types i.e partial and complete. Seed filling is incomplete in head rot infected flower heads.

Management: Considering the soil borne and polyphagous nature of the pathogen, management of disease is still far from satisfaction. Inter specific hybrids of *H. tuberosus* x *H. annuus* and *H. tuberosus* x *H. strumosus* are reported to be resistant to stalk rot (Orellana, 1975). The most promising lines were from crosses including *H. Praecox* ssp.*runyonii*, *H. annuus*, *H. resinosus*, and *H. paradoxus* (Seiler et.al.1993).

Crop rotation for 3 to 5 years with non host crops reduces the sclerotia in the soil. Wide plant spacing allows air movement within crop canopy thus hastening drying of the soil and reduces the chances of infection to sunflower stalks and flower heads by ascospores. Destruction of the disease infected crop debris reduces the build-up of inoculum.

Seed treatment with carbendazim 0.1% or thiophanate-methyl 0.1% or thiram 0.25 % are found to be effective for the management of the disease. Spraying of benomyl at 0.1% two times at the budding and early flowering stage give better control of the disease. Pre sowing flooding for 30 days, seed treatment with carbendazim at 0.2% followed by addition of *Trichoderma harzianum* at 2g/kg soil and spraying of carbendazim at 0.2% after appearance of wilt proved highly effective for management of *Sclerotinia* wilt (Singh and Tripathi, 1997).

8. Head rot (*Rhizopus arrhizus* Fischer (*R. oryzae* Went, *R. nodosus* Man)

The head rot disease infects flower head of the sunflower plant. In wet weather conditions are favourable for head rot disease in sunflower and it also causes yield loss (Agrawat *et al*, 1978). Head rot disease increases as its age advances and maximum rotting is noticed at the soft dough stage. Head rot disease occurs with the help of injury present in the flower head. Injury made by hail, birds and insects to the flower head is necessary for infection. The *Heliothis armigera* larvae causes wounds to the head through it fungus *Rhizopus* enters into it.

Initial symptom appears as brown irregular water soaked spots on the back of ripening head usually adjacent to flower stalks (Fig. 10.9). Spots gradually

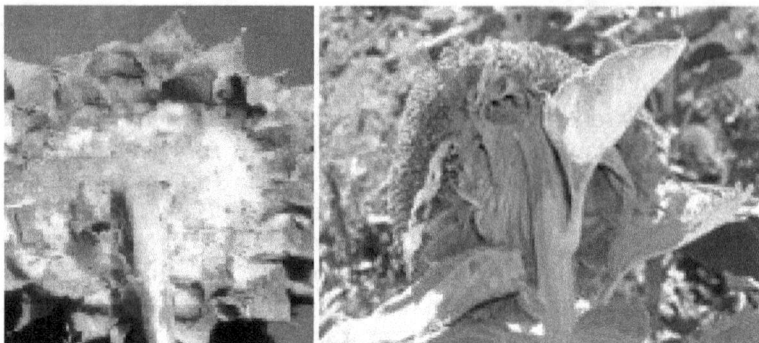

Fig 10.9. Head rot

enlarge and become soft and pulpy and get covered with superficial white mycelium, which later becomes black due to formation of sporangia. The head rot increases from the bud stage up to the full bloom and ripening stages. Higher amount of free fatty acids are present in the oil extracted from *Rhizopus* head rot infected seed (Thompson and Rogers, 1980). In warm humid weather conditions, head rot disease increases very fast. When disease severity is more, seeds looks like a black powdery mass. Wind acts as vector for the fungal spores transfer from one host to other.

Management: Injury to the head due to mechanical operation should be avoided as far as possible. Spraying of endosulphan 0.05% or diazinon 0.03% three times at 15 days interval at the onset of bloom stage is found efficient control of larvae, which causes damage to the head. Spraying fenthion 0.1% plus thiovit 0.2% at the time of head initiation will be effective in controlling the disease. Spraying of copper-oxychloride 0.4% or mancozeb 0.3% or dichloran at completion of flowering stage reported to protect the flower head from head rot.

9. Sunflower leaf curl Virus disease (Su LCV)

Leaf curl viral disease reported two years back from Raichur on sunflower crop and it was recorded in 23.4% of the total field's surveyed. The disease incidence was in the range of 0.5% - 1.2% in Andhra Pradesh., 2.5% - 14.3% in Karnataka and 0.1% - 0.7% in Maharastra. The causal virus of SuLCV was partially purified and identified as Gemini virus of Begamovirus group through electron microscopy. The virus was also detected through PCR based molecular studies using two degenerate specific primers. The phylogenetic analysis of core CP sequence of SuLCV showed close identify to Tomato leaf curl virus (Luc & Ban 2).

The insect vector whitefly which transmits the disease was identified as *Bemisia tabaci* and confirmed.The begamovirus was recorded naturally on cotton, tomato, cowpea, mungbean, marigold and pumpkin but not on castor, safflower, sorghum, and groundnut. Weeds parthenium and azeratum were found positive to infection (Anonymous, 2013). Under the experimental host range studies of SuLCV 12 plant species comprising vegetables, legumes and weeds were tested. The results indicated that among the tested, the susceptible hosts are 85.7% (6 out of 7) in vegetable crops, 42.9% (3 out of 7) in oilseed crops, 100% (2/2) in weeds.

10. Storage microflora

Association of microflora with developing sunflower heads or when the sunflower seeds are kept in storage is well documented in Rajasthan, Punjab, Andhra Pradesh, Uttar pradesh and Haryana (Agarwal and Singh, 1974; Singh *et al.*, 1981). Generally, the storage fungi invade the sunflower seeds stored at 20^0 C and moisture content of the seeds above 7%. *Aspergillus glaucus* group, *A. restrictus, Rhizopus* spp. and *Alternaria* spp are some of the predominantly growing fungi on sunflower seeds in storage. Deterioration in the oil content and oil quality takes place. Production of Aflatoxin B1 to the extent of 33 and 11 ppm has been reported to occur in response to infection of *Aspergillus flavus* and *A. parasiticus* in broken and unbroken seeds respectively.

Management: The losses due to biodeterioration can be reduced by maintaining the temperature and moisture of sunflower seeds at optimum levels. The moisture of stored sunflower seeds should not exceed 6.5 % and relative humidity of storage containers should not exceed 70% (Christensen, 1972).Seed treatment with carbendazim 0.1% reduces the seed microflora and maintains the viability of seeds.

Integrated disease management

- Deep ploughing during summer followed by double harrowing.
- Clean cultivation – Removal and destruction of infected plant residues.
- Compulsory crop rotation of sunflower for 3-4 years.
- Addition of organic amendements like oat straw or castor cake or neem cake and biocontrol agent like *Trichoderma harzianum* to the soil.
- Growing of sorghum / pearl millet / maize 5-7 rows around sunflower as border crop.
- Choosing any resistant or tolerant hybrid cultivar for cultivation.
- Use of healthy and clean seed for sowing.
- Seeds should be treated with thiram or captan 3g/kg or carbendazim 1g/kg (for leaf spot, *Sclerotium* wilt) or Apron 35 SD 6g /kg (for downy mildew) or imidacloprid 5g/kg (for necrosis).
- Adoption of optimum spacing of 60 x 30cm or 45 x 30cm.
- Intercropping of sunflower with groundnut in the ratio of 6:2.
- Application of recommended dose of FYM to the field.
- Application of balanced dose of NPK fertilizers.
- Avoid poorly drained field or those with excess water stagnation.
- Removal of weeds (*Parthenium* ,collateral hosts) from the field and surrounding areas.
- Removal and destruction of disease infected plants.
- Spraying of confidor at 0.05% at 15, 30 and 45 days after sowing for necrosis or mancozeb 0.3% at 10 days interval for *Alternariaster* leaf spot and rust or ridomyl MZ for downy mildew.

Presently downy mildew is restricted to Central and South India, domestic quarantine measures can be adopted to prevent it from spreading to North India where it could pose a threat during rabi and spring seasons.

REFERENCES

Agarwal, V. K., and Singh, O. V. 1974. Fungi associated with sunflower seeds. Indian Phytopath. 27: 240-241.

Agrawat, J. M., Vaish, O. P., Mathur, S. J., and Chkipa, H. P. 1978. Some observations on *Rhizopus* head rot of sunflowers in Rajasthan, India. Abstr.8th Int. Sunflower Conf., Minneapolis, Minn., USA: 23.

Allen, S. J., Kochman, J. K., and Brown, J. F. 1981. Losses in sunflower yield caused by *Alternarai helianthi* in southern Queensland. Aust. J. Exp. Agric., Anim. Husba. 21: 98.

Alves, J.L., Woudenberg, J.H.C. , Duarte1, L.L., Crous P.W. and Barreto, R.W.2013. Reappraisal of the genus *Alternariaster* (*Dothideomycetes*) Persoonia 31: 77 -85.

Anilkumar, T.B. and Gowda, K.T.P. (1983). Possible use of *Trichoderma harzianum* for the control of Sclerotium rolfsii. J. Soil Biol. Ecol., 3: 59-61.

Anonymous. 2001. Annual Report (2001-2002). Directorate of Oilseeds Research, Hyderabad, India. 55-59.

Anonymous. 2006. Research achievements in sunflower. All India Coordinated Research Project on Sunflower, Directorate of Oilseeds Research, Hyderabad, India. 1-4.

Anonymous. 2003. DOR Annual Report: 2002-2003. Directorate of Oilseeds Research, Hyderabad. p. 120.

Anonymous. 2012. Agriculture statistical division, Directorate of Economics and statistics. Ministry of agriculture.

Anonymous. 2013. DOR Annual Report: 2013-2014. Directorate of Oilseeds Research,. Hyderabad. pp. 18-19.

Appaji, S., Raoof, M. A., Chattopadhyay, C., and Prasad, M. V. R. 1996. Sunflower downy mildew- a threat in India. Indian Farming, 46: 6-8.

Balasubramanyam, N., and Kolte, S. J. 1980b. Effect of different intensities of *Alternaria* blight on yield and oil content of sunflower. J .Agric. Sci, Camb. 94: 749.

Baldini, M., Danuso, F., Turi, M., Sandra, M., and Raranciuc, S. 2006. Main factors influencing downy mildew (*Plasmopara halstedii*) infection in high-oleic sunflower hybrids in Northern Italy. Crop Prot. 27: 590–599.

Basappa, H.and Santha Lakshmi Prasad M. 2005. Insect Pests and Diseases of Sunflower and their Management, Ed. Hegde DM, Hyderabad, Directorate of Oilseeds Research, pp. 30–33.

Bhaskaran, R., and Kandaswamy, T. K. 1978. Production of a toxic metabolite by *Alternaria helianthi in vitro* and *in vivo*. Madras Agr. J. 65: 801

Bhat, A. I., Kumar, A., Jain, R. K., Rao, S. C., Ramaiah, M. 2001. Development of serological based assays for the diagnosis of sunflower necrosis virus disease. Ann. Pl. Prot. Sc. 9: 292–296.

Bhat, A. I., Jain, R. K., Chaudhary, V., Krishna Reddy, M., Ramiah, M., Chattannavar, S. N., Varma, A. 2002. Sequence conservation in the coat protein gene of tobacco streak virus isolates causingnecrosis in cotton, mungbean, sunflower and sunn-hemp in India. Indian J. Biotech. 1: 350–356.

Borkar, S. G., and Patil, B. S. 1995. Epidemiology of Alternaria leaf spot disease of sunflower. Indian phytopath. 48: 84-85.

Chander Rao, S., Raoof, M. A., and Singh, H. 2000. Sunflower necrosis disease-Preliminary studies on transmission (extended summary) In: National Seminar on Oilseeds and Oils-Research and Development needs in the millennium. February 2-4,2000.hyderabad, 285-286.

Chander Rao, S., Prasada Rao, R. D. V. J., Singh, H., and Hegde, D. M. 2002. Sunflower necrosis disease and its management. Information Bulletin, Directorate of oilseeds Research, Hyderabad, India.

Chander Rao, S. and Ranganatha, A.R.G. 2003. Evaluation of new fungicides for effective management of Alternaria leaf blight in sunflower. *Indian J. Pl. Prot.,* **31**:135-136.

Chattopadhyay, C. 1999. Yield loss attributable to Alternaria blight of sunflower (*Helianthus annuus* L.) in India and some potentially effective control measures. Inter. J. Pest Manage. 45: 15-21.

Christensen, C.M. 1972. Moisture content of sunflower seeds in relation to invasion by storage fungi. Plant Dis.Rep., 56: 173.

Datar, V. V. and Bindu , K. J. 1974. Collar rot of sunflower (*Helianthus annuus*). A new host record. Curr.Sci. 43: 496.

Dinesh, B. M., Kulkarni, S., Harlapur, S. I., Benagi,V.I., and Mallapur. C. P. 2010. Prevalence of powdery mildew in sunflower growing areas in northern Karnataka. Karnataka J. Agric. Sci. 23: 521-523.

Fick, G. N., and Zimmer, D. E. 1975. Influence of rust on performance of near isogenic sunflower hybrids. Plant Dis. rep.59: 737.

Gautam, M. and Kolte, S.J. 1979. Control of *Sclerotium* wilt of sunflower through organic amendments of soil. Plant and Soil, 53: 233.

Göre, M. E. 2009. Epidemic outbreaks of downy mildew caused by Plasmopara halstedii on sunflower in Thrace, part of the Marmara region of Turkey. Plant Path. 58, 396.

Gulya, T. J. 2006. The downy mildew situation on the 2005.North Dakota sunflower crop. In: Proceedings of the NSA Sunflower Research Workshop 2006, Fargo, ND.

Gulya, T., Rashid, K. Y., and Masirevic, S. M. 1997. Sunflower diseases (In) *Sunflower technology and production* (edt) Schneiter, AA. ASA, CSSA, SSSA, Madison, Wisconsin, USA, 263-379.

Hiremath, P. C., Kulkarni, M. S., and Lokesh, M. S. 1990. An epiphytotic of Alternaria blight of sunflower in Karnataka. Karnataka J. Agril., Sci. 6: 68-69.

Islam, U., and Maric, A. 1980. Contribution to the studies on the biology, epidemiology and resistance of sunflower to *Alternaria helianthi* (Hansf.) Taub.Nish. Zastita Bilja. 31: 35.

Jain, R. K., Bhat, A. I., Byadgi, R. S., Nagaraju., Singh, H., Halker, A. V., Anahosur, K.,and Varma, A. 2000. Association of a tospo virus with sunflower necrosis disease. Current Sci. 79, 1703–1705.

Jain, R. K., Bhat, A. I., and Varma, A. 2003. Sunflower Necrosis Virus Disease – An Emerging Viral Problem. Tech Bulletin-1Indian Agricultural Research Institute, New Delhi, India.

Jasbir, K., Chahal, S. S., and Aulokh, K. S. 1991. Effect of different crop seasons on the influence of *Alternaria helianthi* in sunflower seeds. Plant Dis. Res.6: 83-84.

Kolte, S. J., Balasubramanyam, N., Tewari, A. N., and Awasthi, R. P. 1979. Field performance of fungicides in the control of *Alternaria* blight on sunflower. Indian J. Agril. Scie. 49: 555.

Kolte, S. J., and Tewari, A. N. 1977. Note on effect of planting dates on occurrence and severity of sunflower diseases. Pantnagar J.Res.2: 236.

Lavanya, N. M., Ramiah, R., Ankaralingam, A., Renukadevi, P., and Velazhahan, R. 2005. Identification of hosts for ilarvirus associated with sunflower necrosis disease. Acta Phytopath. Entomol. Hung. 40: 31–34.

Lehman, R. H., and Rice, E. L. 1972. Effect of deficiencies of nitrogen, potassium and sulfur on chlorogenic acid and scopoletin in sunflower. Am.Midl.Nat. 87: 71.

Lokesh, B., Nagaraju, K., Jagadish, K. S., and Hadakshari, Y. S. 2005. Screening of sunflower germplasm against sunflower necrosis virus disease and its thrips vector. Environ. Ecol. 23S, 14.

Mathur, S. J., Prasad, N., Agrawat, J. M., and Chippa, H. P. 1978. Estimation of losses due to Alternaria blight of sunflower.Indian J. Myco. and Pl. Path. 8: 15.

Mayee, C. D. 1992. New approaches to disease management in *kharif* oilseed crops. J. Oilseeds Res. 9: 119-126.

Mayee, C. D., and Patil, M. A. 1987. Downy mildew of sunflower. Indian Phytopath. 38: 314.

Meliala, C., Vear, F., and Tourvieille de Labrouhe, D. 2000. Relation between date of infection of sunflower downy mildew (Plasmopara halstedii) and symptoms development.Helia, 23: 35–44.

Miah, M. A. J., and Sackston, W. E. 1970. Genetics of rust resistance in sunflower. Phytoprotection. 51: 1-16.

Nagaraju., Jayarame G., Gangappa, E., and Virupakshappa, K. 1994. Integrated management of Alternaria leaf spot of sunflower. In *Sustainability of oilseeds*. Prasad, M.V.R.et.al.(Eds.) Indian Soci. Oilseeds Res. Hyderabad, 283-286.

Orellana, R.G. 1975. Photoperoid influence on the susceptibility of sunflower to *Sclerotinia* stalk rot. Phytopathology, 65: 1293

Patil, M. A. and Mayee, C. D. 1988. Downy mildew of sunflower in India. Trop. pest Manag. 33: 81-82.

Patil, M.A., Mayee, C.D. and Phad, H.B. 1992. Sunflower downy mildew. Information Bulletin, ORS, MAU, Latur. pp: 76.

Prasadarao, R.D.V.J., Reddy, A.S., Chander Rao, S., Varaprasad, K.S., Thirumala Devi, K., Nagaraju, Muniyappa,V. and Reddy, D.V.R. 2000. Tobacco streak ilarvirus as causal agent of sunflower necrosis disease in India. J. Oilseeds Res., 17: 400-401.

Punja, Z. K. 1985. The biology, ecology and control of *Sclerotium rolfsii*. Ann. Rev.Phytopathol. 23: 97-127.

Raoof, M. A. 1999. Integrated management of sunflower downy mildew In Integrated Pest management in sunflower.(Eds) Basappa,H., Harvir Singh and Chattopadhyay, C. Directorate of oilseeds Research, Hyderabad, India. 21-24.

Ramaiah, H., Bhat, A. I., Jain, R. K., Pant, R. P., Ahlawat, Y. S., Prabhakar, K., and Varma, A. 2001. Isolation of an isometric virus causing sunflower necrosis disease in India. Plant Disease 85: 443.

Ramanth, A. K., Lambat B. N., Mukewar., and Indrarani. 1981. Interceptions of pathogenic fungi on imported seeds and planting material. Indian Phytopath. 34: 282-286.

Ravi, K. S., and Zehr, U. B. 2001. Tobacco Streak Virus – A Threat to Crop Plants in India. Mahyco Research Center, Jalna, Aurangabad, Maharashtra, India.

Reddy, P. C., and Gupta, B. M. 1977. Disease loss appraisal due to leaf blight of sunflower incited by *Alternaria helianthi*. Indian Phytopath. 30: 569-570.

Reddy, A. S., Prasada Rao, R. D. V. J., Thirumala Devi, K., Reddy, S. V., Mayo, M. A., Roberts, I., Satyanarayana, T., Subramaniam, K., Reddy, D. V. R. 2002. Occurrence of Tobacco Streak Virus on peanut (*Arachis hypogaea*) in India. Plant Dis. 86: 173–178.

Sakr, N., Ducher, M., Tourvieille, J., Walser, P., Vear, F., and Tourvieille de Labrouhe, D. 2009). A method to measure aggressiveness of Plasmopara halstedii (Sunflower DownyMildew). J Phytopathol. 157: 133–136.

Sahu, B., Ghemawat, M. S., and Agrawat, J. M. 1991.Susceptibility of sunflower plants to *Alternaria helianthi* as influenced by plant age. J.Plant Dis.Prot. 98: 103-106.

Santha Lakshmi Prasad, M., Sujatha, K., and Chander Rao, S. 2010. Seed Transmission of *Alternaria helianthi*, Incitant of Leaf Blight of Sunflower. J. Mycol. Pl. Pathology. 40: 63-66.

Sackston, W. E. 1960. Studies on sunflower rust.II.Longevity of uredospores of *Puccinia helianthi*. Can. J Bot. 38: 883.

Seiler, G.J., Gulya, T.J. and Lofgren, J. 1993. Evaluation of interspecific sunflower germplasm for Sclerotinia resistance. *In:* Proc.15th Sunflower Res. Workshop, Fargo, ND. pp.4-9.

Shane, W. W., Baumer, J. S., and Sederstrom, S. G. 1981. *Alternaria helianthi*, a pathogen of sunflower new to Minnesota. Plant Dis. 64: 269-271.

Sharma, S. C., Ghemawat, M. S., and Agrawat, J. M. 1993.Toxin production by *Alternaria helianthi*, the leaf spot and blight pathogen of sunflower. Acta phytopathol.Entomol. Hung.28: 13-19.

Sharman, M., Persley, D. M., and Thomas, J. E. 2009. Distribution in Australia and seed transmission of Tobacco streak virus in *Parthenium hysterophorus*. Plant Dis. 93: 708-712.

Shirshikar, S. P. 1997. Survey of downy mildew disease of sunflower in Marathwada region. J.Maharashtra Agric. Univ. 22: 135-136.

Shirshikar, S. P. 2003. Influence of different sowing dates on the incidence of sunflower necrosis disease. Helia. 26: 109-116.

Shopov, T. 1979 Investigations on some biological features of downy mildew of sunflowers. Abstr. Rev. Pl. Pathology. 59: 5986.

Simmons E. G. 2007. Alternaria: An identification manual. CBS Biodiversity Series 6: 667 -668. Centraalbureau voor Schimmelcultures, Utrecht, Netherlands.

Singh, U. P., and Singh, R. B. 1983.The effect of soil texture,soil mixture,soil moisture and depth of soil on carpogenic germination of *Sclerotinia sclerotiorum* .J. Plant Dis.Prot.90: 662-669.

Singh, B. K., Sinha, M. S., and Prasad, T. 1981. Storage effect on sunflower seed viability. Indian Phytopath. 34: 120.

Singh, S. J., Nagaraju., Krishnareddy, K. M., Muniyappa, V., and Virupakshappa, K. 1997. Sunflower Necrosis –A new virus disease from India. Abstracts of symposium on "Economically important diseases of crop plants". Indian Phytopathological Society South Zone Annual Meeting 18-20, Bangalore, India. 24.

Singh, H. 2005. Thrips incidence and necrosis disease in sunflower,*Helianthus annuus* L. J. Oilseeds Res.22: 90-92.

Singh, R. and Tripathi, N.N. 1997. Management of *Sclerotinia* rot of sunflower by integration of cultural, chemical and biological methods. J. Mycol. Pl. pathol., 27: 67-70.

Shivaprakasam, K., Pillayar Swamy, K., Jagannathan, R. and Navardhnam, L. 1977. Efficacy of some Chemicals in CEntral of Sunflower rust. Madras Agric. J. 64:197.

Sonali, S., Kalyani, G., Kulkarni, N., Waliyar, F., and Nigam, S. N. 2005.Mechanism of transmission of tobacco streak virus by *Scirtothrips dorsalis, Franklinella schulzai* and *Meglurothrips usitatus* in groundnut,*Arachis hypogaea* L. J. Oilseeds Res.22: 215-217.

Spring, O. 2008. Transition of secondary to systemic infection of sunflower with *Plasmopara halstedii*—An underestimated factor in the epidemiology of the pathogen. Fungal Ecol. 2: 75–80.

Spring, O., and Zipper, R. 2000. Isolation of oospores of sunflower downy mildew, *Plasmopara halstedii* and microscopical studies on oospore germination. J Phytopathol. 148: 227–231.

Sujatha, M., Prabakaran, A.J. and Chattopadhyay, C.(1997).Reaction of wild sunflowers and certain interspecific hybrids to *Alternaria helianthi*. Helia 20: 15-24.

Thompson, T. E., and Rogers, C. E. 1980. sunflower oil quality and quantity as affected by *Rhizopus* head rot. J. Am. Oil Chem.Soc.57: 106.

Tourvieille de Labrouhe, D., Walser, P., Serre, F., Roche, S., and Vear, F. 2008. Relations between spring rainfall and infection of sunflower by Plasmopara halstedii (downy mildew). In: Velasco L. (ed.) Proceedings of the 17th International Sunflower Conference. 1: 97–102). Cordoba, Spain.

Van Regenmortel, M. H. V., Fauquet, C. M., Bishop, D. H. L., Carstens, E. B., Estes, M. K., Lemon, S. M., Maniloff, J., Mayo, M. A., McGeoch, D. J., Pringle, C. R., and Wickner, R. B. 2000. Virus taxonomy. In: 7th Report of the International Committee on Taxonomy of Viruses. Academic Press, New York, 1162.

Venkataramanamma,K.Madhusudhan,P.Neelima,S.and Narasimhudu, Y. 2014. Field evaluation of fungicides for the management of Alternaria leaf blight of sunflower. Indian J Plant Prot. 42:165-168.

Ventslavovich, F. S., Novotelnova, N. S., and Gurzhiev, G. A. 1971. Altered metabolism in sunflower plants during infection by downy mildew. Abstr. Rev. Plant Pathol. 40: 2413.

Virupakshappa, K., Vijay Singh, Basappa, H., Prasad, Y. G., and Chattopadhyay, C. 1999. Current status of pest problems in oilseed crops and their management. Indian J Plant Prot. 27: 65-86.

Williamson, A. J. P.1979. Five states give us the current new on sunflower.

Qd.Sunflower. 3: 19.

Wehje, G., Littlefield, L. T., and Zimmer, D. E. 1979. Ultrastructure of compatible and incompatible reactions of sunflower to *Plasmopara halstedii*. Can.J.Bot. 57: 315.

Zad, J. 1978. Transmission of sunflower downy mildew by seed. Iran. J. Plant pathol. 14: 1.

Zimmer, D. E. 1975. Some biotic and climatic factors influencing sporadic occurrence of sunflower downy mildew. Phytopathology. 65: 751.

Zimmer, D.E., Kinman, M.L. and Fick, G.N. 1973. Evaluation of sunflowers for resistance to rust and *Verticillium* wilt. *Plant Dis.Rep.*, 57 (6) : 524.

Diseases of Field and Horticultural Crops
Editor-in-Chief: **P. Chowdappa**
Published by: **Daya Publishing House, New Delhi**

Pages **298-341**

11

Diseases of Castor, Safflower and Sesame

Prasad, R.D. and Raoof, M.A.

(A) Castor

Castor is known to be attacked by several pathogens at every stage of crop growth, which include mostly fungi and bacteria (Table 11.1). However, only few of them can cause diseases of economic importance at different crop growth stages, depending upon the seasonal conditions (Raoof and Nageshwar Rao, 1999). In India, up to mid seventies, rust, powdery mildew, leaf spots and blights were recorded as important diseases on castor (Rangaswamy, 1979). In the past three decades, cultivation of early maturing, fertilizer and irrigation responsive varieties and hybrids of either multi or regional importance established castor as a commercial crop, registering seven-fold increase in castor production and four-fold increase in castor productivity in the country, However, this increase is becoming increasingly unsustainable due to change in the disease scenario such as wilt, root rot and grey rot that have become serious problems, particularly in high yielding varieties and hybrids. Wilt (20-50%) and grey rot (5-85%) in Andhra Pradesh while wilt (20-60%) and root rot (5-50%) in Gujarat are the serious diseases in the castor growing areas causing economic yield losses. Among other diseases of regional and seasonal importance are seedling blight, *Alternaria* blight, Cercospora leaf spot and bacterial blight.

1. Fusarium Wilt (*Fusarium oxysporum* f. sp. *ricini* Nanda & Prasad)

The disease was first reported in Morocco in 1953 (Reiuf, 1953). It is prevalent in Russia, Brazil, Taiwan and Nepal. The extent of disease incidence was up to 80 % in Russia (Moshkin, 1986). In India, it was recorded for the first time from

Rajasthan in 1974 (Nanda and Prasad, 1974) and later from Gujarat, Andhra Pradesh and Karnataka. The extent of seed yield loss depends on the stage at which plants wilt, 77% at flowering stage, 63% at 90 days old crop and 39% at later stages on secondary branches (Pushpavati, 1995). Losses in yield were realized in all cultivated castor hybrids in Gujarat (Dange *et al.*, 1997) and as high as 85 % wilt incidence has been reported under North Gujarat conditions (Dange, 2003).

1.1. Symptoms: The disease appears mostly in patches and affects all the growth stages of crop. Disease generally appears at flowering and spike formation stage and becomes more prominent in later stage of the crop. At seedling (2-3 leaves) stage, discolouration of hypocotyl and loss of turgidity of top leaves are the main symptoms (Fig. 11.1). Within 2-3 days, leaves wilt and dry out without changing the green colour. In few plants, spot or blight symptoms appear along the edge of cotyledons and the first true leaves. Infected stem shows blackish lesions above the collar region and further these lesions spread up to a distance of 15 to 20 cm above the ground level (Rieuf, 1953). But sometimes, dark stripe may cover the whole stem and is formed up to the infected leaves (Raoof and Nageshwar Rao, 1999).

Fig. 11.1. Fusarium wilt in castor

At pre-flowering stage, leaves turn yellow, marginal and intervenial necrosis starts with complete senescence of lower leaves, ultimately irreversible wilting with bend apices occurs. Plants infected at flowering, spike formation and capsule development stages appear sick and leaves become yellow with marginal necrosis. Later, the entire leaf becomes necrotic and shrivels. Lower leaves drop away due to senescence except few top leaves followed by irreversible wilting of plant (Nanda and Prasad, 1974). The branches of infected plants also become shriveled and discoloured. Roots show blackening and necrosis. In case of partial wilting, only one side of root system is observed blackish and necrotic and the other side remains healthy. Transverse and longitudinal sections of the affected roots reveal the presence of the fungus in vascular tissue and in the xylem parenchyma. Tylose formation is also observed in xylem vessels of the infected roots. When the stem is split open, browning, blackening of xylem tissues could be seen. The infected stem tissue shows intercellular mycelium in vessels and hypertrophy of xylem parenchymatous cells (Nanda, 1975). Pinkish overgrowth is also visible on the stem due to formation of sporodochia of the fungus under moist conditions. Infected plants seeds and such seeds are deformed and light in weight

1.2. Causal organism: *Fusarium oxysporum* f. sp. *ricini* Nanda & Prasad. Kingdom: Fungi, Division: Eumycota, Subdivision: Deuteromycotina, Class: Hyphomycetes, Order: Moniliales, Family: Tuberculariaceae. It produces white fluffy mycelial growth on Potato Dextrose Agar (PDA), which turns pinkish by keeping in day light. Microconidia are formed which are hyaline, round to ovoid, single celled or 1 septate. One celled type measures 3.66 x 6.4 µm and the two celled ones measure 15.29 x 3.76 µm in size. Macroconidia are less in number, 2-6 septate (mostly 3 septate), straight, spindle as well as sickle shaped and measure 17.50 - 70.00 x 3.50 - 5.25 x 3.93 µm (Desai *et al.*, 2003). Both terminal and intercalary chlamydospores are present and measure 8.7 x 4.44 µm. Generally, sporodochia develop in two-week-old cultures (Kolte, 1995). Differential reaction of varieties and breeding lines over the years and across the locations have indicated prevalence of variants in the pathogen. Wide variation in cultural, morphological characteristics and pathogenicity among different isolates of the pathogen has been observed (Nanda and Prasad, 1974 and Desai, *et al.* 2003)

1.3. Epidemiology: The pathogen is mainly soil borne though it is also reported to be seed-borne to an extent of 10-20% (Chattopadhyay, 2000, Raoof, *et al.*, 2003). Seeds from wilted castor plants carry inoculum at the micropylar end in 2-19 % seeds and seed infection is confined to testa, tegmen and endosperm (Naik, 1994). The fungus survives for long periods in soil/crop debris as thick walled resting structures called chlamydospores. Under favourable conditions, chlamydospores germinate and produce mycelia along with two types of spores (one celled microconidia and fusiform septate macroconidia). Under moist conditions, the spores germinate in soil and infect the plants when they come into contact with roots. They enter the root tips and older roots through natural wounds, nematode feeding punctures, and other abrasions. After penetration, the fungi grow into the water-conducting vessels (xylem tissue), invade other parts of the plant, plug up the vessels, and produce typical wilt symptoms (Moshkin, 1986). Plants attacked with reniform nematode *Rotylenchulus reniformis* Linford and Oliviara were found to be predisposed to the infection of wilt pathogen *Fusarium oxysporum* f. sp. *ricini*. Nematodes were involved in increased disease severity and breakdown of resistance in castor hybrid GCH-4 (Pathak, 2003).

Table 11.1. List of diseases/pathogens recorded on castor

Disease	Causal organism	Reference
Fungal diseases		
Wilt	*Fusarium oxysporum* f.sp. *ricini* Nanda & Prasad	Nanda and Prasad, 1974
Grey rot	*Botrytis ricini* Godfrey	Godfrey, 1923
Root rot Goid	*Macrophomina phaseolina* (Tassi)	Uppal, 1934
Seedling blight	*Phytophthora parasitica* Racib	Dastur, 1913
Alternaria blight	*Alternaria ricini* (Yoshii) Hansford	Yoshii, 1929
Bacterial leaf spot	*Xanthomonas campestris* pv. *ricini* (Yoshii & Takimoto) Young	
Anthracnose	*Colletotrichum ricini* Petch	
Bacterial wilt	*Pseudomonas solanacearum* (Smith) Yabuuchi	
	Ralstonia solanacearum(Smith) Yabuuchi	
Capsule rot	*Cladosporium oxysporum* Berk and Cart.	
Collar rot	*Scelrotium rolfsii* Sacc.	Siddaramaiah *et al*, 1980
	Rhizoctonia solani Kuhn.	Chattopadhyay, 1995
Damping-off	*Pythium aphanidermatum* (Edson) Fitzp.	Vasudeva 1959
Die-back	*Botryosphaeria quercuum* (Schwein.) Sacc.	
Fruit rot	*Phytophthora colacasiae* Racib.	Patel *et al.*, 1949
Leaf blight	*Phytophthora palmivora* (Butler) Butler *Phytophthora nicotainae* var. *parasitica* Dastur	
Leaf spot	*Cercospora ricinella* Sacc. and Beslase	Vasudeva, 1959
	Cercospora coffeicola Berk.	
	Alternaria tenuissima (Kunze) Wiltshire	
	Leptosphaerulina ricini Karan	
	Sclerotinia sclerotiorum (Lib.) de Bary	
	Myrothecium roridum Tode ex Fr.	
	Corynespora cassiicola (Berk.& Curt.) Wei.	
	Botryodiplodia ricinicola (Sacc.) Petr.	
	Cochliobolus lunatus Nelson & Haasis	
	Sphaceloma ricini Jenkins & Cheo	
	Periconia byssoides Pers.	
	Phyllosticta bosensis Bose & Mathur	Rao 1965
	Phyllosticta ricini E. Rastr	
	Phytophthora nicotianae Breda de Haan	

contd...

Table 11.1. *contd...*

Disease	Causal organism	Reference
Powdery mildew	*Leveillula taurica* (Lev) Arm.	Vasudeva, 1959
Rust	*Melampsora ricini* (Biv-Bern) de Toni	Vasudeva, 1959
Seed rot	*Cephalosporium curtipes* Sacc.	
Stem canker	*Botryosphaeria dothidia* (Moug.: Fr.) Ces. & De Not.	
Stem rot	*Physalospora ventricosa* (Dur. & Mont.) Cooke	
	Phytophthora nicotianae Breda de Haan	
Twig blight	*Glomerella cingulata* (Stonem) S.S.	
Twig rot	*Diaporthe arctii* (Lasch) Mitschke	
Verticillium wilt	*Verticilium alboatrum* Reinke and Berthold	
Bacterial diseases		
Bacterial leaf spot	Xanthomonas axonopodis pv, ricinicola (Elliott) Dowson	
	Xanthomonas campestris pv. *ricini* (Yoshii & Takimoto) Young	
	Pseudomonas syringae	
Bacterial wilt	*Ralstonia solanacearum*(Smith) Yabuuchi	
	Pseudomonas solanacearum (Smith) Yabuuchi	
Viral diseases		
Mosaic and blisters on leaves	*Cucumber mosaic virus*	
Yellowish Vein netting and mottling of leaves	*Olive latent virus 2*	
Tobacco ring spot	*Tobacco ring spot virus*	

The disease generally appears in the months of October-November when the crop is about 3-4 months old and becomes more prominent during February-March when the crop is in seed formation stage (Nanda and Prasad, 1974). The pathogen infects the plant at 13-15°C and the symptoms appear in their full form at 22-25°C (Andreeva, 1979). Infected seeds play an important role in the perpetuation and spread of the pathogen.

1.4. Management: Castor being a monotypic genus, several workers identified few resistant sources to castor wilt by screening large number of diverse germplasm accessions and breeding lines in sick plot and artificial inoculation conditions in pot cultures (Pushpavathi, 1995; Pathak, 2003). Cultivation of resistant castor varieties/hybrids is the cheapest and the best way to manage wilt. Two varieties Jyoti (DCS9) and Jwala (48-1) possessing high degree of wilt resistance were released by DOR, Hyderabad for cultivation in rainfed areas. First wilt resistant hybrid GCH-4 was released from Gujarat in 1986. Subsequently other hybrids GCH-5, GCH-6, DCH-32, DCH-177, RHC1, TMVCH1, PCH1 and DCH-519 were released from different centres for cultivation in wilt endemic areas.

Integration of host resistance, cultural practices, sanitation and use of bioagents is needed for the effective management of castor wilt.

- Avoid cultivation in low-lying and ill drained lands.
- Rotate the crop for 2-3 years with non-host plants like pearl millet, finger millet or other cereals.
- Resort to deep ploughing in summer or soil solarization for 6 weeks during peak summer (Raoof and Nageshwar Rao, 1996).
- Seed treatment with *Trichoderma viride* 10g/kg or carbendazim 2g/kg (Raoof and Rambhadra Raju, 2003).
- Rouge out disease affected plants regularly.
- Cultivate wilt resistant varieties *viz.*, Jyothi, Jwala, and hybrids *viz.*, DCH-32, DCH-177, GCH-5 and GCH-4.

2. *Botrytis* grey rot (*Botrytis ricini* Godfrey)

The disease was first reported from Florida in the year 1918. It is also recorded from USA (Mississippi, Louisiana, Texas), Cuba, Brazil and Bulgaria. In India the disease occurred first in Karnataka and appeared in epidemic form during 1987 in Andhra Pradesh (Moses and Ranga Reddy, 1989). The disease is confined to few states of India *viz.*, Andhra Pradesh, Tamil Nadu, Karnataka and Orissa. The disease can cause 10 to 100% seed yield loss (Godfrey, 1923; AICRP, 1988).

2.1. Symptoms: The disease appears predominantly on spikes. Staminate/ male flowers below the pistillate flowers are highly susceptible. Initially pale blue blotch-like spots are formed from which drops of yellow liquid exude (Godfrey, 1923). Later fungal threads grow from these spots and transform into characteristic dense woolly growth (Maiti *et al.*, 1988). The growth of the fungus may vary from pale to olive grey in colour. Severely infected capsules rot and fall. The immature seeds become soft and mature ones hollow, exhibiting discoloured seed coat.Circular to sub-circular or irregular lesions develop on leaves. They enlarge and coalesce to form blighted areas either marginally or apically near or around the leaf base where dew or rain usually collects (Bheema Raju, 1999). Abundant sporulation can also be seen on leaves. Water soaked blackish lesions develop on tender shoots usually when infected spikes fall on them leading to rot and breakage (Moses and Ranga Reddy, 1989).

2.2. Causal organism: *Botrytis ricini* Godfrey (Anamorph); *Sclerotinia* (=*Botryotinia) ricini* (Godfrey) Whatzel (Teleomorph), Kingdom: Fungi; Division: Eumycota; Subdivision: Deuteromycotina; Class: Hyphomycetes; Order: Moniliales; Family: Moniliaceae. The fungus causing grey mould of castor in Florida was first described as a new species of *Botrytis* by Godfrey (1919) and named it as *Sclerotinia ricini*. In Europe the disease is attributed to *Botrytis cinerea* Pers. ex Fr. (Golenia, 1955 and Zarzicka, 1958). Later, called the fungus causing leaf blight of castor beans as *Botryotinia ricini* (Godfrey) Whetzel. In India, Moses and Ranga Reddy (1989) identified the causal fungus as *Botrytis ricini* Godfrey (*Sclerotinia ricini*). It is host specific and not reported to infect crops other than Euphorbiaceae.

The fungus as described by Mosses and Reddy (1989) produced submerged mycelium that sporulated in greyish masses on PDA. The colonies are diffused and grey in colour. Conidiophores are erect, septate, 11-23 µm thick, blackish-brown with several projections at the tip from which conidia are formed singly on very fine warts, finally becoming lateral. The conidia stand so thickly on the projections that thick beads are produced, which soon fall off. Conidia are globose, 8-9 µm in diameter with almost hyaline slightly brownish wall. Sclerotia were formed on PDA, acidified PDA, V_8 juice medium, and host leaf extract medium and on infected seeds upon incubation (Bheema Raju, 1999). Sclerotia are black, rough, elongate, irregular 1-25 mm in length (usually 3-9mm) suberumpent to superficial developing on axes of old inflorescence and on stalks (Godfrey, 1919).

2.3. Epidemiology: The fungus survives in soil through mycelia/sclerotia in crop debris. It can survive in infected crop debris on the soil surface for 6 months whereas under laboratory conditions the survival was for 9 months (DOR, 2002). Sclerotia germinate to produce mycelium or mycelium with spores. Though carpogenic germination (formation of apothecia) is reported from other countries (Godfrey, 1923), there are no reports from India. Spores penetrate the host tissue by mechanical means and after penetration, host tissue gets disorganized completely. A succession of several continuously wet days lead to the development of abundant sporulation. Spores formed on the infected spikes get readily disseminated by wind and rain, which facilitate the secondary spread of infection (Godfrey, 1923).

The fungus also survives in seeds especially in the caruncle and beneath the seed coat (Godfrey, 1923). Latent infection was recorded from apparently healthy seeds to an extent of 13% (DOR, 2002). The pathogen has a wide host range and can infect plants in family Euphorbiaceae (*E. pulcherrima, E. hirta, E. geniculata, E. prostrata) Jatropha* spp. (*J. curcus, J. multifida, J. gossypifolia, J. podagrica*) (Mehtab Yasmeen *et al.*, 2003). A succession of several (5 days) continuously wet days with high relative humidity (>90%) and cool temperature (25-28 ^{0}C) during flowering and capsule formation is essential for disease development (Godfrey, 1923).

2.4. Management: Adjust sowing time in such a way that the crop maturation occurs during dry season (Raoof and Nageshwar Rao, 1999).

- Adopt wider spacing (90 x 60 cm).
- Remove diseased spikes and destroy them.
- Jwala (48-1) a spineless variety with loose spike is less susceptible to grey rot disease (Raoof and Nageshwar Rao, 1999).
- Spray carbendazim / thiophanate methyl 1g/l before the onset of cyclonic rains based on weather forecast followed by second spray soon after disease appearance.
- Spray of *Trichodema viride* (10^{10} spores/ml) and *Pseudomonas fluorescens* (10^8 CFU/ml) recorded 52 and 62% disease reduction respectively. *Trichoderma* could over grow *B. ricini* and thus, not only involved in controlling the disease but also restricting the dispersal of spores by hyperparasitising the pathogen (Raoof *et al.*, 2003).

3. Root rot (*Macrophomina phaseolina* (Tassi) Goid)

The disease has been recorded from Russia, Morocco, Peru, Italy, Yugoslavia, Pakistan, China, Phillipines, Palestine, Sri Lanka, East Africa, West Indies and eastern United States. The disease was first reported from India in 1932 (Uppal, 1934) and is widely distributed in the states of Andhra Pradesh, Maharashtra, Gujarat, Bihar and Tamil Nadu. Das and Prasad (1989) reported yield loss of 194 and 138 kg/ha in hybrid GAUCH-1 and variety Aruna respectively.

3.1. Symptoms: Different symptoms are observed *viz.*, seedling blight/die-back (Rieuf, 1953; Maiti and Raoof, 1984), spike blight, stem blight, collar rot, root rot and twig blight (Moses and Ranga Reddy, 1989). Initially, the plant shows signs of water shortage, later leaves and petioles droop down and the entire plant dries up (Fig. 11.2). Dark black lesions appear on the stem region near the ground level, this phase (collar rot) is observed at 30-40 days after sowing. Tap root also dries up and root bark shreds off easily. Micro sclerotial bodies are seen scattered in the shredded root bark and on the surface of the roots. Pycnidia develop as minute black spots on the surface of woody tissues. Elongated, irregular greyish to brownish black spots also develop on the stem, petiole and branches (Chaudhuri and Patel, 1992). Later, aerial infection is seen in the form of small brown depressed lesions on or around the nodes. The lesions increase in size and result in 2-20 cm necrotic area (Maiti and Raoof, 1984). Several such lesions often coalesce and girdle the stem causing wilt and leaf drop leaving a bare yellow stalk, which ultimately becomes black. In severe infection entire branch or top of the plant withers away. Wilting of leaves starts at the apex and progresses downwards. Pith region shows brown discolouration due to formation of sclerotia. The affected spikes are discoloured and turn black and infected capsules due to drying up drop off easily. Infected capsules become discoloured and drop off easily. Pycnidia are produced on outersurface of capsule, whereas sclerotia are produced inside the capsule (Maiti and Raoof, 1984; Moses and Ranga Reddy, 1989).

Fig. 11.2. Macrophomina root rot caused by Macrophomina phaseolina

3.2. Causal organism: *Macrophomina phaseolina* (Tassi) Goid (Pycnidial Stage); *Rhizoctonia bataticola* (Taub.) Butler (Sclerotial stage), Kingdom: Fungi; Division: Eumycota; Subdivision: Deuteromycotina; Class: Gonomycetes; Order:

Aganomycetales; Family: Aganomycetaceae. The pathogen was found associated with aerial infections. Mycelial hyphae are filiform or thick, up to 5 to 7 μm or irregular diameter, septate, frequently ramose, secondary branching at right angles and anastomosing at first hyaline then becoming dark reddish to black, penetrating deep into the matrix and forming microsclerotia, immersed or erumpent, scattered or gregarious, typically spherical, 50 to 100 μm in diameter, formed of angular parenchymatous cells, abundantly filled with refractive oily droplets (Dhingra and Sinclair, 1978).

Pycinidia are erumpent, globose, dark brown to black and 140-200 μm in diameter with an ostiole. Conidia are formed from hyaline, ampulliform phialides from the inner cells of pycinidia. These conidia are hyaline, oval to oblong, straight, single or two celled, round at each and measure 10-16 x 3.5 μm in size (Kulkarni *et al.*, 1966). The root rot phase is caused by *Rhizoctonia bataticola* (Taub.) Butler. The sclerotia formed on the affected roots are dark, black and roughly globose and measure 100 μm in diameter (Sarwar, 1974).

3.3. Epidemiology: The pathogen is soil borne and survives for long periods as sclerotia in the crop debris. The fungus harbouring in the roots and stems of crop debris also plays an important role in the disease initiation in the field. Sclerotia germinate under favourable conditions and mycelia enter through roots causing infection. Later, as the disease progresses pycnidia are formed on stem near the inter-nodes. Pycnidial bodies germinate and produce pycnidiospores that help in the secondary spread of the disease (Maiti and Raoof, 1984). The pathogen survives on a wide range of hosts *viz.*, maize, soybean, sunflower, groundnut, mustard and cowpea (Moshkin, 1986). Disease appears when there is hot and dry climate. Soil temperature (above 32°C) is crucial for the disease development, increase in 1°C soil temperature results in 0.7-1 % increase in diseased plants (Das and Srivastava, 1992). Rain after prolonged dry spell predisposes the plants to the disease.

3.4. Management: Use disease free seeds and follow seed treatment with *Trichoderma viride* 4g/kg or thiram/captan 3g/kg seed.

- Burn crop debris along with sclerotia (Chaudhari and Patel, 1992).
- Use farmyard manure or green manure (Chaudhari and Patel, 1992). Application of organic amendments in soil is helpful in reducing the disease in other host crops like soybean (Purkayastha and Menon, 1985; Narasimhalu and Bhaskaran, 1987; Muthusamy and Mariappan, 1994), sesamum (Waghe and Lanjewar, 1981) and cowpea (Ratnoo and Bhatnagar, 1993).
- Rotate with non-host crops (Chaudhari and Patel, 1992).
- Grow tolerant and resistant varieties like Jwala and JHB-665.
- Treat the seed with thiram / captan 3g/kg or carbendazim 1g/kg (Das and Shankar, 1990).
- Topsin M-70 (thiophanate methyl) is also found to be effective (Sarwar and Raju, 1985).

4. Seedling blight (*Phytophthora parasitica* (Dastur) Waterhouse)

The disease is distributed in Argentina, Brazil, Bulgaria, China, Malaysia, Sumatra (Thompson, 1929; Liang, 1964; Kolte, 1995). In India, it is known to exist since 1889 and authentically recorded for the first time from Pusa (Bihar) in 1909 (Dastur, 1913) and later from Assam, Andhra Pradesh, Uttar Pradesh, Karnataka and Gujarat. It can cause plant loss to an extent of 10% (Maiti *et al.*, 1988) though 30 to 40 % seedling mortality was also observed (Dastur, 1913).

4.1. Symptoms: Infection starts from the basal portion of the plant. Symptoms appear in the form of dull green colour, roundish patches on both the surfaces of cotyledonary leaves of young seedlings (15-20 cm high). Soon, the cotyledonary leaves droop down and the infection gradually spreads from the leaf to the base of petiole, point of attachments and then to the stem and growing point ultimately killing the seedling.

Symptoms on older plants are restricted to leaf lamina. Yellow irregular spots develop which turn into light to dark brown concentric zones with yellowish green halo on the upper surface of the leaves and brownish grey border on the under surface. The spots coalesce to cover almost the entire leaf and under humid conditions a very fine whitish growth of the fungus occurs on the under surface of the leaves. Ultimately leaves get blighted and shed prematurely (Dastur, 1913).

4.2. Causal organism: *Phytophthora parasitica* (Dastur), Kingdom: Fungi; Division: Eumycota; Subdivision: Mastigomycotina; Class: Oomycetes; Order: Pernosporales; Family: Pythiaceae. The pathogen develops inside the host tissues as intracellular and intercellular myclium and does not form haustoria. The sporangiophores are slender, unbranched, 35-500 µm long and emerge from the lower epidermis of the leaf through stomata. A single colourless, ovoid or roundish sporangium is formed at the tips of sporangiophores. The sporangia are papillate, pear shaped, hyaline and measure 25-50 x 2-40 µm. Under dry and dark conditions mature sporangia germinate directly by producing germ tube but in the presence of light and water the sporangia germinate by liberation of zoospores (Mundkur, 1959). Number of zoospores in each sporangium varies from 5-45 (av. 30). The zoospores are reniform, 8-12 x 5-8 µm in size, biflagellate and settle within 20 minutes to 2 hours after their liberation from the sporangium. They germinate readily by single germ tube or rarely by two. The fungus also forms chlamydospores in culture or on inoculated plants. They are hyaline smooth, round, thick walled yellow bodies measuring 20-60 µm in diameter. These are formed freely in hot and dry months when spornagia are scanty and retain the power of germination for many months. Oospores (sexual spores) measure 13-24 µm in diameter are also produced on artificial culture media.

4.3. Epidemiology: The disease is soil-borne and fungus survives as chlamydospores for long periods in crop debris and also on alternate hosts like potato, tomato, brinjal, sesame, cotton and guava. Chlamydospores germinate under favourable conditions and sporangia are formed (Dastur, 1913). Sporangia germinate readily on the cotyledonary leaves by producing zoospores, which in turn produce germ tubes and initiate infection producing disease spots within 24 hours. Production of next crop of sporangia follows within two days. Sporangia are easily disseminated by wind and help in the spread of secondary infection (Mundkur, 1959).The disease appears at seedling stage during rainy season under moist and water-logged conditions (Dastur, 1913).

4.4. Management:

- Provide proper field drainage and avoid damp and low-lying areas (Butler, 1918).
- Adopt seed dressing with thiram / captan, 2-3 g/kg or metalaxyl 3g/kg (AICRP, 1970).
- Need based foliar spray with Bordeaux mixture 1% or copper oxychloride 0.3% (AICRP, 1970).
- Field sanitation, summer ploughing, soil solarization and other cultural practices should be taken up.

5. *Alternaria* leaf spot (*Alternaria ricini* (Yoshii) Hansford)

The disease was first reported from Korea in 1929 (Yoshii, 1929) and later from Japan, Bulgaria, Brunei, Egypt, Jamaica, Niger, Sarawak, South Africa, Southern Rhodesia, Sudan, Yugoslavia, Tanzania and USA. In India, it is distributed in Uttar Pradesh (Singh, 1955), Madhya Pradesh, Andhra Pradesh, Rajasthan, Maharashtra and Tamil Nadu. It can cause 35-85% seed yield loss, 1961).

5.1. Symptoms: All the aerial parts of the plant are affected but symptoms are more pronounced on leaves. In the seedling stage, light brown circular spots appear on cotyledonary leaves, which later become angular and coalesce to cause foliage blight. Severe infection results in death of the seedlings. On adult plants brown, zonate, irregular spots of variable sizes with concentric rings surrounded by yellow halos appear scattered on leaves. Sometimes, bluish green spore mass could also be seen. In severe infections, several spots coalesce to form big patches resulting in premature defoliation of the plant.

Under humid conditions, disease appears on inflorescence and capsules in the form of sooty growth. Two types of symptoms may be observed on capsules one involving sudden wilt of half mature capsule with purple or dark brown discolouration and the other characterized by development of sunken spots on fully developed capsules on one side which gradually enlarge to cover the whole capsule with fungal growth (Stevenson, 1945). On shaking of infected spikes black clouds of spores are released. Such capsules are smaller in size and contain under developed or wrinkled seeds with little or no oil content Young racemes and even flower primordia are killed during severe infection.

5.2. Causal organism: *Alternaria ricini* (Yoshii) Hansford. Kingdom: Fungi; Division: Eumycota; Subdivision: Deuteromycotina; Class: Hyphomycetes; Order: Moniliales; Family: Dematiaceae. The young mycelium on PDA is fluffy in growth and of spreading type. The hyphae are hyaline, septate and branched irregularly. As the hyphae get older the colony colour changes from white to olive grey. The conidiophores are erect, simple, straight or flexous almost cylindrical or rather thicker towards the base, septate, pale brown, smooth with one or few conidial scars up to 80 μm long and 5-9 μm thick. Conidia are solitary or occasionally in chains of two, straight or curved, obclavate or with the body of the conidum ellipsoidal, tapering abruptly to a very narrow beak which is equal in length up to twice as long as the body, reddish brown in colour smooth, sometimes constricted at the septa, overall length 70-170 (140 µm) x 13-15 µm thick in the broadest part. Beak is simple and non-septate (Ellis and Holiday, 1970). The optimum temperature for the spore germination lies between 25 and 30°C. Low temperature (20°C) favours the production of spores. The fungus requires organic sources of nitrogen like asparagine, aspartic acid and carbon compounds like xylose, maltose and raffinose. The optimum pH lies between 5 and 6 and the amount of aerial mycelium produced in culture is greater in acidic than in alkaline reaction.

5.3. Epidemiology: The pathogen is externally and internally seed borne (Stevenson, 1945). Primary infection spreads through infected seeds. The fungus also survives in crop debris. The pathogen produces spore mass on the infected seeds or crop debris under humid and moist conditions. Spores germinate by producing germ tubes, penetrate the epidermis and grow within the leaf. Later conidiophores emerge from the diseased leaf. Spores produced on the conidiophores are spread by wind and splashing rain. Newly formed spores cause leaf and capsule infections. The pathogen can also infect other plants like *Jatropha pundurifolia* and *Bridelia hamiltoniana*, which may act as alternate hosts (Kolte, 1995). Humid and cloudy weather with low temperature (16-20°C) favours the disease development (Maiti *et al.*, 1988).

5.4. Management:

- Seed treatment with thiram 3g/kg is effective in the initial phase of infection (Maiti *et al.*, 1988).
- Three sprays with mancozeb 0.025% or iprodione 0.1% at an interval of 15 days starting from 90 days of crop growth are effective (Maiti *et al.*, 1988).
- Judicious application of nitrogenous fertilizers reduces the disease development.

6. *Cercospora* leaf spot (*Cercospora ricinella* Sacc. and Beslese)

The disease was recorded from Australia, Brazil, Colombia, Dominican Republic, Indonesia, Jamaica, Mozambique, Nicaragua, Portugal, Sierraleone, Somalia, Tanzania, Tongo, USA, and Venezuela. In India, it was first recorded from Hyderabad (Andhra Pradesh) in 1957 (Salam and Rao, 1957) and subsequently from Bihar and Uttar Pradesh.

The incidence and severity of the disease is variable and losses are considerable, where the leaves are fed to the eri-silk worm (Kolte, 1995).

6.1. Symptoms: Initially minute, circular to angular pale yellow watery lesions with pale green halo appear on the mature leaves. Later, the lesions enlarge and turn circular to irregular in shape with dark brown borders and greyish-white centres. The fructifications of the fungus are visible as tiny black dots in the white central region. The spots are visible on both the surfaces of the leaf. When the spots coalesce, the intervening leaf tissue withers and large brown patches of dried leaf result. Severe infection also leads to leaf blighting (Raoof *et al.*, 2003).

6.2. Causal organism: *Cercospora ricinella* Sacc. and Beslese. Kingdom: Fungi; Division: Eumycota; Subdivision: Ascomycotina; Class: Dothiodeomycetes; Family: Mycosphaerellaceae. The hyphae collect beneath the epidermis to form very small stomata. Clusters of conidiophores emerge through the epidermis and form fructifications of the fungus. The conidiophores are brown below and lighter towards the tips, septate, unbranched and measure 24-70 x 3 x 6.5 μm in size. The upper part of the conidiophores is characteristically knobby or flexed. Conidia are obclavate, sub-hyaline, thin walled, broad at the base and narrow at the tips, straight or slightly curved, multi septate (up to 7 transverse septa) and measure 50-105 x 4-6.5 μm.

6.3. Epidemiology: The disease cycle of *Cercospora* leaf spot begins in the spring with conidia produced from the surviving pathogen in lesions on crop debris remaining on the soil surface or as dormant mycleia in seeds (Singh, 1948). Conidia on leaves germinate, penetrate through stomata and produce hyphae. The hyphae get collected beneath the epidermis and form stroma. Cluster of conidiophores may arise through stomata or by piercing through the epidermis followed by production of conidia (Kolte, 1995). Conidia are primarily wind-dispersed and spread by air currents to the newly expanded leaves. Free water on leaf surface triggers conidial germination and causes secondary infection of leaves. The disease is favoured by high temperature and high humid conditions.

6.4. Management

- Spray copper oxychloride 0.3% or mancozeb 0.25% two to three times at 10-15 days interval.
- Spray of Bordeaux mixture (5: 5: 50) checks the spread of the disease.

7. Bacterial leaf spot (*Xanthomonas campestris* pv. *ricini* Yoshii and Takimoto)

The disease was first reported under the name 'Black rot' from Sudan in 1927 (Archibald, 1927) later it appeared in Russia, Zambia, Hongkong, USA, Uganda, Saudi Arabia, Poland, Korea, Japan, Brazil, South Africa and Taiwan. In India, it was first reported from Gujarat (Patel *et al.*, 1951) and subsequently from Andhra Pradesh, Rajasthan and Tamil Nadu. It can cause up to 25% loss in seed yield (Poole, 1954).

7.1. Symptoms: Initially, minute water soaked spots with dark margins develop in large number on the leaves. These spots appear first on the undersurface and then on the upper surface of the leaves. Cotyledonary leaves also get affected. Later, the spots enlarge, become angular and several spots coalesce to form irregular brown to black patches covering the veins and blade. Bacterial ooze in the form of encrustation is seen on the affected tissue in the morning hours. Leaves become blighted and fall off. Affected petioles and young branches show dark elongated lesions (Desai and Shah, 1963).

7.2. Causal organism: *Xanthomonas casmpestris* pv. *ricini* (Yoshi and Takimoto), Kingdom: Prokaryotae; Division: Gracilicutes; Class: Proteobacteria; Order: Pseudomonadales; Family: Pseudomonadaceae. The bacterium is gram negative, rod shaped, 1.5 x 0.7 μm with a single polar flagellum, capsulated, non-acid fast and non-spore forming. The isolate liquifies gelatin, hydrolyses starch, digest casein, produces ammonia, acid and H_2S, peptonises milk and reduces litmus. It is negative to Methyl Red and Voges-Proskauer tests; it does not hydrolyze cellulose and it is strictly aerobic. Both yellow and white strains of this pathogen have been reported (Sabet, 1959; Jindal and Patel, 1972).

7.3. Epidemiology: The pathogen is seed-borne and also known to survive for 120 days in dry soil (Khalil and Sabet, 1971). In young capsules, the bacteria spread through pericarp to the seed integument and gains entry through hydathodes. Rain spreads the disease within cropping season (Desai and Shah, 1963). The pathogen also infects plants like *Cassia, Crotolaria juncea* and *Cymopsis tetragonoloba* that may act as collateral hosts.Disease is favoured by warm and high humid conditions (Desai and Shah, 1963), poor drainage and heavy rains accompanied by high wind.

7.4. Management:

- Seed-borne infection is eliminated by hot water treatment of seeds at 50-60°C for 10 min (Poole, 1954), seed treatment with streptocycline and Sandos 6334 (Singh *et al.*, 1976).
- Soil amendment, crop rotation, use of healthy seed and disinfestations of implements is also recommended.
- Need based spray of streptocycline 1g/10 l of water or paushamycin 0.025% + copper oxychloride 0.3%.

B. Safflower

About 50 diseases caused by fungi, bacteria, viruses and mycoplasma are reported to infect safflower crop although only a few of them are of economic importance (Kolte, 1985). Diseases like wilt (*Fusarium oxysporum* f.sp. *carthami*), *Alternaria, Ramularia* and Cercospora leaf spots, rusts, powdery mildew and root rot diseases though reported on safflower but were not causing economic losses till early eighties. With introduction of new high yielding cultivars during late eighties, intensification of cultivation of safflower in large areas as a sole crop with irrigation and fertilizers application in major safflower growing areas, wilt, root rot and leaf spot diseases started appearing in severe form. Though viral diseases like cucumber mosaic virus, alfalfa mosaic virus, lettuce mosaic virus, aster yellows, tobacco streak virus and few bacterial diseases are reported on safflower, they have never caused economic losses.

Table 11.2. Fungal, viral and bacterial diseases recorded on safflower in India

Disease	Pathogen	Reference
Fungal diseases		
Fusarium wilt	Fusarium oxysporum Schelecht. f. sp. carthami	Singh et al., 1975
Sclerotinia wilt	Sclereotinia sclerotiorum (Lib.) de Bary, Sundararaman and Ramakrsihnan	Rangaswami, 1993
Alternaria leaf blight	Alternaria carthami Chowdhury	Chowdhury, 1944
Phytophthora root rot	Phytophthora drechsleri Tucker	Balakrishnan and Krishnamurthy, 1947
Brown leaf spot	Ramularia carthami Zaprometov	Mohanty and Das, 1958
Cercospora leaf spot	Cercospora carthami (H. and P. Sydow)	Sundararaman and Ramakrishnan, 1928
Root rot	Rhizoctonia bataticola (Taub.) Butler	AICRP, 1982
Powdery mildew	Erysiphe cichorecearum DC	Saluja and Bhide, 1962
Rust	Puccinia carthami Corda.	Prasada, 1947
Pythium root rot	P. aphanidermatum (Eds.) Fitz	Ranganathan et al.,1973
Anthracnose	Colletotrichum capsici (Syd.) But. and Bisby	Ramakrishnan, 1941
Leaf spot	Alternaria alternata (Fr.) Keissler Alternaria zinniae	Kore and Zinjurde, 1981 Krishna Prasad, 1988
Root rot	Fusarium equiseti (Corda) Sacc. Fusarium solani Rhizoctonia bataticola (Taub) Butler, pycnidial stage Macrophomina phaseolina (Tassi.) Goid	Singh et al., 1975 Shukla and Bhargava, 1977 Kore and Deshmukh, 1981; Singh and Bhowmik, 1979; Quadri and Deshpande, 1982
	Phytophthora palmivora (Butler) Phytophthora sp.	Knowles, 1955 Balakrishnan and Krishnamurthy,1947
	Rhizoctonia sp.	AICRP, 1982
Storage rots	Aspergillus, Rhizopus, Penicillium Alternaria, Cladosporium	Chandra et al., 1981
Viral diseases		
Mosaic	Cucumber mosaic virus (CMV)	Thangamani et al., 1970
Necrosis disease	Tobacco streak virus	Chander Rao et al., 2003
Bacterial diseases		
Leaf spot	Pseudomonas syringae van Hall	Gnanamanickam and Kandaswamy, 1978

1. Wilt

Wilt is a major soilborne disease in safflower particularly where safflower is cultivated without crop rotation. The disease was first reported in India by Singh *et al.,* (1975).

1.1. Distribution and losses: Fusarium wilt of safflower was first observed in the Sacramento Vallley of California in USA during 1962 (Klisiewicz and Houston, 1962), in Egypt and in India in 1975 (Singh *et al.,* 1975). Wilt is widely distributed in safflower growing areas of Maharashtra state with an incidence of 25 to 40% (Nirmal *et al.,* 1989; Deokar *et al.,* 1991). The disease was also reported in Uttar Pradesh (Chakrabarti and Basuchaudhary, 1978). The yield losses due to safflower wilt are not quantified precisely in any country. In India it has appeared as a serious threat to safflower cultivation, destroying up to 25% of plants, amounting to considerable yield loss in the Gangetic valley (Chakrabarti, 1980). The disease caused yield losses ranging from 7.2 to 100% (Sastry and Rama Chandran 1994). Fusarial mycotoxins have been reported to be produced in infected seeds in quantities sufficient to cause mycotoxicosis (Ghosal *et al.*, 1977).

1.2. Symptoms: In seedling stage, cotyledonary leaves show small brown spots either scattered or arranged in a ring on the inner surface and they may be shrivelled, rolled or curved (Chakrabarti, 1980). Symptoms become distinct when plants are in 6th to 10th leaf stage. Unilateral infection on branches and leaves, golden yellow discolouration of leaves, epinasty, vascular browning appearing only on one side of root or stem with unilateral yellowing and drying of plants (Fig. 11.3). The symptoms develop in acropetalous succession. In older plants, lateral branches die, while the other portion of the plant remains apparently healthy. The severely affected plants produce small sized flower heads, which are partially blossomed. On infected side of stem, necrosis of bark initiates above collar region which progresses upwards resulting in drying of branches and splitting of bark. A streak is also seen on infected stem. The vascular regions of infected plants show reddish brown discolouration (Klisiewicz and Houston, 1962; Holdeman and McCarthey, 1964; Singh *et al.*, 1975.

Fig. 11.3. Safflower wilt caused by *Fusarium oxysporum* f.sp. carthamai

1.3. Causal organism: The pathogen is *Fusarium oxysporum* f. sp. *carthami* Klisiewicz and Houston. The fungus can be easily isolated from diseased stem and root portions on potato dextrose agar (PDA). The mycelium is pink or white usually with a purple tinge, sparse to abundant, branched, and septate.

Microconidia are borne on simple phialides arising laterally on the hypha or on short sparsely branched conidiophores, abundant, oval to elliptical, 1-celled, slightly curved, and measuring 5 to 16 x 2.2 x 3.5 µm. The macroconidia are hyaline, may be up to 5 septate, borne in sporodochia, straight or curved, often pointed at the tip with round base, and measure 10 to 36 x 3-6 µm, mostly 28 x 4 to 5 µm. Chlamydospores are one-celled, smooth, faintly coloured, and 5 to 13 x 10 µm. They are formed abundantly and are terminal or intercalary, usually solitary but occasionally formed in chains (Singh *et al.*, 1975). The fungus is specific in its pathogenicity on safflower and six other species of *Carthamus* (Klisiewicz and Houston, 1963).

Physiological races exists in *Fusarium oxysporum* f. sp. *carthami* and 4 races were identified in United States based on reaction of 14 pathogen isolates on safflower varieties Gila, Nebraska 6, UC-31 and US Biggs (Klisiewicz and Thomas, 1970; Klisiewicz, 1975). In India, 3 physiological variants are observed among *Fusarium oxysporum* f. sp. *carthami* isolates. At DOR, Hyderabad, 4 races were identified based on reaction of 6 differentials to 12 isolates (DOR, 1993; Kalpana Sastry and Chattopadhyay, 2003). The number of pathogen isolates considered in these studies is very limited and may not give clear pattern of wilt pathogen race distribution in safflower growing countries. Intensive efforts are needed to study race pattern using exhaustive collection of pathogen isolates from safflower growing areas. Molecular analysis of genetic variability using RAPD, Microsatellite and ITS-RFLP markers revealed 2 distinct groups among 12 isolates of *Fusarium oxysporum* f. sp. *carthami* (Prasad *et al.*, 2004; DOR, 2004). Further, principal coordinate analysis of similarity values generated by RAPD markers of 50 isolates of the pathogen showed 3 groups (DOR, 2005).

1.4. Epidemiology: The fungus enters the host cells by mechanical means, is easier when the plants are in the seedling stage and tissues are soft (Chakrabarti, 1980). Once inside the host, the fungus multiplies and moves to the xylem vessels, where the mycelium is formed in abundance to partially or completely plug the vessels, interrupting the transport of nutrients to the above ground plant parts, thus stunting plant growth. The fungus produces enzymes like polygalacturonase, pectin-methyl esterase, cellulase, protease and amylase, which help in degradation of host cell components and necrosis. Continued dislocation of the flow of plant nutrients lead to plant wilt. The fungus induces the production of tyloses, which, when present in abundance in the xylem vessels, prevent the movement of nutrients. Also a toxic substance called fusaric acid has been shown to be at least partially responsible for the wilt symptoms. The disease is severe in acidic soils with high nitrogen and warm moist weather. The fungus survives in seed (Klisiewicz, 1963), in soil as chlamydospores (Chakrabarti, 1980; Klisiewicz and Thomas, 1970) and in infected plant debris.

1.5. Management

1.5.1. Host resistance: Sources of resistance to wilt in wild and cultivated *Carthamus* species are available. Many resistant lines are identified from cultivated safflower germplasm lines and wild germplasm maintained at DOR, Hyderabad, India and some exotic accessions from US and other countries. Wild safflower species like *Carthamus oxyacantha, C. lanatus, C. glaucus, C. creticus* and *C. turkestanicus* are immune to wilt. Many lines from crosses of oxyacantha-tinctorius and tinctorius-turkestanicus advanced breeding lines showed immunity to wilt. The breeding line 96-508-2-90 showed immune reaction to wilt and is also high yielder (Anjani *et al.,* 2005). Among released Hybrids DSH-129, NARI-NH-1, NARI-H-15 and varieties A1, PBNS-40 and NARI-6 are tolerant to wilt.

1.5.2. Cultural: Deep summer ploughing and burning of infected trash would reduce disease incidence. Chickpea, lentil, pea and wheat, usually grown as mixed crop in India, have been found to decrease the disease incidence by releasing some inhibitory substances and also encouraging microflora antagonistic to pathogen (Chakrabarti, 1976, 1979).

1.5.3. Biological: Many workers reported the effect of antagonists like *Trichoderma viride, Trichoderma harzianum, Trichoderma virens* and *Aspergillus fumigatus* against wilt disease in safflower (Chattopadhyay and Sastry, 1997; Gaikwad and Behere, 2001; Prasad, 2003; Patibanda and Prasad, 2004). *T. harzianum* and *T. viride* seed treatment @ 10g/kg seed with moderately susceptible variety A1 was found to be very effective in reducing wilt incidence and increasing seed yield under field conditions.

1.5.4. Chemical control: Cersan wet, benomyl and vapam check the growth of *Fusarium oxysporum* f. sp. *carthami* (Chakarbarti and Basuchaudary, 1975). Brestan-60, captaf-83 W, ceresan wet, dithane M-45 and thiram + bavistin (1: 1) @ 0.25% are effective in eradication of seed mycoflora of safflower (Singh, *et al.,* 1987). Soaking seeds in calixin or bavistin for 18 hrs is effective in reducing wilt disease and also soaking seeds in a mixture of ceresan + bavistin or ceresan + calixin (1:1) can completely eradicate the seedborne inoculum of the pathogen. Soaking of safflower seed in 0.2% a. i. cersean wet solution for 24 hrs can give good protection against the safflower wilt. Application of benomyl in furrows (2 mg a.i/ 2kg soil) is also highly effective against the pathogen. The fungus is eradicated from the soil when vapam (7.62 ml a.i/ 2kg soil) was applied (Chakarbarti and Basuchaudary, 1978). Carbendazim is effective against *Fusarium* spp. followed by benomyl and thiobendazole and tolerance to *Fusarium oxysporum* f. sp. *carthami* was most rapidly induced with carbendazim and slowly with benlate. Pedgaonkar and Mayee (1991) tested thiram, carbendazim and mancozeb *in vitro* against the pathogen inducing safflower wilt and found that the seed germination is increased and pre emergence mortality is reduced. A new seed dressing formulation of a fine particle carbendazim with a sticker at 0.1% is effective in eliminating seedborne pathogens of safflower (Kalpana sastry and Jayaraman, 1993).

2. Alternaria leaf spot/ leaf blight

2.1. Distribution and losses: Alternaria blight is first reported from India by Chowdhury (1944) subsequently from USSR (Nelen and Vasileva 1960), United States (USDA, 1961), Ethiopia (Ellis and Holliday, 1970), Kenya (Ellis and Holliday, 1970), Australia (Irwin, 1975), Pakistan (Stovold, 1979) and Italy (Zazzerini and Buonaurio, 1981). Available reports indicate considerable yield loss from the disease in USA (Zimmer, 1963; Crowell, 1973) and Ausatralia (Irwin, 1976). In India, the disease is reported to cause 25-60% yield loss every year (DOR, 1984). Survey on intensity of Alternaria leaf blight of safflower in northern India revealed 27-90% yield loss when the disease appears at early stage of crop growth (Krishna Prasad, 1988)

2.2. Symptoms: The initial symptoms of leaf blight in safflower incited by *A. carthami* appears during seedling stage (30 days old) as small isolated light brown dark brown circular spots 1-2 mm in diameter on the lower leaves each gradually spreading to upper leaves (Fig. 11.4). The diameter of the spot gradually increases to about one cm. As the infection progresses, these spots enlarge and coalesce forming bigger spots. There is brown dot in the centre of spot and is surrounded by many dark concentric rings alternating with light ones. In mature spots, shot holes usually appear in the infected area. The disease spreads very fast in the susceptible varieties and the entire plant turns brown and dries without producing any seed (Fig. 11.5). In others, the disease symptoms develop on the leaves occupying the lower half of the stem. They do appear on the upper leaves, but they are old in the form of small-scattered ones. In several cases, the symptoms were also observed on stem as elongated dark brown to black spots and later crack develop in the stem. In floral parts, the symptoms were observed as minute dark brown spots first appearing at the base of involucral bracteoles. These spots later enlarged and spread to other pots of the captiulum. The unopened capitula shrivel and dry up (Krishna Prasad, 1988).

Fig. 11.4. Safflower Leaf Spot caused by *Alternaria carthami* Chowdhury

Fig. 11.5. Field view of Safflower crop effected with Alternaria Leaf Spot

2.3. Causal organism: The pathogen involved is *Alternaria carthami* Choudhury. The conidia are straight or curved, light brown and translucent in shade, and possess a long beak (Chowdhury, 1944). They measure 36 to 171 μm (with beak) and 36 x 99 μm (without beak) in length, and 12 to 28 μm in width. The spores have 3 to 11 transverse septa and up to 7 longitudinal septa

2.4. Epidemiology: The pathogen survives in seed and infected plant debris. Primary infection develops from infected seed; secondary infection takes place through airborne conidia. The inoculum is mainly airborne, spreading by means of conidia, which infect the leaf to cause leaf blight. It is also seedborne externally. Seed infection may take place in advanced stages of kernel formation; such infections becoming deep seated, to constitute a source of internal seedborne infection. The fungus sporulates; the spores thus produced becoming airborne to serve as a source of secondary infection. Secondary infection takes place through airborne conidia. Continuous cloudy and wet weather conditions are conducive for disease development and spread of the disease. The disease becomes particularly severe in irrigated crop and in warmer areas where periods of dew or frequent showers occur.

2.5. Management

2.5.1. Host resistance: The wild species *viz.*, *Carthamus oxyacantha*, *C. lanatus*, *C. glaucus*, *C. creticus* and *C. turkestanicus* are immune to Alternaria blight. Many interspecific derivatives having resistance to the disease were identified (DOR, 2004). A variable degree of resistance is observed in safflower lines EC 32012, NS 133, 1015, 1016, 1021 and 181866A, GMU lines 822, 1921, 163, 487, 624, 645, 1824, 1825 (AICRP, 1982). The hybrid NARI-NH-1 and variety NARI-6 are moderately resistant to the disease.

2.5.2. Cultural: As the pathogen is seedborne, seeds from infected fields should be avoided. Recommended date of sowing should be followed and avoid growing crop in low-lying areas.

2.5.3. Chemical: Seed treatment with thiram or captan @ 0.25%, and need based sprays with carbendazim 0.1% or copper oxychloride 0.25% are effective in the management of the disease (Ayyavoo and Shanmugam, 1984; Indi *et al.*, 1987). Two sprays of dithane Z- 78 and dithane M-45 (0.02%) at an interval of 15 days from 30 days after sowing are also effective in controlling Alternaria blight.

3. Phytophthora root rot (*Phytophthora drechsleri* Tucker)

3.1. Distribution and losses: Phytophthora root rot of safflower is first reported in 1947 in USA (Classen *et al.*, 1949). It is also reported to occur in most of the safflower growing countries including India (Weiss, 1971; Balakrishnan and Krishnamurthy, 1947). Yield losses up to 3% have been reported in US due to Phytophthora root rot and during epidemic years up to 80% of plants have been reported to be killed (Weiss, 1971). In Tamilnadu, India Phytophthora root rot occurred in serious proportions and caused 90-95% loss in yield (Balakrishnan and Krishnamurthy, 1947) and in Karnataka it caused 50% seedling mortality in irrigated areas.

3.2. Symptoms: On succulent plants of 2-3 weeks of age, water soaked spots develop and cortical tissue of lower leaves collapses. Stem softens and plants fall, shrivel and die. On older plants near the bloom stage, black necrotic lesions encompass the roots and some times extend 2 to 5 cm above the ground on the lower parts of the stems. In advanced stages, the vascular tissue and pith also become necrotic and dark-coloured (Erwin, 1952). Leaves of such plants sometimes turn yellow and the entire plant then wilts. The tap and lateral roots of affected plants totally rot (Klisiewicz, 1977).

3.3. Causal organism: The pathogen involved is *Phytophthora drechsleri* Tucker. The mycelium is hyaline, non-septate, and branched. The sporangia are thin walled, non-papillate, pyriform to ovate and measure 24 to 38 µm x 15 to 24 µm. The zoospores measure 10 to 20 µm in diameter. Variation in pathogen and occurrence of physiological races was reported in U.S (Schneider and Zimmer, 1965; Pratt *et al.*, 1974; Thomas and Klisiewicz, 1963).

3.4. Epidemiology: Safflower root exudates stimulate the germination and growth of zoospores. Primary infection takes place when an optimum temperature and humidity condition prevails. The infection hyphae penetrate roots directly (Klisiewicz and Johnson, 1968; Berkenkamp, 1962). Massive intercellular and intracellular spread of the hyphae occurs in cortex and vascular region of hypocotyls and causes cell disruption and collapse of cells. Secondary infection is spread by the sporangia and zoospores, disseminated by irrigation water and windborne rains. The fungus is capable of living saprophytically in the soil, surviving adverse climatic conditions by means of oospores. Induction of water stress predisposes safflower plants to infection by the pathogen (Duniway, 1975, 1977; Davia *et al.*, 1981; Davia and Knowles, 1978). Soil temperature is an important factor influencing the pathogenicity of the *P. drechselri* to safflower. The optimum temperature for disease development is 25 to 30°C (Thomas, 1952). The effect of such water stress conditions can actually be seen under natural conditions when drought conditions prevail before irrigating the crop, followed by the severe development of the disease after irrigation (Fig. 11.6).

Fig. 11.8. Symptoms of Phytophthora blight in Safflower

3.5. Management

3.5.1. Host resistance: No attempts are made in India to develop resistant varieties as disease is of minor importance. But in U. S., many moderately resistant varieties are evolved (Erwin, 1952; Thomas, 1956, 1964; Rubis *et al.*, 1958; Thomas *et al.*, 1960; Beech, 1969). The Biggs safflower variety is found resistant to all races of the pathogen under even heavy flood irrigated conditions (Thomas and Klisiewicz, 1963; Thomas and Allen, 1970; Johnson, 1970). Several safflower germplasm lines in UC 150 and UC 164 series have been reported to be resistant to the disease (Davia *et al.*, 1981).

3.5.2. Cultural: Growing safflower in raised beds, not permitting water to stand in the field after irrigation, following crop rotation and not growing safflower in disease affected fields are some measures to reduce the damage due to the disease

3.5.3. Chemical: Captan seed treatment and soil drenching @ 0.2% is found to reduce pre emergence mortality due to seedling blight and alliette (fosetyl-Al) can inhibit the pathogen (Siddaramaih *et al.*, 1978).

4. Ramularia leaf spot

4.1. Distribution and losses: The disease was reported to occur in several countries including India. In USSR the disease caused severe yield losses. Ramularia leaf spot is one of the major diseases of safflower under irrigated conditions in Maharashtra, which can cause a yield loss of 18-23%. Epiphytotic occurrence of the disease was reported at Phaltan in the Maharashtra state (AICRP, 1982).

4.2. Symptoms: Greyish-chestnut to brown spots of 2 to 10 mm diameter appears on lower leaves (Minz *et al.*, 1961; Rathaiah and Pavgi, 1977). The underside of the spot may show the presence of white growth of the fungus (Fig. 11.7). The spots may coalesce to cause withering of large area of the leaf.

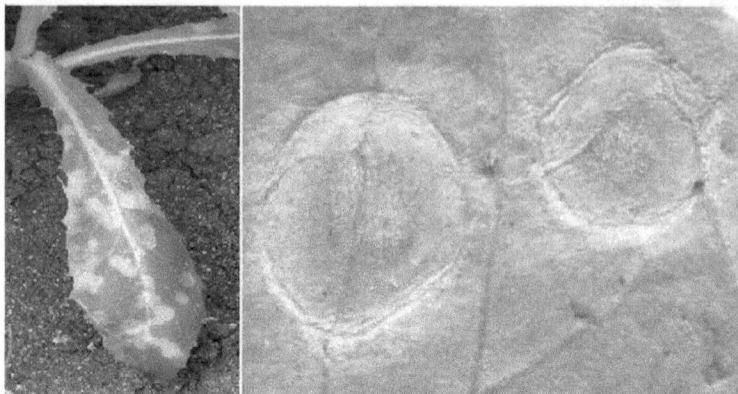

Fig. 11.7. Symptoms of Ramularia leaf spot in Safflower

4.3. Causal organism: The pathogen is *Ramularia carthami* Zaprometov. The hyphae are hyaline, septate, 2 to 3 μm in diameter. The conidiophores are hyaline, unbranched, and measure 15 to 81 x 3 to 5 μm. The conidia are one or two-celled, rarely three-celled, hyaline, cylindrical with rounded apices and measure 14 to 25 x 4.5 to 6 μm (Mohanty and Das, 1958; Khan, 1960). The occurrence of perfect state of the pathogen is also suspected and spermagonia may develop in the old conidial stomata.

4.4. Epidemiology: The pathogen is not carried through seed. The pathogen is viable in debris up to 84 days. The pathogen survives in the form of sclerotia. The inoculum is mainly airborne spreading by means of conidia. The fungus invades the leaf by penetration of stomata. Once inside the host, the sclerotia are formed by continued multiplication of cells of a single hyphal branch just below the epidermis. Disease development was favoured at a temperature of >28°C coupled with high humidity (Patil and Hegde, 1988). The disease incidence is greater under irrigated conditions than rainfed conditions. Disease increases with increased number of irrigations.

4.5. Management

4.5.1. Host resistance: The wild safflower lines *C. oxyacantha* and *C. flavescens* are resistant to Ramularia leaf spot. In India, safflower lines NS133, 140E, 999 and 1021 are reported to be moderately resistant to brown leaf spot.

4.5.2. Chemical: Spraying the crop with copper oxychloride (0.3%) or mancozeb (0.25%) has been found to give satisfactory control of the disease. Three sprays of mancozeb (0.2%) and carbendazim (0.05%) at intervals of 15 days starting at 55 days after sowing is also effective in controlling the disease.

5. Cercospora leaf spot

5.1. Distribution and losses: The Cercospora leaf spot of safflower is worldwide in occurrence and appears in severe form when safflower is grown in a large area as sole crop. It is reported to occur in Ethiopia, India, Israel, Kenya, Philippines, USSR and US. Cercospora leaf spot first reported in India on safflower in 1924

(McRae, 1924). The disease is observed in epidemic forms at Coimbatore during 1921, 1924 and 1925. The disease was recorded in Maharashtra and Karnataka states where safflower is grown on large scale (Karve, 1980).

5.2. Symptoms: Symptoms on leaves are characterized by formation of circular to irregular brown sunken spots measuring 3 to 20 mm in diameter. The spots have a yellowish tinge at the border and they are some- time zonate. In advanced stages, the leaves turn brown and distorted. Under moist conditions, the spots have a velvety greyish-white appearance caused by sporulation of the fungus. Flower buds and bracts are affected in severe cases. Minute black fructifications of the pathogen may be seen on spots on both surfaces. Stems and nodes may also be affected. Flower buds turn brown and dry. The entire capitulum may also be affected without formation of seed.

5.3. Causal organism: The pathogen is *Cercospora carthami* (H. and P. Sydow) Sundararaman and Ramakrishnan. The mycelium is hyaline, septate, branched, and collects in the stromatal areas where stromata are formed. The conidiophores emerge separately or in fascicles (tufts of 12 to 20 conidiophores) on both surfaces of leaves. The conidia are hyaline, linear, 2 to 20 septate, and are borne on the conidiophores acrogenously. They are broad at the base and taper towards the end in a whip-like manner, measuring 2.5 to 5 μ x 50 to 300 μ. C. tinctorius is the lone known host for *C. carthami* (Rathaiah and Pavgi, 1973).

5.4. Epidemiology: The pathogen perpetuates through seed, as vegetative saprophytic mycelium and through viable stromata embedded in crop debris, which serves as primary sources of inoculum for infection (Sundararaman and Ramakrishnan, 1928). Secondary spread is through airborne conidia. Infection of the bracts, stem and node causes withering of the plant parts. They enter the susceptible plants through the stomata on the underside of the leaves. Wet and humid weather favours disease development.

5.5. Management: Spraying the crop with copper oxychloride (0.3%) or mancozeb (0.25%) has been found to give satisfactory control of the disease. The disease can also be controlled by spray application of 1 % Bordeaux mixture and 0.1 % carbendazim.

6. Root rot

6.1. Distribution and losses: The Rhizoctonia root rot caused by *Rhizoctonia bataticola* (Taub) Butler is sporadic all over the country and the pathogen has got wide host range. The disease regularly causes 1-10% yield losses.

6.2. Symptoms: Symptoms are seen as wilt in patches in the field conditions. If such plants are pulled out, rotting of roots and stems at crown level is seen. The tissues are weakened and break of easily. Roots become black and shreds. In affected stem, black microsclerotia are found in the pith that rots and becomes hollow. In advanced cases sclerotial bodies may be seen scattered on the affected tissues.

6.3. Causal organism: The pathogen involved is *Rhizoctonia bataticola* (Taub) Butler. The pycnidial stage of the pathogen is *Macrophomina phaseolina* (Tassi.) Goid. The fungus invades the host both inter- and intra- cellularly. It grows rather fast, covering large areas of the host tissues and eventually killing them in a short time. It produces numerous sclerotial bodies on the host tissue, which measure about 110-130μ in diameter. Often, the conidial or pycnidial stage is produced on the host. The pycnidia are dark brown, ostiolate and of varying size, depending upon the host. The pycniospres are elliptical, thin walled, single celled, hyaline and measure 10 – 42 × 6-10μ

6.4. Epidemiology: Infected soil and plant debris forms the source of inoculum. The fungus is saprophytic and survives in soils in the absence of host plants as sclerotia or as mycelium colonizing plant debris. Infection begins soon after seeds are planted. After emergence, the primary point of invasion may be in the hypocotyl area at the soil line. If seedlings survive the damping-off phase, infections may continue to develop into root rot when lesions expand down to the root system. Soil temperature of 35°C and soil moisture in the range of 15-20% are favourable for disease development. In *R. bataticola* infested soil, seedling infection gradually decreases with increasing soil moisture. Infection was 100% at 30% soil moisture, but none occurred at 90%. Saprophytic activity of the fungus is highest at low soil moisture.

6.6. Management

6.6.1. Host resistance: Resistance sources are not available either in cultivated or wild safflower. Germplasm lines GMU-8, 15, 23, 38, 42, 48, 60, 66, 69 and 73 are tolerant to the disease (Deokar *et al.*, 1991).

6.6.2. Cultural: Removal and destruction of infected plants will help in minimizing the inoculum load in the field. Deep ploughing during summer months will expose resting structures of the pathogen to high temperature and inactivate them. Avoiding water stress during crop growth can reduce the disease incidence. Crop rotation with non-host crops of the pathogen will reduce inoculum load of the pathogen in the soil. Using recommended dose of fertilizers reduces the infection levels. Ammonical form of nitrogen and phosphorus as single super phosphate are effective in managing the disease.

6.6.3. Biological: Seed treatment with biological control agent *Trichoderma viride* formulations at 4g/kg seed will also be helpful in managing root rot.

6.6.4. Chemical: Practically no field control is available by using chemicals. However, to minimize seedborne inoculum, thiram seed treatment at 2-4g/kg seed is recommended.

7. Powdery mildew (*Erysiphe cichoracearum* DC)

7.1. Distribution and losses: Powdery mildew of safflower caused by *Erysiphe cichoracearum* DC is reported from Afghanistan, France, India, Israel, USSR and US. No information is available on yield losses caused by this disease in safflower.

7.2. Symptoms: The disease appears in the form of dirty white fluffy patches on both the sides of leaves, the initial infection starts on the lower side of the leaves. Under favourable environmental conditions, the entire stem and leaves are covered with the fungus. Early infected plants remain small in size and produce less number of capsules.

7.3. Causal organism: A biological form of *Erysiphe cichoracearum* DC affecting safflower has been named as *E. cichoracearum* f. sp. *carthami* Milovtzova (Milovtzova,

1937). The mycelium is greyish white, septate, and profusely branched, but disappears when perithecia are formed. Mature perithecia are black with numerous hypha-like appendages, spherical or slightly depressed, and measure 103 to 154 μ with an average of 136 μ. The asci are enclosed in perithecia, stalked, hyaline, narrow to broadly ovate, and measure 47 to 70 x 23.5 μ to 38.5 μ. The ascospores are hyaline to light yellow and measure 14 to 32.5 μ x 11 to 21 μ.

7.4. Epidemiology: The pathogen survives the off-season in the cleistothecial stage. The ascospores released during the crop season bring about primary infection, and secondary infection is from the windborne conidia. As the crop advances to maturity, the pathogen produces perithecia, which remain on the debris after harvest to provide the primary source of inoculum for the next season. High temperature, low humidity low or no rainfall with wind currents are favourable for disease development.

7.5. Management: The disease can be managed by one or two sprays of wettable sulphur (0.2%) or karathane 0.1%.

8. Rust (*Puccinia carthami* Corda)

8.1. Distribution and losses: The rust is one of the important diseases of safflower in many safflower growing countries. It is first reported by Corda attacking *Carthamus tinctorius* L. in Bohemia in 1840 (Arthur and Mains, 1922). Occurrence of this disease is now reported in all safflower growing countries. The disease is more serious in countries where safflower is grown year after year. Severe epidemics of this disease are reported from US (Schuster and Christiansen, 1952), Mexico (Zimmer and Jensen, 1970) and France (Bernaux, 1953; Darpoux, 1946). Field trials with rust resistant and susceptible safflower varieties have shown that rust infected but resistant varieties exhibit stand loss of 26%. But the surviving plants of such resistant varieties have growth compensation ability and loss in yield remains non-significant as compared to the stand loss of 55 to 97% in susceptible varieties with a significantly reduced yields. The average annual yield loss due to safflower rust in the U.S. has been estimated to be about 5%.

8.2. Symptoms: Rust is identified by the formation of orange yellow pustules on seedlings. On leaves many such pustules coalesce to give burnt appearance of the leaves. Pustules are also seen on stems, flowers, and fruits (Fig. 11.8). In mature plants, girdling of the invaded area due to collapse of the tissue is a very characteristic symptom. Such plants remain erect, but leaves are generally in a wilted condition. The plants often break at girdled areas. Presence of the rust pustules has been reported on the underground parts like tap root and lateral roots (Sackston, 1953). The teleutospores are formed in the uredopustules when safflower plant matures giving dark brownish colour to the rust affected plant parts

Fig. 11.8. Rust (Puccinia carthami) in Safflower

8.3. Causal organism: The pathogen is *Puccinia carthami* Corda. *P. carthami* is an obligate pathogen with an autoecious life cycle on *Carthamus* spp. It is macrocyclic rust. The uredosori are found scattered on both sides of the leaves usually near pycnia. The uredosori contain numerous globoid or broadly ellipsoid uredospores measuring 21 to 27 µ x 21 to 24 µ in size. The teleutosori are formed in uredosori. The teliospores are bicelled, ellipsoid, 36 to 44 x 24 to 30 µ in size. Pycnia are formed in groups, are subepidermal, flask-shaped or spherical, 80 to 100 µ diameter. A large number of pycniospores are produced in pycnia. The heterothallic nature of this rust has been proved (Siddiqui and Prasada, 1959). Different races of the rust have been reported in the U.S. and host differentials identified (Zimmer and Leininger, 1965).

8.4. Epidemiology: Though the fungus is autoecious, it repeats the uredial stage. The teleutospores remain dormant on the seed and debris and perhaps play a significant role in the perpetuation of the fungus and initiation of primary infection. The wild plant *C. oxyacantha* may also carry the fungus through the off-season and teleutospores from the wild safflower also initiate primary infection. A week after primary infection by sporidia, orange spots containing spermagonia appear on cotyledons and after two or three days, primary uredosori develop around them. These infect first leaves and serves as primary foci of infection. The secondary spread of the disease is mainly through wind blown uredospores. The optimum temperature for germination of teleutospore is 12 to 18°C (Prasada and Chotia, 1950; Klisiewicz, 1972; Calvert and Thomas, 1954). The optimum temperature for uredospore germination is between 18 and 20°C. Cool temperature and high relative humidity favour the infection.

8.5. Management:

8.5.1. Host resistance: Sources of resistance are available in cultivated safflower and wild safflower. Several commercial varieties have been developed utilizing different sources of resistance in breeding programmes (Ashri, 1971; Singh and Singh, 1974). The variety Sagar Muthyalu (APRR-3) is resistant and JSI-73 is moderately resistant to rust.

8.5.2. Cultural: Early sowing of the crop (September 20) will help in minimizing loss due to rust (Desai and Radder, 1984). Incidence and severity of the disease can be significantly reduced by flooding the soil with water as teleutospores viability is reduced under submerged conditions. Seeds from disease free areas should be used as the disease is seedborne

8.5.3. Chemical: Seed dressing with captafol or thiram @ 0.2% and two to three sprays with systemic fungicides like oxycarboxin or benomyl at an interval of 15 days are found to manage the rust (AICRP, 1982).

Viral Diseases

9. Mosaic (*Cucumber mosaic virus* (CMV)

Safflower mosaic is reported from safflower growing areas in India, Israel, Iran, Morocco and the U.S. Exact information on yield losses is not available. The causal virus is *Cucumber mosaic virus* (CMV). The virus is spherical in shape. The virus is sap transmissible and also transmitted by the aphid *Myzus persicae*. Young leaves show irregular yellow or light green patches alternating with normal green areas. Leaves may become blistered and distorted and infected plants are stunted in growth. In few plants, primary leaves are produced, forming rosette of leaves exhibiting mosaic mottling and from centre of this, the axis bearing secondary leaves is produced. Wild hosts serve as primary source of infection. The virus can be transmitted to healthy safflower crop through sap or aphids. Further, the disease is spread by many repeated cycles of secondary infection through the process of insect transmission. Moist and cloudy weather is more favourable for disease development.Rouging and destruction of the infected plants and spraying of systemic insecticides, monocrotophos 1.6ml or dimethoate 2ml per litre of water for control of aphid vector can manage the disease.

C. Sesame

Sesame is the most ancient oilseed crop of India. India is the largest producer of the seed in the world cultivated over an area of 24.2 lakh ha, with an annual production of 8 lakh MT and 331 kg/ha productivity. Under the present situation there is need to increase production vertically by taking into account the 54% available from exploited yield reservoirs. In this background, crop protection in a given agro climatic situation gains importance. An estimated 40% yield loss has been reported due to various biotic stresses in sesame. There are mainly due to pests and diseases. Among different disease attacked by sesame charcoal rot, Phytophthora blight, wilt, Cercospora leaf spot and bacterial leaf spot are important. To manage these diseases the following control measures are given here. Sesame is the oldest known oilseed crop in India. It is grown during kharif, rabi and summer seasons, but kharif crop is prone to several diseases because of high temperature and relative humidity. The crop is affected by many diseases of which some are most limiting in successful cultivation of the crop. Currently stem and root rot, phtophthora blight, bacterial blight and phyllody are the important diseases.

1. Stem or root rot/ charcoal rot

The disease is caused by *Macrophomina phaseolina* (Tassi) Goid occurring in all sesame growing areas of India (Mehta, 1951) and also reported from Bangladesh (Khan *et al.*, 1976), Cyprus (Nattrass, 1934), Myanmar (Small,1927), Greece (Sarejanni and Cortzas, 1935), Iraq (Al-Ani, 1970), Korea (Yu and Park.,1990), Palestine (Reichert, 1930), Pakistan (Prasad, 1944), Srilanka (Small, W. 1927.), Turkey (Bremer., 1944),Nigeria (Abd-El-Ghany, *et al.*, 1970), Uganda (Ashby,1927) and USA (Altsatt, 1944) It is a very destructive occurring in all sesame growing areas of India. In kharif 1993 and 1994 there was very high incidence of the disease in the states of Rajasthan, Maharashtra and Tamilnadu. Yield loss has been estimated up to 100% when there was 66% infection. High soil temperature of 2530 oC and moisture favour the spread of the pathogen and increase the disease severity. Periods of drought between heavy rains favour disease development. The pathogen has a wide host range and is seed and soil borne.

Seed treatment with carbendazim @ 0.1% a.i., Thiram + Captan or with biocontrol agents like *Gliocladium virens* and *Trichoderma viride* are found effective. Seed treatment with botanicals viz., neem leaf extract @ 1% (w/v) before sowing also can reduce the disease significantly. Soil solarization by covering the soil with transparent polyethylene mulch of 50 u for six weeks during hot summer after ploughing and irrigation helps in controlling the disease and increasing yield under Indian conditions. The strategy is also found to be ecofriendly due its minimal effect on the beneficial microflora of the soil . Inter cropping sesame with moth bean at 1: 1 ratio is effective in managing the disease. Clean cultivation in well drained soil is advisable. There is no source of resistance against the disease located in the available germplasm till date.

2. Phytophthora blight

The disease is caused by *Phytophthora parasitica* (Dastur) var. sesami (Prasad). Its incidence is reported to be up to 100% with up to 79.8% yield loss The disease is known to cause heavy crop losses in the Tikamgarh area of Madhya Pradesh state. The disease can attack at all stages of the plant growth after it attains 10 day age. Initial symptoms are water soaked spots on leaves and stems. Dieback rotting develops very fast and causes blight. Under favourable conditions these brown discoloured spots enlarge rapidly and may coalesce with each other. Heavy rains for two weeks and >90% RH for three weeks favour disease development. The disease is favoured by a soil temperature of 2830°C with the activity of the pathogen decreasing on further rise in temperature. The disease is severe in areas of heavy soil with high rainfall. The pathogen is soil and seed borne. Pathogen mycelium and chlamydospores can survive up to 50 and 52°C respectively. Association of plant parasitic nematodes viz., *Rotylenchulus reniformis* are reported to increase incidence of the disease.

Seed treatment with Apron 35 SD @ 6g/kg is useful. Application of Bordeaux mixture at 3: 3: 50 ratio, copper oxychloride or Ridomil MZ as spray @ 0.3% or even extract of Lawsonia may be used when conditions favourable for the disease prevail with 3 sprays at seven day interval. Soil amendment with mustard cake is reported to cause significant disease reduction. Clean cultivation, destruction of crop residues and crop rotation is advisable. There are some sources of resistance against the disease in the germplasm screened in India, viz., TKG22.

3. Phyllody

The disease caused by phytoplasma. The pathogen is transmitted in a persistent manner by a leaf hopper *Orocius albicintus*. It is a serious disease capable of causing heavy yield losses; 1% increase in disease incidence reduces yield by 8.36 kg/ha.

Presently, its incidence ranges up to 20% in India. Affected sesame plants express symptoms depending on the growth and time of infection. Early infection results in stunting to about twothirds of a normal plant and the entire plant may be affected. The inflorescence is totally replaced by a growth consisting of short, twisted leaves closely arranged on a stem with very short internodes. Infected plants show excessive branching and shortening of internodes. Such plants do not generally bear capsules. Under Coimbatore conditions maximum population of inoculative insects (60%) was reported in midFebruary and highly significant correlation was noted between hopper population and phyllody incidence. Early flowering disease resistant lines need to be used. *Sesamum mulayanum* has been reported to be a source of resistance to the pathogen. Late sowing with soil treated with phorate 10 G @ 10 kg/ha at sowing time and spraying of oxydemeton methyl 0.05 %. Monocrotophos (0.01% a.i.) at flowering are effective. Also spraying of 500 ppm of tetracycline or neem oil (1%) at flowering is effective.

4. Alternaria leaf blight

The disease is distributed worldwide and is among the most destructive diseases of the crop. It was first reported in India by Mohanty and Behera (1958), and is widely prevalent in Maharashtra and Orissa (Deshopande and Shinde, 1976). It is also reported from EI Salvador, Ethiopia and the USA (Kinman and Martin, 1954).The pathogen attacks all parts of the plant at all stages. Small, dark brown water soaked, round to irregular lesions, 1-8mm in diameter with target-board look appear on the leaves and spread fast to cover the entire stem length, giving a blighted appearance. The lesions may also appear on the midrib and veins of the leaves without showing the typical leaf spotting. Milder attacks cause only defoliation while in severe cases the plants may die.

The disease is caused by Alternaria sesame (Kawamura) Mohanty and Behera. Conidiophores are pale brown, cylindrical, simple, erect, 0-3 septate, measure 30-54 x 4-7μ and bear conidia at the apex. Single or chains of conidia are straight to curved, obclavate, yellowish brown to dark olivaceous brown, 30-120 x 9-30μ with 4-12 transverse and 0-6 longitudinal septa. Physiology of the pathogen has been studied (Mohapatra *et al.*, 1977) and variations among the geographical isolates of the pathogen have been indicated (Sehgal. and Daftari,, 1966). The pathogen is seed borne (Leppik and Sowell, 1964) and is favoured by cool high humid conditions (Mehta and Prasad, 1976). The pathogen infects the seed embryo and even sporulates inside the seed; the dormant mycelium occurs in the sub-epidermal layer of the seed coat (Singh *et. al.*, 1980). Seed treatment with thiram @ 3g/kg seed or captan (0.25%) and foliar spray of mancozeb @ 0.2% a.i. (Abraham, *et al.*, 1976) at 35, 50 and 65 days after sowing helps to reduce the disease spread. Use of resistant varieties viz., RT-46 and HT-2 and intercropping with castor at 1: 4 ratio is suggested.

5. Powdery mildew

The disease starts with small whitish spots on the upper leaf surface. These spots coalesce to form larger patches finally covering the entire leaf surface with the powdery fungal growth. In severe cases the entire plant may be affected. The disease is caused singly or in combination(s) of *Oidium erysiphoides* Fr. [reported from Ethiopia (Ciccarone, 1940), India (Mehta, 1951), Iraq, Japan (Uozumi and Yoshii,1952.), Tanzania (Weiss,1971) and Uganda (Snowden, 1927)], *Sphaerotheca fuliginea* (Schlecht.) Pollacci [reported from India (Gemawat, and Verma, 1972), Malawi (Lawrence, 1951) and Sudan (Tarr, 1954)] and *Leveillula taurica* (Lev.) Trnaud. [reported from India (Patel,*et al.*, 1949), Sicily and Venezuela.

The disease generally appears in the later stages of the crop and hence do not pose any yield loss. Thus, under such conditions no control measure is taken. However, if the disease appears early and congenial conditions prevail, foliar application of wettable sulfur @ 0.2% or dusting sulfur @ 20kg/ha (Shanmugam, *et al.*, 1976) or miltox application (Abraham,1976) is found to be effective.

6. Bacterial leaf spot

The disease is of worldwide distribution and is reported from India (Durgapal and Rao, 1967) and also from Brazil (Pereira,1967), China (Bradbury, 1981), Greece (Zachos and Panagopoulos,1960), Japan (Nakata, 1930), Korea (Bradbury,1981), Mexico (Rodriguez, S. H., 1972), Somalia (Curzi, 1934), South Africa (Gorter, 1977), Sudan (Tarr,1954), Tanzania (Welsford,1932), Turkey (Bremer, *et al.*, 1974), Uganda (Emechebe,1975), the USA (Brown, and Streets, 1934), Yugoslavia (Sutic and Dowson,1962) and Venezuela (Urdaneta and Mazzani,1976). Yield losses from the disease have been reported to be 21.07-27.12% in susceptible cultivars in Rajasthan (Vajavat and Chakravarti, 1978).

Symptoms appear on all above ground parts of the plant. Initially light brown angular spots with dark purple margins appear. The spots are located between the leaf veins. They may merge to cover a large area. Defoliation may occur. Severe stem infection results in death of the plant. Spots on capsules are sunken and shiny. Early capsule infection renders them black and seedless. The pathogen is *Pseudomonas syringae* Van Hall pv. Sesame (Malkoff) Young, Dye and Wilkie. The bacteria is a gram negative rod, motile by 2-5 polar flagella, 0.6-0.8 x 1.2-3.8μ. There are two races of the pathogen in the US (Culp, 1964). The pathogen is seed borne. High rainfall and persistent humidity favours the disease. Spraying urea @ 0.2% results in increased disease severity. An epidemic of the disease was reported from Uttar Pradesh (Singh, 1968).Cultivars "Dulce" and "Margo" were reported to be resistant to the disease (Culp, 1964). White seeded early varieties are more resistant. Seed soaking in 250 ppm streptomycin or 250 ppm agrimycin + 500 ppm ceresin wet provides good control (Durgapal, *et al.*, 1969). Streptocycline spray at 0.3g in 25 gallons water also gives effective control. Hot water seed treatment at 52°C for 10 min controls seed brone infection (Singh, 1969).

7. Bacterial blight

The disease is reported from India (Rao,1962), Sudan (Sabet and Dowson, 1960) and Venezuela (Malaguti,1971). Plants of all ages may be attacked. Small water soaked lesions may develop on the cotyledonary leaves. About 4% mortality due to the disease in 4-6 week old plants is reported (Rao,1962). If the seedlings survive, similar symptoms appear on true leaves that turn angular, olive green to black colour. Several spots may coalesce to form large patches. Leaves may dry and turn brittle. Petioles and capsules may also be attacked. *Xanthomonas campestris* (Pamel) Dowson pv. *Sesami* (Sabet and Dowson, 1960). Dye is the pathogen. Bacterium is rod shaped, 0.4-0.6 x0.8-1.6μ, motile by a single polar flagellum. Two strains of the pathogen are reported from India. One to four plasmids of varying mobility with molecular weights of 48.5 kb, 12.9 kb, 3.0 kb and 1.97 kb have been identified (Sheela., *et al.*, 1994).The pathogen is seed borne (Rao and Durgapal,1966) and secondary spread occurs through spattering of rain water. High temperature and humidity favour the disease incidence. Seedling infection is severe at 20°C soil temperature and 30-40% soil moisture with 75-87% RH. Plants with higher stomatal frequency and higher nitrogen content are more susceptible (Shukla, *et al.*, 1975).T-58 appears to be resistant (Jain and Kulkarni,1967). Copperoxychloride (0.5%) and captafol (0.16%) are useful in disease control. Other chemical control measures are same as bacterial leaf spot.

REFERENCES

Abd-El-Ghany, A.K., Ess-El-Rafei, M., Bekhit, M.R. and El-Yamany, T., 1970. Studies on root rot wilt disease of sesame, Agric. Res. Rev. Cario 48: 85.

Abraham, E. V., Shanmugham, N., Natrajan, S. and Ramakrishnan, G. 1976. An Integrated programme for controlling pests and diseases of sesamum. Madras. Agric. J. 63: 532-536.

AICRP. 1970. Annual Progress Report: Castor. Directorate of Oilseeds Research, Hyderabad, India.

AICRP. 1982-2010. Annual Progress Reports: Safflower, Directorate of Oilseeds Research, Hyderabad.

AICRP. 1988. Annual Progress Report: Castor. Directorate of Oilseeds Research, Hyderabad, India.

Akashe, V. B., Deokar, C. D., Patil, M. W., and Shewale, M. R. 1995. Seasonal incidence of safflower aphid, *Uroleucon compositae* in safflower. Madras Agric.J., 82: 232-233.

Al-Ani, H.Y., Natour, R.M. and El-Behandli, A.H. 1970. Charcoal rot of sesame in Iraq. Phytopath Mediterr, 9: 50.

Altsatt, G.E. 1944. Ashy stem blight and a bacterial leaf spot of sesame in Texas. Plant Dis. Reptr. 28: 1104.

Andreeva, L.T. 1979. *Fusarium* wilt of *Ricinus communis*. Zashchita Rasternil 7, 22.

Anjani, K., Singh, H., and Prasad, R. D. 2005. Performance of multiple resistant line of safflower (*Carthamus tinctorius* L.). Indian J. Agric. Sci. (in press).

Archibald, R.G. 1927. The castor oil plant (*Ricinus communis* L.)-Black rot in the Gazira, *Tropical Agriculture* (Trinidad): 4. Pp 127.

Arthur, J. C., and Mains, E. B. 1922. Bullaria, N. Am. Flora, 7: 482.

Ashby S.F. 1927. Macrophomina phaseoli (Maubl) Comb. Nov.,the pycnidial stage of *Rhizoctina bataticola* (Taub) Bult. Trans. Brit. Mycol. Soc. 12: 141-147.

Ashri, A. 1971. Evaluation of the world collection of safflower, *Carthamus tinctorius*L. I. Reaction to several diseases and associations with morphological characters in Israel, Crop Sci. 11: 253.

Ayyavoo, R. and Shanmugam, N. 1984. Control of Alternaria leaf spot of safflower. Pesticieds 18: 60-61

Balakrishnan, M. S., and Krishnamurthy, C. S. 1947. Seedling blight of safflower (*Carthamus tinctorius* L), *Curr. Sci.,* (India), 16: 291.

Beech, D. F. 1969. Safflower. Fld.Crop. Abstr., 22: 107.

Beranaux, P. 1953. A contribution to the study of safflower rust (*Puccinia carthami*) corda. Bull. Soc. Mycol. Fr., 68: 327.

Berkenkamp, B. B. 1962. Studies concerning *Phytophthora* root rot of safflower (*Carthamus tinctorius* L.). Diss. Abstr., 22, 12: 4153.

Bheema Raju, A. 1999. Biology and Management of *Botrytis* Grey Mould of castor (*Ricinus communis* L.). M.Sc Thesis. Acharya, N.G. Ranga. Agricultural University,

Bradbury, J. F. 1981. *Pseudomonas syringae* pv. *Sesami. CMI* Descriptions of Pathogenic Fungi and Bacteria No. 696, Commonwealth Mycol. Inst., Kew, Surrey, England.

Bremer, H. 1944. On wilt diseases in South west Anatolia. Istinbul Yaz. 18: 40.

Bremer, H., Ismen, H., Karel, g., Ozkan, H. and Ozkan, M. 1974. Contribution to knowledge of the parasitic fungi of Turkey. Rev. Fac. Sci. Univ. Istanbul. Ser. B. 13: 122-172.

Brown, J. G. and Streets, R. B. 1934. Diseases of field crops in Arizona. Ariz. Univ. Agric. Exp. Stn. Bull. 141: 181.

Butler, E.J. 1918. Fungi and disease in Plant. Thacker Spink & Co. Calcutta.

Calvert, O. H., and Thomas, C. A. 1954. Some factors affecting seed transmission of safflower rust, Phytopathology 44: 609.

Chakrabarthi, D. K., and Basuchaudhary, K. C. 1975. Laboratory evalution of fungicides against a *Fusarium oxysporum f.sp. carthami*. Proc. Indian Sci. Cong. (Ag. Sci.) 71-71. (Abstract).

Chakrabarti, D. K. 1976. Cultural control of safflower wilt, Intensive Agric.14: 17.

Chakrabarti, D. K. 1979. Non –host crops, their efferct on *Fusarium oxysporum* f. sp. *carthami,* the incitant of safflower wilt, Proc. Natl. Sci. Acad. 45: 633.

Chakrabarti, D. K. 1980. Survival of *Fusarium oxysporum* f sp *carthami* in soil. Sci.& Cul. 46: 65.

Chakrabarti, D. K., and Basuchaudhary, K. C. 1978. Incidence of wilt of safflower caused by *Fusarium oxysporum* f.sp *carthami* and its relationship with a age of the host, soil and environmental factors. Pl. Dis. Report., 62: 776-778.

Chander Rao, S., Prasada Rao, R. D., Manoj Kumar, V. J., Raman, D. S., Raoof, M. A., and Prasad, R. D. 2003. First report of tobacco streak virus infecting safflower (*Carthamus tinctorius* L.). Plant Dis. 87: 1396.

Chandra, S., Marang, M., and Srivastava, R. K. 1981. Changes in associated mycoflora and viability of seeds of some oilseeds in prolonged storage. Geobios 8: 200-204.

Chattopadhyay, C. 1995. Collar Rot, *Rhizoctonia solani,* on castor bean (*Ricinus cummunis* L.) plant- A new record.

Chattopadhyay, C. 2000. Seed borne nature of *Fusarium oxysporum* f.sp. *ricini* and relationship of castor wilt incidence with seed yield. Journal of Mycology and Plant Pathology. 30: 265.

Chattopadhyay, C., and Kalpana Sastry, R. 1997. Effect of bioagents on safflower wilt caused by *Fusarium oxysporum* f. sp. *carthami*. IV International Safflower Conference, Bari. 2-7, June 1997, pp 299-302.

Chaudhari, S.M. and Patel, I.D. 1992. Management of castor diseases. Indian Farming, 2: 5-6.

Chowdhury, S. 1944. An Alternaria disease of safflower. J. Indian Bot. Soc. 23: 59.

Ciccarone, A. 1940. Italian East Africa, Plant diseases reported in 1939. Intl. Bull. Plant Prot. 15: 117.

Classen, C. E., Schuster, M. L., and Ray, W. W. 1949. New diseases observed in Nebraska on safflower. Pl. Dis. Rep. 33: 73.

Crowell, C. 1973. Safflower power. Montana State University Co- operative Extension service Bulletin 9: 8-10

Culp, T. W. 1964. Race 2 of *Pseudomonas sesami* in Mississippi. Plant Dis. Reptr. 48: 86

Curzi, M. 1934. Of African fungi and diseases. I. Concerning *Pseudomonas* parasitic on plants in Italian Somaliland. Boll. R. Staz. Pat. Veg. N.S. 12: 173-184.

Dange, S.R.S. 2003. Wilt of castor-An overview. Journal of Mycology and Plant Pathology. 33: 333-339.

Dange, S.R.S., Desai, A.G. and Patel, D.B. 1997. Management of wilt of castor in Gujarat State in India. In: *Proc. Int. Confr. on Integrated Plant. Dis. Mangt for Sustainable Agriculture*, 10-15 Nov. 1997, IARI, New Delhi, India, pp. 107.

Darpoux, H. 1946. A contribution to the study of the diseases of oleiganous plants in France, Abstr. Rev. Appl. Mycol. 25: 416.

Das, N.D. and Prasad, M.S. 1989. Assessment of yield losses due to root rot, *Macrophomina phaseolina* (Tassi) Goid. of castor in water shed areas. Indian Journal of Plant Protection, 17: 287-290.

Das, N.D. and Shankar, G.R.M. 1990. Effect of root disease (*Macrophomina phaseolina*) of castor (*Ricinus communis*) by statistical modeling of cultural practices and carbendazim application. Indian Journal of Mycology and Plant Pathology, 20: 234-240.

Das, N.D. and Srivastava, N.N. 1992. Effect of some environmental factors on root/ charcoal rot disease (*Macrophomina phaseolina*) of castor. Indian Journal of Plant Protection, 20: 232-233.

Dastur, J.F. 1913. *Phytophthora parasitica sp. Nova*, a new disease of castor oil plant. Mem Dep. of Agric., India Botany Series 7-231.

Davia, D. J. and Knowles, P. F. 1978. The effect of water stress on resistance to *Phytophthora*, root rot of safflower, Agron. Abstr. 93.

Davia, D. J., Knowles, P. F. and Klisiewicz, J. M. 1981. Evaluation of the world safflower collection for resistance to *Phytophthora*. Crop Sci. 21: 226.

Deokar, C. D., Shinde, P. B. and Akashe, V. B. 1991. Screening of safflower germplasm against Rhizoctonia root rot under lab conditions. Sesame and Safflower Newsletter, 7: 72-75.

Desai, A.G., Dange, S.R.S., Patel, D.S. and Patel, D.B. 2003. Variability in *Fusarium oxysporum* f.sp. *ricini* causing wilt of castor. Journal of Mycology and Plant Pathology. 33: 37-41.

Desai, M.V. and Shah, H.M. 1963. Bacterial leaf blight of castor beans. Current Science, 32: 474.

Desai, S. A. and Radder, G. D. 1984. Relative performance of sunflower and safflower varieties with reference to date of planting and disease incidence. Pl. Path. Newsletter, 8: 12.

Deshpande, G.D. and Shinde, D.D. 1976. Occurrence of cultural strains of *Alternaria sesame* (Kawamura) Mohanty and Behera in Maharashtra. J. Maharashtra Agric. Univ., (India) 1: 124-125.

Dhingra, O.D. and Sinclair, J.B. 1978. Biology and pathology of *Macrophomina phaseolina* Universidade Federal De Vicosa. Minas Gerais, Brasil pp. 166.

DOR. 1984. Annual Report: Directorate of Oilseeds Research, Hyderabad, pp..

DOR. 1993. Annual Report: Directorate of Oilseeds Research, Hyderabad, pp.106.

DOR. 2000. Annual Report. 2000. Directorate of Oilseeds Research, Hyderabad, India.

DOR. 2002. Annual Report. 2002. Directorate of Oilseeds Research, Hyderabad, India.

DOR. 2004. Annual Report: Directorate of Oilseeds Research, Hyderabad, pp.120

DOR. 2005. Annual Report: Directorate of Oilseeds Research, Hyderabad, (in press).

DOR. 2005. Package of practices for increasing production of safflower. Directorate of Oilseeds Research, Hyderabad, pp. 21.

Duniway, J. M. 1975. Water relations in safflower during wilting induced by *Phytophthora* root rot. Phytopathology 65: 886.

Duniway, J. M. 1977. Predisposing effect of water stress on the severity of *Phytophthora* root rot in safflower. Phytopathology 67· 884.

Durgapal, J. C. and Rao, Y. P. 1967. Bacterial leaf spot of sesamum (*Sesamum orientale* L.) in India. Indian Phytopath. 20: 178-179.

Durgapal, J. C. and Rao, Y. P. and Singh, R. 1969. Eradication of infection of Pseudomonas sesami from sesamum seeds. Indian Phytopath. 22: 400-402.

Ellis, M. B., and Holliday, P. 1970. Description of pathogenic and bacteria. No 241, Kew Surrey, England: Common wealth Mycological Institute.

Emechebe, A. M. 1975. Some aspects of crop diseases in Uganda, Kampala. Uganda Makerere Univ.

Erwin, D. C. 1952. *Phytophthora* root rot of safflower. Phytopathology 42: 32.

Gaikwad, S. J., and Behere, G. T. 2001. Biocontrol of wilt of safflwer caused by *Fusarium oxysporum* f. sp. *carthami*. In: Proceedings of Vth International Safflower Conference, 63-66pp.

Gemawat, P. D. and Verma, O. P. 1972. A new powdery mildew of *Sesamum indicum* incited by *Sphaerotheca fuligenea*. Indian J. Mycol. Plant Pathol. 2: 94.

Ghosal, S., Chakrabarti, D. K., and Chaudhary, K. C. B. 1977. The occurrence of 12,13-epoxytrrichothecenes in seeds of safflower infected with *Fusarium oxysporum* f.sp. *carthami*. Experientia 33: 574.

Gnanamanickam, S. S., and Kandaswamy, T. K. 1978. Bacterial leaf spot and stem blight- a new disease of safflower in India caused by *Pseudomonas syringae*, Curr. Sci. 47: 506.

Godfrey, G.H., 1919. *Sclerotinia ricini* n. sp. parasitic on castor (*Ricinus communis*), Phytopathology. 9: 565.

Godfrey, G.H., 1923. Gray mold of castorbean. Journal of Agricultural Research, 23 (9): 679.

Golenia, A. 1955. Sclerotia of greymold (*Botrytis cinerea* Persoon) castorbean. Acta Microbio Polonica., 4: 153.

Gorter, G. J. M. 1977. Index of plant pathogens and the diseases they cause in cultivated plants in South Africa. Sci. Bull. Dep. Agric. Technol. Services Republic South Africa 392: 177.

Holdeman, Q. I., and Mc Carthey, W. O. 1964. Diseases of safflower, Bureau of Plant pathology, Dep. Agric., Californis, 26.

Indi, D. V., Lukade, G. M., Shambharkar, D. A., and Patil, P. S. 1987. Compartive bioefficacy of some fungicides against *Alternaria* leaf spot of safflower (*Alternaria carthami*) in dryland conditions. Pesticides 21: 24-26.

Irwin, J. A. G. 1975. *Alternaria carthami* on safflower. Australian Path. Soc. News Lett. 4: 24-25.

Irwin, J. A. G. 1976. *Alternaria carthami*, a seed borne pathogen of safflower. Australian J. Exper. Agri. 16: 921-925.

Jain, A. C. and Kulkarni, S. N. 1967. Varietal reaction of sesamum (*Sesamum orientale* L.) to bacterial blight JNKVV Res. J. (India) 1: 181-182.

Jindal, J.K. and Patel, P.N. 1972. New albino strain of *Xanthomonas ricinicola* (Elliott) Dowson, an incitant of bacterial leaf spot disease in castor. Indian Phytopathology. 25: 152-154.

Johnson, L. B. 1970. Symptom development and resistance in safflower hypocotyls to *Phytophthora drechsleri*. Phytopathology 60: 534.

Kalpana Sastry, R., and Chattopadhyay, C. 2003. Development of Fusarium wilt reistant genotypes in Safflower (*Carthamus tinctorius*). Eur. J. Pl. Path. 109: 147-151.

Kalpana Sastry, R., and Jayaraman, J. 1993. Eradication of wilt fungus from heavily infected safflower seeds. J. Oilseeds Res. 10: 227-231.

Kalpana Sastry, R., and Ramchandram, M. 1992. Differential genotypic response to progessive development of wilt in safflower. J. Oilseeds Res. 9: 297-305.

Karve, A. D. 1980. Resistance of safflower (*Carthamus tinctorius* L.) to Insect and diseases *Final Technical report of N.A.R.I.* Phaltan. Annual Linseed and Safflower Reaesrach Worker's Group Meeting, Ranchi, September 16-20.

Khalil, O. and Sabet, K.A. 1971. A study of certain etiological aspects of bacterial leaf-blight disease of castorbean. Sudan Agriculture Journal, 6: 26.

Khan, A.L., Fakir, G.A and Thirumalachar, M.J. 1976. Comparative pathogenicity of two strains of *Macrophomiona phaseolina* from sesame. Bangladesh. J. Bot. 5: 77-81.

Khan, S. A. 1960. A leaf spot of safflower. *Pak.J. Sci.Res.,* 12: 130.

Kinman, M.L. and Martin, J.A., 1954. Present status of sesame breeding in the united States. Agron. J. 46: 24.

Klisiewicz, J. M. 1972 Effect of host plant materials and temperature on germination of teliospore of *Puccinia carthami*. Phytopathology 62: 436.

Klisiewicz, J. M. 1977. Identity and relative virulence of some heterothallic *Phytophthora* species associated with root and stem rot of safflower. Phytopathology 67: 1174.

Klisiewicz, J. M., and Houston, B. R. 1962. *Fusarium* wilt of safflower. Plant Dis. Rep. 46: 748.

Klisiewicz, J. M., and Houston, B. R. 1963. A new form of *Fusarium oxysporum*, Phytopathology 53: 241.

Klisiewicz, J. M., and Johnson, L. B. 1968. Host- parasite relationship in safflower resistant and susceptible to *Phytophthora* root rot. Phytopathology 58: 1022.

Klisiewicz, J. M., and Thomas, C. A. 1970. Pathogenic races of *Fusarium oxysporum* f.sp. *carthami*. Phytopathology 60: 83.

Knowles, P. F., Klisiewicz, J. M., and Hill, A. B. 1968. Safflower introductions resistant to Fusarium wilt. Crop Sci. 8: 636-637.

Kolte, S. J. 1985. Diseases of Annual Edible Oilseed Crops. III. Sunflower, Safflower and Nigerseed Diseases, 97-126.

Kolte, S.J., 1995. Castor diseases and crop improvement. Shipra Publications, Delhi, pp. 119.

Kore, S. S., and Deshmukh, R. W. 1981. Charcoal rot of safflower caused by *Macrophomina phaseolina*, Abstr., 3rd Int. Symp. on Plant Pathol., New Delhi, December 14 to 18.

Kore, S. S., and Zinjurde, D. D. 1981. Investigation on leaf spot disease of safflower (*Carthamus tinctorius*) caused by *Alternaria alternata*, Abstr., 3rd Inst. Symp. On Plant Pathol., New Delhi, December 14 to 18 , 172.

Krishna Prasad, N. V. 1988. Studies on the alternaria leaf blight of safflower (*Carthamus tinctorius* L.) Ph.D., Thesis, Varanasi: Benaras Hindu University. P.168.

Kulkarni, N.B., Patil, B.C. and Ahmed, L. 1966. Studies on the pycnidial formation by *Macrophomina phaseolina* (Maubl.) Ashby Development of pycinidia and pycnidiospores. Mycopatho. Mycol. Appl. 28: 337-341.

Lawrence, E. 1951. Report of the Acting Director of Agriculture. Report of the department of Agriculture, Nyasaland, for the year 1949.

Leppik, E. E. and Sowell, G. 1964. *Alternaria sesami*, a serious seed borne pathogen of world-wide distribution. FAO Plant Prot. Bull. 12: 13-16.

Liang, P.Y. 1964. Identification of *Phytophthora* species causing boll rot and castor bean blight in North China. Acta Phytopath. Sinic. 7: 11-20.

Maiti, S. and Raoof, M.A. 1984. Aerial infection of *Macrophomina phaseolina* on castor. Journal of Oilseeds Research (1): 232-233.

Maiti, S., Hegde, M.R. and Chattopadhyay, S.B. 1988. Handbook of annual oilseed crops. Oxford and IBH, New Delhi.

Malaguti, G. 1971. A severe bacterial disease of sesame in Venezuela. Agron. Trop. 21: 333-336.

McRae, W. 1924. Economic Botany Part III. Mycology Repoty for the year 1922-23, Board Scientific Advice, India, 31pp.

Mehta, P. R. 1951. Observations on new and known diseases of crop plants of the Uttar Pradesh. Plant Prot. Bull. New Delhi 3: 7-12.

Mehta, P. R. and Prasad, R. N. 1976. Investigations on the *Alternaria sesami* causing leaf blight of til. Proc. Bihar Acad. Agril. Sci. 24: 104-109.

Mehta, P.R. 1951. Observations and new and known diseases of crop plants of the Uttar Pradesh. Plant Prot. Bull., New Delhi 3: 7-12.

Mehtab Yasmeen, Raoof, M.A and Rana Kausar. 2003. Host range of *Botrytis ricini* Godfrey, castor grey mold pathogen. National symposium on plant pathogens diversity in relation to plant health, Osmania University, Hyderabad January 16-18, 2003 (*Abst*).

Milovtzova, M. O. 1937. New species of fungi on the medicinal and essential oil plants of the Ukraine, Trav. Inst. Bot. Univ. Kharkoff, 2: 7.

Minz, G., Chorin, M., and Solel, Z. 1961. Ramularia leaf spot of safflower in Israel. Bull. Res. Coun. Israel, Sect. D. 10: 207.

Mohanty, N. N., and Das, S. N. 1958. Leaf spot of safflower. Sci. Cult. 24: 284.

Mohanty, N.N. and Behera, B.C. 1958. Blight of sesame (*Sesamum orientale* L.) caused by *Alternaria sesame* (Kaw). Current Sci. 27: 492-493.

Mohapatra, A., Mohanty, A. K. and Mohanty, N. N. 1977. Studies on the physiology of the sesamum leaf blight pathogen, *Alternaria sesami*. Indian Phytopath. 30: 432-434.

Moses, G.J. and Ranga Reddy, R. 1989. Grey rot of castor in Andhra Pradesh. Journal of Research APAU, 17: 74-75.

Moshkin, V.A. 1986. Castor. Amerind Publishing Co. Pvt. Ltd. New Delhi. pp. 315.

Muhammad, S.V. and Dorairaj, M.S. 1967. Inheritance studies in *Ricinus communis* L. 1. Resistance to jassid infestation. Madras Agricultural Journal, 54: 323-325.

Mundkur, B.B. 1959. Fungi and plant disease. Macmillan & Co. Ltd., New York. pp.256

Muthusamy, S. and Mariappan, V. 1994. Effect of soil amendments on the incidence of charcoal rot disease of soybean. In: *Crop Disease, Innovative Techniques and Management*, (Sivaprakasan, K. and Seetharaman, K., Eds.), Kalyani Publishers, New Delhi, pp. 233-235.

Naik, M.K. 1994. Seed borne nature of *Fusarium* in castor. Journal of Mycology and Plant Pathology. 24: 62-63.

Nakata, K. 1930. Comparative studies of *Bact. sesami* with *Bact. solanacerarum* and *Bact. sesamicola*. Annl. Phytopath. Soc. Japan 2: 229-243.

Nanda, S. 1975. Studies on *Fusarium* wilt of castor. Ph.D. thesis, University of Udaipur, Udaipur. pp 216.

Nanda, S. and Prasad, N. 1974. Wilt of castor – A new record. Indian Journal of Mycology and Plant Pathology, 4: 103-105.

Narasimhalu, G. and Bhaskaran, R. 1987. Effect of organic amendments on root rot of soybean caused by *M. phaseolina*. In: Workshop on biological control of plant diseases (Abstr.), March 10-12, 1987 at Tamil Nadu Agricultural University, Coimbatore, pp.3.

Nattrass, R.M. 1934. Annual Report of the Mycologist for the year 1933. Dep. Agric. Cyprus.

Nelen, E. S., and Vasileva, L. N. 1960.The pathogenic fungus flora of the flowering plant in the Far-East Botanical Garden. Rev. Appl. Microbiol. 39: 91.

Nirmal, D. D., Kulthe, K. S., and Patil, K. S. 1989. Occurrence of *Fusarium* wilt of safflower in Marathwada. Indian Phytopath. 42: 344.

Patel , M.K., Kulkarni, Y.S. and Dhande, G.W. 1951. Bacterial leaf spot of castor. Current Science. 20: 20.

Patel, M. K., Kamat, M. N. and Bhide, V. P. 1949. Fungi of Bombay, Supplement 1. Indian Phytopath. 2: 142-155.

Pathak, H.C. 2003. Emerging trends in castor seed development. In: Proc. of Nat. Seminar on Castor seed, castor oil and its value added products. 22nd 2003. Solvent Extract Association of India. Ahmedabad, India.pp54-62.

Patibanda, A. K., and Prasad, R. D. 2004. Screening of *Trichoderma* isolates against wilt pathogen of Safflower, *Carthamus tinctorius* L. J. Biol. Control, 18: 103-106.

Patil, M. S., and Hegde, R. K. 1988. Epidemiological studies on safflower leaf spot caused by *Ramularia carthami* Zaprometov under irrigated conditions. J. Oilseeds Res. 5: 45-48

Pedgaonkar, S. M., and Mayee, C. D. 1991. Screening of safflower genotypes and efficacy of seeds dressing fungicides against safflwer wilt. J. Oilseed Res. 265-270.

Pereira, A. L. G. 1967. A study of *Pseudomonas sesami*, causal agent of a bacteriosis of sesame. Arq. Inst. S. Paulo 34: 113-125.

Poole, D.D. 1954. Disease observations on castorbeans in 1953. Plant Disease Reporter 38 (3): 218.

Prasad, N. 1944. Studies on root rot of cotton in Sind. II. Reduction of root rot of cotton with root rot of other crops. Indian J. Agric. Sci. 14: 388-391.

Prasad, R. D. 2003. Potential of *Trichoderma* species as biocontrol agents of safflower wilt. In: Extended summaries: ISOR National seminar on Stress management on Oilseeds for attaining self-reliance in vegetable oils. January 28-30, 2003. Indian Society of Oilseeds Research, Hyderabad pp 134-135.

Prasad, R. D., Sharma, T. R., Jana, T. K., Prameela Devi, T., Singh, N. K., and Koundal, K. R. 2004. Molecular analysis of genetic variability in *Fusarium* species using microsatellite markers. Indian Phytopath.57: 272-279.

Prasada, R. 1947. The rust of safflower, *Puccinia carthami* (Hutz.,)Corda. Curr. Sci.16: 292.

Prasada, R., and Chothia, H. 1950. Studies on safflower rust in India. Phytopathology, 40: 363.

Pratt, B. H., Heather, W. A., and Shepherd, C. J. 1974. Pathogenicity to three agricultural plant species of *Phytophtora drechsleri* isolates from Australian forest communities. Aust . J. Bot. 22: 9.

Purkayastha, R.P. and Menon, U. 1985. Extraneous supply of organic additives affecting nodulation, phytoalexin synthesis and disease susceptibility of soybean. International Journal of Tropical Plant Diseases. 2: 9-18.

Pushpavati, B. 1995. Role of *Fusarium oxysporum* f. sp. *ricini* (Wr.) Gordon in castor wilt complex. M.Sc. (Ag) thesis, Andhra Pradesh Agricultural University, Rajendranagar, Hyderabad. pp. 107.

Quadri, S. M. H., and Deshpande, K. S. 1982. Studies on root /collar rot of safflower caused by *Rhizoctonia bataticola*. Indian Bot. Rep. 1: 141.

Ramakrishnan, T. S. 1941. Studies in the genus *Colletotrichum*. I. Saltation in *Colletotrichum capsici* (Syd.), Proc. Ind. Acad. Sci., Sect. B., 13: 60.

Ranganathan, K., Shanmugam, N., and Marimuthu, T. 1973. A new root rot of safflower in India, Sci. and Cult., (India), 39: 354.

Rangaswamy, G. 1979. Disease of crop plants in India. 2nd Edition, Prentice Hall of India, New Delhi, pp. 520.

Rangaswamy, G. 1993. Diseases of crop plants in India, 3rd Ed. pp.315-336.

Rao, V. G. 1965. The genus *Phyllosticta* in India. Sydovia 19: 117-120.

Rao, Y. P. 1962. Bacterial blight of sesamum (*Sesamum orientale* L.). Indian Phytopath. 15: 297.

Rao, Y. P. and Durgapal, J. C. 1966. Seed transmission of bacterial blight disease of sesamum (*Sesamum orientale* L.) and eradication of seed infection. Indian Phytopath. 19: 402-403.

Raoof, M.A and Rama Bhadra Raju. 2003. Seed treatment for the management of castor wilt. National Seminar on "Stress management in oilseeds for attaining self-reliance in vegetable oils" held at Directorate of Oilseeds Research, Hyderabad during January 28-30,2003 (*Abst*).

Raoof, M.A. and Nageshwar Rao, T.G. 1996. A simple screening technique for early detection of resistance to castor wilt. Indian Phytopathology, 48(4): 82-84.

Raoof, M.A. and Nageshwar Rao, T.G. 1999. Diseases of castor and their integrated management in IPM systems in Agriculture (Eds. Rajeev, Upadhyay, Mukerji, K.G. and Rajak, R.L.) Vol.5 (Oilseeds). Aditya Books Pvt. Ltd., New Delhi. pp 559-574.

Rathaiah, Y., and Pavgi, M. S. 1973. Perputation of speices of *Cercospora* and *Ramularia* parastic on oilseed crops. Ann. Phyotpath. Soc. Jpn. 39: 103.

Rathaiah, Y., and Pavgi, M. S. 1977. Development of sclerotia and spermogonia in *Cercospora sesamicola* and *Ramularia carthami*. Sydowia 30: 148.

Ratnoo, R.S. and Bhatnagar, M.K. 1993. Effect of plant age on susceptibility of cowpea for ashy gray stem blight disease caused by *Macrophomina phaseolina*. Indian Journal of Mycology and Plant Pathology. 23: 205-207.

Reichert, I. 1930. Tilestine: root disease caused by *Rhizoctonia bataticola*. Intnl. Bull.Plant Prot. 4: 17.

Reiuf, P. 1953. A note on wilt of the castor plant. Rev. Path. Veg. 32: 120-129.

Rodriguez, S. H. 1972. Enfermedades parasitarias de los cultivos agricols en Mexico. Folleto Miscelaneo INIA 23: 58.

Rubis, D. D., and Black, D. S. 1958. Gilla, A new safflower variety. Ariz. Agric. Exp. Stn. Bull., 301.

Sabet, K. A. and Dowson, W. J. 1960. Bacterial leaf spot of sesame (*Sesamum orientale* L.). Phytopath. Z. 37: 252-258.

Sabet, K.A. 1959. Studies in bacterial diseases of crops. II. Bacterial leaf blight disease of castor (*Ricinus communis* L.). Annals of Applied Biology. 47: 49.

Sackston, W. E. 1953. Foot and root infection by safflower rust in Manitoba. Plant Dis. Rep. 37: 522.

Salam, M. A.. and Rao, P.N. 1957. Fungi from Hyderabad (Deccan). India. Journal of Indian Botanical Society 36 (3): 421.

Saluja, V. K., and Bhide, V. P. 1962. Powdery mildew of safflower (*Carthamus tinctorius* L.) caused by *Erysiphe cichorecearum* DC in Maharashtra State. Indian Phytopath. 15: 291.

Sarejanni, J.A. and Cortzas, C.B. 1935. A note on the parasitism of *Macrophomiona phaseolina* (Maubl) Ashby. Ann. Inst. Phyto path. Benaki, Greece 1: 38.

Sarwar, H.A.K. 1974. Studies on the root and stem rot disease of castor caused by *Rhizoctonia bataticola* (Taub.) Butl. M.Sc. Thesis submitted to the Andhra Pradesh, Agricultural University, Hyderabad, 49 pp.

Sarwar, H.A.K. and Raju, D.G. 1985. Topsin M-70, the most effective fungicide for the control of *R. bataticola* (Taub) Butler, The root and stem rot disease causing pathogen in castor. Pesticides 19: 56-57.

Sastry, K. and Ramchandran 1994. Effect of wilt on yield attributes of safflower. Indian Phytopathol. 47: 108-110.

Schneider, C. L., and Zimmer, D. E. 1965. Pathogenicity and growth rates of *Phytophthora drechsleri* isolates from safflower and sugarbeet. Plant Dis. Rep. 49: 293.

Schuster, M. L., and Christiansen, D. W. 1952. A foot and root disease of safflower caused by *Puccinia carthami* Cda. Phytopathology 42: 211.

Sehgal, S. P. and Daftari, L. N. 1966. A new leaf spot disease of sesamum. Current Sci. 35: 416.

Shanmugam, N. Natrajan, S. and Ramakrishnan, G. 1976. Fungicidal control of powdery mildew of *Sesamum indicum* L. Madras. Agric. J. 63: 420.

Sheela, P., Amuthar, G. and Mahadevan, A. 1994. Plasmids in *Xanthomonas campesteris* p.v. *sesami*. Zeitschrift fuer Pflanzenkrnkheiten und Pflanzenschutz 105: 482-486.

Shukla, B. N., Chand, J.N. and Kulkarni, S. N. 1975. Effect of leaf age on bacterial blight of sesamum. Indian Phytopath. 28: 3014.

Shukla, D. N., and Bhargava, S. N. 1977. Some studies on *Fusarium solani* (Mart) Sacc. isolated from different seeds of pulses and oilcrops. Proc., Natl., Acad. Sci. 47: 199.

Siddaramaiah, A. L., Desai, S. A., and Bhat, R. P. 1980. Eradication of *Alternaria carthami* Chowdhury, a seed born pathogen of safflower. Pesticides 14: 22-23.

Siddaramaiah, A. L., Hegde, R. K., and Prasad, K. S. K. 1978. Control of seedling blight of safflower. Curr. Res. 11: 92.

Siddiqui, M. R., and Prasada, R. 1959. Heterothallism in *Puccinia carthami* (Hutz.) Corda, the rust of safflower. Indian Phytopath. 12: 59.

Singh, A. K., Chakarbarthi, D. K., and Basuchaudhary, K. C. 1975. Two new diseases of safflower from India. Curr. Sci. 44: 397.

Singh, A., and Bhowmik, T. P. 1979. Occurrence of charcoal rot of safflower in India. Indian Phytopath. 32: 626.

Singh, B. P., and Singh, P. P. 1974.Incidence of rust *Puccinia carthami* on safflower *Carthamus tinctorius* varieties at Jabalpur, Punjabrao krishi vidhyapeeth . J., 3: 81.

Singh, D., Mathur, S. B. and Neergard, P. 1980. Histological studies of *Alternaria sesamicola* penetration in sesame seed. Seed Sci. Tech. 8: 85-93.

Singh, J., Banerjee, A.K. and Swarup, J. 1976. Efficacy of fungicides in controlling bacterial leaf spot and blight of castor and varieties resistant to it. Indian Journal of Farm Science. 4: 125-127.

Singh, R. N. 1968. Outbreaks and new records. FAO Plant Prot. Bull. 17: 138-142.

Singh, R. N. 1969a. Control of bacterial diseases of sesame (*Sesamum orientale* L.) by streptocycline. Telhan Patrika (India) 1: 1-2.

Singh, R.S. 1955. *Alternaria* blight of castor plants. Journal of Indian Botanical Sciences, 34: 130.

Singh, S. N., Agrawal, S. C., and Khare M. N. 1987. Seedborne mycoflora of safflower, their significance and control. Seed Res. 15: 190-194.

Singh, U.B. 1948. Phytopathological examination of seeds by Ulster method. Current Science, 17 (9): 266.

Small, W. 1927. Further occurrence of *Rhizoctonia bataticola* (Taub.) Bulter. Trop. Agric. 69: 202.

Snowden, J. D. 1927. Report of the Acting Mycologist for the period November 10, 1925 to September, 30, 1926. Annual Report Uganda Dep. Agric. for the year ending 31 Dec. 1926.

Stevenson E.C. 1945. *Alternaria ricini* (Yoshii) Hansford the cause of serious disease of the castorbean plant (*Ricinus communis* L.) in the United States, Phytopathology 35 (4): 249.

Stovold, G. E. 1979. The incidence of *Alternaria carthami* on safflower seed in New South Wales and its relation to crop production methods. Rev. Pl. Path. 58: 491.

Sundararaman, S., and Ramakrishnan, T. S. 1928. A leaf spot disease of safflower (*Carthamus tinctorius* L.) caused by *Cercospora carthami* sp. nov. Agric. J. India. 23: 383.

Sutic, D. and Dowson, W. J. 1962. Bacterial leaf spot of sesamum in Yugoslavia. Phytopath. Z. 45: 57.

Tarr, S. A. J. 1954. Diseases of economic plants in Sudan. II. Fibres, oilseeds, coffee and tobacco. FAO Plant Prot. Bull. 2: 161-165.

Thangamani, G. S., Sellammal, I. P., Janki, T. T., Kandswamy, V., Mariappon, and Padmanabhan, C. 1970. A mosaic disease of safflower (*Carthamus tinctorius* L.) in Tamilnadu, Madras Agric. J. 57: 326.

Thomas, C. A. 1952. A greenhouse method of evaluating resistance in safflower to *Phytophthora* root rot. Phytopathology 42: 219.

Thomas, C. A. 1956. The effect of rust on yield and oil content of safflower, Abstr. Phytopathology 46: 229.

Thomas, C. A. 1964. Registration of US 10 safflower. Crop Sci. 4: 446.

Thomas, C. A. 1976. Resistance of VFR-1 safflower to *Phytophthora* root rot and its inheritance. Plant Dis. Rep., 60: 123.

Thomas, C. A., and Kilisiewicz, J. M. 1963. Selective pathogenesis within *Phytophthora drechsleri*. Phytopathology 53: 368.

Thomas, C. A., and Allen, E. H. 1970. An antifungal polyacetylene compound from *Phytophthora*- infected safflower, Phytopathology 60: 261.

Thomas, C. A., Rubis, D. D., and Black, D. S. 1960. Development of safflower resistant varieties resustant to *Phytophthora* root rot, Phytopathology 50: 129

Thompson, A. 1929. *Phytophthora* species in Malaya. Malayan Agric. J. 17: 53

Uozumi, T. and Yoshii, H. 1952. Some observations on the mildew fungus affecting the cucurbitaceous plant. Annl. Phytopath. Soc. Japan. 16: 123-126.

Uppal, B. N. 1934. Appendix K, Summary of the work done under the plant patholgist to Government, Bombay Presidency, Poona for the year 1932-33, pp 171.

Urdaneta, U. R. and Mazzani, B. 1976. Effectiveness of different chemicals for the control of bacteriosis of sesame (*Sesamum indicum* L.) in Venezuela. Agron. Trop. 26: 47-54.

Vajavat, R. and Chakravarti, B. P. 1978. Survival of *Pseudomonas sesami* and effect of an antagonistic bacterium isolated from seeds on the control of the disease in the field. Indian Phytopath. 31: 286-288.

Valaskove, E. 1983. Reaction of *Fusarium* species to simultaneous use of Benzimidazoles prepartion and resistant and cross-resistance. Drchark Roslin. 19: 37-38.

Vasudeva, R.S. 1959. Diseases of castor. In: Castor (ed. L.G. Kulkarni). Indian Central Oilseeds Committee, Hyderabad, 1952, pp.107.

Waghe, S.V. and Lanjewar, R.D. 1981. Effect of different oilcakes amendments on seedling survival against wilt, root rot and stem rot disease of sesamum. Oilseeds Journal. Pp. 69-73.

Weiss, E. A. 1971. *Castor, Sesame and Safflower*. Leonard Hill, An Intertext Publisher, London, 709 pp.

Weiss, E. A. 1971. Castor, Sesame and Safflower. Leonard Hill, An Intertext Publisher, London.

Welsford, E. J. 1932. Diseases of simsim. Mycol. Circ. Tanganyika Dep. Agric. 3: 4.

Wiess, E.A. 1983. Oilseed crops: Safflower pests, Publication Longman, UK, pp. 660.

Yoshii, H. 1929. A leaf spot or blight disease of *Ricinus communis* L. caused by *Macrosporium ricini* sp., Bull. Sci. Fak. Terkult Kjusu Imp. Univ. Fukuoka, Japan. 3: 327.

Yu, S.H. and Park,J. S.1990. *Macrophomiona phaseolina* detected in seeds of *sesamum indicum* and its Pthogenicity Korean J. Plant Prot. 19: 135.

Zachos, D. G. and Panagopoulos, C. G. 1960. The bacterium *Pseudomonas sesami* Malkoff in Greece. Ann. Inst. Phytopathol. Benaki N.S. 3: 60-64.

Zaprometov, N. G. 1927. Materials for the mycoflora of middle Asia. Abstr. Rev. Appl. Mycol.,6: 123.

Zarzica, H. 1958. Odpornosc niektorych odmian Racznika (*Ricinus communis*) na grzyb *Botrytis cinerea* Pers. Acta Agrobot. 7: 117-130.

Zayed, M. A., Yehia, A. H., El-Sabaey, M. A., and Gowily, A. M. 1980. Studies on the host parasite relationships of safflower root rot disease caused by *Fusarium oxysporum* Schlect. Egypt. J. Phytopath., 12: 63.

Zazzerini , A., and Buonaurio, R. 1981. Diseases of safflower: a leaf spot due to *Alternaria* spp. Informatore Fitopatologico, 31: 7-10

Zazzerini , A., and Cappelli, C. 1982. Safflower diseases in Italy: rust *Puccinia carthami* Cda.) Abstr., Rev. Pl. Path. 61: 345.

Zimmer, D. E. 1963. *Alternaria* leaf spot of safflower in 1962. Pl. Dis. Rep. 47: 643-648.

Zimmer, D. E., and Jensen, J. J. 1970. Seedling rust and yield of safflower. Plant Dis. Rep.54: 364.

Zimmer, D. E., and Leininger, L. N. 1965.Sources of rust resistance in safflower. Plant Dis. Rep. 49: 440.

Zimmer, D. E., and Urie, A. L. 1967. Influence of irrigation and soil infestation with strains of *Phytophthora drechsleri* on root rot resistance of safflower. Phytopathology 57: 1056.

Diseases of Field and Horticultural Crops
Editor-in-Chief: **P. Chowdappa**
Published by: **Daya Publishing House, New Delhi**

Pages **342-385**

12

Diseases of Apple and Pear

I.M. Sharma, H.S. Negi, Durga Prashad and Ved Ram

I. Apple

Apple (*Malus domestica* Borkh.) is the most important fruit of temperate regions of the world. It belongs to the family Rosaceae. Although, it is usually said to derive from *M. pumila* Mill., a small-fruited species that occurs naturally in eastern Europe and western Asia, yet it might have originated more directly from *M. sieversii* (Ledeb.) M. Roem., a diverse species from the mountains of central Asia (Stebbins and Aldwinckle, 1990). It is the most important temperate fruit crop of Indian Himalayan region. In India, the commercial cultivation of apple is largely confined to the states of Jammu and Kashmir, Himachal Pradesh, Uttrakhand and Arunachal Pradesh. Being a golden crop of these hilly states, it has boosted the economic returns per unit area thereby greatly helped in improving the socio-economic status of the farmers.

In India, apple is being grown over an area of 311.5 thousand ha, with an annual production of 1915.4 thousand MT (Anonymous, 2013a). Jammu and Kashmir is the leading state in area and production of apple with highest productivity followed by Himachal Pradesh, Uttrakhand and Arunachal Pradesh (Table 12.1).

Table 12.1. State wise area, production and productivity of apple in India

State	2008-09			2009-10			2010-11			2011-12			2012-13		
	Area (000' ha)	Produ-ction (000' MT)	Pdy. (MT/ha)	Area (000' ha)	Produ-ction (000' MT)	Pdy. (MT/ha)	Area 000' ha)	Produ-ction (000' MT)	Pdy. (MT/) ha)	Area (000' ha)	Produ-ction (000' MT)	Pdy. (MT/ ha)	Area (000' ha)	Produ-ction (000' MT)	Pdy. (MT/ ha)
Jammu & Kashmir	133.7	1332.8	10.0	138.1	1373.0	9.9	141.7	1852.4	13.1	170.6	1775.0	10.4	157.28	1348.2	8.6
Himachal Pradesh	97.2	510.2	5.2	99.6	280.1	2.8	101.5	892.1	8.8	103.6	275.0	2.7	106.23	412.43	3.9
Uttrakhand	32.7	132.3	4.1	32.4	114.0	3.5	33.0	135.9	4.1	33.7	122.7	3.6	33.76	133.2	3.7
Arunachal Pradesh	10.8	9.8	0.9	12.8	10.0	0.8	12.8	10.0	0.8	13.9	30.5	2.2	14.07	31.0	2.2
Others	0.0	0.1	1.4	0.1	0.2	2.2	0.1	0.2	2.7	0.1	0.1	4.3	0.2	0.6	3.0
Total	274.4	1985.1	7.2	282.9	1777.2	6.3	289.1	2890.6	10.0	321.9	2203.4	6.8	311.5	1915.4	6.1

Table 12.2. List of important diseases of apple

S. No.	Name of the disease	Causal organism
Fungal diseases		
1	Apple scab	*Venturia inaequalis* (Cke.) Wint. (Ana.: *Spilocaea pomi* Fr.: Fr.)
2	Premature leaf fall	*Marssonina coronaria* (Ell. and J. J. Davis) J. J. Davis (Tel.: *Diplocarpon mali* Harada and Sawamura)
3.	Cankers	
a.	Smoky blight canker	*Botryosphaeria obtusa* (Schw.) Shoemaker (Ana.: *Sphaeropsis malorum* Berk.)
b.	Pink canker	*Corticium salmonicolor* Berk. and Br. (Ana.: *Necator decretus* Mass.)
c.	Stem brown canker	*Botryosphaeria dothidea* (Moug.: Fr.) Ces. and de Not. (Ana.: *Dothiorella mali* E. and E.)
d.	Nail head canker	*Nummularia discreta* (Schw.) Tul.
e.	Silver leaf canker	*Chondrostereum purpureum* (Pers.: Fr.) Pouzar
f.	Stem black canker	*Coniothecium chomatosporum* Corda
g.	Cryptosporiopsis canker	*Cryptosporiopsis corticola* (Tel.: *Pezicula malicorticis* (H. Jacks.) Nannf.)
h.	European canker	*Nectria galligena* Bres. (Ana.: *Cylindrocarpon heteronemum* (Berk. and Br.) Wollenweb.)
i.	Valsa canker	*Valsa caratosperma* (Tobe.: Fr.) Maire (Ana.: *Cytospora sacculus* (Schwein) Gvritishvili
4.	White root rot	*Dematophora necatrix* Hartig. (Tel *Rosellinia necatrix* Berl.: Prill.)
5.	Collar rot	*Phytophthora cactorum* (Leb. and Cohn) Schroet.
6.	Seedling blight	*Sclerotium rolfsii* Sacc. (Tel.: *Athelia rolfsii* (Curzi) C. C. Tu and Kimber.)
7.	Powdery mildew	*Podosphaera leucotricha* (Ell. and Ev.) Salm. (Ana.: *Oidium farinosum* Cooke)
8.	*Alternaria* leaf spot and blight	*Alternaria mali* Robt.
9.	*Mycosphaerella* leaf spot	*Mycosphaerella pyri* (Auersw.) Boerema (Ana.: *Septoria pyricola*)
10.	*Phyllosticta* blotch	*Phyllosticta solitaria* Ellis and Everhart
11.	Flyspeck	*Schizothyrium pomi* Mont.: Fr. (Ana.: *Zygophiala jamaicensis* Masson)
12.	Sooty blotch	*Gloeodes pomigena* (Schw.) Colby.
Bacterial diseases		
1.	Crown gall	*Agrobacterium tumefaciens* (Smith and Townsend) Conn.
2.	Hairy root	*Agrobacterium rhizogenes* (Riker *et al.*) Conn.
Viral and viroid diseases		
1.	Apple chlorotic leaf spot	*Apple chlorotic leaf spot virus*
2.	Apple mosaic	*Apple mosaic virus*
3.	Apple stem grooving	*Apple stem grooving virus*

Contd...

Contd...

S. No.	Name of the disease	Causal organism
4.	Apple stem pitting	*Apple stem pitting virus*
5.	Dapple apple	*Dapple apple viroid*
6.	Apple scar skin	*Apple scar skin viroid*
Post-harvest diseases		
1.	Alternaria rot	*Alternaria alternata* (Fr.) Keissler
2.	Bitter rot	*Glomerella cingulata* (Stonem.) Spauld. and Schrenk (Ana.: *Colletotrichum gloeosporioides* (Penz.) Sacc.)
3.	Black rot and white rot	*Botryosphaeria obtusa* (Schw.) Shoem. (Ana.: *Sphaeropsis malorum* Berk.) *Botryosphaeria dothidea* (Moug.: Fr.) Ces. and de Not. (Ana.: *Dothiorella mali* E. and E.)
4.	Blue mould rot	*Penicillium expansum* Link.
5.	Brown rot	*Monilinia fructigena* (Aderh. and Ruhl.) Honey *M. laxa* (Aderh. and Ruhl.) Honey (Ana.: *Monilia* spp.)
6.	Core rot and mouldy core	*Alternaria alternata*; *Fusarium* spp.; *Trichothecium roseum*; *Aspergillus niger*; *Penicillium funiculosum*; *Phomopsis mali*; *Phoma* sp.
7.	Pink mould rot	*Trichothecium roseum* Link.
8.	Fusarium rot	*Fusarium* spp.
9.	Mucor rot	*Mucor piriformis* Fischer.
10.	Rhizopus rot	*Rhizopus stolonifer* (Ehrenb.: Fr.) Lind.
11.	Scab	*Venturia inaequalis* (Cke.) Wint. (Ana.: *Spilocaea pomi* Fr.: Fr.)
12.	Flyspeck	*Schizothyrium pomi* Mont.: Fr. (Ana.: *Zygophiala jamaicensis* Masson)
13.	Sooty blotch	*Gloeodes pomigena* (Schw.) Colby.
14.	Grey mould rot	*Botryotinia fuckeliana* (de Bary) Whetzel (Ana.: *Botrytis cinerea* Pers.: Fr.)
15.	Phytophthora rot	*Phytophthora cactorum* (Leb. and Cohn) Schroet.

1. Apple scab

1.1. Occurrence and distribution: Apple scab is one of the most destructive diseases of the apple prevalent throughout the world. In India, it was first reported in 1930 on the native cultivar "Ambri" in Kashmir valley (Nath, 1935) where it appeared in epidemic form during 1973 (Joshi *et al.*, 1975). In 1977, it was first noticed in localized pockets in apple growing districts of Himachal Pradesh (Gupta, 1978). Due to its uncontrolled spread in the following 3-4 years, it appeared in severe epidemic form during 1983 (Gupta, 1987). Thereafter, the disease was also reported from other apple growing states including Uttrakhand and Sikkim (Mallick *et al.*, 1984a), Arunachal Pradesh (Mallick *et al.*, 1984b) and Nilgiri hills of Tamil Nadu (Gupta, 1987). At present, the disease is prevalent in all the major apple growing states of India (Fig. 12.1) except Himachal Pradesh, where it is confined only to Janjhehli in district Mandi, Sangla and Nichar in district Kinnaur and Barshaini in district Kullu (Fig. 12.2) (Sharma, 2012a).

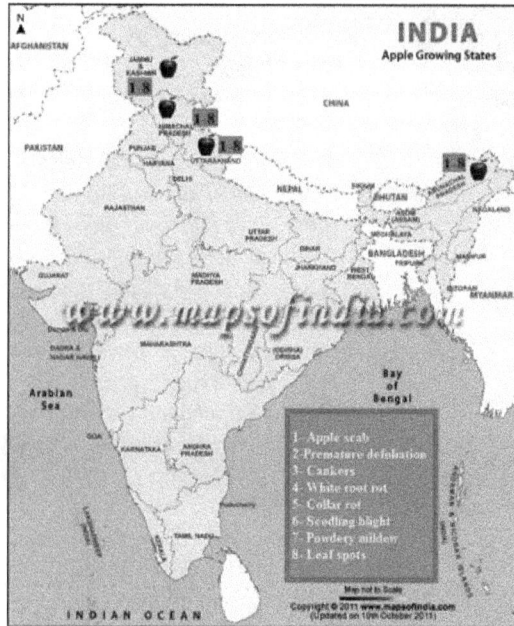

Fig. 12.1. Distribution of apple diseases in India

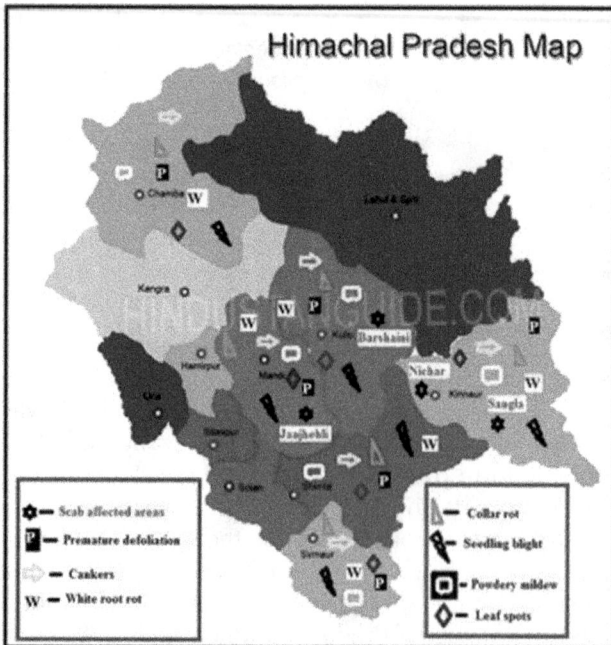

Fig. 12.2. Distribution of apple diseases in Himachal Pradesh

1.2. Crop Losses: Both direct and indirect losses due to scab disease is through

(i) Reduction in crop

    (a) Reduction in fruit set due to attack on flower stalks.

    (b) Pedicel infection and early fruit infection cause premature fruit drops

(ii) Devaluation in grade

    (a) Spoil the shape and appearance of fruits

    (b) Impairs the keeping quality of fruits

    (c) Scabbed fruits encourage more storage rot

(iii) Foliage loss

    (a) Less CO_2 assimilation

    (b) Restricted leaf growth

    (c) Premature defoliation

    (d) Stunted or reduced growth of young infected plants

(iv) High expenditure on pesticides, sprayers and spray operations

(v) Ancillary industries like sawmills and transport are also badly affected

In 1973, apple scab appeared in epidemic form in the Kashmir valley wherein about 70,000 acres orchard area got infected. It resulted in a loss of rupees 54 lakhs in a season (Joshi *et al.*, 1975). In Himachal Pradesh during 1983 epidemic, 10 per cent of apple crop (30,000 out of 3,00, 000 MT) were made unfit for market and had to be destroyed on the spot resulting in a loss of Rs. 15 millions (Gupta *et al.*, 1984). Kumar and Singh (1989) reported 17.0-100.0 per cent incidence of apple scab in 1986 from Uttarkashi district of the Uttrakhand state. Singh and Kumar (1999) recorded incidence up to an extent of 47.92-74.33 per cent in Bhatwari fruit belt of the Uttrakhand from 1994 to 1998. Kaul (1985) recorded 20-35 per cent storage losses of apple due to storage scab and other rots during epidemic year of 1983-84 in Himachal Pradesh.

A loss of about Rs. 18.2 to 21.0 million resulted in Himachal Pradesh in the year 1996, in spite of the considerable expenditure (Rs. 30.6 millions) incurred for the control of this disease. Thus, the accumulated loss by the disease was around Rs. 50 millions (Thakur and Sharma, 1999). Nevertheless, the expenditure occurred in fungicide spray to check disease, which further increases the cost of cultivation. Himachal Pradesh government has been providing the subsidy of crores of rupees on fungicides to manage this and other important diseases. During 2012, the State Govt. purchased fungicide (368MT) of worth Rs. 13.14 crores and provided subsidy to the tune of 4.02 crores to the farmers to control apple diseases including scab (Anonymous, 2012).

1.3. Symptoms: Scab symptoms can be observed predominantly on the leaves and fruits. The symptoms first appear on the lower exposed surfaces of the emerging leaves. As the leaves unfold further, symptoms appear on both the surfaces. Initially, velvety brown to olive green or mousy black lesions having feathery and indistinct margins appear on the leaves (Fig. 12.3a & b). Later, these lesions coalesce and the tissues surrounding them thicken resulting in deformed leaves. Sometimes, the entire leaf surface gets covered with scab and such type of symptoms are called as 'sheet scab' (Fig. 12.3c). Leaves showing sheet scab symptoms often fall prematurely. Lesions on young fruit are similar to those on leaves. As the fruits enlarge, the lesions turn brown and corky. Early infection results in appearance of cracks on the fruit skin and flesh and the fruit may become deformed. Although the entire fruit surface is susceptible to infection but majority of infection starts from the calyx end (Fig. 12.3 d, e & f).

Fig. 12.3. Symptoms of apple scab on leaves and fruits

1.4. Causal organism: The disease is caused by the fungus, *Venturia inaequalis* (Cke.) Wint. with its imperfect stage *Spilocaea pomi* Fr.: Fr. *Venturia inaequalis* produces ascocarps (pseudothecia) of 90-150 µm diameter in a stroma in overwintered leaves or fruit on the orchard floor. Asci (55-75×6-12 µm) about 50 to 100 per pseudothecium are fasciculate, cylindrical, short-stipitate, and each contain eight ascospores (Fig. 12.4a & b). Ascospores (11-15×5-7 µm) are yellowish green to tan and unequally two celled with upper cell shorter and wider than the lower cell. The flame shaped conidia (12-22×6-9 µm) are produced on conidiophores in spring and summer and are the source of secondary infection (Fig. 12.4c).

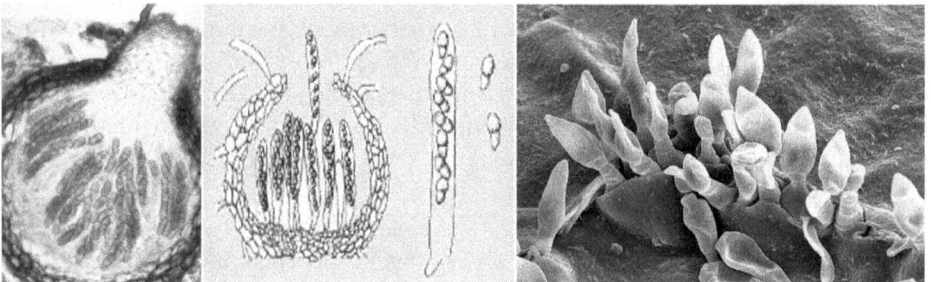

Fig. 12.4. a and b. Pseudothecium containing asci and ascospores,
c. Conidia of the fungus

1.5. Epidemiology: Weather is the most important factor in the development of apple scab, which is particularly severe in areas of high rainfall and relative humidity and also known as wet weather disease. A minimum leaf wetness of 9 hours is required for ascospores infection and 5.9 hours for conidial infection and symptoms would appear after 9 days at optimum temperature range of 18.2-23.8°C (Mills, 1944). In the presence of adequate moisture, the overwintering pseudothecia may be initiated at temperature below 0°C but their density is maximum at temperature between 4 to 8°C (Gupta, 1979; Gadoury and MacHardy, 1982) while, a temperature of 15-20°C is favourable for ascospore maturation. Maximum discharge of ascospores takes place only under wet conditions at full bloom to petal fall stage (Gupta and Lele, 1980). The secondary infection takes place by conidia during summer and early autumn. Atmospheric humidity up to 90 per cent and temperature around 16 to 20°C favours the conidial production.

Pseudothecia matures during March and discharges the ascospores on wetting. They are carried by rain splashes and air currents to the emerging plant tissues during spring to start primary infection. The conidia produced on these infected tissues are mostly washed by rain and sometimes even blown by wind (Sutton *et al.*, 1981; Kumar and Gupta, 1986).The apple scab occurred in epiphytotic form during 1973 in Kashmir valley and during 1983 and 1996 in Himachal Pradesh. Thereafter the disease has been appearing in moderate to severe form in all apple growing states but in Himachal Pradesh, disease suddenly disappeared during the crop season of 1999 (Sharma, 2003a). It was again noticed in 2003 at Janjhehli in district Mandi and presently confined only to four highly wet and humid places (Sharma, 2012a).

2. Premature defoliation

2.1. Occurrence and distribution: Premature defoliation of apple is one of the major causes in declining apple production in apple growing states of India. The disease was first reported in Kullu valley of Himachal Pradesh in 1994 (Sharma and Bhardwaj, 1994) and then it appeared in large proportion during 1996 and 1997 in all the apple growing areas of the state (Sharma and Bhardwaj, 1997). The disease is also prevalent in Jammu and Kashmir, Uttrakhand and Arunachal Pradesh states of India (Fig. 12.1). The appearance of the disease is due to the indiscriminate sprays of EBI fungicides to control scab and powdery mildew which provided an opportunity to a weak pathogen (*Marssonina coronaria*) to build up and cause severe defoliation.

2.2. Crop losses: The loss occurs through severe defoliation in apple plants thereby adversely affecting the fruit size, colour and quality. The spots on the fruits lower its market acceptability and price as well. In 1996, the disease affected more than 80 per cent apple orchards of the Himachal Pradesh causing huge losses to the growers (Sharma and Verma, 1999). In 1997, 91.6 per cent orchards in Shimla district of Himachal Pradesh were found affected with the disease with 47-60 per cent premature defoliation by the end of September (Sharma *et al.*, 2004). In Kalpa and Nichar block of Kinnaur district of Himachal Pradesh, its incidence varied from 10 to 100 per cent along with fruit infection of about 20-30 per cent (Sharma *et al.*, 2011b).

Sharma and Bhardwaj (2006) reported that in plant with 90 per cent early defoliation (*M. coronaria*) the net decrease in the fruit weight was to the tune of 25.7 per cent and it was only 2.5 per cent at 15-20 per cent disease index. They also observed the decrease in quality parameters like TSS, acidity, fruit pressure and total and reducing sugars with the increase in defoliation levels. It was to the tune of 12.8°B, 0.26, 9.6 Kg/cm^2, 10.1 and 7.2 per cent in healthy plants, respectively whereas these were only 10.9 °B, 0.34, 8.4Kg/cm^2, 8.6 and 5.9 per cent in plants with 90 per cent early defoliation. Dong *et al.* (2011) also observed a decrease in the soluble solid contents of the fruit of Fuji on M 9 rootstock when the defoliation per centage by *Marssonina* blotch was over 10 per cent before the end of September and fruit weight decreased when per centage of defoliation was over 30 per cent. Colour and starch contents of fruit tend to decrease as per centage of defoliation near the fruit development increased. Return bloom, fruit weight, and shoot growth in the following year tend to decrease as per centage of defoliation increased. They further suggested that the reduction of fruit quality as a consequence of defoliation is mainly due to decrease of starch contents in fruit, which was caused by the lower photosynthetic rates of leaves due infection of *Marssonina* blotch in August.

2.3. Symptoms: Disease symptoms appear both on leaves and fruits. Initially green circular spots of 5-10 mm size develop on upper leaf surface on mature leaves present on lower portion of the plant mostly in the first fortnight of June. These patches later become necrotic and turn brown. Small pin head size, black acervuli are visible on the surface of diseased area. Later, these spots coalesce to form large dark brown blotches and surrounding areas turn yellow (Fig. 12.5a). These symptoms are followed by severe premature falling-off of leaves and only fruits hanging on the tree are seen (Fig. 12.5c). Sometimes, in the affected trees, the flowering occurs with the onset of winters in October when some of fruits are still on the tree (Fig. 12.5d). On fruits initially circular (4-5 mm in diameter) spots brown in colour develop on its suface. Later these spots become oval, depressed and dark brown with age and almost black at harvest time (Fig. 12.5b). The surface of the fruit is somewhat indented and small black acervuli are visible in the lesions.

(a)   (b)   (c)   (d)

Fig. 12.5. Symptoms of premature defoliation disease on a. Apple leaf, b. Tree c. Flowering while fruits still on tree d. Apple fruits.

2.4. Causal organism: This disease is caused by a mitosporic fungus *Marssonina coronaria* (Ell. and J.J. Davis) J.J. Davis Syn.: *M. mali* (P. Henn.) Ito. The fungus produces asexual fruiting bodies called acervuli which are 100 to 200 μm in diameter, subcuticular and irregularly rupturing the cuticle (Fig. 12.6a). Conidia are borne on small clavate conidiophores and are single septate, constricted at the septum and hyaline guttulate and 20-24 × 6.5-8.5 μm in size (Fig. 12.6b). Spermatia are produced in acervuli, solely or mixed with conidia and are rod shaped to elliptical, continuous, hyaline and 4-6 × 1-2 μm in size. *Diplocarpon mali* Harada and Sawamura is the perfect stage of the fungus (Fig. 12.6c) (Harada *et al.*, 1974).

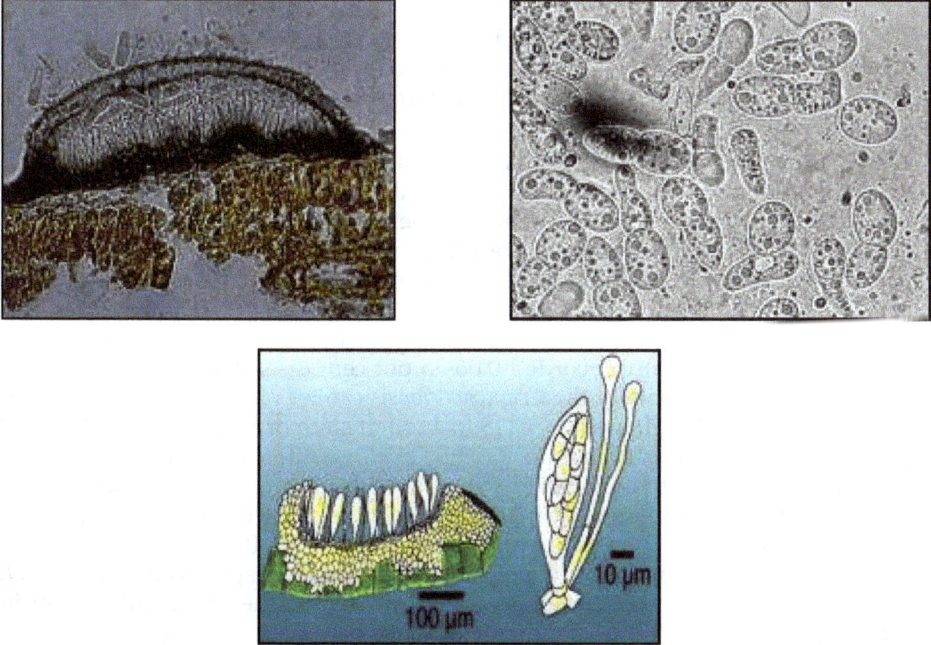

Fig. 12.6 a. Acervulus containing conidia b. Conidia c. Apothecia and ascus containing ascospores of Diplocarpon mali anamorph Marssonina coronaria syn. M. mali

2.5. Epidemiology: The fungus overwinters in the form of apothecia on fallen leaves. Primary infection has been reported to initiate from ascospores produced in the apothecia in diseased leaf litter (Harada *et al.*, 1974) at just before bloom stage. Rain is necessarily required for spore release and infection of the leaf tissues. In India, the fungus overwinters on the infected fallen leaves in the form of acervuli and remained viable till the next season (Nirupma *et al.*, 2009). So, most of the primary and secondary infection occurs through conidia. A temperature ranging from 20 to 25°C has been found suitable for conidial infection of leaves in the inoculation experiments (Takahashi and Sawamura, 1990; Sharma, 2003b). Initiation of disease at lower temperature requires longer leaf wetness of 8 hrs while at temperature of 20 - 25°C, it started only after 4 hrs of leaf wetness and for severe infection it requires a minimum of 40 hrs of leaf wetness (Nirupma *et al.*, 2009).

Predominantly conidia are source of both primary and secondary infection and spread through rain splashes and air currents. The disease appeared in epiphytotic form first in Himachal Pradesh during 1996, 1997 and 1999 (Sharma, 2003b). Again in 2005 and 2006 cropping season, it occurred in epiphytotic form (Nirupma *et al.*, 2011).

3. Canker diseases

Canker diseases are prevalent throughout the apple growing areas of the world irrespective of their causal agents. In India, many fungi are associated with the canker symptoms depending upon prevalent environmental factors. Fourteen different cankers have been reported occurring in apple growing states of the country in more or less severe form (Sharma and Bhardwaj, 1999; Sharma, 2006) (Fig. 12.1). The canker diseases of economic importance causing severe losses are discussed below.

3.1. Losses caused by canker diseases: Canker causes losses by girdling and eventually killing shoots, limbs and main trunk. These cankers result in lowering of yield and early decline of orchard. The infection on fruits makes them unmarketable and during storage it results in rotting of fruits (Sharma, 2006). Of the various cankers, stem black canker is of common occurrence (upto 60%) in Kumaon hills of Uttrakhand (Dey and Singh, 1939). Sharma (2006) recorded 2.5-10.4 per cent incidence of bull's eye rot of apple in 2003 from Kullu district of Himachal Pradesh. He also recorded 0.06-35 per cent disease incidence and 0.02-45 per cent disease severity of the *Cryptosporiopsis* canker phase during 2002. Khan *et al.* (2010) observed high incidence (18 to 41%) of stem brown canker in Kashmir valley of Jammu and Kashmir state. Krishna *et al.*, (2010) found more than 90 per cent of apple orchards to be infected with stem brown canker in Kumaon hills of Uttrakhand. Sharma and Ram (2010) recorded the severity of three major cankers from Kullu district of H.P. and it varied between 0.1- 18, 0.3- 22 and 0.2- 16 per cent in smoky blight, pink and *Cryptosporiopsis* canker, respectively. Khan *et al.* (2011) have reported *Valsa* canker as a new threat for the apple cultivation in Jammu and Kashmir with its incidence upto 25 per cent.

3.1.1. Smoky blight or black rot

3.1.1.1. Occurrence and distribution: Smoky blight or black rot canker is prevalent in severe form in all the apple growing regions of the world. Paddock (1899) reported the fungus *Sphaeropsis malorum* causing black rot of apple fruit and frog eye leaf spot to be associated with the canker symptoms. In India, Mundkur and Kheswalia (1943) recorded the presence of this fungus with canker affected branches on imported apple plants. Later, Agarwala and Gupta (1971) reported occurrence of the canker disease on apple trees in Himachal Pradesh caused by the same fungus *Sphaeropsis malorum* and named it smoky blight canker, because of the characteristic appearance of affected stem.

3.1.1.2. Symptoms: The disease appears in three phases i.e. leaf spot (frog eye leaf spot), fruit rot (black rot) and canker (smoky blight). Leaf symptoms usually first appear 1 to 3 weeks after petal fall as small purple flecks, which later enlarge to form circular lesions of 4 to 5 mm diameter. The margin of the lesion remains purple and centre becomes tan to brown giving the lesion a frog eye appearance (Fig. 12.7a). Severely infected leaves become chlorotic and defoliate. On fruits, initially reddish flecks appear after petal fall, which later develop into purple pimples of 0.1-1.0 mm diameter with the initiation of fruit maturity. The spot area becomes irregular in shape, black in colour and is surrounded by red halo. These lesions enlarge and form a series of concentric rings alternating from black to brown colour. Affected fruits show early development of colour and ripen 3 to 6 weeks prior to healthy fruits. They mummify and may remain attached to the tree (Fig. 12.7b).

Fig. 12.7. Symptoms of smoky blight canker on a. leaf, b. fruit and c. tree

The most serious phase of the disease is the development of cankers on limbs, trunks and branches. On larger limbs, cankers usually develop on the upper surface as reddish brown, sunken lesions which turn smoky with series of alternate rings. The canker develops lengthwise more rapidly and becomes elliptical and completely girdles the affected limb (Fig. 12.7c). The canker also appears as superficial roughening of the bark and callus is formed around the wound. The wood below is stained reddish brown. The spur, branches and twigs above the canker are killed. Numerous pimples like protuberances usually appear over the bark of blighted twigs or along the margins of canker (Klingner and Pontis, 1974).

3.1.1.3. Causal organism: The disease is caused by *Botryosphaeria obtusa* (Schw.) Shoemaker (Syn.: *Botryosphaeria quercum* (Schw.) Sacc. with imperfect stage *Sphaeropsis malorum* Berk. (Stevens, 1933; Shoemaker, 1964). The pycnidia are

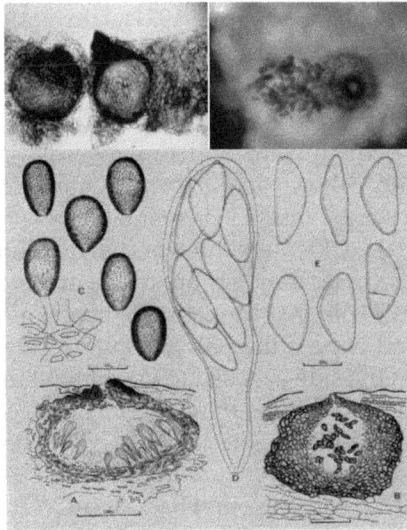

Fig. 12.8 Fruiting bodies and spores of *Botryosphaeria obtusa*. **(I)** Pycnidium. **(II)** Dark, asexual conidia coming out from pycnidium. **(III)** A, Cross section of an ascocarp. B, Cross section of a pycnidium. C, Conidiogenous cells and conidia. D, Ascus. E, Ascospores

globose, solitary or botryose and stromatic with papillate ostiole. Conidia (20-26 × 9-12 µm) are ellipsoidal, brown and one celled. The pseudothecia are upto 3 mm wide, solitary, botryose, stromatic, dark brown to black in colour with ostioles. Asci (90-120 × 17-23 µm) are bitunicate and eight spored. Ascospores (25-33 × 7-12 µm) are fusiform and rarely septate (Fig. 12.8).

3.1.1.4. Epidemiology: High humidity and relatively longer period of low temperature (6-16°C) favours the release of ascospores and conidia. Maximum spore release takes place during rains occurring at night (Holmes and Rich, 1969). A temperature between 23 to 27°C and relative humidity ranging between 96 to 100 per cent is most suitable for spore germination (Arauz and Sutton, 1989). The fungus is primarily a wound parasite (Gupta, 1975). However, various types of injuries resulted through limb rub, hailing, spray of chemicals and frost also offer an opening for the pathogen to enter. The orchards situated at lower elevations (900-1200 m.a.s.l.) and plants above 3 years of age are more prone to the disease (Shandilya *et al.*, 1973). Drought conditions prevalent in summer and rainy season predispose the plants to infection (Holmes and Rich, 1969). Higher doses of nitrogenous fertilizer increase the susceptibility of plants to infection whereas; phosphorus and potash decrease it (Heladze, 1968). Conidia are primarily waterborne whereas, ascospores are airborne and are disseminated by rain drops and rarely by windblown mist (Sutton, 1981).

3.2.2. Pink canker

3.2.1.1. Occurrence and distribution: The disease is also known as pink disease, twig blight or limb blight. It is fairly important disease in southern states of America and Japan (Sakuma, 1990), where it causes limb blight. In India, the disease caused an extensive damage to apple in Kumaon (Singh, 1943b) and Himachal Pradesh (Agarwala and Gupta, 1971). Currently it is present in all the apple growing regions of the country (Fig. 12.1).

3.2.2. Symptoms: The disease develops on trunk, stems and twigs causing canker, blight and dieback symptoms. Infection mostly starts from the forks of thick branches and proceed both upward and downwards. Usually the lesions are sunken, dull brown with cobweb like growth. Oftenly the mycelium remains superficial and transforms into pinkish incrustation during rainy season. Sometimes mycelium penetrating the bark enters the wood and results in death or blight of terminal parts. The bark becomes minutely broken by numerous small protuberances consisting of white or buff coloured fungal structures which later expand to form star shaped openings and expose the pale or pink to salmon coloured pustules. The cankerous lesions on the side exposed to bright light develop nodular, orange red pustules which burst through the bark. They later become erumpent and are separated into constituent cells. This is called as "Necator stage" of the fungus. In dry conditions, the growth of mycelium in the bark is checked and it results in the formation of a typical canker. The outer bark of such canker flakes off and inner bark falls out revealing the dead wood and the irregular cankerous formation with the callus (Fig. 12.9). The fungus exists in four forms namely; cob-web form, pustular form, necator stage and pink incrustation (Bakshi *et al.*, 1970; Pathak, 1986).

Fig. 12.9. Pink canker on stem of apple tree

3.2.3. Causal organism: The disease is incited by *Corticium salmonicolor* Berk. and Br. [Syn. *Botryobasidium salmonicolor* (Berk. and Br.) Venkatanarayan, *Pellicularia salmonicolor* (Berk. and Br.) Dastur] with its imperfect stage *Necator decretus* Mass. Necator stage consists of nodular orange red conidial pustules. The conidia are irregularly oval or spherical, hyaline and measure 4-20 × 8-10 μm (Fig. 12.10). The perfect stage contain thin, effused, rose to pink coloured fruiting bodies, which fade rapidly and crack.

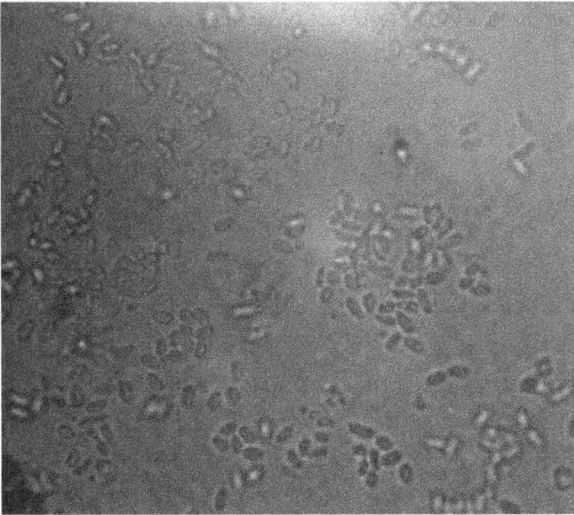

Fig. 12.10. Conidia of the *Necator decretus*

3.2.4. Epidemiology: Fungus survives and spreads by means of necator spores and deep seated mycelium (Singh, 1943; Verma, 1988). Maximum germination of conidia and basidiospores takes place at a temperature of 25°C and rain water also helped in their germination (Verma, 1988). Maximum progress and spread of infection takes place during rainy season. Temperature of 28°C coupled with 90 per cent relative humidity favoured the formation of cobweb form, basidial and sterile nodular stages during rainy season resulting in maximum disease incidence (Verma, 1991). Average temperature of 24°C with 75 per cent relative humidity during post-rainy season supported the formation of necator pustules. Overcrowding, poor drainage, high humidity and comparatively warm weather favour the spread of disease.

3.3. Stem brown canker

3.3.1. Occurrence and distribution: Stem brown disease is also known as Botryosphaeria canker, Dothiorella canker, bark canker, Botryosphaeria rot, white rot or Dothiorella rot. The disease was first reported from South Africa by Putterill (1919) and the causal agent was identified as *Botrysphaeria mali*. In India, the disease was reported from Kumaon hills (Singh, 1942), Kashmir (Malik, 1967), and Himachal Pradesh (Agarwala and Gupta, 1971).

3.3.2. Symptoms: The disease appears in two phases i.e. stem canker and fruit rot. The canker usually develops on sun burnt or wounded surfaces of limbs and twigs as a small, sunken, reddish brown lesions with purplish margins. The infected bark become depressed and sometimes blisters are formed on it, which often exude watery liquid on the surface of the lesion. During spring season a few pimple like elevations develop on the bark of the lesion as a result of the formation of pycnidia and stromata underneath it. The small shoots above the cankered lesion show dieback and become wrinkled. The bark is loosened, becomes tan to burnt orange or brown and papery, which sometimes peels back. Wood below the bark is necrotic and stained dark brown. It turns slimy and develops fissures (Fig. 12.11). Large limbs are also girdled by the fusion of cankers (Gupta and Agarwala, 1973; Pathak, 1986, Sharma and Bhardwaj, 1999).

Fig. **12.11.** Symptoms of stem brown canker on apple tree **A**, Cross section of an ascocarp. **B**, Cross section of a pycnidium. **C**, Part of a pycnidial wall and conidiogenous cells. **D**, Macroconidia. **E**, Microconidia. **F**, Ascus. **G**, Ascospores

Lesions on fruit become noticeable 4 to 6 weeks prior to harvest as small, circular, sunken, light green spots encircled by reddish brown halo on yellow skin varieties. The spots later become depressed and brown surrounded by alternate light green and brown margins. In some of these spots series of concentric rings are formed (Fig. 12.12). As lesion expands in diameter the rotted area extends in a cylindrical manner in cross-section. The entire fruit rots and flesh become mushy (Fulkerson, 1960). Infection in mature fruit develops rapidly under warm conditions and the fruit becomes tan to light brown, soft and watery. Under cooler areas the rotted fruits are firmer, deeper tan and usually drop but some shrivel and remain attached to tree (Sutton, 1990).

Fig. 12.12. White rot symptoms on fruits

3.3.3. Causal organism: Stem brown canker is caused by *Botryosphaeria dothidea* (Moug.: Fr.) Ces. and de Not. (Syn.: *Botryosphaeria ribis* Gross. and Dug.) with imperfect stage of *Fusicoccum aesculi* Corda (Punithalingam and Holliday, 1973;

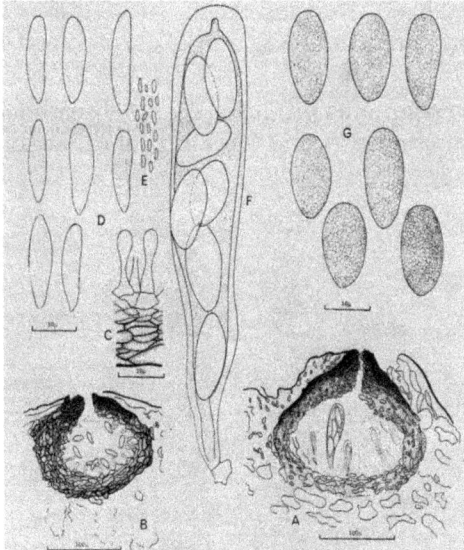

Fig. 12.13. Fruiting bodies and spores of Botryosphaeria dothidea A, Cross section of an ascocarp. B, Cross section of pycnidium. C, Part of a phcinidial wall and conidiogenous Cell. D, macroconidia, E, Microconidia. F, Ascus. G, Ascospores.

Sutton, 1990) and *Dothiorella mali* E. & E. (Birmingham, 1924). In India, only *Dothiorella mali* stage has been recorded (Gupta and Agarwala, 1973). Pycnidia appear on cankered bark in the spring, borne in stroma and produce fusoid, hyaline conidia (16-31 × 4.5-4.7 µm). In autumn and winter the pseudothecia are found in stromatic bodies of *Dothiorella.* The pseudothecia are embedded in a black erumpent stroma. Asci are club shaped, 8 spored with numerous filiform paraphyses. Ascospores are elliptical, unicellular and hyaline (16-23 × 5-7 µm) (Fig. 12.13).

3.3.4. Epidemiology: The fungus survives the winter as mycelium, pycnidia and pseudothecia in canker in both live and dead limbs and mummified fruits. Both spore production and release is dependent upon rainfall. Number of spores released during rain is related to amount and duration of rainfall (Sutton, 1981). Optimum temperature for germination of ascospores and conidia is 28 to 32°C and also require free water for their germination (Kohn and Hendrix, 1982; Travis *et al.*, 1992). Kohn and Hendrix (1983) found that infection in fruit occurs during last six to eight weeks of growing period when the total soluble solids in fruit approaches 10.5 per cent. Period of moisture stress or summer drought following infection favours the canker development rapidly (Shey and Witterly, 1954). Other predisposing factors are frost injury, poor nutrition, high temperature injury and poorly managed orchards. Plants aged between 11 to 14 years and orchards situated at an elevation of 900-1200 m.a.s.l. are most susceptible (Shandilya *et al.*, 1973). Conidia are primarily waterborne i.e. spread through rain splashes whereas, ascospores are airborne in nature (Sharma and Bhardwaj, 1999).

3.4. Stem black canker

3.4.1. Occurrence and distribution: This disease is also known as blister disease, branch blister or rough scab and of common occurrence (upto 60%) in Kumaon hills of Uttrakhand (Dey and Singh, 1939). It is of minor importance in Himachal Pradesh and Jammu and Kashmir (Agarwala and Gupta, 1971; Sharma and Bhardwaj, 1999).

3.4.2. Symptoms: Symptoms produced by the disease include fruit russeting cracking and dieback of branches and formation of canker on main branches and

Fig. 12.14. Apple twigs showing stem black canker symptoms

limbs. On the infected branches purplish, raised blisters are formed which later become dry. Long vertical cracks containing black powder develop in the bark resulting in blackening of branches which later die away. On larger branches a jet black streak appears in a conical fashion and slowly surrounds the whole branch. In advanced stage canker is formed and the affected branches are killed outrightly (Fig. 12.14).

3.4.3. Causal organism: The disease is incited by *Coniothecium chomatosporum* Corda. The mycelium is poorly developed (short hyphae) and is dark in colour. Conidia are irregular, muticate, non-catenulate, often coalescent and variable (8.4-36.9 × 5.6-30.8 µm). Chlamydospores are formed and the pycnidial stage has not been observed (Dey and Singh, 1939; Agarwala and Gupta, 1971).

3.4.4. Epidemiology: The fungus enters through wounds in the wood and through lenticels in fruits. Disease development is favoured by drought conditions in summer. Light soils with poor water holding capacity and deficient in potash promote disease incidence (Sharma and Bhardwaj, 1999).

3.5. *Cryptosporiopsis* canker

3.5.1. Occurrence and distribution: The disease is also known as anthracnose or perennial canker of apple in different parts of the world. In India, the disease was reported from Kullu valley of Himachal Pradesh (Sharma, 2006).

3.5.2. Symptoms: Initial symptoms comprise oozing of cell sap from lenticels of the branches. Later on, well defined necrotic lesion forms on branches, if hot and dry weather prevails during June-July. The fungus also causes fruit rot known as bull's eye rot with the occurrence of is excessive rain and high humidity from fruit colour development to pre-harvest stage (Sharma, 2006) (Fig. 12.15).

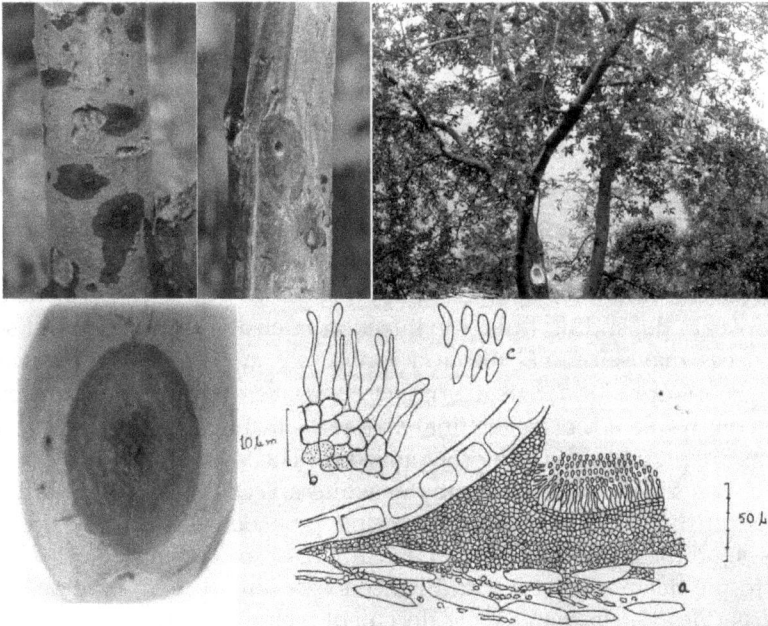

Fig. 12.15. *Cryptosporiopsis corticola* Canker

3.5.3. Causal organism: In India, disease is caused by *Cryptosporiopsis corticola* (Tel.: *Pezicula malicorticis* (H. Jacks.) Nannf.), however perfect stage could not be found from the country (Sharma, 2006). Acervuli (0.2-1 mm in diameter) are subepidermal and erumpent. Numerous simple or branched, hyaline conidiophores arises from the surface of acervulus. Macroconidia (15-24 × 4-6 µm) are unicellular, hyaline, sickle to U-shaped (occasionally straight), and coarsely granular to guttulate. Microconidia (5-8 × 1.5-2.5 µm) are produced in culture.

3.5.4. Epidemiology: Grove (1985) mentioned that canker phase on twigs and branches of the apple tree are favoured by hot and dry weather during summers and early rainy season. Whereas, excess rainfall and high humid conditions during this period favours fruit rot phase (bull's eye rot) of the disease.

4. White root rot

4.1. Occurrence and distribution: White root rot is one of the most serious soil-borne diseases causing heavy losses both in nurseries as well as grown up orchards. In India, the disease was first observed in Uttar Pradesh hills (now Uttrakhand) by Singh in 1939 (Singh, 1943) and later on in 1961 from Himachal Pradesh (Agarwala, 1961). The fungus causing white root rot is distributed throughout the world but its seriousness on apple has been observed in India, Israel, U.S.A., Brazil and Britain whereas in other countries it is serious on other host plants. In India, it is prevalent in almost all the apple growing states (Fig. 12.1).

4.2. Losses: Agarwala and Sharma (1966) recorded 8-33 per cent incidence of the disease in apple nurseries of different apple growing areas of Himachal Pradesh and also estimated the annual loss of Rs. 1.3 million in India due to this disease. Sharma (1999) reported that the incidence of this disease varied between 10-25 per cent in apple orchards of Himachal Pradesh and losses were estimated up to the tune of 10-12 lakhs per year.

4.3. Symptoms: The earliest above ground symptom of the disease is diminution in the leaves size, scattered leaves and a stunted tree growth resulting in the progressive decline in the vigour as whole or certain branches. These symptoms are usually associated with a heavy blossom and fruiting next year. In the succeeding years, few leaves emerge and much of the fruit fail to reach maturity and dying back of branches is quite evident (Fig. 12.16a). Infected trees often persist for 2 to 3 years depending upon the infection intensity in the roots but severely infected trees may succumb within a single season (Jain, 1961; Agarwala and Sharma, 1966; Gupta, 1977; Sztejnberg *et al.*, 1987). Below ground expression of the pathogen on the plant roots is the final diagnostic feature as similar above ground symptoms are produced by other root maladies. The lateral roots turn dark brown and are covered with a greenish grey or white mycelial mat having a flocculent web of whitish strands or ribbons during monsoon season. As the disease progresses, all the roots are attacked and fibrous root system almost disappears leading to the death of the tree and hence expressing disease on above ground plant parts (Fig. 12.16b & c). White flocculent web disappears after the death of the host plant, leaving the root surface dotted with small round black sclerotia.

Fig. 12. 16. Apple tree showing symptoms of root rot, a. Above ground symptoms, b and c. Below ground symptoms

4.4. Causal organism: The disease is caused by *Rosellinia necatrix* Berl.: Prill. (Ana.: *Dematophora necatrix* Hartig). The perfect stage of the fungus is not known to occur in India. The diagnostic feature of *D. necatrix* is the pear shaped swellings at each septum of the fungal hyphae (Fig. 12.17). The fungus produces scattered black, microsclerotia (138-98 µm) and does not infect xylem tissues of apple roots but bark is completely damaged. Two types of mycelium are present in the fungus. The greenish grey to white rhizomorph like, having fan shaped structure marking the site of destroyed roots and the diffused fine white mycelium occupying the soil cavities. The later is responsible for the infection of new plants (Gupta and Sharma, 1999).

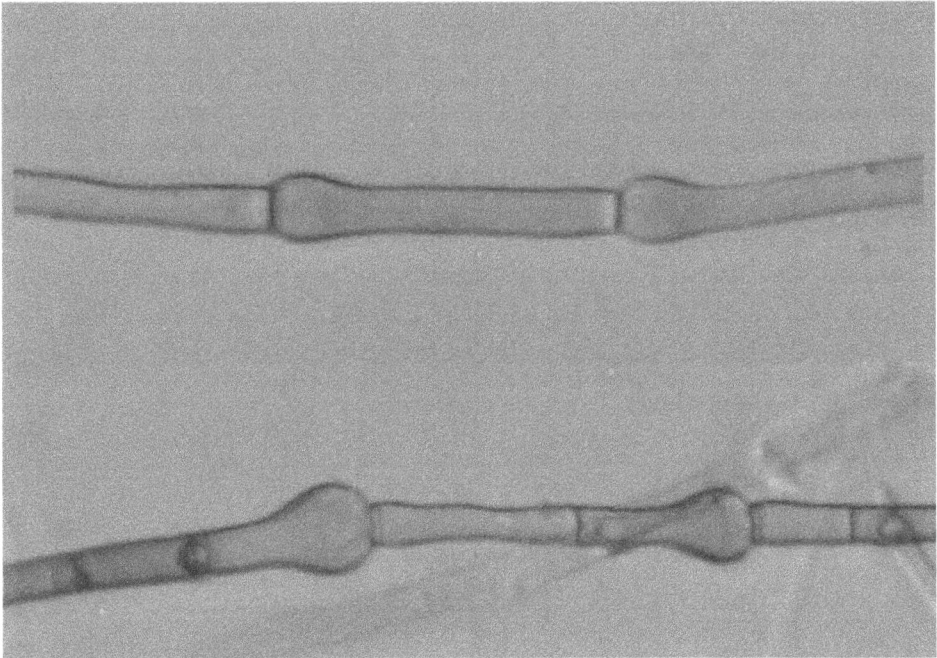

Fig. 12.17. Pear shaped swellings at the septa of hyphae of *Dematophora necatrix*

4.5. Epidemiology: The disease is most serious in water-logged, clay loam and acidic soils at pH 6.1 to 6.5. Maximum disease spread during July-August under moderate temperature (15-25°C) and high moisture (Agarwala and Gupta, 1981). The mycelium spreads from infected to healthy roots through infested soil and implements, and rain water or through contact of new plant roots with old dead roots, infecting the feeder roots first.

5. Collar rot

5.1. Occurrence and distribution: Collar rot of apple is prevalent in all the apple growing regions of the world. Baines (1939) first described the disease on apple. In India, collar rot of apple was first reported from Himachal Pradesh in 1960 (Anonymous, 1960), while the fruit rot phase of the disease was recorded by Bose and Mehta (1951). The disease is serious in all apple growing regions of the country (Fig. 12.1).

5.2. Crop losses: Losses from the disease is severe as the pathogen attacks apple trees at soil level and girdles the stem/crown completely leading to death of the plant. It is a serious problem in few apple growing pockets of the Himachal Pradesh.Incidence of collar rot of apple in nursery and orchards of Himachal Pradesh was reported varying between 2.5-24.5 and 0.2-77.5 per cent, respectively (Sharma *et al.* 2011a). Tree mortality of 30-40 per cent has been reported in Kinnaur district of Himachal Pradesh due to the disease (Sharma *et al.*, 2011b).

5.3. Symptoms: The pathogen attacks the collar/crown region of the tree and produces small necrotic lesions. As it girdles through the bark, a moist rot is produced and the necrotic tissues turn dark brown. The rotting can go quite high on stem and may sometimes bleed (Baines, 1939). In many cases, the tree may be completely girdled before its condition is noticed. During late rainy season to early fall the symptoms of the disease appears as chlorosis followed by purplish red colouration of veins and margins of leaves giving a reddish-bronze to purple appearance of the affected tree (Fig. 12.18a). Further, there is reduction in size of leaves and terminal growth. These symptoms are directly correlated with the rotting of stem near soil line (Gupta and Sharma, 1999) (Fig. 12.18b).

Fig. 12.18 Symptoms of the collar rot on apple plant,
a. Reddish-bronze appearance of apple tree, b. Rotted collar region of the tree

5.4. Causal organism: Collar rot of apple is mainly caused by *Phytophthora cactorum* (Leb. and Cohn) Schroet. though other pythiaceous fungi are also capable of causing similar symptoms (Sharma and Gupta, 1989). *P. cactorum* is semi-aquatic, homothallic fungus with a complex life cycle consisting of an asexual phase in which motile zoospores are produced from sporangia and a sexual phase resulting in the formation of oospores (Fig. 12.19a & b). The mycelium is slender and aseptate (Fig. 12.19c). Sporangia measures 35-65 μm in length and produces about 40-90 zoospores.

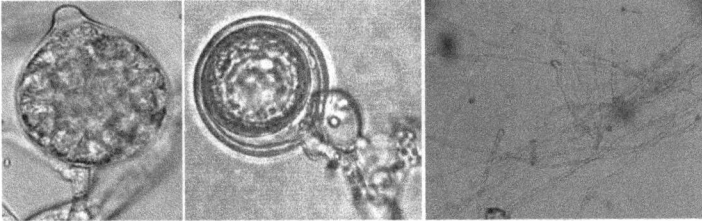

Fig. 12.19. *Phytophthora cactorum* a. Sporangium b. Oogonium and antheridium c. Aseptate mycelium

5.5. Epidemiology: Waterlogged soil with temperature of 12 to 20°C and pH 5 to 6 has been found to be the best for disease spread. Soil temperature of 18 to 22°C along with high soil moisture is favourable for sporangial production but oospores are produced at low moisture level at the same temperature (Sneh and McIntosh, 1974; Rana and Gupta, 1983). Flooded or saturated soil predisposes collar region to infection and colonization of bark tissues is greatest between pink bud stage of tree phenology and beginning of shoot elongation (Gupta, 1996). Disease is mainly spread through mycelium and floating zoospores in water from one tree to another. Infested soil and farm equipments also carry the pathogen from diseased to healthy plant.

6. Seedling blight

6.1. Occurrence and distribution: The disease is mostly prevalent in nurseries, affecting 2-3 years old apple seedlings. In India, the disease was first reported from Mysore (Anonymous, 1937) and later in Himachal Pradesh by Jain (1962). It is present in almost all the apple nursery producing states of the country (Fig. 12.1).

6.2. Crop losses: Losses in nurseries are due to the death of apple seedlings. Bhardwaj and Agarwala (1986) recorded 40 per cent losses in nurseries of Himachal Pradesh however; Sharma and Bhardwaj (2002) recorded these losses between 30-40 per cent.

6.3. Symptoms: Blightening of the foliage upon root infection is the common symptom of the disease (Fig. 12.20a). The above ground symptoms are generally confused with the collar and white root rot but its identity can be confirmed by examining the vicinity of dead seedling, where mustard like sclerotia of the fungus can be seen (Fig. 12.20 b & c).

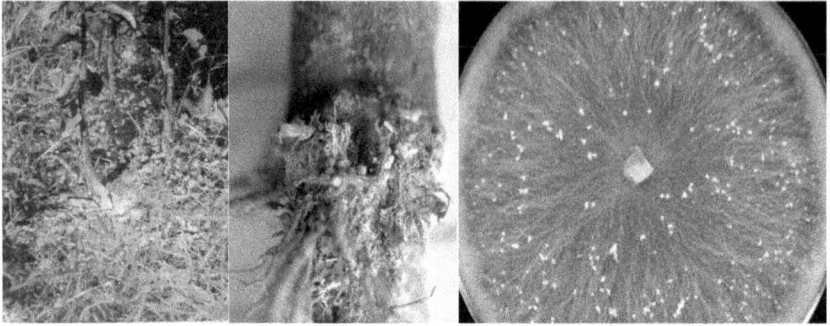

Fig. 12.20 a. Dead apple seedling, b. Sclerotia on root, c. Sclerotia on culture

6.4. Causal organism: The disease is caused by *Sclerotium rolfsii* Sacc. (Tel.: *Athelia rolfsii* (Curzi) C. C. Tu and Kimber.) The fungus produces sclerotia which are round mustard seed sized and tan to reddish brown to dark brown coloured. The asexual spores are not produced. The pathogen is more frequent in warm moist conditions. The sclerotia are the surviving structures and can germinate to infect new plants under favourable conditions.

6.5. Epidemiology: Sclerotia are the main source of primary infection in nurseries. A soil temperature of 30-33°C, pH 6.0 and above 38 per cent soil moisture is favourable for the disease development (Bhardwaj and Agarwala, 1986). Sandy loam soil is more suitable for the infection than loamy or clay loam soil. Disease spreads through sclerotia with water and infected implements.

7. Powdery mildew

7.1. Occurrence and distribution: Powdery mildew is a major foliar disease of apple and prevalent in all apple growing countries of the world. The disease was first noticed on apple seedlings in Iowa, USA in 1871 (Bessey, 1877). In India, the disease is prevalent in all the apple growing regions (Fig. 12.1).

7.2. Crop losses: Sharma (1985) reported the incidence of powdery mildew between 5-80 and 10-100 per cent in orchards and nurseries, respectively at different locations of Himachal Pradesh. Losses to the tune of 8-10 per cent occur every year due to the severe destruction of seedlings in nursery and also through the infection of grown up trees particularly; on pollinizer cultivars, resulting in loss of fruit yield and quality.

7.3. Symptoms: The pathogen infects foliage, blossom, twigs, and fruits of most of the commercial varieties. The symptoms first appear on lower surface of leaf as small whitish felt like patches of fungal structures. The mycelium and powdery coating of spores soon become evident on the entire leaf (Fig. 12.21a). The infected foliage becomes crinkled, curled, hard and brittle and is frequently killed. Severe infection leads to defoliation of mildewed shoot with subsequent curtailment of its growth and a reduction in the number of potential fruiting buds. Diseased twigs are either stunted with shortened internodes or shows dieback symptoms along with silvery appearance (Fig. 12.21b). Lateral buds become purplish, elongated and bunched and usually open 5 to 8 days later than healthy buds. The petals of infected flowers are pale yellow or light green, distorted and infected flowers fail to set. In some varieties, the tree shows stunted growth and typical russeting (Fig. 12.21c). In nurseries, the fungus is more destructive and spreads to all developing leaves leading to severe defoliation along with stunting of terminal growth.

Fig. 12.21. Symptoms of powdery mildew of apple

7.4. Causal organism: The disease is caused by *Podosphaera leucotricha* (Ell. and Ev.) Salm. (Ana.: *Oidium farinosum* Cooke). It is an ascomycetous heterothallic fungus. Conidia (20-30 x 12-15 μm) are ellipsoidal, truncate and hyaline (Fig. 12.22a). Cleistothecia are subglobose, dark brown embedded in external felt like growth of fungus and have apical and basal appendages. One ascus is present in each cleistothecium (Fig. 12.22b). Ascospores are hyaline, subglobose or oblong, single celled, measuring 22-26 x 12-14 μm. Puttoo (1972) reported the occurrence of *Erysiphe heracoli* (D.C.) St Am. on apple seedlings in Kashmir.

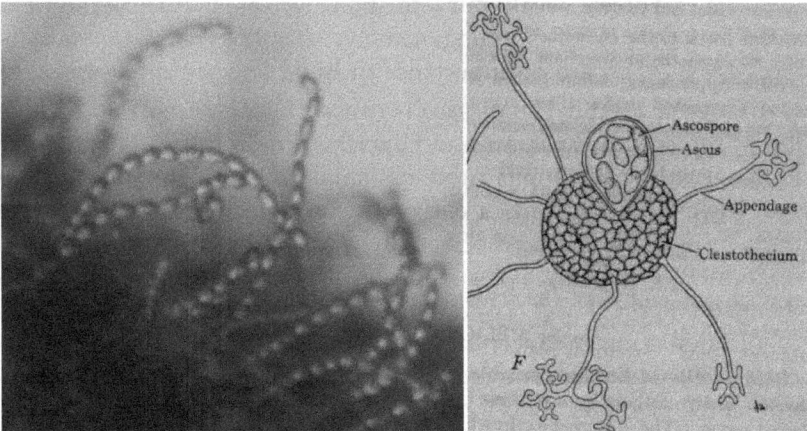

Fig. 12.22 a. Conidia of *Oidium farinosum* b.*leistothecium*

7.5. Epidemiology: *P. leucotricha* is an obligate parasite and overwinters in the form of dormant mycelium in diseased terminal and 3-4 lateral vegetative buds. During spring or summer conidial germination takes place at temperature ranging between 5 to 30°C with maximum at 20°C and high relative humidity approaching 100 per cent favoured maximum germination (Sharma and Gupta, 1991). Burr (1980) reported that prevalence of relative humidity more than 90 per cent and temperature ranging between 19 to 22°C favoured the powdery mildew infection. He also established that although high relative humidity is needed for infection but the spore will not germinate if immersed in water. Leaf wetting is therefore not conducive to powdery mildew development. Dry weather favoured the rapid disease development and wet conditions appear to restrict it. Increasing light intensity raised photosynthetic activity and inhibited disease development (Pathak, 1986). During spring or summer large numbers of conidia produced in infected tissues are dispersed by wind causing secondary infection. Conidial dispersal is low during night than day and is interrupted by heavy prolonged rains (Stephan, 1988).

8. Leaf spots

Any blemish produced on leaf resulting in chlorotic or necrotic lesion is generally termed as leaf spot. In apple, these spots are produced in different times of the season and cause premature defoliation thereby leading to reduction of photosynthetic efficiency. Consequently, it results in reduced viability and vigour of the plant leading to lowered fruit production mainly through Losses occurs due to premature defoliation leading to reduction of photosynthetic efficiency which results in reduced viability and vigour of the plant leading to lowered fruit production mainly through weak spur formation, decrease in fruit size and excessive pre-harvest fruit drop. Incidence of leaf spot can be upto 60-80 per cent depending upon favourable climatic conditions causing losses upto 15-20 per cent through reduced fruit size and its quality due to early defoliation and reduction in photosynthetic area of leaf, weak spur formation, decrease in fruit size and excessive pre-harvest fruit drop. Leaf spot fungi namely, *Alternaria mali, Phyllosticta solataria* and *Mycosphaerella pyri* are of importance in India and are discussed as under:

8.1. *Alternaria* blight: The disease is caused by *Alternaria mali* Robt. and in India it was reported in Kashmir and Himachal Pradesh (Gupta and Agarwala, 1968; Puttoo, 1987). The fungus produces small, oval to irregular, dark brown lesions on leaves in late spring or early summer, which enlarge and coalesce to form irregular patches (Fig. 12.23). Sometimes round, blackish brown and sunken spots are formed on growing twigs and fruits. In severe case the affected twigs are killed. Lenticels on shoot also occasionally appear swollen as a result of infection. Primary infection occur in late spring and increases rapidly during rainy season (Yoon *et al.,* 1989). Optimum temperature for infection, mycelial growth, sporulation and germination of spores ranges between 25 to 30°C. Filajdic and Sutton (1992) found that increase in wetness (2-48 hours) and temperature ranging between 12 to 28°C also increases the infection.

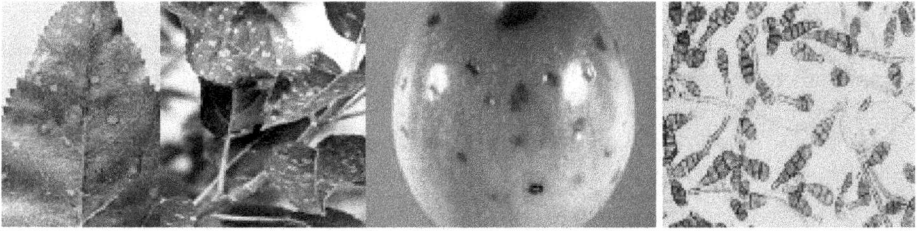

Fig. 12.23. Symptoms of Alternaria blight

8.2. *Mycosphaerella* leaf spot: The disease occurs more frequently in regions with frequent rainfall and is incited by *Mycosphaerella pyri* (Auersw.) Boerema (Ana.: *Septoria pyricola)*. Chona *et al.*, (1956) reported *M. pyrina* causing similar leaf spot of apple in India. It produces greyish white spots with purplish margins which contain small black scattered pycnidia in their centre (Fig. 12.24a). When disease is serious, premature defoliation occurs in late summer (Singh, 1944). The fungus overwinters as perithecia formed in the dead leaves. Perithecia discharge ascospores in spring and lesions are developed two weeks later. Secondary infection takes place through conidia formed in pycnidia (Fig. 12.24b).

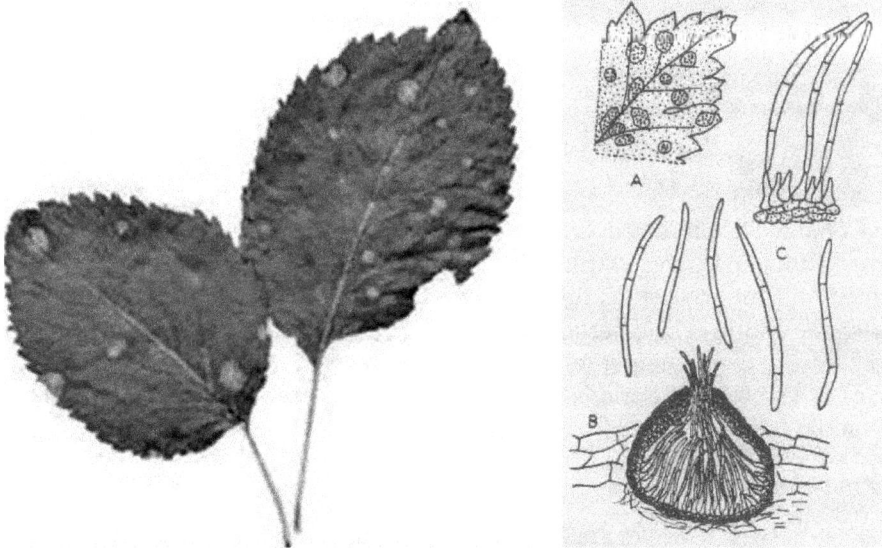

Fig. 12.24 Symptom of Mycosphaerella leaf spot on apple leaf

8.3. *Phyllosticta* blotch: Disease is incited by *Phyllosticta solitaria* in United States (Ellis and Everhart, 1896) and India (Rao, 1965). Other species *of Phyllosticta* recorded in India causing leaf spot of apple are *P. pirina* (Roy *et al.* 1985), *P. mali* (Sud and Agarwala, 1972) in Himachal Pradesh and *P. pyrina* (Singh, 1944) in Kumaon hills. The disease causes damage to leaves, twigs, small branches and fruits. Two distinct types of blotch are produced on leaves. The more severe type consists of elongated, sunken light tan or buff lesions mostly appearing on the veins of lower leaf surface and petiole. The second less severe type is pin head sized, yellowish green to brownish spots form in interveinal areas (Fig. 12.25a). In each spot a small black speck (pycnidium) develops. The lesions may appear on the leaf petiole resulting in early defoliation. Symptoms are also produced on twigs, small branches and fruits. Lesions on twigs and small branches are dark, purplish or black and raised or blister like. Later these lesions become roughened and bark sloughed off. Lesions on fruits appear as shiny blotches which are slightly raised or sunken and have irregularly lobed edges. They often crack as the fruit enlarges.

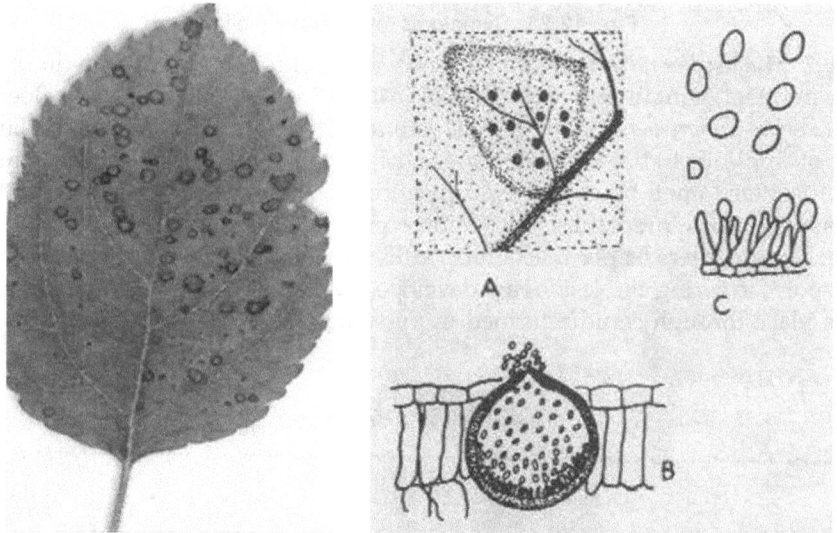

Fig. 12.25. Symptom of Phyllosticta leaf spot on apple leaf

Conidia are produced during early spring and cause primary infection on leaves, fruits and twigs starting from petal fall to about 4 weeks later. Pycnidia formed on primary lesions produce conidia which serve as secondary inoculum during late summer and early fall (Fig. 12.25b). The disease is favoured by warm and wet weather. Temperature ranging between 25 to 30°C, heavy rains and extended leaf wetting periods promote exudation, dissemination and germination of conidia (Guba, 1924).

9. Post-harvest diseases

Post-harvest losses in apple have been estimated to the tune of 20-30 per cent in India and 5 to more than 20 per cent in developed countries (Janisiewicz and Korsten, 2002). Apples are often stored after harvest and are subject to attack by numerous post-harvest pathogens. Many serious post-harvest diseases affects apple but, most of the losses in India are contributed by *Alternaria, Penicillium, Monilinia, Botrytis* and *Trichothecium* sp. (Fig. 12.26). Most of the post-harvest rots are caused by wound invading fungi like *Penicillium expansum, Botrytis cinerea, Rhizopus stolonifer, Aspergillus* spp. and *Monilinia* spp. Some post-harvest rots also result from pre-harvest latent infections by fungal pathogens (*Trichothecium roseum, Botrytis cinerea, Monilinia laxa, M. fructigena, Glomerella cingulata, Botryosphaeria dothidea, B. obtusa, Sphaeropsis malorum, Alternaria mali, A. alternata,*

Rhizopus arrhizus, Phytophthora cactorum and *Cryptosporiopsis corticola*) in the field at flowering or early fruit developmental stages when the calyx end provides a better mode for pathogens to enter and survive. The initial symptoms of this latent infection start appearing in the month of June-July (colour development stage). These stages of infections also coincide with maturity and release of spore mass for primary and secondary infection of most of these fungi (Sharma and Kaul, 1999). Sharma and Bhardwaj (2000) identified eleven different fungi associated with fruit rots during storage and five namely: *P. expansum* (25.5%), *T. roseum* (16.8%), *Botrytis cinerea* (15.8%), *M. laxa* (10.6%) and *G. cingulata* (8.0%) were most prevalent whereas; others namely; *A. mali, B. obtusa, B. dothidea* and *R. arrhizus* occurred in low frequencies (3.7-7.2%).

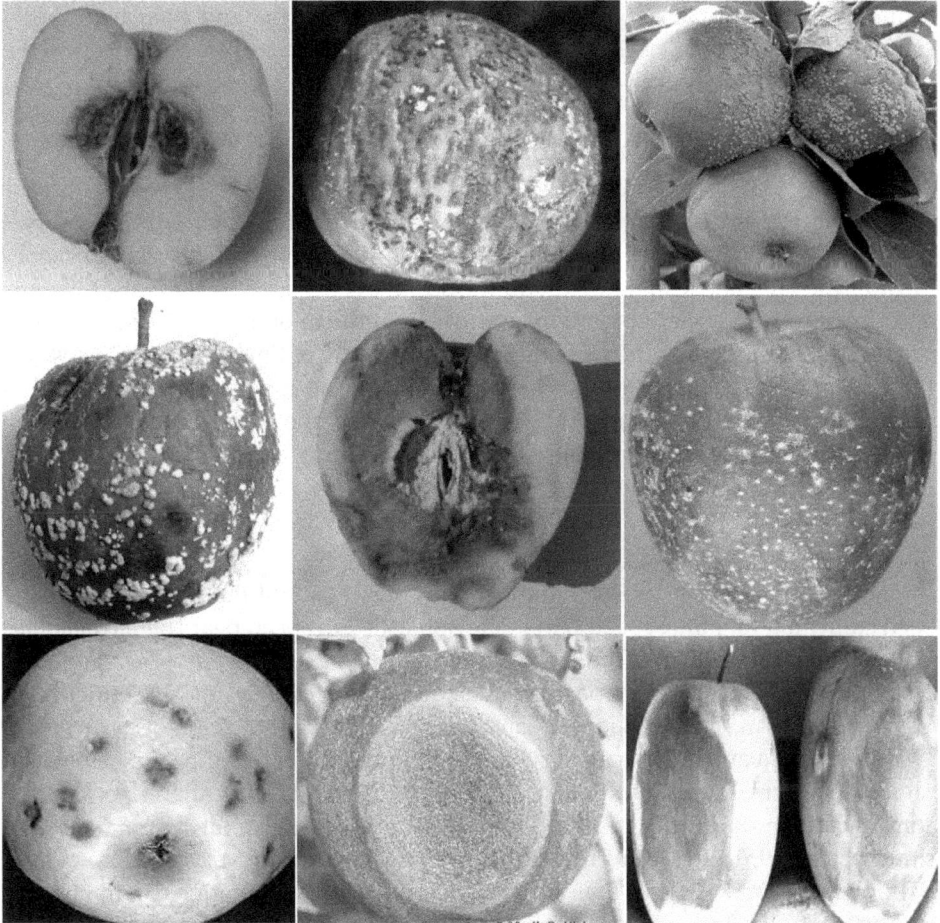

Fig. 12.26. Post-harvest rots of apple. **a.** *Alternaria mali*: Core rot, **b.** *Penicillium expansum*: Blue mould rot, **c & d.** Monilinia laxa: Brown rot, **e.** *Trichothecium roseum*: Pink mould rot, **f.** *Botrytis cinerea*: Grey mould rot, **g & h.** *Glomerella cingulata* Ana.: *Colletotrichum gloeosporioides*: Bitter rot, **i.** *Botryosphaeria ribis* Ana.: *Dothiorella gregaria*: White rot.

In apple, fruit rotting fungi cause post-harvest and storage losses to the tune of 20-30 per cent. However, Kaul and Munjal (1982) recorded 10.3 to 18.0 per cent losses in apple due to post-harvest fungal pathogens from Himachal Pradesh. Storage scab and other rots have been reported to cause post-harvest and storage losses of apple upto 20-35 per cent in Himachal Pradesh (Kaul, 1985). Incidence of core rot in field in the form of pre-harvest fruit drop was reported from 0.2 to 20.3 per cent while the incidence of storage rot was 0.8-8.5 per cent during 2008-2011 (Sharma, 2012b). Ratnam and Nema (1967) observed 12.4-26.7 per cent loss in apples from the stores and markets of Jabalpur (M.P.). Dasgupta and Mandal (1989) reported 19.7 per cent loss of apple and pear fruits from Calcutta fruit markets. Singh (2002) reported 14.0-34.6 per cent loss of apples and 28.8-33.1 per cent loss of crab apples from Jammu fruit market. Sharma and Bhardwaj (2005) estimated the pre and post-harvest losses in apple due to fruit rots were up to the tune of 15-18 per cent.

Socio-economic impact due to the losses caused by various diseases on apple

As the apple diseases lead in lowering the fruit yield thereby has a direct impact on economic returns. In the absence of sufficient money in pocket to spend, most of the farmers hesitate to attend the social events such as festivals, fairs etc. and delays others social obligations (marriages, education of the children, others functions) thereby lower their socio-economic status. Continued disease losses will force them to take loan from the rich persons of the society or from banks and in case of not repaying the loans in time will also affect the social status of a person concerned. In extreme cases, the affected person may abandon the orcharding profession and move to the city for livelihood.

2. Pear

Pears are one of the oldest and most patronized fruit of the world. They have been cultivated for almost 3000 years. The domestic European pear (*Pyrus communis* L.) has been derived from the large fruited wild species of *P. caucasica* Fed. and *P. nivalis* Jacq. The domestic pears of Asia, have been derived mostly either from *P. pyrifolia* (Burm.) Nak. (Japanese sand pear) or from the selection of ussuri pear, *P. ussuriensis* Maxim. In India, pear is found right from cold dry temperate Himalayan region to warm humid sub-tropical plains in the form of scattered plantation comprising the states of Himachal Pradesh, Jammu and Kashmir, Uttrakhand and some parts of Punjab. Pear cultivation is also practiced in Mizoram, Nagaland, Manipur and some parts of Tamilnadu. Pear crop has a wide range of climatic and soil adaptability and can be grown even in marginal lands due to which pear plantation is gaining popularity among the orchardists (Verma and Sharma, 1999). Pears are being grown in India in an area of 42000 ha with a production of 382000 MT (Anonymous, 2011b). In Himachal Pradesh, it is the second most important temperate fruit crop after apple and occupies an area of 7370 ha with an annual production of 32075 metric tonnes (Anonymous, 2011c).

Table 12.3. List of important diseases of pear

S. no.	Name of the disease	Causal organism
Fungal diseases		
1.	Pear scab	*Venturia pirina* Aderh.
		(Ana.: *Fusicladium pyrorum* (Lib.) Fucket)
2	Black rot	*Botryosphaeria obtusa* (Schw.) Shoemaker
		(Ana.: *Sphaeropsis malorum* Berk.)
3.	Phyllosticta leaf spot	*Phyllosticta* sp.
4.	*Mycosphaerella* leaf spot	1. *Mycosphaerella pyri* (Auersw.) Boerema
		(Ana.: *Septoria pyricola* (Desmaz.) Desmaz.)
		2. *M. chaubattiensis* Bose and Roy
5.	Powdery mildew	1. *Podosphaera leucotricha* (Ell. and Ev.) Salm.
		2. *Phyllactinia guttaca* (Wallr.: Fr.) Lev
6.	European pear rust	*Gymnosporangium fuscum* R. Hedw. in DC
7.	Cankers	
a.	European canker	*Nectria galligena* Bres.
		(Ana.: *Cylindrocarpon heteronemum* (Berk. and Br.)
		Wollenweb.)
b.	Diaporthe canker	*Diaporthe tanake* Kobayashi and Sukuma
		(Ana.: *Phomopsis tanake*)
c.	Silver leaf canker	*Chondrostereum purpureum* (Pers.: Fr.) Pouzar
8.	Japanese pear black spot	*Alternaria altornata* (Fr.) Keissler
9.	Flyspeck	*Schizothyrium pomi* Mont.: Fr.
		(Ana.: *Zygophiala jamaicensis* Masson)
10.	Sooty blotch	*Gloeodes pomigena* (Schw.) Colby.
Bacterial diseases		
11.	Bacterial blossom blast	*Pseudomonas syringae* pv. *syringae* Van Hall
Post-harvest diseases		
1.	Alternaria rot	*Alternaria alternata* (Fr.) Keissler
2.	Bitter rot	*Glomerella cingulata* (Stonem.) Spauld. and Schrenk
		(Ana.: *Colletotrichum gloeosporioides* (Penz.) Sacc.)
3.	Black rot	*Botryosphaeria obtusa* (Schw.) Shoem.
		(Ana.: *Sphaeropsis malorum* Berk.)
4.	Blue mould rot	*Penicillium expansum* Link.
5.	Brown rot	*Monilinia* sp.(Ana.: *Monilia* spp.)
6.	Pink mould rot	*Trichothecium roseum* Link.
7.	Fusarium rot	*Fusarium* spp.
	Scab	*Venturia pirina* Aderh.
		(Ana.: *Fusicladium pyrorum* (Lib.) Fucket)
8.	Mucor rot	*Mucor piriformis* Fischer
9.	Rhizopus rot	*Rhizopus stolonifer* (Ehrenb.: Fr.) Lind.
10.	Flyspeck	*Schizothyrium pomi* Mont.: Fr.
		(Ana.: *Zygophiala jamaicensis* Masson)
11.	Sooty blotch	*Gloeodes pomigena* (Schw.) Colby.
12.	Cylindrocarpon rot	*Nectria galligena* Bresad.
		(Ana.: *Cylindrocarpon mali* (Allesch.) Wollenw.
13.	Grey mould rot	*Botryotinia fuckeliana* (de Bary) Whetzel
		(Ana.: *Botrytis cinerea* Pers.: Fr.)
14.	Phytophthora rot	*Phytophthora cactorum* (Leb. and Cohn) Schroet.

1. Pear scab

1.1. Occurrence and distribution: Pear scab is a serious disease in almost all pear growing countries including India (Fig. 12.27). In India, the disease was first reported from Himachal Pradesh by Chona *et al.* (1956) and subsequently from Kashmir valley by Puttoo and Chaudhary (1984). It is a serious problem of Chinese, Japanese and European pears such as Bartlett, Beurre Bose, Comice, d'Anjou and Winter Nelis (Verma and Sharma, 1999).

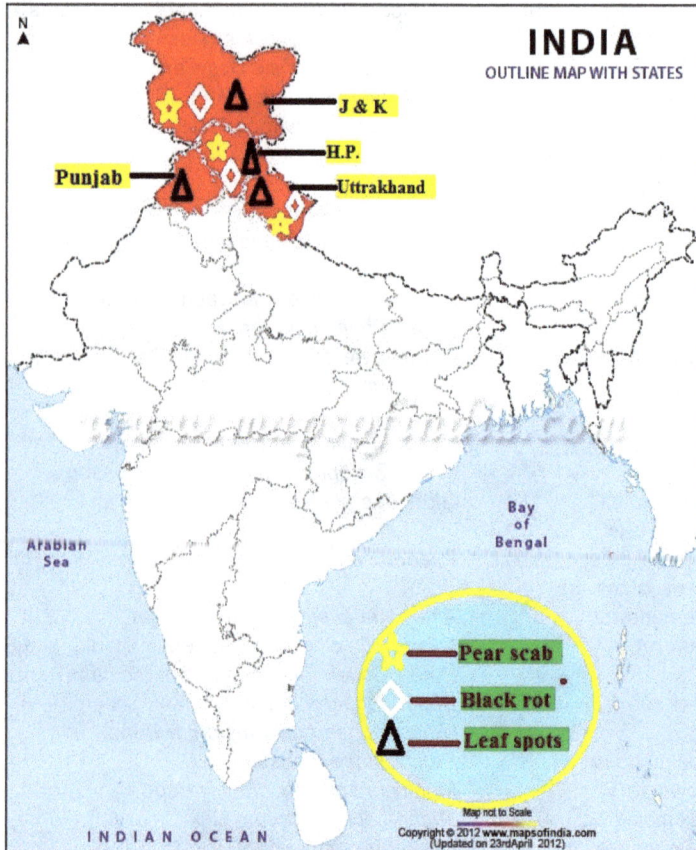

Fig. 12.27. Distribution of major pear diseases in India

1.2. Crop losses: The losses occurred due to mid season defoliation, failure of fruit bud formation, weakening of trees, reduction in fruit production and devaluation of fruit quality (Ahmad and Sagar, 2010). Pear scab has been reported to cause an annual loss up to 30 per cent (Gaumann, 1927).

1.3. Symptoms: On leaves, circular, rough and olive green coloured lesions develop. The lesions vary from 3-8 mm in size (Fig. 12.28a). Lesions on actively growing branches of susceptible cultivars are brown and velvety. Later on these lesions become canker like with few spores (Putto and Chaudhary, 1984). On fruits, the lesions appear on the calyx-end. Infected fruit often become misshapen having coalesced dark brown to black patches (Fig. 12.28b).

Fig. 12.28. Symptoms of pear scab a. leaves b. fruits

1.4. Causal organism: Pear scab is incited by *Venturia pirina* Aderh. (Ana.: *Fusicladium pyrorum* (Lib.) Fucket), which produces globose to conical pseudothecia in the overwintered leaf tissues. Asci are oblong and bitunicate with eight ascospores. Ascospores are smooth, oblong, 2-celled and slightly constricted at the septum. The smaller cell is oriented towards the base of ascus. Conidiophores are brown coloured, erect and septate, bearing distinct scars indicative of detachment sites of conidia. Conidia are olive brown, smooth and fusiform, mostly one-celled but sometimes bi-celled (Fig. 12.29), measuring 15-40 × 6-10 μm in natural habitat (Doliveira, 1937).

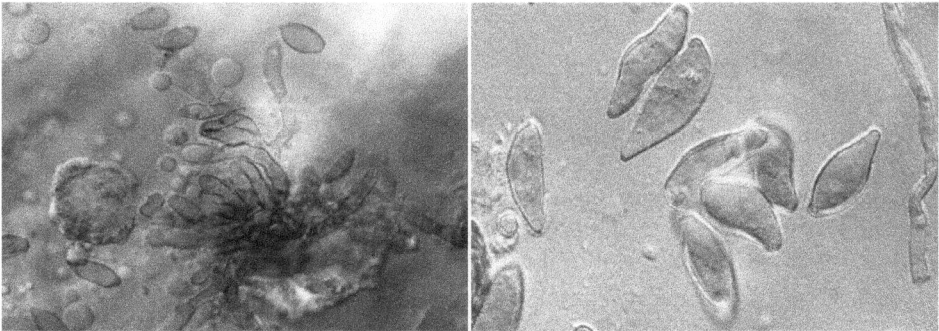

Fig. 12.29 Conidia of *Venturia pirina*

1.5. Epidemiology: The period of wetting and favourable temperature required for infection is quite similar to that of apple scab. If the leaf surface is wet, the optimum temperature for infection is 16-20°C. The duration of wetting period required for infection depends on the temperature. The incubation period for lesion development varies from 8 days in young leaves to more than 60 days in older leaves (Ogata *et al.*, 1994). Conidia produced in primary lesions lead to secondary infection. Young fruits are most susceptible, but mature fruits may be infected if wetting period exceeds 24 hours. Increased nitrogen fertilization increases the disease incidence (Unemoto, 1991). The conidia produced from infected twigs are disseminated by rain. Light stimulated the discharge of ascospores (Spotts and Cervantes. 1994).

2. Black Rot

2.1. Occurrence and distribution: Black rot of pear is an important disease and has been reported from India, Australia, New Zealand, North and South America and Zimbabwe. In India, it appears in more or less severe form depending upon the environmental factors of the particular state (Fig. 12.27).

2.2. Crop losses: Losses from the disease occur through fruit rot, twig cankers and weakening of trees through premature defoliation however the exact estimated has not been reported.

2.3. Symptoms: Sepal infection is the most common form of initial infection which appears as minute red specks later turning purple and bordered by red ring. Later in the season the sepal infection results in blossom end rot. After petal fall, infections on young fruit begin as reddish flecks which develop into purple lesions. These lesions remain static until fruit begin to mature. As lesions enlarge, a series of concentric rings alternating from black to brown are formed. Pycnidia appear scattered on the lesion surface. Mummified fruits remain attached to the tree. In India, the pathogen has also been reported to cause fruit rot (Srivastava and Tandon, 1969) (Fig. 12.30a & b). Leaf symptoms usually appear 1 to 3 weeks after petal fall. The centre of the lesion is tan to brown and margin is purple, giving the lesion a frog eye appearnace. Twig cankers begin in the bark as slightly sunken reddish brown areas. Branches damaged by climatic stresses are often attacked by this pathogen (Fig. 12.30c). Sometimes infected branches are girdled and killed (Jones, 1990).

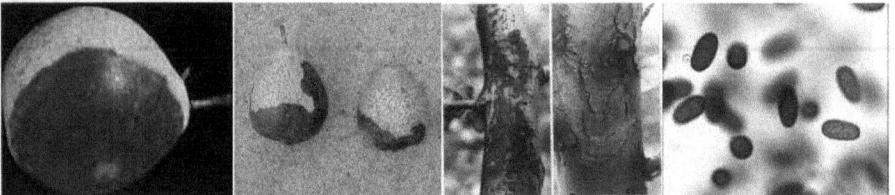

Fig. 12.30. Symptoms of black rot on pear a and b on fruits c. on stem. d. Conidia of the Sphaeropsis malorum

2.4. Causal organism: Black rot is caused by *Botryosphaeria obtusa* Schwein (Ana.: *Sphaeropsis malorum* Berk.). It produces pycnidia commonly on infected twigs and fruit, and rarely on leaves (Kawai *et al.*, 1992). The pycnidia are globose, stromatic with papillate ostioles. Conidia are non-septate, ovoid with rough echinulations (Fig. 12.30d). Pseudothecia are usually produced on infected twigs which are interspersed with pycnidia. The pseudothecia are dark brown to black, up to 3 mm wide, with ostioles darker around the neck region. Asci are 8 spored, bitunicate, and ascospores are single celled, sometimes two celled and fusiform (Punithalingam and Waller, 1973).

2.5. Epidemiology: The pathogen overwinters in cankers and mummified fruits. Conidia are released from pycnidia throughout the year. Ascospores are released during rainfall usually after bud break. However, maximum conidia are released before mid May to late August and ascospores during late April to mid June (Ko and Sun, 1995). The rate of conidia and ascospores germination decline with the decrease in relative humidity, more than 80 per cent germinate at 100 per cent relative humidity within 2 hours incubated at 28° to 32°C. No germination occurs at relative humidity less than 92 per cent. Infection occurs through stomata on leaves, sepals and in early season fruit. However, infection on mature fruits takes place through lenticels, cuticle cracks and wounds. Conidia are primarily dispersed by rain water whereas ascospores are airborne as well as waterborne.

3. Leaf spots

Many leaf spot diseases of pear probably cause more damage than they appear to. Leaf spot diseases caused damage by reduction in active photosynthetic area. They are known to cause moderate to severe premature defoliation in mid-season and hinder the photosynthetic efficiency of the trees, thereby affecting the quality and yield of the fruit produce. Many fungal pathogens are involved in causation of leaf spots on pear but, in India economic losses are caused mainly by *Mycosphaerella* spp. and *Phyllosticta* spp. (Verma and Sharma, 1999) (Fig. 12.27).

3.1. Losses: Leaf spot diseases cause losses by reducing the active photosynthetic area and heavy infestation resulting in premature defoliation thereby reducing the quality and yield of the fruit produce. *Phyllosticta* leaf spot of pear occurred in epiphytotic form in 1986 in Solan and Bilaspur districts of Himachal Pradesh with 100 per cent incidence by the end of July (Gautam, 1987). Sharma and Bhardwaj (1998) reported 90 per cent incidence of this disease in district Kullu of Himachal Pradesh and estimated the losses up to an extent of 21.4 per cent. Parihar (2006) recorded average disease incidence of 82.33 per cent and 52.15 per cent disease severity of pear leaf spots (*Phyllosticta pyricola* and *Mycosphaerella pyri*) from different pear growing regions of Himachal Pradesh.

3.2. *Mycosphaerella* leaf spot: This disease is of great significance in nurseries. The disease also occurs on pear trees in an areas receiving frequent rains. Initially spots of about 3 mm diameter appear on the upper leaf surface. The spots are greyish white with purplish margin and contain small, black pycnidia in their centre (Fig. 12.31a & b). Sometimes shot-hole symptoms are also produced. Higher incidence results in premature defoliation of such affected leaves.

Fig. 12.31 Mycosphaerella leaf spot of pear a and b. Symptoms on pear leaf. c. Conidia of *Septoria* sp. Tel. *Mycosphaerella* sp

The disease is caused by *Mycosphaerella pyri* (Auersw.) Boerema (Ana.: *Septoria pyricola* Desmaz.), which produce perithecia and pycnidia. The perithecia are black with a long ostiole and occur in clusters on fallen leaves. The asci are clavate and ascospores are pyriform and 2-celled (Tzavella Klonari and Tamoutseli, 1986). The pycnidia are embedded in the leaf tissues and similar to perithecia in size and shape. The conidia are filiform and usually have two septa (Fig. 12.31c). From India, *Mycosphaerella* leaf spot has been reported to be caused by *M. pirina* Ell. and Ev. and *M. chaubattiensis* Bose and Roy (Bose and Roy, 1970). *Mycosphaerella pyri* overwinters in infected leaves as perithecia. The ascospores are discharged in spring and lesion development takes place in about 2 weeks. After about 1½ months of infection pycnidia are formed. High relative humidity favours the epidemic development (Aloj *et al.,* 1994).

3.3. *Phyllosticta* leaf spot: Lobik (1928) identified two species of *Phyllosticta* namely *P. piricola* and *P. pirina.*reported from pear leaves from North Caucasus. In India, leaf spot disease caused by *Phyllosticta piricola* was reported by Sharma *el al.,* (1986) from Himachal Pradesh. Disease symptoms occur on the both sides of the leaves, but are more pronounced on the upper surface. The leaf spots are initially small and circular in size and later enlarged and formed irregular necrotic spots. The centre of the spot became round and greyish in colour at later stages of infection. Outer necrotic area enlarged and coalesced with each other and ultimately covered most part of the leaf. The affected leaves drop prematurely. Numerous dots like structures called pycnidia are noticed on greyish to brown spots of the leaves in the month of July.(Fig. 12.32a). The disease is caused by *Phyllosticta piricola* Sacc. Pycnidia are dark coloured, ostiolated and immersed and measures 80.5 µm to 140 µm in diameter. Pycnidiospores are single celled, small, hyaline, oval to oblong and measuring 3.5 µm to 6.9 µm in size. (Fig. 12.32b).

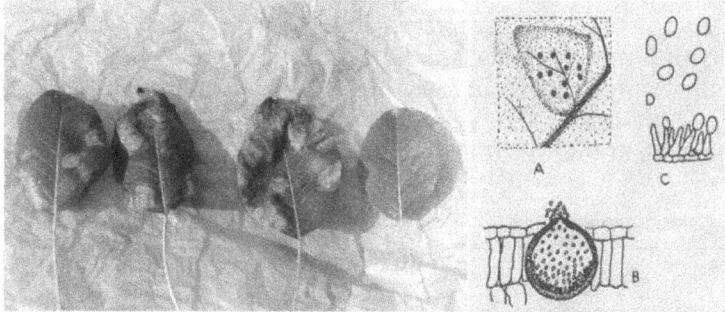

Fig. 12.32. a. Symptom of Phyllosticta leaf spot. b. Pycnidium, conidiophore

Gupta and Gautam (1989) reported that leaf spot of pear caused by *Phyllosticta piricola* appeared mostly with the onset of rain in July and its intensity increased in rainy season with 80-90 per cent relative humidity at 25-27°C temperature. The disease spreads through rain splashes. Sharma and Bhardwaj (1998) established that disease appeared in May with the occurrence of mean temperature of 22°C accompanied with RH of 53.2-62 per cent. It spread at a faster speed between third week of June to mid August with prevalence of the temperature and RH ranging between 23-25C, 73.2-82.4 per cent, respectively

4. Post-harvest diseases

Post-harvest diseases are important in reducing the supply of pears in the market for sale and are responsible for losses that occur during picking, packing, transportation and storage of fruits. Sharma and Agarwala (1985) recorded incidence of bitter rot of pear up to an extent of 16 per cent in Himachal Pradesh. Sharma (1991) recorded 2.1 to 19.7 per cent storage rot of China pear due to different fungi from Himachal Pradesh. Of these, grey mould rot (33.1%) caused by *Botrytis cinerea* and brown rot caused by two species of *Monilinia* viz. *M. laxa* (20.6%) and *M. fructigena* (4.0%) were most prevalent. Ratnam and Nema (1967) observed 10.3-23.4 per cent loss in pears from the stores and markets of Jabalpur (M.P.). Dasgupta and Mandal (1989) reported 19.7 per cent loss of apple and pear fruits from Calcutta fruit markets. Singh (2002) reported 13.1-25.9 per cent loss of pears from Jammu fruit market.. The major fungi responsible for post-harvest losses of pears in India are *Botrytis, Monilinia, Penicillium* and *Glomerella* sp. (Fig. 12.33).

Fig. 12.33. (From left to right) a. Grey mould rot. b. Brown rot. c. Blue mould rot. d. Bitter rot

REFERENCES

Agarwala, R. K. 1961. Problem of root rot of apple in Himachal Pradesh and prospectus of its control with antibiotics. Himachal Hortic. 2: 171-178.

Agarwala, R. K., and Gupta, G. K. 1971. Canker disease complex of apple trees. Second International Symposium on Plant Pathology. IPS, New Delhi, Jan. 1971. 160p. (Abst.).

Agarwala, R. K., and Gupta, V. K. 1981. Effect of seasonal changes on the occurrence of white root rot of apple and production of macerating enzymes in disease syndrome. Third International Symposium on Plant Pathology. 1981. IPS, New Delhi. pp. 113.

Agarwala, R. K., and Sharma, V. C. 1966. White root rot disease of apple in Himachal Pradesh. Indian Phytopath. 19: 82-86.

Ahmad, S. and Sagar, V. 2010. Pear scab. In: The Pear: production, post-harvest management and protection. (R. M. Sharma, S. N. Pandey and V. Pandey, eds.): 623-626. IBDC Publishers, Lucknow, India.

Aloj, B., Nanni, B., Marziano, F., and Noviello, C. 1994. Severe and unusual occurrence of leaf fleck on pear trees in Campania. Inf. Fitopatol. 44: 25-29.

Anonymous. 1960. Annual Report of Regional Research Station, Mashobra, H.P. for the year 1959. 60p.

Anonymous. 1937. Some diseases of apple. Mysore Agric. Cal. pp. 25, 33.

Anonymous. 2013a. Indian Horticulture Database, National Horticulture Board, Gurgaon, India.

Anonymous. 2011b. http: //www.fao.org

Anonymous. 2011c. Area and Production of Fruit Crops in H.P. Directorate of Horticulture, Navbahar, Shimla (HP).

Anonymous. 2012. Directorate of Horticulture, Navbahar, Shimla (HP).

Arauz, L. F., and Sutton, T. B. 1989. Temperature and wetness duration requirements for apple infection by *Botryosphaeria obtusa*. Phytopathology 79: 440-444.

Baines, R. C. 1939. Phytophthora trunk canker or collar rot of apple trees. J. Agric. Res. 59: 159-184.

Bakshi, B. K., Ram Reddy, M. A., Singh, S., and Pandey, P. C. 1970. Disease situation in Indian forests-1. Stem diseases of some exotic plants due to *Corticium salmonicolor* and *Monochaetia unicornis*. Indian For. 96: 826-829.

Bessey, C. W. 1877. On injurious fungi: the blight (*Erysiphe*). Iowa State College of Agric. Bienn. Rep. pp. 185-204.

Bhardwaj, L. N., and Agarwala, R. K. 1986. Effect of some important edaphic factors on the incidence of seedling blight of apple caused by *Sclerotium rolfsii* Sacc. In: Advances in Research on Temperate Fruits. (T. R. Chadha, V. P. Bhutani and J. L. Kaul, eds.): 403-406. UHF, Nauni.

Birmingham, W. A. 1924. A canker of apple tree due to a fungus *Dothiorella mali* E. and E. *Agric. Gaz.* N.S.W. 35: 525-527.

Bose, S. K., and Mehta, R. B. 1951. Records of disease occurrence. Plant Prot. Bull., New Delhi 3: 46-48.

Bose, S. K., and Roy, A. J. 1970. Leaf spot disease of pear in Kumaon caused by *Mycosphaerella chaubattiensis* sp. nov. Prog. Hortic. 1: 57-62.

Dey, P. K., and Singh, U. B. 1939. The stem black disease of apple in Kumaon. Indian J. Agric. Sci. 9: 703-710.

Doliveira, B. 1937. Indications for thestudy of the genus *Fusidladium* III conidia fructification of *Fusidladium dendreticum* and *F. ericbetrye*. Rev. Agric. Lisboa 25: 140-164.

Dong, H. S., Hung, J. K., Yang, Y. S., Moo, Y. P., Jong, C. N., Soek, B. K., and Sang, G. L. 2011. Influence of defoliation by Marssonina blotch on vegetative growth and fruit quality in 'Fuji'/M.9 apple tree. Korean J. Hortic. Sci. 29: 531-538.

Ellis, J. B., and Everhart, B. M. 1896. A new species of fungi from various localities. Proc. Acad. Nat. Sci. Phila. 1895: 430.

Filajdic, N., and Sutton, T. B. 1992. Influence of temperature and wetness duration on infection of apple leaves and virulence of different isolates of *Alternaria mali*. Phytopathology 82: 1279-1283.

Fulkerson, J. F. 1960. *Botryosphaeria ribis* and its relation to rot of apple. Phytopathology 50: 394-398.

Gadoury, D. M., and MacHardy, W. E. 1982. Effects of temperature on the development of pseudothecia of *Venturia inaequalis*. Plant Dis. 66: 468.

Gaumann, E. 1927. The economic significance of our principal plant diseases. *Landw. Jhrb. Der Schweiz.* 12: 319-324.

Gautam, H. R. 1987. Studies on Phyllosticta leaf spot disease of pear. M. Sc. Thesis, Dr. YS Parmar University of Horticulture and Forestry, Nauni, Solan. 69p.

Grove, G. G. 1985. Anthracnose and perennial canker. In: Compendium of Apple and Pear Diseases. (A. L. Jones and H. S. Adwinckle, eds.): 36-38. APS Press, St. Pual Minnesota.

Guba, E. F. 1924. Phyllosticta leaf spot fruit blotch and canker of apple: its etiology and control. Phytopathology 14: 234-237.

Gupta, G. K. 1975. Canker of apple trees due to *Sphaeropsis malorum* Berk. and its control. Himachal J. Agric. Res. 3: 44-52.

Gupta, G. K. 1978. Present status of apple scab (*Venturia inaequalis*) in Himachal Pradesh and strategy for its control. Pesticides 12: 13-14.

Gupta, G. K. 1979. Some observations on apple scab in Himachal Pradesh. Indian phytopath. 32: 172.

Gupta, G. K. 1987. Apple scab and its management. Indian Hortic. 32: 48-52.

Gupta, G. K. 1996. Fungal diseases of temperate fruits in India. In: Advances in Diseases of Fruit Crops in India. (S. J. Singh, ed.): 429. Kalyani Pub., Ludhiana.

Gupta, G. K., and Agarwala, R. K. 1968 Alternaria blight of apple. *FAO* Plant Prot. Bull. 16: 32.

Gupta, G. K., and Agarwala, R. K. 1973. Canker diseases of apple in Himachal Pradesh. Indian J. Mycol. Pl. Pathol. 3: 189-192.

Gupta, G. K., and Lele, V. C. 1980. Morphology, physiology, and epidemiology of the apple scab fungus, *Venturia inaequalis* (Cke.) Wint. in Kashmir valley. Indian J. Agric. Sci. 50: 51-60.

Gupta, G. K., Verma, K. D. and Pal, J. 1984. Some observations on the prevalence and severity of apple scab in Himachal Pradesh during the year 1978 to 1983. Indian J. Mycol. Pl. Pathol. 14: 12-14 (Abst.).

Gupta, V. K. 1977. Root rot of apple and its control by carbendazim. Pesticides 11: 49-52.

Gupta, V. K., and Gautam, H. R. 1989. Control of Phyllosticta leaf spot of pear. In: Tree Protection, Indian Society of Tree Scientists, Solan. pp. 42.

Gupta, V. K., and Sharma, S. K. 1999. Soil borne diseases of apple and their management. In: Diseases of Horticultural Crops. (L. R. Verma and R. C. Sharma, eds.): 89-104. Indus Publishing Co., New Delhi.

Harada, Y., Sawamura, K., and Konno, K. 1974. *Diplocarpon mali* sp. nov., the perfect state of apple blotch fungus *Marssonina coronaria*. Ann. Phytopathol. Soc. Jpn. 40: 412-418.

Heladze, V. S. 1968. Black canker of apple tree. Zashch. Rast. (Mosc.). 13: 56.

Holmes, J., and Rich, A. E. 1969. The control of black rot canker on apple trees in New Hampshire. Plant Dis. Rep. 53: 315-318.

Jain, S. S. 1961. Root and collar rot diseases in Himachal Pradesh. Himachal Hortic. 2: 19-23.

Jain, S. S. 1962. Studies on apple seedling blight *Sclerotium rolfsii* Sacc. in Himachal Pradesh. Proc. Indian Sci. Cong. 3: 247.

Janisiewicz, W. J., and Korsten, L. 2002. Biological control of post-harvest diseases of fruits. Ann. Rev. Phytopathol. 40: 411-441.

Jones, A. L. 1990. Diplodia canker. In: Compendium of Apple and Pear Diseases. (A. L. Jones and H. S. Adwinckle, eds.): 100. APS Press, St. Paul Minnesota.

Joshi, N. C., Mallik, A. G., Kaul, M. L., and Anand, S. K. 1975. Some observations on the epidemic of scab disease of apple in Jammu and Kashmir during 1973. Indian Phytopath. 28: 288-289.

Kawai, Y., Saito, Y., and Iuima, A. 1992. The time of infection of pear *Physalospora* canker on fruit and the disease development. Proc. Kanto Tosan Plant Prot. Soc. 39: 143-144.

Khan, N. A., Ahmad, M., and Ghani, M. Y. 2010. *Botyrosphaeria dothidea* associated with white rot and stem bark canker of apple in Jammu & Kashmir. Appl. Biol. Res. 12: 69-73.

Khan, N. A., Ahmad, M., Ahmad, K., and Beig, M. A. 2011. Etiology and occurrence of valsa apple canker in Jammu & Kashmir state. Appl. Biol. Res. 13: 48-50.

Klingner, A. E., and Pontis, R. E. 1974. Apple canker caused by *Sphaeropsis malorum* in Mendoza. Rev. Fac. Cienc. Agrar. Univ. Nac. Cuyo. 18: 57-68.

Ko, Y., and Sun, S. K. 1995. Physiological characteristics and population dynamics of *Botryosphaeria dothidea*. *Plant Prot*. Bull. (Taipei) 37: 281-293.

Kohn, F. C. Jr., and Hendrix, F. F. 1982. Temperature and free moisture and inoculum concentration effect on incidence and development of white rot of apple. Phytopathology 72: 313-316.

Kohn, F. C. Jr., and Hendrix, F. F. 1983. Influence of sugar contents and pH on development of white rot of apple. Plant Dis. 67: 410-412.

Kaul, J. L. 1985. Storage scab and its management. In: Principal and Concepts of Apple Scab Disease Management. (R. K. Agarwala, J. L. Kaul and L. N. Bhardwaj, eds.): 96-98. Haryana Agricultural University Press, Hissar.

Kaul, J. L., and Munjal, R. L. 1982. Apple losses in Himachal Pradesh due to post-harvest fungal pathogens. *Indian J. Mycol. Pl.* Pathol. 12: 209-210.

Krishna, H., Das, B., Attri, B. L., Grover, M., and Ahmad, N. 2010. Suppression of *Botryosphaeria* canker of apple by arbuscular mycorrhizal fungi. Crop Protection 29: 1049-1054.

Kumar, J., and Gupta, G. K. 1986. Influence of host response and climatic factors on the development of conidial stage of apple scab fungus (*Venturia inaequalis*). Indian J. Mycol. Pl. Pathol. 16: 123-135.

Kumar, J., and Singh, U. S. 1989. Status of apple scab in U.P. hills. Pesticides 23: 39-42.

Lobik, A. I. 1928. Fungi parasitizing fruit trees and their control. Terek Reg. Plant Prot. Stn. News 1-2: 20-83.

Malik, R. A. 1967. The canker that damage apple in Kashmir. Indian Hortic. 11: 25-26.

Mallick, F., Shukla, N. B., and Bhutia, U. 1984a. Occurrence of apple scab in Sikkim. Plant Prot. Bull. 36: 121.

Mallick, F., Singh, A. K., and Shukla, C. P. 1984b. Occurrence of apple scab in Arunachal Pradesh. Plant Prot. Bull. 36: 123.

Mills, W. D. 1944. Efficient use of sulphur dust and sprays during rain to control apple scab. Cornell Univ. Ext. Bull. 630: 4.

Mundkur, B. B., and Kheswalia, K. F. 1943. A canker of apple tree in Mysore. Mysore J. Agric. Sci. 13: 397-398.

Nath, P. 1935. Studies in the disease of apples in Northern India. 11. A short note on apple scab due to *Fusicladium dendrilicum* Fuckel. J. Indian Bot. Soc. 14: 121-124.

Nirupma, S., Thakur, V. S., Mohan, J., Paul Khurana, S. M., and Sharma, S. 2009. Epidemiology of *Marssonina* blotch (*Marssonina coronaria*) of apple in India. Indian Phytopathol. 62: 348-359.

Nirupma, S., Thakur, V. S., Sharma, S., Mohan, J., and Paul Khurana, S. M. 2011. Development of *Marssonina* blotch (*Marssonina coronaria*) in different genotypes of apple. Indian Phytopath. 64: 358-362.

Ogata, T., Ito, K., and Ochiai, M. 1994. Analysis of a regional characteristic for Japanese pear scab development by utilization of the weather data obtained from AMEDAS (Automated Meteorological Data Acquisition System). Annu. Rep. Soc. Plant Prot. North Jpn. 45: 114-116

Paddock, W. 1899. The New York apple tree canker. New York Agric. Exp. Stn. Bull. (Geneva) 163: 173-206.

Parihar, A. S. 2006. Studies on leaf spot diseases of pear and their management. M.Sc. thesis, Dr. YS Parmar University of Horticulture and Forestry, Nauni, Solan, 86p.

Pathak, V. N. 1986. Diseases of Fruit Crops. Oxford and IBH Publishing Co., New Delhi. 309 p.

Punithalingam, E., and Holliday, P. 1973. *Botryosphaeria ribis*. Description of Pathogenic Fungi and Bacteria No. 395, CM1, Kew, Surrey, England.

Punithalingam, E., and Waller, J. M. 1973. *Botryosphaeria obtusa*. Descriptions of Pathogenic Fungi and Bacteria No. 394, CMI. Kew, Surrey, England.

Putterill, V. A. 1919. A new apple tree canker. S. Afr. J. Soc. 16: 258-272.

Puttoo, B. L. 1972. Two new fungi of apple *(Malus sylvestris)* in Kashmir. Indian J. Mycol. Pl. Pathol. 2: 193.

Puttoo, B. L. 1987. A new defoliation disease of apple in Kashmir. Indian J. Mycol. Pl. Pathol. 17: 109-110.

Puttoo, B. L., and Chaudhary, K. C. B. 1984. A study of pear scab-symptomatology and morphology of the pathogen. Indian Phytopath. 37: 699-701.

Rana, K. S., and Gupta, V. K. 1983. Effect of different soil factors and amendments on the production of sporangia and zoospores in *Phytophthota cactorum*. Proc. Natnl. Acad. Sci. B. 49: 706-710.

Rao, V. G. 1965. The genus Phyllosticta in India. Sydowia. 19: 117-120.

Ratnam, C. V., and Nema, K. G. 1967. Studies on market diseases of fruits and vegetables. Andhra Agric. J. 14: 60-65.

Roy, A., Seth, J. N., and Shah, A. 1985. Comparative effect of fertilizers and different fungicidal treatments on the incidence of leaf spot disease of apple var. Red Delicious. Prog. Hortic. 17: 241-244.

Sakuma, T. 1990. Canker disease of deciduous fruit trees in Japan. Pestic. Inf. 57: 12-14.

Shandilya, T. R., Thakur, M. S., and Agarwala, R. K. 1973. Effect of age and altitude on the incidence of canker diseases of apple in Himachal Pradesh. Indian J. Mycol. Pl. Pathol. 3: 102-103.

Sharma, I. M. 1999. White root rot of apple: problem and management. Divya Himachal (Daily Newspaper). July 28.

Sharma, I. M. 2003a. Probable reasons for zero level of apple scab incidence in Kullu valley of Himachal Pradesh. J. Mycol. Pl. Pathol. 33: 96-100.

Sharma, I. M. 2003b. Influence of environmental factors on the development of pre-mature defoliation disease caused by *Marssonina coronaria* in apple and its management. J. Mycol. Pl. Pathol. 33: 89-95.

Sharma, I. M. 2006. Outbreak of a new canker disease of apple in Himachal Pradesh and its control. J. Mycol. Pl. Pathol. 36: 332-335.

Sharma, I. M. 2012a. Personal communication. Dr. Y.S. Parmar University of Horticulture and Forestry, Nauni, Solan (HP).

Sharma, I. M. 2012b. Diagnosis and management of core rot in apple: A new emerging disease in Himachal Pradesh, India. Proceedings of 3rd Global Conference on Plant Pathology for Food Security Jan. 10-13, 2012 at MPUAT, Udaipur, pp. 172 (Abst.).

Sharma, I. M., and Bhardwaj, S. S. 1994. Annual Progress Report. Dr. Y.S. Parmar University of Horticulture and Forestry, Regional Horticultural Research Station, Bajaura (Kullu). 63p.

Sharma, I. M., and Bhardwaj, S. S. 1997. Diagnosis and management of premature defoliation disease of apple in Kullu valley of Himachal Pradesh. Proceedings of International Conference on Integrated Plant Disease Management for Sustainable Agriculture 10-15 Nov. 1997, by IPS and ICAR, New Delhi. pp. 329 (Abst.).

Sharma, I. M., and Bhardwaj, S. S. 1998. Leaf spot diseases of pome and stone fruits In: Annual Progress Report, Regional Horticultural research Station of Dr. Y.S. Parmar university of Horticulture and Forestry, Bajaura, Kullu, p.54-57.

Sharma, I. M., and Bhardwaj, S. S. 1999. Canker and Foliar Diseases of Apple. In: Diseases of Horticultural crops-Fruits. (L. R. Verma and R. C. Sharma, eds): 15-33. Indus Publishing Co., New Delhi.

Sharma, I. M., and Bhardwaj, S. S. 2000. Use of plant extracts and yeast antagonists in management of storage scab, and rots of apple fruits. J. Biological Control. 14: 17-23.

Sharma, I. M., and Bhardwaj, S. S. 2002. Management of seedling blight of apple through eco-friendly methods. In: Proceedings of National Symposium on Perspective in Integrated Plant Disease Management at NRC for Citrus, Nagpur, Feb. 13-14. pp. 82 (Abst.).

Sharma, I. M., and Bhardwaj, S. S. 2005. Disease management in apple-progress, challenges and strategies. In: Proceedings of Centenary Symposium on Plant Pathology at CPRI Shimla, April 7-8. pp. SIV-3 (Abst.).

Sharma, I. M., and Bhardwaj, S. S. 2006. Effect of early defoliation disease of apple on yield, quality and next year's fruit set. Pl. Dis. Res. 21: 185-186.

Sharma, I. M., and Gupta, V. K. 1989. Screening of apple germplasm for susceptibility to *Pythium ultimum* Trow during different seasons. Scientia Hort. 40: 63-69.

Sharma, I. M., and Ram, V. 2010. Occurrence and management of important cankers in apple. J. Mycol. Pl. Pathol. 40: 213-218.

Sharma, I. M., Rathore, R., Gupta, B., and Bhardwaj, S. S. 2011a. Development of ecofriendly integrated management strategy against collar rot (*Phytophthora cactorum* (Leb. and Cohn) Schroet.) in apple. Proceedings of International Workshop, Seminar and Exhibition on Phytophthora Diseases of Plantation Crops and their Management, Sept. 12-17, 2011 at RRI, Kottayam, Kerela. pp. 144-146.

Sharma, J. N., and Verma, L. R. 1999. Pre-mature leaf fall in apple. In: Diseases of Horticultural Crops-Fruits (L. R. Verma and R. C. Sharma, eds): 80-89. Indus Publishing Co., New Delhi.

Sharma, J. N., Sharma, A., and Sharma, P. 2004. Out-break of *Marssonina* Blotch in warmer climates causing premature leaf fall problem of apple and its management. Acta Hort. 662: 405-409.

Sharma, K. K. 1985. Epidemiology and control of powdery mildew of apple. PhD thesis, Dr. YS Parmar University of Horticulture and Forestry Nauni, Solan (HP). 121p.

Sharma, K. K., and Gupta, V. K. 1991. Effect of temperature and relative humidity on conidial germination and germ tube length of *Podosphaera leucotricha* on glassslide. Plant Dis. Res. 6: 91-95.

Sharma, R. C., Gupta, V. K., and Agarwala, R. K. 1986. Epidemiology and control of Phyllosticta leaf spot of pear. In: National Symposium on Researches in Social Forestry for Rural Development held at Delhi University, Delhi, 1-2, jan. 1986. pp. 12 (Abst.)

Sharma, R. C., and Kaul, J. L. 1999. Post harvest diseases of temperate fruits and their management. In: Diseases of Horticultural Crops-Fruits (L. R. Verma and R. C. Sharma, eds): 582-623. Indus Publishing Co., New Delhi.

Sharma, R. L. 1991. Prevalence of storage rots of China pear in Himachal Pradesh. Pl. Dis. Res. 6: 86-88

Sharma, S., and Agarwala, R. K. 1985. Incidence of bitter rot of pear caused by *Glomerella cingulata* in Himachal Pradesh. Indian J. Mycol. Pl. Pathol. 15: 103-104.

Sharma, S. K., Kumar, A., Singh, J., and Kumar, P. 2011b. Emerging disease problems of temperate fruits in cold desert and dry temperate regions of Kinnaur-the strategic issue. National Symposium on Strategic Issues of Plant Pathological Research, Nov. 24-25, CSK HP Krishi Vishvavidyalaya, Palampur (HP). pp. 76(Abst.).

Shey, J. R., and Witterly, W. R. 1954. Botryosphaeria canker of apple. Phytopathology 44: 504.

Shoemaker, R. A. 1964. Conidial state of some *Botryosphaeria* species on *Vitis* and *Quercus*. Can. J. Bot. 42: 1297-1301.

Singh, K. P., and Kumar, J. 1999. Efficacy of different fungicidal spray schedules in combating apple scab severity in Uttar Pradesh Himalayas. Indian Phytopath. 52: 142-147.

Singh, U. B. 1942. Stem brown disease of apple in Kumaon. Indian J. Agric. Sci. 12: 368-380.

Singh, U. B. 1943. Control of fruit diseases in Kumaon. Indian Farm. 4: 411-412.

Singh, U. B. 1943b. Pink disease of apple in Kumaon. Indian J. Agric. Sci. 13: 528-530.

Singh, U. B. 1944. Leaf spot diseases of apple in Kumaon. Indian Farm. 5: 566-567.

Singh, Y. P. 2002. Studies on the market diseases of some pome fruits and their management. Ph.D. Thesis. University of Jammu, Jammu.

Sneh, B., and McIntosh, D. L. 1974. Studies of the behavior and survival of *Phytophthora cactorum* in soil. Can. J. Bot. 52: 795-802.

Spotts, R. A., and Cervantes, L. A. 1994. Factors affecting maturation and release of ascospores of *Venturia pirina* in Oregon. Phytopathology 84: 260-264.

Srivastava, M. P., and Tandon, R. N. 1969. Physalospora rot of *Pyrus communis.* Indian Phytopath. 33: 476-477.

Stebbins, R. L., and Aldwinckle, H. S. 1990. Introduction. In: Compendium of Apple and Pear diseases. (A. L. Jones and H. S. Aldwinckle, eds.): 1-5. APS Press, St. Paul Minnesota.

Stephan, S. 1988. Studies on the sporulation and spore dispersal of apple powdery mildew [*Podosphaera leucotricha* (Ell. & Ev.) Saim.]. Arch. Phytopathol. Pflanzenschutz. 24: 491-501.

Stevens, N. E. 1933. Two apple black rot fungi in United States. Mycologia 25: 536-548.

Sud, V. K., and Agarwala, R. K. 1972. New records of pome and stone fruit diseases from Himachal Pradesh. II. Indian Phytopath. 25: 580-582.

Sutton, T. B. 1981. Production and dispersal of ascospores and conidia by *Physalospora obtusa* and *Botryosphaeria dothidea* in apple orchards. Phytopathology 71: 584-589.

Sutton, T. B. 1990. White rot. In: Compendium of Apple and Pear Diseases. (A. L. Jones and H. S. Aldwinckle, eds.): 17-18. APS Press, St. Paul Minnesota.

Sutton, T. B., James, J. R., and Nardacci, J. F. 1981. Evaluation of a New York ascospore maturity model for *Venturia inaequalis* in North Carolina. Phytopathology 71: 1030-1032.

Sztejnberg, A., Freeman, S., Chet, I., and Katan, J. 1987. Control of *Rosellinia necatrix* in soil and in apple orchard by solarization and *Trichoderma harzianum.* Plant Dis. 71: 365-369.

Takahashi, S., and Sawamura, K. 1990. *Marssonina* blotch. In: Compendium of Apple and Pear Diseases. (A. L. Jones and H. S. Aldwinckle, eds.): 33. APS Press, St. Paul Minnesota.

Thakur, V. S., and Sharma, R. D. 1999. Apple scab and its management. In: Diseases of Horticultural Crops-Fruits. (L. R. Verma and R. C. Sharma, eds.): 54-79. Indus Publishing Co., New Delhi.

Travis, J. W., Rytter, J. L., and Hickey, K. D. 1992. Environmental factors affecting the infection of white rot on apple wood under laboratory conditions. *Pa.* Fruit News 72: 32-35.

Tzavella Klonari, K., and Tamoutseli, D. 1986. The development and structure of the spermogonia and ascocarps of *Mycosphaerella sentina.* Cryptogam Mycol. 7: 267-273.

Unemoto, S. 1991. Relationship between Japanese pear scab development and nutrient content in leaves on trees given different amounts of nitrogen fertilizer. Ann. Phytopathol. Soc. Jpn. 55: 623-628.

Verma, K. S. 1988. Role of necator stage of *Corticium salmonicolor* in initiation of pink canker disease of apple in Himachal Pradesh. Plant Dis. Res. 3: 226-228.

Verma, K. S. 1991. Factor affecting the development of pink canker of apple. Plant Dis. Res. 6: 40-45.

Verma, L. R., and Sharma, R. C. 1999. Fungal and bacterial diseases of pear. In: Diseases of Horticultutal Crops-Fruits. (L. R. Verma and R. C. Sharma, eds): 140-166. Indus Publishing Co., New Delhi.

Yoon, J. T., Lee, J. T., Park, S. D., and Park, D. O. 1989. Effect of meteorological factors on the occurrence of Alternaria leaf spot caused by *Alternaria alternata* f.sp. *mali*. Korean J. Plant Pathol. 5: 312-316.

Diseases of Field and Horticultural Crops
Editor-in-Chief: **P. Chowdappa**
Published by: **Daya Publishing House, New Delhi**

Pages **386-411**

13

Diseases of Stone Fruits and Walnuts

Ved Ram, Durga Prashad, I.M.Sharma and H.S.Negi

India produces all deciduous fruits including pome fruits (apple and pear) and stone fruits (peach, plum, apricot and cherry) in considerable quantity. These are mainly grown in the North-Western Indian States of Jammu and Kashmir (J&K), Himachal Pradesh (H.P.) and in Uttar Pradesh (U.P.) hills. The North-Eastern Hills region, comprising of the States of Arunachal Pradesh, Nagaland, Meghalaya, Manipur and Sikkim also grows some of the deciduous fruits on a limited scale. Due to introduction and adaptation of low chilling cultivars of crops like peach, plum and pear, they are also now being grown commercially in certain areas of the north Indian plains. Out of all the deciduous fruits, apple is the most important in terms of production and extent. Sweet cherry was introduced from Europe before India's independence in 1947, while commercial cultivars of sour cherry have been brought mainly from USA in more recent years. The European and Japanese plum varieties are grown both in high and low hill areas. A plum variety 'Santa Rosa' reported to be a hybrid between Japanese and American species predominates (70-80%) plantations in the hills. Low chilling cultivars of peach and nectarine such as Flordasum, Flordared, and Sunred nectarine are successful introductions to the north-Indian plains. Some local selections of peach (Shan-e-Punjab, Sharbati), plum (Jamuni, Alubhokhara) and sand pear (Patharnakh) are also cultivated on a commercial scale in sub-tropical - marginal chilling areas of north India.

Pest and diseases have great impact on crop yield and quality, and also reduce resource-use efficiency. Improved crop protection strategies to prevent such damage and loss are required to increase production and make a substantial contribution to food security. It is fact that substantial crop is lost annually from the field, even in crops where pesticides and cultivars with improved genetic resistance to pests and diseases are used. The losses may be substantially greater in subsistence agriculture, where crop protection measures are often not applied. The key issues facing crop protection scientists in the 21st century: (1) to devise pest and disease control systems that are sustainable and not compromised by the evolution of pest and disease strains able to overcome crop resistance or chemicals. (2) To develop appropriate crop protection technologies, as well as mechanisms for their use, in lower-input farming systems and (3) more effective, efficient and durable crop protection measures are therefore priority. Fruit crop losses due to plant disease have been mentioned earlier but the latest data on stone fruit losses is lacking. Some of the losses in stone fruits are mentioned below

PEACH AND NECTRINE

Peach (*Prunus persica* Batsch) is the third most important fruit crop grown in the world. In India peach is grown on commercial scale in states of H.P, U.P, J&K. Peach require relatively warmer climate than other temperate fruits. In India peach cultivation extends from Northern plains to an elevation of 2000 MSL (Soodan *et al.*, 1994). During the last four decades there has been a remarkable increase in area under peach & nectarine cultivation in India and in H.P, which is estimated to be around 39100 ha with an annual production of 286500 MT (FAO, 2010-11) and 5182 ha with annual production of 9527 MT respectively (Anonymous, 2010-11). The diseases of peach and nectrine are given in Table 13.1 and their geographical distribution is shown in Fig. 13.1.

Table 13.1. Diseases of peaches and nectrine

Diseases		Casual organisms
Fungal diseases		
1	Peach leaf curl	*Taphrina deformans* (Berk.) Tul.
2	Brown rot	*Monilinia fructicola* (Wint.) Honey
3	Peach blight/ Shot hole of peach	*Wilsonomyces carpophilus* (Lev.)
4	Peach Scab/Freckles/black spot	*Venturia carpophila* Fisher Anamorph- *Fusicladium carpophilum* (Thuem.)
5	Peach Rust	*Tranzschelia discolor* (Fuckel)Tranzschel
6	Silver leaf	*Stereum purpureum* (Pers.ex Fr.)
7	Fusicoccum canker	*Fusicoccum amygdali* Delacr.
8	Powdery mildew	*Sphaerotheca pannosa* (Wallr. Ex Fr)
9	Perennial / Cytospora canker	*Valsa cincta* (Fr.ex Fr.)
10	Frosty mildew	*Cercospora persica* Sacc.
Bacterial diseases		
11	Bacterial canker/ Gummosis	*Pseudomonas syringae* pv. *syringae* Van. Hall
12	Bacterial spot	*Xanthomonas arboricola* pv. *pruni* (Smith) Vauterin *et al.*
Post-harvest diseases		
13	Alternaria rot	*Alternaria alternata* (Fr.) Keissler
14	Bitter rot/ anthracnose	*Glomerella cingulata* Spauld. **&** v. Schrenk
15	Blue mould rot	*Penicillium expansum*
16	Grey mould rot	*Botrytis cinerea* Pers.
17	Rhizopus rot	*Rhizopus stolonifer* (Ehrenb. ex Fr.) Vuill
Viral diseases		
18	Peach yellows	Peach yellows Virus, vector- Plum leaf hopper (*Macropsis trimaculata*)
19	Phony Peach (Phony disease)	*Xylella fastidiosa*
20	Peach – X	Phytoplasma

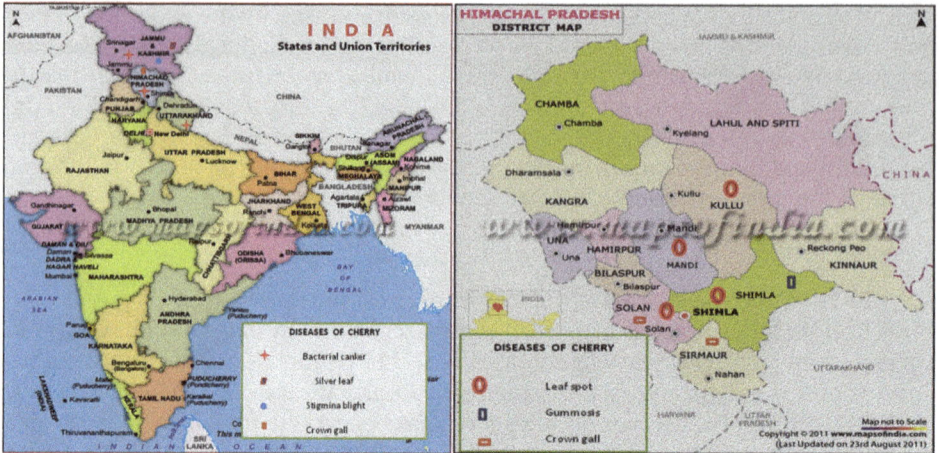

Fig. 13.1. Geographical distribution of cherry diseases in India and Himachal Pradesh

1. Peach leaf curl

Peach leaf curl is common disease of all the stone fruits, but peach is the main host for fungus, which causes it. Disease causes defoliation of peach trees, which may lead to small fruit or fruit drop. Different species of *Taphrina* cause leaf, flower and fruit deformation on stone fruits and forest trees. Thus *Taphrina deformans* causes leaf curl on peach and nectarine (Fig. 13.2b), *T. communis* and *T. pruni* cause plum pocket on American and European plums, respectively, *T. cerasi* causes leaf curl and witches broom on cherries and *T. coerulescens* causes leaf blister of oak. The most common losses, however, are those caused primarily on peach, nectarine and sometimes on plum.

Curling and puckering of leaves (Fig. 13.2a) occurs due to hypertrophy and hyperplasia of the cells resulting abnormal, irregular leaves which affect the photosynthetic activity, hence reduced fruits. The fungus causing peach leaf curl was first de-scribed by Berkeley in 1857 as *Ascomyces deformans,* but the name was changed in 1869 by Fuckel to *Exoatcus deformans.* In 1866 Tulasne de-scribed the organism as *Taphrina deformans,* which is the binomial generally employed by current writers. Asci of the fungus break through the cuticle of the distorted leaf.

Fig. 13.2. a. Curling and puckering of leaf, b. Naked asci of *T. deformans*

They appear as a powdery grey felt like area on the thickened leaf. The mature asci on being ruptured release ovoid ascospores. Under favourable conditions, ascospores will produce conidia continually until the entire ascospore content is depleted. The budding of newly formed bud conidia is also of common occurrence and is the common vegetative type of growth and multiplication.

Leaf curl is damaging when the weather is cool and moist. Rain is necessary for infection while low temperatures are considered to retard maturation of leaf tissue, thus prolonging the time of infection. The fungus can penetrate young leaves at temperatures between 10 and 21.1°C but only weakly below 7.2°C. The disease severity can be predicted if average maximum temperature during later half of February and March fluctuates in between 15 to 20°C (18°C optimum) along with high humidity or precipitation (Agarwala *et al.* 1966). It is indicated in literature that peach leaf curl aphid *(Brachycaudatus helichrysii)* acts as vector for the spread of peach leaf curl. The fungus overwinters as ascospores or thick walled conidia on the tree, perhaps on the bud scales. Ascospores are drifted to new tissue by air currents. In India, it is prevalent in Himachal Pradesh, Jammu and Kashmir, Assam, Bihar and Kumaon hills of Uttar Pradesh (Jain, 1962; Ferraris, 1928; Sydow and Butler, 1916).

2. Stigmina blight/ Shot Hole

This disease is also known as Coryneum blight. California peach blight, fruit spot, winter blight and pustular spot. In India, Munjal and Kulshrestha (1968) first reported the occurrence of this disease on peach, apricot and almond. Shot hole of peach is also called as 'peach blight", which also attacks other one fruits, but peach is the main host. The fungus attacks the twigs, blossoms, leaves, fruits and unopened buds. Small circular deep purple spots appear on fruits. On leaves, dark brown scattered lesions enlarge rapidly and loss of the dead areas results in the formation of shot holes.

The disease is caused by a fungus *Wilsonomyces carpophilus* (Lev.) Adaskaveg *et al,* (Adaskaveg *et al.,* 1990). The earlier generic names for this fungus include *Clasterosporium, Coryneum* and *Stigmina.* The fungus produces black dot like fruiting bodies on the necrotic cankers and twigs. These fruiting bodies produce numerous conidia which are thick walled, ellipsoidal or fusiform, three to five transverse double walled septa and infect buds and twigs of trees in winter with the onset of rains. Disease spread is continuous as the day temperature rises and it appears in epidemic form with the occurrence of frequent rains in subsequent months of spring and summer.The disease has been reported in severe form in lower hills of Kullu Valley, stone fruit areas of Mandi, Solan and Shimla districts of Himachal Pradesh (Gupta *et al.* 1973). As conidia gets splashed with rain drops and thus spread to adjacent branches. Spots on fruits make them unmarketable, whereas the hole in the leaves reduces photosynthetic activity.

3. Brown rot

The term brown rot refers primarily to the discolouration and decays of maturing fruits on trees of apple, pear, peach, cherry, plum, apricot and almond, is caused by many species of *Monilinia* genus. However, *Monilinia fructicola* and *M. fructigena* are primarily involved in its cause.

Causal Organism: *Monilinia fructicola* (G. Wint.)

Brown rot disease is caused by *Monilinia fructicola* (Wint.) Honey. *M. fructigena* (Aderh. & Ruhl.) Honey, *M. laxa* (Aderh. & Ruhl.) Honey, and *M. laxa* (Aderh. & Ruhl.) Honey f. sp. *mali* Wormald sensu Harrison. Gupta and Byrde (1988) found *Monilinia laxa* to cause blossom wilt of apricot and almond in North India. The mycelium produces chains of *Monilia-type* conidia on hyphal branches arranged in groups or tufts (sporodochia). The sexual stage (apothecium) originates from pseudosclerotia formed in mummified fruit partly or wholly buried in soil or debris. Warm, wet, humid weather is particularly favourable for brown rot. At 20°C, 3 to 5 hours of wetting may lead to significant infection. With 24 hours of wetting may leads to severe blossom infection may result regardless of temperature. Optimum temperature for blossom infections of peach is 25°C (Biggs and Northover, 1988a). The pathogen overwinters as mycelium in mummified fruit on the tree and in cankers of affected twigs or as pseudosclerotia in mummies in the ground. In the spring, the mycelium in mummified fruit on the tree and in the twig cankers produces new conidia. While the pseudosclerotia in mummified fruit buried in the ground produce apothecia, which form asci and ascospores. Conidia are windblown or may be carried by rain water and splashes or insects to floral parts and initiate infection within a few hours. Brown rot pathogen (*M. laxa*) associated with blossom wilt, an important phase of brown rot disease of stone fruits in Kullu Valley of Himachal Pradesh in late March 1982. Whereas direct infection to blossoms of peach, apricot and plum incited by *M. Laxa* were found extensively for the first time in the spring of 1983 in Rohru (Shimla distt.) and Kullu areas (Sharma and Kaul, 1988). Blossom blight phase reduces the number of flowering buds, whereas the decaying of fruits at maturity imparts poor quality.

4. Bacterial canker or Gummosis

The disease is also known as gummosis, blossom blast, and dieback, spur blight and twig blight. It occurs in all major fruit growing areas of the world. Tree losses from 10 to 75 per cent have been observed in young orchards. The disease also kills buds and flowers of trees, resulting in yield losses of 10 to 20 per cent but sometimes up to 80 per cent (Agarwala, 1961).Severe gummosis on fruits reduces market value of the produce. Three fungal species namely *Botryosphaeria dothidea*, *B. obtusa* and *B. rhodina* have been shown to cause gummosis in peach trees.The destructive phase of this disorder occurs when trees are attacked by *Pseudomonas syringae* pv. *syringae*. This disease is serious on stone fruit trees throughout Himachal Pradesh, Kashmir Valley and Kumaon and Garhwal hills of Uttar Pradesh (Agarwala, 1961; Durgapal, 1974).

The pathogens causing canker are *Pseudomonas syringae* pv. *syringae* Van. Hall and the more specialized *P. syringae* pv. *morsprunorum* (Wormald). This is restricted predominantly to plum. *P. syringe* pave. *syringe* produces the phytotoxic syringomycin, which appear to play a role in the virulence of the pathogen. The bacteria overwinter in active cankers, infected buds and systemically in apparently healthy tissues of the tree. Infection occurs in the autumn through leaf scars and in spring through scars left by bud scales. In spring, bacteria from cankers and infected buds are disseminated by rain onto emerging tissues. Infection to developing blossoms leaves and fruit occurs through natural openings and wounds (Ram Ved and Bhardwaj, 2004). Leaves and blossoms are probably more susceptible when injured by frost. Crosse (1966) reported that infection of stone fruits by *P. syringae* is favoured by wet weather and rain slashes. It was registered 21-24°C as optimum temperature for canker development. Periods of frequent rainfall, cool temperatures and high speed winds are most favourable for early season infection and spread. Gummosis of stone fruits has been reported in benign and severe form on all varieties of stone fruits infecting trunks, branches, twigs, spurs, leaves and fruits in Himachal Pradesh (Agarwala, 1961). The incidence of bacterial canker /gummy lesion was maximum (89) with peach/plum combination followed by peach/sloh (47 and 17) respectively, as carried out during 1981-82 at college orchard of Punjab Agricultural University, Ludhiana (Verma and Singh, 1986). Similarly gummosis of stone fruits was noticed in the orchard of University of Horticulture and Forestry, Nauni, Solan and Regional Fruit Research Station Mashobra during March, 1988 and in 1989 respectively,(Durgapal, 1974).

5. Scab

Peach scab is also known as freckles or black spot. The disease occurs primarily in the warmer peach producing areas. It is worst where a good spray schedule is not followed early in the season. Besides peach, the disease occurs on apricot, nectarines and cherry. Baruah *et al.* (1980) reported occurrence of this disease on peach from India.

The disease is caused by *Venturia carpophila* Fisher, whose imperfect stage is known as *Fusicladium carpophilum* (Thuem.) Oudem or *Cladosporium carpophilum* (Thuem).The fungus overwinters in lesions on the twigs. Conidia are airborne and water borne and are most abundant 2 to 6 weeks after the shuck split stage of development. If weather conditions are favourable, infection begins to occur at about shuck fall. The fruit remain susceptible until harvested. Spores from the fruit reinfect the twigs and leaves (Lawrence and Zehr, 1982).

6. Rust

This disease is of worldwide occurrence, but severe only in warmer areas. It often causes less damage as it appears late in the season after the trees have attained sufficient growth for the production of the following years. The disease was first reported from Australia in 1890 (Dunegan. 1938). Later on was reported from China. Japan, Brazil, New Zealand and is now widely distributed in Europe and North America also. The disease has been observed to be very serious on apricot, peach and plum (Fig. 13.3) in northern parts of India (Sharma *et al.* 1988b: Bhardwaj and Shyam. 1986; Waraich and Khatri, 1977).

Persoon (1801) first named the causal agent as *Puccinia*. However, Arthur (1906) erected a new genus *Tranzschelia* on the basis of teliospore morphology. *Tranzschelia* is distinguished from *Puccinia* by virtue of the fasciculate bunching of teliospores attached by their pedicels to a common basal cell and also by the ready separation of two cells of the spore. However, in India only *T. discolor* (Fuckel) Tranzschel & Litv. (Waraich and Khatri, 1977) and *T. pruni-spinosae* (Pers. ex Pers.) Dietel (Waraich and Khatri, 1977; Bhardwaj and Shyam. 1986) have been recorded. The infection can take place at 15 to 23°C and the incubation period is 7 and 70

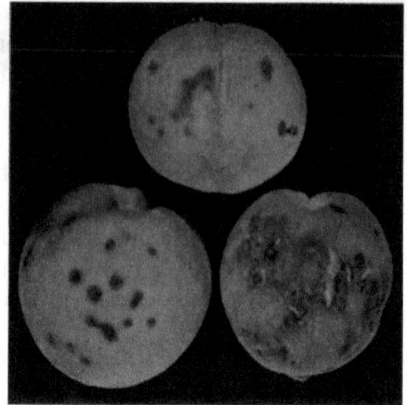

Fig. 13.3. Rust on peaches and nectrine

days at higher and lower temperatures, respectively. Leaf and fruit infection occurs under high humidity (Goldworthy and Smith. 1931) which is favoured by abundant rains during August (Vitanov, 1976). Epidemics are favoured by rainy period in summer when temperature is around 20 to 30°C which is most favourable for disease development (Simcone *et al.*, 1985). Stone fruit rust overwinters in twigs as uredosori and produce uredospores in spring (Jafar, 1958). However, *Anemone coronaria* has been reported to be the alternate host. Jafar (1959) further showed that canker developed on young wood also produce spores in late spring, thus serving as source of infection. The extent of overwintering inoculum is dependent on twig infection in the late season which remains dormant till spring. Perennial mycelium in 1 to 2 years old shoots has also been shown as the perpetual sources of infection (Nadazdin and Nadazdin, 1977).

7. Rhizopus Rot

Rhizopus rot is one of the most important post-harvest diseases of stone fruits, and can be responsible for heavy losses, especially in peaches, nectarines and cherries. The common name "whiskers" is derived from the fact that the fruit is covered with the mycelium of the fungus (Fig. 13.4), and the black fruiting bodies extend out into the spaces between the fruits and even on to the sides of the container. The fungi are widespread and records on stone fruits include the USA, Australia, India, South Africa and European countries. This disease causes more than 60 per cent loss of peach fruits in Ludhiana, during storage and transit in years 1979 and 1980 (Singh and Prashar, 1984).Quality of the final produced will be reduced to many folds as it appears during storage and transit.

The fungi exist on dead plant material and their spores (sporangiospores) are very common in the atmosphere. Immature fruits are generally resistant to attack: there are a few reports of infection of mature fruits in the orchard. Much more

Fig. 13.4. Rhizopus rot on peaches and plum

commonly, however, infection takes place via injuries, such as are sustained during harvesting and handling. Once established, these fungi can invade adjacent sound fruits, causing extensive damage to the stored fruits and cause huge losses. *R. stolonifer* grew between 15- 35°C with optimum at 25°C, followed by 20°C. Sporulation of pathogen was found excellent at 20 and 25°C, and poor at 15°C. As regarding, fruit infection, optimum fruit rot due to *R. stolonifer* developed at 20 and 25°C followed by 30°C. Fruit rot progressed fast at and above 80 per cent RH (Singh and Prashar, 1984). Soft-rotting by these fungi is accomplished by means of enzymes, some of which are heat-tolerant and survive the sterilization process in fruit-canning factories, leading to break-down of the canned product. Rhizopus rot is a serious rot of stone fruits, especially peaches in transit and storage. The fungus spreads rapidly through the container. Often a single infected peach results in the rotting of most of the fruit within a few days if the temperature is above 50 or 60°F. Singh and Prashar (1984) observed more than 60 per cent loss of peach fruits due to *Rhizopus* species during storage and transit in years 1979 and 1980 in the Department of Plant Pathology, Punjab Agricultural University, and Ludhiana.

CHERRY (*Prunus avium* L.)

Sweet cherry belongs to family *Rosacae* is an important crop of temperate region of the world. Leading cherry producing countries are USSR, USA, West Germany, Italy, France and India. USA ranks first in area and production of cherry in the world. In India, cherries are grown to a limited scale in the state of J&K; H.P. Cherries are grown at higher altitudes ranging from 2100 to 2700 a.m.s.l. In India it is cultivated in an area of 3600 ha with annual production of 15300 MT (FAO, 2010-11), whereas it occupies an area of 480 ha with annual production of 1039 MT in H.P alone, (Anonymous, 2010-11). The diseases that affect cherry cultivations are listed in Table 13.2.

Table 13.2. Diseases of cherry

Diseases		Causal organisma
Fungal Diseases		
1	Stamina blight	*Wilsonomyces carpophilus* (Lev.) Adaskaveg *et al.*
2	Cercospora leaf spot	*Cercospora circumscissa*
3	Blumeriella Leaf Spot	*Blumeriella jaapii*
4	Leaf curl	*Taphrina deformans* (Burk.).
5	Brown rot	*Monilinia fructicola* & *M. laxa*
6	Cytospora canker	*Cytospora cincta*
7	Powdery mildew	*Podosphaera clandestine*
Soil Borne Diseases		
8	Phytophthora root rot and Crown rot	*Phytophthora megasperma, P. drechsleri* P.cryptogea
9	Armillaria root rot	*Armillaria mellea* (Vahl.) Qu
10	Verticillium wilt	*Verticillium albo-atrum, V. dahlia*
Bacterial Diseases		
11	Bacterial Canker	*Pseudomonas syringae* pv. *syringae*
12	Crown gall	*Agrobacterium tumefaciens*
Virus and Phytoplasma diseases		
A	Cherry disease spread by pollen or seed	
13	Lace leaf, Necrotic leaf spot, Necrotic ring spot, Rugose mosaic	Prunus necrotic ring spot virus (PNRSV)
14	Blind wood, chlorotic ringspot, shot hole,	Prunus dwarf virus (PDV)
B Diseases spread by air borne vectors		
15	Little Cherry Disease (LCD)	Vector- Apple mealy bug (*Phenacocus aceris*)
16	Mottle leaf disease (MLD)	Cherry bud mite (*Eriophyes inaequalis*)
17	X- disease	Phytoplasma, transmitted by leafhopper
18	Rusty Mottle Disease (RMD)	Unknown vector
19	Twisted Leaf Disease (TLD)	Unknown vector
C Diseases spread by soil borne vectors		
20	Nepoviruses: Cherry leaf roll (CLRV)	Nematode transmission
21	Tomato bushy stunt virus (TBSV)	No vector known

1. Stigmina Blight

The disease affects all the above ground parts of the plant causing blight of apical twigs, blossom, leaves and unopened buds. Besides sweet cherry, blight affects wild cherries (*P. serotina. P. virginiana* and *P. padus*). The cherry laurel (*P. laurocerasus)* and *P. davidiana*. In addition, the natural hosts to this disease are peach, apricot, nectarines and almond. In India, Stigmina blight of cherry has been reported from Kashmir valley where 80 per cent of leaves and 60 to 70 per cent fruits were affected and yield was reduced to one-fourth of the normal crop (Purtoo and Razdan, 1991). The disease is also commonly known as Coryneum blight, shot hole, corynosis fruit spot and pustule spot. The lesions on fruits are small circular, deep purple, sunken and may render poor quality fruits which ultimately reduce its market value.

The blight is caused by *Stigmina carpophila* (Lev.) Ellis. Perfect stage of the fungus is still unknown; only one type of propagative structure is regularly produced. Conidiophores arise from mycelial cushion which are simple and short. Conidia are 4 to 6 celled, ovoid and yellow in colour.The fungus passes its life cycle on the tree as mycelium and conidia that remain alive in the diseased buds and twigs. After rains begin, conidia develop on the surface of twig lesions and new conidia develop inside diseased buds. Moving air is ineffective in detaching the conidia from the conidiophores. For infection and germination, conidia require a film of water on the host surface and no infection occurs during dry weather. Infection initiates when susceptible plant parts are wet for longer time. Conidia germinate, and germ tube penetrates the host tissue. Incubation period is determined by the temperature. In warm spring weather lesions develop 5 to 6 days after rainy period, whereas it takes 15 to 18 days in winter. Conidia are disseminated by rain splashes. This disease was reported as occurring in California as a destructive disease as early as 1900. In North America it is most commonly found in the western states. In India, stigmina blight of cherry appeared in severe form in Kashmir valley during March-May 1987.The incidence of disease was recorded in Harwan area (Distt. Srinagar) during 1987 and found to be as high as 80 per cent on leaves and 60-70 per cent on developing fruits. When infection occurs on the petiole, the entire leaf is soon killed. Severely infected fruits dried in spurs. Undeveloped and dried fruits remained attached to the branches for a longer time. Infected buds turned dark and flowers withered before these were fully expended.

2. Brown Rot

Brown rot disease is widespread in occurrence and affects cherry in almost all the cherry growing regions of the world viz. North America. Australia, New Zealand. South America, South Africa, North-Eastern Japan and India. The disease also affect other temperate stone fruits like apricot, nectarines, plums and peaches throughout the world where sufficient rainfall occurs during ripening period of the fruits. Sour cherries are comparatively resistant than sweet cherries. The disease has been considered as one of the main factors interfering in the way of commercial cultivation of cherry in Himachal Pradesh (Chopra and Singh 1972). Losses from brown rot result primarily by rotting of the fruit in the orchard, transit and storage. Yields are also reduced by destruction of flowers during blossom blight stage of the disease In severe infection in the absence of good protection measures 50 to 75 per cent of the fruit may rot causing enormous losses.Fruit infection also takes place after harvest in storage and in transit. The infected fruit continues to rot after harvest (Fig. 13.5). Mycelium and conidia infect wounded healthy fruits during storage. As a result of infection quality of fruits will drastically reduced.

Brown rot is caused by different *Monilinia* species. *M. fructicola* (G. Wint.). The mycelium of the fungus is hyaline and multinucleate. It produces chains of elliptical conidia on hyphal branches which are not distinct conidiophores, but are arranged in groups or tufts (sporodochia) as asexual fruiting bodies. The sexual stages (apothecia) develop on the surface of the mummified fruit which are partly or wholly buried in the soil or debris. Apothecia are fleshy, cup shaped, light brown in colour measuring 3 mm in diameter at maturity. The upper surface of the apothecia is lined by numerous cylindrical asci interspersed with paraphyses. The pathogen overwinters as mycelium or conidia on the mummified fruits. Overwintering on twig cankers are rare in *M. fructicola* (American brown rot) but very common in *M. laxa* which overwinters in cankers and

Fig. 13.5. Brown rot on sweet cherry

conidia are produced the following spring around bloom time if adequate moisture is present. Warm, wet and humid weather is most favourable for brown rot development. The hours of wetting necessary for blossom infection decreases from 18 hours at 10°C to 5 hours at 25°C (Wilcox, 1989). Infection occurs more slowly at temperatures above 27°C and below 12.7°C, but may continue at as low as 4.4°C. The conidia are wind borne or may be carried by rain splashes or insects but the ascospores are forcibly discharged by the ascus and carried by air currents on to the flowers. Brown rot of stone fruits in Himachal Pradesh occurred during the years1982, 1983 and 1984. Blossom wilt, an important phase of brown rot disease was observed in three stone fruits orchard in Kullu valley in the late March 1982 (Gupta and Byrde, 1988).

3. Bacterial Canker

Bacterial canker occurs wherever cherries are grown. It is particularly severe on sweet cherry and less severe on sour cherry. The disease affects tree by causing cankers on branches and main trunks, killing young trees and reducing the yield or killing older ones. More than 83 per cent cherry nurseries in Pakistan were found to be infected with this disease and incidence varied from 20 to 100 per cent (Wicks *et al..* 1993). Losses are difficult to assess because of serious damage to trees as well as reduction of yields. Tree losses which vary from 10 to 75 per cent have been observed in young orchards. Bacterial canker and gummosis also kills buds and flowers of trees, usually resulting in yield losses of 10 to 20 per cent but sometimes up to 80 per cent leaves and fruits are also attacked resulting in weaker plants and in low quality or unsalable fruit (Agarwala,1961; Durgapal, 1974, Jindal and Rana, 1992)

Two closely related bacteria, *Pseudomonas syringae* pv. *syringae* van Hall and *P.s.* pv. *morsprunorum* (Wormald) (Young, 1987) cause bacterial canker. Bacterium is rod shaped 2.5 μm long and 0.7 μm wide. It has one to several polar flagella and these occur singly, in pairs and occasionally in short chain. The bacterium is aerobic and produces a capsule. A green fluorescent pigment is produced by the bacterium in culture. *Pseudomonas syringae* is also known to produce a toxin in culture which causes severe damage to plant varieties susceptible to the pathogen but has little effect on resistant varieties. From India, *P. syringae* pv. *morsprunorum* was found to be the cause of canker in wild cherry (Durgapal and Singh, 1980). The two pathovars are distinguished from each other primarily by some biochemical tests. Pathovar *morsprunorum* shows more host specialization than pathovar *syringae*. *P.s* pv *morspru-norum* has been reported from sweet cherry, sour cherry and plums, while *P.s.* pv. *syringae* reported on all stone fruit crops and several ornamental *Prunus* species.Growth and dissemination of the bacteria are favoured by cool and wet weather. Both pathovars are found as residents on symptomless blossoms and on leaf surfaces from anthesis through to leaf fall (Latorre and Jones, 1979). Free water on leaf surfaces and high relative humidity are required for at least 24 hours before significant infection to leaves following violent storms. Symptoms appear about 5 days later at temperatures of 20° to 25°C. No significant correlation exists between resistance to frost and *Pseudomonas syringae* (Fischer *et. al.*, 1995). The bacteria are capable of surviving from one season to the next in cankers, in apparently healthy buds and in the vascular tissues of cherry trees (Sundin *et al.*, 1988). Bacteria from these overwintering sites colonize blossoms and leaf tissue as they unfold from buds in the spring. In autumn, the bacteria may be sucked into leaf scars, where they overwinter. Outbreaks of bacterial canker are relatively rare, but they can be very serious when they occur. Disease outbreaks are often associated with late spring frost or with severe storms that injure the emerging blossom and leaf tissues (Sule and Seemuler, 1987).Bacterial canker or gummosis of stone fruits is serious disease in Himachal Pradesh, Kashmir Valley, and Kumaon and Garwal hills of Uttar Pradesh (Agarwala, 1961; Durgapal, 1974, Jindal and Rana, 1992)

4. Crown Gall

Crown gall has been reported from most countries of the world and is worldwide in distribution it affects many woody and herbaceous plants belonging to 140 genera of more than 60 families. Crown gall is characterized by the formation of rumors or galls of varying sizes and forms. It is common on the roots and shoots of various nursery seedlings. Plants with tumors at their crowns or on their main roots grow poorly and produce reduced yield. Severely infected plants may die.

Crown gall is caused by *Agrobacterium tumefaciens* (Smith & Townsend) Conn. The bacterium is rod shaped and measure about 1-3 x 0.4-0.8 μm. It occurs singly or in short chains and has 2 to 4 flagella at the same pole. It forms a capsule. The pathogen is sensitive to sunlight and drying, but in soils with sufficient moisture it remains viable and virulent for many years. The bacterium is soil-borne and enters the roots and crown through wounds produced when caring for and handling the nursery stock. The free water is essential for the growth and development of bacterium. Bacteria are disseminated to new plants and planting sites by splashing rain or irrigation water. The incidence of crown gall during 1999 to 2001 varied from 5.70 to 70.00 and 5.00 to 100 .00 per cent on peach and cherry, respectively in different stone fruits nurseries in Himachal Pradesh (Gupta *et al.*, 2005). Crown gall caused by *Agrobacterium tumefaciens* is the major limiting factor in raising healthy stone fruit plants in nurseries. The incidence of crown gall varies from 4 to 97.5 per cent at different locations resulting in outright rejection of an average of 30 per cent stone fruit plants in nurseries (Gupta *et al.* 2010).The incidence of crown gall varies from 4 to 97.5 per cent at different locations causing an estimated loss of Rs. 6 million, annually. However, peach and cherry are worst affected by crown gall. Almost in every block of Rajgarh and Sangrah (H.P) known as peach bowl of India, the incidence on peach is as high as 97.5 per cent (Gupta *et al.*, 2010).

ALMOND (*Prunus amygdalus* Batsch)

Almond is the oldest nut fruit crops of the world. It is native to central Asian mountain areas (India, Pakistan and Iran). It is also found in South Africa, America and Australia. Production of unshelled almond is around one million tons per year of which the USA and Spain produce more than 50 per cent of the world production. In India its cultivation is confined to hilly areas of the J&K and cool and dry regions of Himachal Pradesh. Besides its cultivation is also catching up in mid and low hills of H.P mainly for use as green almond. The commercial cultivation of this nut in some area could not spread much on account of some intrinsic problems like damage by rains, winds, frost and hail storms during the blossoming time, lack of proper pollination and outbreak of pests and diseases are the main limiting factors in almond cultivation. Almond tree is usually the earliest deciduous fruit crop to bloom in spring because of its low chilling requirement and ready response to warm growing temperature. Almond occupies an area of about 5578 ha with an annual production of 1795 MT in H.P, (Anonymous, 2010-11), whereas the data on area and production of almond in India is lacking. The diseases that affect the productivity and production are presented in Table 13.3.

Table 13.3. Diseases of Almond

Major Diseases	Common Name	Causal Organism
Fungal diseases		
1	Blossom blight/ Blossom wilt and Brown rot	*Monilinia laxa, M. fructigena*
2	Cytospora canker / Valsa canker	*Cytospora cincta*
		C. leucostoma
3	Almond rust	*Tranzschelia discolor*
Minor Diseases		
4	Vericillium wilt	*Verticilluim albo-atrum*
5	Fusicoccum canker	*Fusicoccum amygdale*
6	Shot hole	*Wilsonomyces carpophilus*
7	Root rot	*Armillaria mellea*
8	Powdery mildew	*Podosphaera tridactyla*
		Sphaerotheca pannosa
Bacterial diseases		
9	Almond leaf scorch	Xylem limited fastidious walled bacteria
Virus and virus like diseases		
10	Almond bud failure	Almond bud failure virus
11	Almond yellow bud mosaic	Almond yellow bud mosaic virus

1. Blossom blight/wilt and Brown rot

The disease is present in all the almond growing areas of the world. When blossom blight is severe, reduction in fruit yield may be expected, because infection often progress into twigs, encircle them and cause death of the twigs. Under such situations, 50 to 75 per cent of the bloom get blighted and fail to set the fruit. Late infection leads to poor fruit development. Such fruit become shrivelled turn black and remain hanging as mummies on the infected spurs. Blossom wilt is an important disease of almond and apricot, which may cause 30-40 losses in certain localities.

The conidia of *M. fructigena* are the largest of the three species. These conidia are produced in sporodochia in monilioid chains without disjunctors. The apothecia of *M. fructicola* and *M. fructigena* are similar, light brown, about 3 mm across at maturity, cupulate and stipitate. While the apothecia in *M. laxa* and *M. fructigena* are rare. Asci (112-180 x 9-12 µm for *M. fructigena,* 121-188 x 7.5-11.8 µm for *M. laxa* and 102-215 x3-12 µm for *M. fructicola)* are eight spored.Warm, wet and humid weather is most favourable for brown rot development. At spring temperature of 13-20°C, disease symptoms develop within 3 to 6 days. With 24 hrs of wetting, severe blossom infection may result regardless of temperature. Optimum temperature for blossom infections of peach is 25°C (Biggs and Northover, 1988). Blighted blossoms and twigs produce sporodochia and spores which are released and spread by wind and rain. Conidia produced on blossoms, twigs and fruits infected in the previous season serve as inoculums for the spring blossom infections. Apothecia have not been found in India. Insects can carry spores from flowers and fruits to injured healthy fruits. Sepals are highly susceptible to infection.The fruit rot and canker phase of brown rot of stone fruits were first recorded by Gupta (1988) in an orchard in Kullu valley of HP.

2. Almond rust

Almond rust is of worldwide occurrence but severe in warmer and humid areas where it generally appears late in the season. In India, the disease was reported by Butler and Bisby in 1931. The disease has been observed to be very serious on almonds in northern parts (Sharma *et al.,* 1988b; Bhardwaj and Shyam, 1986; Kaul, 1966; Waraich and Khatri, 1977, Singh *et al.,* 1976).The mid-season defoliation is the most characteristic and cumulative effect of the rust (Fig. 13.6) which leads to gradual decline of the orchard productivity (Sharma *et al.,* 1988b). Direct infection to fruits reduces the market value of the final produce (Dippenaar, 1941)

Fig. 13.6. Rust pustules on almond leaf

The causal agent was first named as *Puccinia* by Persoon (1801). However, Arthur (1906) reported new genus *Tranzschelia* on the basis of teliospore morphology. *Tranzschelia* is distinguished from *Puccinia* by virtue of the fasciculate bunching of teliospores attached by their pedicels to a common basal cell and also by the ready separation of two cells of the spore. In India only *T discolor* (Fuckel) Tranzschel & Litv. Has been recorded (Kaul, 1966; Waraich and Khatri, 1977). Perennial mycelium in 1 to 2 years old shoots has also been shown as the perpetual source of infection (Nadazdin and Nadazdin, 1977). The infection can take place at 15 to 30°C the incubation period is 7 and 70 days at higher and lower temperatures, respectively. Leaf and fruit infection occurs under high humidity (Goldworthy and Smith, 1931) which is favoured by rainy period in summer when temperatures are around 20 to 23°C which is most favourable for disease development (Simcone *et al.*, 1985). Stone fruit rust overwinters in twigs as uredosi and produce uredospores in spring which are drifted by air currents (Jafar, 1958) However, *Anemone coronaria* has been reported to be the alternate host as the rust is heteroecious.The incidence of rust disease in almond was up to 64 per cent in some of the collections in Punjab Agricultural University, Ludhiana during November 1974. Whereas a new rust disease of *Prunus puddum* (*Tranzschelia pruni-spinosae* (Pers) Diet was reported by Bisht *et al.*, from Horticultural Experiments and Training Centre, chaubattia (Almora) Uttarakahand.

3. Almond leaf scorch

Almond leaf scorch disease in India poses a serious threat to cultivation in dry temperate zone of Himachal Pradesh. Leaf scorch disease in almond in India poses a serious threat to cultivation in dry temperate zone of Himachal Pradesh and requires strict quarantine measures to check its further spread. The newly emerged leaves were first symptomless, later exhibited such scorch symptoms (Fig. 13.7), causing premature defoliation and subsequent death of the plant within 2-3 years (Jindal and Sharma, 1987). The association of xylem limited fastidious bacteria (*Xyllela fastidiosa* Wells) was confirmed by the phony peach chemical test (Jindal and Sharma, 1987)

Fig. 13.7. Leaf scroch of almond

Diseases caused by fastidious xylem-limited, gram-negative bacteria are Pierce's disease of grape, citrus ariegation chlorosis, phony peach disease, almond leaf scorch, and plum leaf scald. Leaf scorch is caused by the bacterium *Xylella fastidiosa*, (formerly *Clavibacter xyli subsp. xyli*). Warm, wet and humid climate is considered congenial for the growth and development of the pathogen. Disease spread through irrigation water. The lethal disease was observed at Regional Horticultural Research Sub-station, Tabo in the first week of June (Gupta and Sharma, 1998).

APRICOT (*Prunus armeniaca* L.)

Apricot is one of the most important and delicious fruit of temperate region. It is generally cultivated in the areas situated between 900 to 2000m a.m.s.l. It is commercially cultivated in HP, J&K and hilly areas of UP and to some extent in north eastern parts of the country including Manipur, Arunachal Pradesh, Meghalaya, Mizoram and Nagaland. Among stone fruits, apricot ranks next only to peach and plum in area, production, productivity and varietal status. It occupies an area of 5400 ha with an annual production of 19100 MT in India (FAO, 2010-11) whereas it is grown over an area of 3483 ha with an annual production of 3341 MT in H.P, (Anonymous, and 2010-11). The diseases of apricot in given in Table 13.4.

Table 13.4. Diseases of Apricot

	Disease	Scientific name
A	**Fungal diseases**	
1	Blossom wilt and Brown rot	*Monilinia laxa.*
		Monilinia fructicola
2	Rust	*Transzchelia discolor*
3	Silver leaf	*Chondrostereum purpureum*
4	Verticillium wilt	*Verticillium albo-atrum*
		V. dahlia
B	**Bacterial Diseases**	
5	Bacterial Canker and gummosis	*Pseudomonas syringae* pv.
		syringae
6	Bacterial leaf spot or shot hole	*Xanthomonas pruni*

1. Brown rot

Brown rot caused by *M. fructicola* (Wint.) Honey, appears to be restricted to the America, South Africa, Japan, Australia and New Zealand. *M. fructigena* is widespread in Europe and Asia and has also been recorded in parts of North Africa and South America .The principal phases of the disease are blossom and twig blight. Twig blight occurs during the blossoming period and continues for three to four weeks as the fungus progresses from the dead blossoms girdling the twigs. Shoot blight is more likely to develop in association with green fruit infection than that of ripe fruit rot because in green fruit infection, mummies remain attached to the twigs.

Warm, wet and humid weather is favourable for brown rot development. Blossom infection occurs on petal, stigma and anthers, at spring temperatures of 13 to 20°C.Reduction of yield also occurs as a result of blossom blight, but the more serious aspect of this is the production of twig cankers, thus increasing the possibility of fruit rot later in the season. Cankers, like blossom blight, may cause indirect loss by furnishing inoculum for fruit infections. Before a successful fungicide had been found it was not unusual for growers to report from 50 to 75 per cent loss of the crop in the orchard in a wet season, and the entire crop was often destroyed before it reached the market.

2. Bacterial leaf spot

Bacterial spot has been reported only on species of the genus *Prunus*. Bacterial leaf spot are more commonly seen on apricot and plum but are rarely observed on peach. A shot-hole effect is produced by the early excision and dropping out of the diseased tissue.

Bacterial leaf spot caused by *Xanthomonas campestris* pv. *pruni* is a short rod, with rounded ends, which may occur singly or in short chains. It is motile by one to several polar flagella. The bacteria measure from 0.4 to 1.7 µm long and from 0.2 to 0.8 µm wide. Capsules are present in cultures after about 9 days. The organism stains Gram-negative and aerobic. Colony colour is yellow on culture media. A warm, moderate season with light frequent rains accompanied by fairly high winds and heavy dews is most favourable for severe infections. It has been reported that pathogen survive the winter in fallen leaves. Twig cankers are the main source of primary spring infections.

3. Gummosis

Gumming is of common occurrence on most stone fruits. Trunks, limbs, twigs, and fruit often exude gum when injured by mechanical agents or insect punctures. Cankers as a result of winter injury, insect attack, or fungus invasion also are of common occurrence. It occurs in all major fruit growing areas of the world. The disease is serious on stone fruit trees throughout Himachal Pradesh, Kashmir Valley and Kumaon and Garhwal hills of Uttar Pradesh (Agarwala, 1961; Durgapal, 1974). Most destructive phase of the disease is that on the trunk and branches where gumming symptoms are evident besides canker. The girdled branches fail to develop or die.

The pathogens causing canker are *Pseudomonas syringae* pv. *syringae* Van. Hall and the more specialized *P. syringae* pv. *morsprunorum* (Wormald). *P syringae* pv. *syringae* produces the phytotoxin syringomycin, which appear to play a role in the virulence of the pathogen. The rod-shaped bacterium measures 0.6 by 1.2 to 1.8 µm. It does not form spores, is capsulate, motile by one to several polar flagella, and is Gram-negative. It produces a green fluorescent pigment in culture. In spring, bacteria from cankers and infected buds are disseminated by rain onto emerging tissues. Infection to developing blossoms, leaves and fruit occurs through natural openings and wounds. Leaves and blossoms are probably more susceptible when injured by frost. Periods of frequent rainfall, cool temperatures and high speed winds are most favourable for early season infection and spread. Infection spreads from one branch to another through rain splashes. Gummosis on apricot was noticed in the orchard of University of Horticulture and Forestry, Nauni, during

March, 1988 and similar disease was reported earlier from Regional Fruit Research Station, Mashobra in1989 (Durgapal,1974)

4. Rust

The rust is now world-wide in its distribution but is most commonly found in the warmer sections of the stone fruit areas (Fig. 13.8). The mid-season defoliation is the most characteristic and cumulative effect of rust which leads to gradual decline of the orchard productivity (Sharma *et al.,* 1988b). Direct infection of fruits reduces the market value (Dippenaar, 1941).

The rust on cultivated stone fruits is caused by a heteroecious rust fungus, *Tranzschelia discolor* (Pers.).The infection can take place at 15 to 30°C. The incubation period is 7 and 70 days at higher and lower temperatures, respectively. Spores produced in late spring, are washed or blown to surrounding area. *Anemone coronaria* has been reported to act as alternate host.

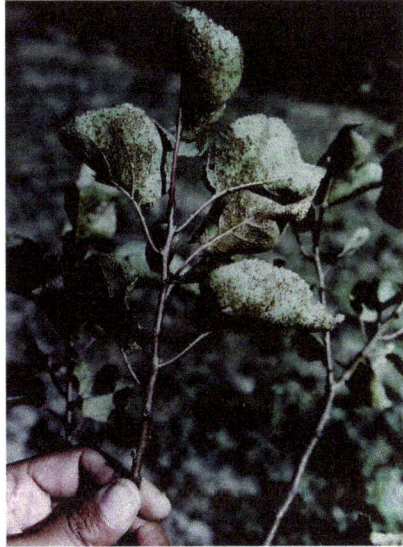

Fig. 13.8. Rusty pustule on apricot leaf

5. Rhizopus rot or whiskers rot

A serious rot of stone fruits, especially in transit and storage. Disease is caused by the bread mold fungus *Rhizopus stolonifer.* It is one of the major post-harvest diseases of stone fruits in Himachal Pradesh. The fungus is a wound parasite, and con-sequently, care should be taken to avoid any break in the fruit surface. Also, as the fungus requires rather high temperatures for rapid growth. Seven post-harvest diseases of apricot were recorded causing 4.5 to 43.1 per cent losses during storage and marketing in Himachal Pradesh (Sharma *et al.,* 1989).

PLUM (*Prunus salicina* Lindl.)

Plum is an important fruit crop of temperate region of the world. It occupies unique position among the stone fruits in world fruit production. Among stone fruits, it ranks next to peaches in economic importance. In India its cultivation is mainly confined to H.P, J&K and Uttarakhand in North India and to a limited extent in Nilgiri hills in South India. Plums are successfully grown in elevation of 1000 to 1600 MSL. During the last four decades there has been a remarkable increase in area under plum cultivation in India and in H.P, which is estimated to be around 27200 ha with an annual production of 236100 MT (FAO, 2010-11) and 8477 ha with annual production of 13717 MT respectively (Anonymous, 2010-11). The diseases of plum is given in Table 13.5.

Table 13.5. Diseases of Plum

	Disease	Scientific name
A Fungal diseases		
1	Brown rot	*Monilinia fructicola M. laxa*
2	Frosty mildew	*Cercosporella perska.*
3	Rust	*Transzchelia discolor*
4	Silver leaf	*Chondrostereum purpureum*
B Bacterial Diseases		
5	Bacterial Canker and gummosis	*Pseudomonas syringae* pv. *syringae*
6	Bacterial leaf spot or shot hole	*Xanthomonas pruni*
Post-Harvest Diseases		
7	Blue mould rot	*Penicillium expansum*
8	Rhizopus rot	*Rhizopus stolonifer*
Viral Disease		
9	Plum pox	Plum Pox Virus

1. Silver leaf

Silver leaf disease is a destructive disease of stone fruit trees, and is present in most temperate zone production areas. The first report of the disease was by Prillieux in France in 1885. The foliage of affected plants show a metallic luster in contrast to the normal green colour of the healthy trees Diseased limbs show dark discolouration of the heartwood. Silver leaf has been reported from most countries of continental Europe and is especially common in England. In New Zealand it is general in all the fruit sections and has been re-ported from South Africa. In India, this disease had been reported from Kashmir Valley during a field survey in the year 1982-83(Chib and Andotra, 1985).

The disease is caused by *Chondrostereum purpureum* (Pers. ex Fr.) Pouzer. Disease causes rapid decline and death of the grown up bearing tree and source of infection-potential to the healthy plants in the orchard. Infection is also favoured when trees are forced into rapid growth by heavy pruning, particularly in nurseries when rootstocks are headed to the scion buds in early spring. The fungus is a wound parasite, usually enters through pruning cuts. The risk of infection and disease is enhanced if trees are pruned during late winter and early spring.

Post-Harvest Diseases

2. Brown rot

In many countries where stone fruits are grown, brown rot is of major importance. *M. fructicola* (Wint.) Honey, appears to be restricted to the America, South Africa, Japan, Australia and New Zealand. *M. fructigena* is widespread in Europe and Asia and has also been recorded in parts of North Africa and South America.

The brown rot fungi survive the winter on mummified fruits and blighted twigs in the orchard. Much more commonly, however, the brown rot fungi invade stone fruits either through pre-harvest wounds caused by insects or adverse weather, or through injuries resulting from harvesting and handling operations subsequently brown rot can spread rapidly through a batch of harvested fruit. There is a direct correlation between the severity of blossom blight and the subsequent fruit rot. Blighted blossoms and twigs, mummies and infected peduncles produce sporodochia. Wet weather during the spring months results in the production and dissemination of spores which are disseminated by rain splash, wind and insects. In warm and wet weather pathogen produces sporodochia and spores that are disseminate by wind, rain splash and insect vector.

WALNUT (*Juglans regia* L.)

Nut trees are a promising potential food resource but this potential is not adequately utilized world-over, because the suitable farm land is either not put under nut crops or disappearing due to population pressure. They are rich in protein and fat as well as in mineral and some vitamins. The walnut is grown extensively in almost all the temperate countries where summers are not too cool. In India, Jammu and Kashmir the principal walnut growing state having monopoly in the production of export quality nuts. India produces nearly twenty-six tonnes of dry walnut every year most of which is grown in Jammu and Kashmir, Himachal Pradesh and Uttar Pradesh hills (Anon, 1997). In India, walnuts with shell occupy an area of 31800 ha with an annual production of about 33400 MT (FAO, 2010-11), whereas it is cultivated on an area of 4619 ha with an annual production of 1632 MT in HP (Annonymous, 2010-11). The diseases that reduces the productivity is listed in Table 13.6.

1. Walnut Blight

More than 50 per cent of the crop is damaged in USA in unprotected orchards in year of epiphytotics (Miller, 1953) In India, it was first reported from UP hills (Adhikari *et al.*, 1988) infecting 16 to 47 per cent nuts of different exotic varieties.The disease is caused by a gram negative bacterium *Xanthomonas arboricola* pv.*juglandis* (Pierce). The bacterium is aerobic, rod shaped (0.2-0.5 x 1.2-2.0 µm) in size, forms no spores or capsules. Colonies on beef agar are circular slightly raised, straw to pale yellow in colour (Adhikari *et al.*, 1988).Moisture is essential for the infection to take place, so high rainfall, fog, dew and sprinkler irrigation favours faster multiplication and spread of the pathogen and thus aggravate the blight problem. Long and frequent rains just before and during blossoming and for about 2 weeks thereafter may bring serious disease outbreaks as the nuts are more susceptible during this period. Subsequent infection caused little damage as the infection remains superficial on the nuts.The bacterium overwinters in dormant diseased buds and disseminated by rain drops. The bacteria enter the tissues of the current growth through stomata and injury. In India, it was first recorded on all 10 exotic cultivars of walnut in the orchard of the Government Fruit Research Station, Pithoragarh, during May and June, 1988. Similarly, disease has also been reported from H.P (Jindal and Dwivedi, 1994). Disease is reported to cause infection of about 16 to 47 per cent in nuts of different exotic varieties (Adhikari *et al.*, 1988). Severe infection leads to defoliation. On the stem and young shoots the disease causes slightly depressed black spots which often girdle the shoots and cause dieback symptoms. Early infection causes nut drop while late infection results in nut shriveling and discolouration.

Table 13.6. Diseases of Walnut

	Disease	Causal organisms
A Fungal diseases		
1	Leaf Blotch and Anthracnose	*Marssonina juglandis* (Lib) Magn. [Tel.: *Gnomonia leptostyla* (Fʀ.) Ces. and De Not.
2	Root and Crown Rot	*Phytophthora cactorum* (Lebert & Cohn) Schrot
3	Shoestring Root and Crown Rot	*Armillaria mellea* (Varhl. ex Fr.) Kumm.
4	Stem cankers and Dieback	*Cytospora leucosperma* (Pers. ex Fr.) Fr. (Syn.:.*Cytospora juglandicola* Ell. & Everh), *Fusarium solani*, *Fusarium incarnatum*, *Nectria galligena* Bres.
5	walnut rot	*Alternaria* sp., *Rhizopus oryzae* & *R. stolonifer*
B Bacterial Diseases		
6	Walnut Blight	*Xanthomonas arboricola* pv.*juglandis* (Pierce) Vauterin *et al.*
7	Crown gall	*Agrobacterium tumefaciens* (Smith & Townsend) Conn.
C Viral Diseases		
8	Black line	Cherry leaf roll virus (CLRV)
9	Mosaic	Walnut mosaic virus
Minor Diseases		
10	Branch wilt *fawuttii* Wilson)	*Hendersonula toruloidea* Nattrass (Syn.: *Exosporina*
11	Ring spot	*Aschochyta juglandis* Boltshauser
12	Downy leaf spot	*Microstroma juglandis* (Berenger)Sacc.
13	Powdery mildew	*Phyllactinia quttata* (Wallr. ex Fr.) and *Microsphaera extensa* Cooke & Peck
14	Heart rot	*Polyporus squamosus* (Huds. Ex Fr.) Fr.

2. Leaf Blotch and Anthracnose

The disease is widespread in occurrence and destructive at young nut stage causing pre-mature nut drop and extensive yield loss. It was observed in almost all the walnut growing localities of Kashmir (Kaul 1962).

The disease is caused by *Marssonina juglandis* (Lib) Magn. [Tel.: *Gnomonia leptostyla* (Fʀ.) Ces. and De Not.]. Acervuli appear in early August as small black specks on the lower surface of the diseased leaves. Conidiophores are hyaline, short, simple, elliptic, and one celled packed together in a small layer bearing a conidium at their tips. The conidia are variously shaped being straight, ovoid, falcate or with only one end rounded and the other pointed. There is one septa, and two cells are unequal with prominent oil globules and measures 15-26 x 2-5 µm. On fallen leaves brown coloured perithecia develop. The body of perithecia is immersed in the leaf tissues while the beak protrudes considerably on to the leaf surface. Infection starts with the onset of spring rains with the release of ascospores discharged forcefully from the perithecia and are carried by wind current to new shoots where they cause primary infection. The subsequent generation of summer spores (conidia) produced on primary and secondary lesions further spread the disease. The process of disease spread continues until late autumn, however, it is most conspicuous during July and August (Kaul, 1962). The fungus overwinters on infected fallen leaves, nuts as well as on infected twigs. The process of disease spread continues until late autumn, however, it is most conspicuous during July and August (Kaul, 1962). Early infection on nuts results in premature fruit drop. It was observed in almost all the walnut growing localities of Kashmir (Kaul, 1962). Disease is most destructive at young nut stage causing premature nut drop and extensive yield loss (Kaul, 1962).

3. Root and Crown Rot

The disease is widespread in occurrence and quite disastrous than other diseases because it kill the tree out rightly at any age. Infected trees may produce low yields of poor quality nuts.

Many species of *Phytophthora* are involved but *P. cactorum* (Lebert & Cohn) Schrot is the most common fungus associated with the disease. Presence of free water is necessary for the release of zoospores from sporangia. Free water carries the zoospores on to the soil surface where they lose their flagella, encyst, germinate and produce hypha which infects the root or collar region. Injury on these parts hastens the infection process. Poorly drained soils with prolonged water stagnation, low level of budding and injuries to roots as well as collar region are predisposing factors. The fungus survives in the soil in form of chlamydospore. Because of widespread in occurrence and quite disastrous than other diseases, wherever it comes causes mortality of plants.Infected trees may produce low yields of poor quality nuts, because nutrition will not be translocated to the aerial parts of the plants.

4. Crown Gall

Crown gall is a common problem in nurseries of Persian walnut, chestnut, hickory, pecan, European hazelnut and almond. The disease has been known since 1853 and is worldwide in distribution. It was also reported on walnut from India by Jindal and Dwivedi (1994).

Crown gall is caused by a bacterium *Agrobacterium tumefaciens* (Smith & Townsend) Conn. The bacterium is a rod-shaped measuring 1-3 x 0.4-0 8 µm. It occurs singly or in short chains and has 2 to 4 flagella at the single pole and forms capsule. On certain media, *A tumefaciens* forms star-shaped arrangements which have been considered as stages in the sexual processes. Various strains of this bacterium are known, and they exhibit different degrees of virulence. The bacterium overwinters in infested soils, where it can live as a saprophyte for several years. When plants of susceptible hosts are grown in such infested soils, the bacterium enters the roots or stems near the ground through fresh wounds caused by cultural practices, grafting and insects. Once inside the tissue, the bacteria stimulate the surrounding cells to divide. One or more groups or whorls of hyperplastic cells appear in the cortex or in the cambial layer depending on the depth of the wound. These cells may contain one to several nuclei They divide at a very fast rate, producing cells that show no differentiation. The cells continue to divide and elongate so fast that 10 to 14 days after inoculation a small tumour can be seen with the naked eye. The bacterium overwinters in infested soils, where it can live as a saprophyte for several years

5. Stem Cankers and Dieback Diseases

These cankers normally do not kill the trees but reduce the quality and quantity of nuts as well as affect the quality of wood adversely. Several canker diseases caused by different fungi have been reported on black and Persian walnuts which varied in their magnitude of distribution and losses. It has also been reported from hills of Uttar Pradesh. More than 84 per cent black walnut plantation in North America was infected with stem canker caused by *Fusarium solani*.

There are several canker causing fungi viz., *Cytospora leucosperma* (Pers. ex Fr.) Fr. (Syn. *Cytospora juglandicola* Ell. & Everh), *Nectria galligena* Bres., *Fusarium solani* (Mart.) Sacc. and *Botryosphaeria dothidea* (Moug. ex Fr.) Ces. & de Not. These canker fungi are generally wound parasites and require wounded or injured host tissues to penetrate. Unprotected pruning cuts, insect punctures and injuries caused during various horticultural operations predispose the host to infection. Asexual or sexual fruiting bodies are produced on the host surface which produce infection propagules specific to the fungus. In India, 63 per cent population of 6 to 9 years old budded plants were found to be infected with Cytospora canker in Kashmir valley (Puttoo and Chaudhary, 1984). These fungi survive in the infected tissues and mycelium starts growing when host growth starts. The propagules are disseminated and initiate the infection in susceptible hosts. Disease cause losses indirectly, as the girdling and die back of stem and branches result in poor translocation of food to aerial parts, hence the nuts with reduced sized, hampering the market value of the nuts.

REFERENCES

Adaskaveg, J. E., Ogawa, J. M., and Buller, E. E. 1990. Morphology and ontogeny of conidia in *Wilsonomyces carpophilus*. gen. nov. and comb nov. causal pathogen of shot hole disease of *Prunus* species. Slycouaon 37: 275-290.

Adhikari, R. S., Bora, S. S., and Singh, S. B. 1988. *Xanthomonas campestris* pv. *juglandis*—A new report from India. Curr. Sci. 57: 728.

Agarwala, R. K., Arora, K. N., and Singh, A. 1966. Effect of temperature and humidity variation on the development of peach leaf curl in mid hills and its control. Indian Phytopath. 19: 308-309.

Agarwala, R. K. 1961. Bacterial gummosis of stone fruits. Himachal Home 2: 49-51.

Anonymous. 1997. World Production. FAO Production Year Book. Vol 51.

Anonymous. 2010-11.Indian Horticulture Database. National Horticultural Board, Gurgaon, India.

Arthur, J. C. 1906. Cultures of uredinae . J. Moycol. 13: 189-205.

Baruah, S. N., Bamah, A., and Bora, K. N. 1980. A fungal disease of peach plant collected from Gauhati University Campus, Assam *(Venturia carpophila)*. Sci Cult 46: 264.

Bhardwaj, L. N., and Shyam, K. R. 1986. Occurrence of rust on Japanese plum. Himachal J. Agric. Res. 12: 62.

Biggs, A. R., and Northover, J. 1988. Influence of temperature and wetness duration on infection of peach and sweet cherry fruits by *Monilinia fructicola*. Phytopathology 78: 1352-1356.

Chib, H. S., and Andotra, P. S.1985.Prevelance of silver leaf disease in Kashmir. Indian Journal of Mycology and Plant Pathology 15 (3): 321.

Chopra, S. K., and Singh, I. I. 1972. Some problems of cherry cultivation in India Himachal Hortic. 12: 1-8.

Crosse, J. E. 1966. Epidemiological Relations of the Pseudomonad Pathogens of Deciduous Fruit Trees. Annual Review of Phytopathology 4: 291-310

Dippenaar, B. R. 1941. Diseases of fruit trees caused by leaf rust, manganese and zinc deficiency and their control. S. Afr. J Sci. 37: 136-155.

Dunegan, J. C. 1938. The rust of stone fruits. Phytopathology 28: 417-426.

Durgapal, J. C. 1974. A preliminary note on bacterial diseases of emperale plants in India. II Bacterial diseases of stone fruits. Indian Phytopath. 24: 379-382.

Durgapal, J. C., and Singh, B. 1980. Taxonomy of Pseudomonads pathogenic to horse-chestnut, wild fig and wild cherry in India. Indian Phytopath. 33. 533-535.

Ferraris, T. 1928. Agriculture of litopathologianel Kashmir. Cur. Plant 6: 61-86.

Fischer, M., Ebert, A., and Hohlfeld, B. 1995. Tests on resistance of sweet cherries *(Prunus avium* L). Pan 2 Resistance against *Pseudomonas syringae* (bacterial canker). Erwerbsobstbau 37. 102-107.

Fuckel, L. 1869-1870. Symbolae mycologicae, *Jahrb. Nassau ver.* Naturk, 23-24: 1-459.

Goldworthy, M. C., and Smith, R. E. 1931. Studies on a rust of clingstone peaches in California. Phytopathology 21: 133-168.

Gupta, A. K., and Sharma, R. C. 1998.Almond leaf scorch- A serious threat to almond cultivation in Himachal Pradesh. Indian Phytopathology 51(2): 203.

Gupta, A. K., Kamal, B., and Kumar, S. 2005. Biovar characterization of *Agrobacterium tumefaciens* isolates from stone fruits and effect of crown gall disease on growth parameters on peach and cherry rootstock-colt. Indian Phytopathology 58(4): 486-488.

Gupta, A. K., Khosla, K., Bhardwaj, S. S., Thakur, Aman., Devi, S., Jarial, R. S., Sharma, C., Singh, K. P., Srivastava, D. K., and Lal, R. 2010. Biological control of crown gall on peach and cherry rootstock colt by native *Agrobacterium radiobacter* Isolates. The Open Horticulture Journal 3: 1-10

Gupta, G. K., and Byrde, R. J. W. 1988. *Monilinia laxa* associated with blossom wilt of apricot and almond in Himachal Pradesh, India. Plant pathol. 37: 591-593

Gupta, G. K., Agarwala, R. K., and Dutt, K. 1973. Blight of peaches in HP can be controlled. Indian Hortic. 17: 13.

Gupta, G. K. 1988. *Monilinia laxa* associated with blossom wilt of apricot and almond in Himachal Pradesh, India. Plant Pathology 37: 591-593

Jafar, H. 1958. Studies on the biology of peach rust *(Tranzschelia discolor* Pers) in New Zealand. Investigations on the control of peach rest N.ZJ. Agric Res I: 642-651.

Jafar, H. 1959. Current research and investigations—Orchards. N.Z.J. Agric. Res. 2: 17-19.

Jain, S. S. 1962. Tests of fungicides for control of peach leaf curl *(Taphrina deformans* Tul.) Himachal Home 3: 11-15.

Jindal, K. K., and Dwivedi, M. P. 1994. Occurrence of crown gall on walnut plants in India. Plant Dis. Res. 9: 67-68.

Jindal, K. K., and Rana, H. S. 1992. Studies on germplasm resistance and chemical control of bacterial canker of apricot. Plant Disease Research 7(1): 7-10

Jindal, K. K., and Sharma, R. C. 1987. Outbreaks and new records. Almond leaf scorch-a new disease from India. FAO. Plant Protection Bulletin 35(2): 64-65

Kaul, T. N. 1966. Diseases of stone fruits in Kashmir. Horticulturist 2. 52-58.

Kaul, T. N. 1962. Occurrence of *Gnomoma leptostyla* (Fr.) Ces. et de Not. on walnut in India. Curr. Sci 31: 349.

Latorre , B. A. and Jones, A. L. 1979. *Pseudomonas morsprunorum* the cause of bacterial canker of sour cherry in Michigan, and its epiphytic association with *P. syringae.* Phytopathology 69 335-339.

Lawrence, E. G. Jr. and Zehr, E. L. 1982. Environmental effects on the development and dissemination of *Cladosporium carpophilum* on peach. Phytopathology 72: 773-776.

Miller, P. W. 1953. Filberts and Persian Walnuts. In: Plant Diseases. The Year Book of Agriculture. 8OO-808. Oxford & IBM Publ. Co., New Delhi.

Munjal, R. L. and Kulshreshta, D. D. 1968. Some dematiaceous hyphomyectes from India V. *Stigmina* species. Indian Phytopath. 21: 309-314.

Nadazdin, M., and Nadazdin, V. 1977. Some morphological and biological features of the pathogens of rust of peach and apricot in Hcrcegovina. Zast Bilja 28: 327-333.

Persoon, C. N. 1801. Synopsis Methodica Fungorum. Goattingae, Dietcrich.

Purtoo, B. L., and Razdan, V. K. 1991. Stigmina blight of cherry- a new record from India. Plant Disease Research 6(1): 60.

Puttoo, B. L., and Chaudhary, K. C. B. 1984. Canker of walnut trees and its management in Kashmir. Pesticides 18: 38-39.

Ram, V., and Bhardwaj, L. N. 2004. Stone fruit disease and their management .*In:* Disease of fruits and vegetables: diagnosis and management. (Naqvi, S. A. M. H. ed.): 2: 485-510.

Sharma, R. L., and Kaul, J. L. 1988a. Occurrence of brown rot *(Monilima* spp.) on stone fruits in Himachal Pradesh Plant Dis. Res. 3: 46-47.

Sharma, R. C., Jindal, K. K., and Gupta, V. K. 1988b. Reaction of some almond cultivars to rust. Plant Dis. Res. 4: 80-81.

Simcone, A. M., Ialonga, M. T., and Corazza, L. 1985. Reaction of some plum varieties of rust *(Tranzschelia pruni-spinosae)* Dietal var. *discolor)* in a locality of Latium Coast *Inf. Fitopatol* 35: 37-41.

Singh, G., Chauhan, J. S., Bhatt, A. S., and Malhi, C. S. 1976. Evaluation of germplasm collections of almond rust in Punjab. Indian J Mycol. Plant Pathol. 6: 81.

Singh, S. R., and Prashar, M. 1984.Studies on rhizopus rot of peach and its control. Indian Journal of Mycol. Pl. Pathol.14(2): 185-187.

Soodan, A. S., Wafri, B. S., and Kaul, A. K. 1994. Peach diversity in Kashmir. Indian Horticulture 32(2): 21-44.

Sule, S., and Seemuller, E. 1987. The role of ice formation in the infection of sour cherry leaves by *Pseudomonas syringae* pv. *Syringae.* Phytopathology 77: 173-177.

Sundin, G. W., Jones, A. L., and Olson, B. D. 1988. Overwintering and population dynamics of *Pseudomonas syringae* pv *syringae* and *Pseudomonas syringae* pv. *Morsprunorum* on sweet and sour cherry trees. Can. J. Plant Pathol. 10. 281-288.

Sydow, H., and Butler, E. J. 1916. Fungi Indiae Orienlalis Part IV. Ann Mycol. 14: 177-220.

Verma, K. S., and Singh, A. 1986. Incidence of bacterial canker of peach in Punjab. J. Res. Punjab Agric. Univ. 23 (1): 74-77.

Vitanov, M. 1976. On some questions of the pathogens of plum rust in Bulgaria and its control Rastit. Zshch 24: 34-36.

Waraich, K. S., and Khatri, H. L. 1977. Occurrence of pink disease and rust on plum in India. Indian J. Mycol. Plant Pathol. 7: 202-206.

Wicks, I. J., Iiaque, M. L., and Doolan, D. W. 1993 Crown gall on stone fruits in the northern areas of Pakistan *f AO* Plant Prot. Bull. 41: 23-28.

Wilcox, W. F. 1989. Influence of environment and inoculum density on the incidence of brown rot blossom blight of sour cherry. Phytopathology 79: 530-534.

Diseases of Field and Horticultural Crops
Editor-in-Chief: **P. Chowdappa**
Published by: **Daya Publishing House, New Delhi**

Pages **412-441**

14

Diseases Bananas and Plantains

R. Selvarajan

Banana is one of the important fruit crops grown in the tropical and sub-tropical regions in the developing countries. India is one of the origins for bananas and it was grown even before the Vedic period. Bananas are known for its antiquity and are interwoven with Indian heritage and culture. The plants are considered as the symbol of 'prosperity and fertility'. Owing to its greater socio-economic significance and multifaceted uses, they are referred as 'Kalpatharu' (Plant of Virtues) and Kalpavriksh. Botanically, banana fruits are a wonder berry, which forms the staple food of millions of people across the globe, providing more balanced food than any other fruit or vegetable. As a dessert, banana is more filling, easy to digest, fat free, rich source of carbohydrate with a calorific value of 90 Kcal per 100 g fruit and free from sodium, making it a salt free food fruit. It contains various vitamins and has therapeutic values for the treatment of many diseases. In North Eastern India, banana powder issued as baby food and being a good source of potassium is good for heart patients. Stem core juice is a well-recognized medicine for dissolving kidney stones. Plantains are rich in vitamin A (beta carotenes) and they aid indigestion. Ripe fruits are being used in the treatment of asthma and bronchitis. Nectar from the flower buds is rich in vitamin, hence given for strengthening of babies. Juice from the male bud provides effective remedy for all stomach problems. The powdered peel of ripe banana has antiseptic properties, hence used in the healing of cuts and wounds. Banana skin is also a rich source of vitamin and is curative to retina related disorders. There has been ample diversification in products. Frozen banana purees are being used in the manufacture of yoghurts, milk shakes, ice-creams, breads, cakes, banana flavored drinks, baby food and sauces. Banana flour obtained by drying and grinding the green fruit is easily digestible than the cereal flour. Banana juices are being used in the production of beer with low alcohol content and rich in vitamin B6 can help to relieve stress and anxiety. Unripe banana and plantain can be dried and made into a meal, which can be used to substitute up to 70-80% of the grain in pig and dairy diets. Starch can be extracted from banana pseudo stem and corms, which is used for making glue, used in the manufacture of cartons for exporting fresh bananas. Banana seeds are used for making necklaces and other ornaments. Use of sap as a dye, fruit as a meat tenderizer and ash in the manufacture of soap and anti-dandruff shampoos. Sap can be used as a natural film forming resin, which along with isopropyl alcohol produces better quality ink for inkjet printers. Dried banana peel contains 30-40 per cent tannin, making it suitable for leather processing. The productions of floor wax and shoe polish from banana peels are also being explored. Under non-edible sector of banana products, fibre plays a major role. Banana yields a fibre, which is used extensively in the manufacture of certain papers especially where greater strength is required. The papers are also being used for making tea bags and currency notes like those of Japanese Yen. It has use in the textile industry, for making cordages and handicrafts. India is the world largest producer of banana, with a total production of 29.78 million tonnes from an area of 0.830 million hectares. In India, banana is well adopted in the regions varying from humid tropics to humid sub tropics and semi-arid subtropics, and from the sea level up to an elevation of 2000 m above mean sea level. Production has increased from 7.79 M tonnes in 1991-92 to 29.78M tonnes in 2011-12 (Table 14.1). Availability of wide genetic diversity, varying production

Table 14.1. Area, Production and Productivity of banana in India

Year	Area (in 000'ha)	Total fruit area (%)	Production (000'mt)	Fruit production (%)	Productivity (Mt/ha)
1991-92	383.9	13.4	7790.0	27.2	20.3
2001-02	466.2	11.6	14209.9	33.0	30.5
2002-03	475.3	12.5	13304.4	29.4	28.0
2003-04	498.6	10.7	13856.6	30.4	27.8
2004-05	589.6	11.9	16744.5	34.0	28.4
2005-06	569.5	10.7	1887.8	34.1	33.2
2006-07	604.0	10.9	20998.0	35.3	34.8
2007-08	658.0	11.2	23823.	36.3	36.2
2008-09	709.0	11.6	26217.0	38.3	37.0
2009-10	770.3	12.2	26469.5	37.0	34.4
2010-11	830.0	13.0	29780.0	39.8	35.9

system, its suitability on wide range of agro-climatic conditions is attributed for wide range adoption of banana. It is grown as homestead garden as well as commercial plantation. The productivity has increased phenomenon is attributed to nature of soil and availability of water. However, cultivation has also been extended to areas where water sources have been created. Banana has assumed more significance nowadays, as it is the source of regular income for small and marginal farmers, and responds very well to management system. With the adoption of advanced technologies like drip irrigation, use of quality tissue culture plants, fertigation and integrated plant protection technologies lead the country to be number one in area and production. The major banana growing states are Andhra Pradesh, Assam, Bihar, Gujarat, Karnataka, Kerala, Madhya Pradesh, Maharashtra, Orissa and West Bengal. (Table 14.2). However, banana cultivation continues to face several pests and diseases which affect the production and productivity constantly. This most valuable fruit crop is vulnerable for many fungal, bacterial and viral pathogens (Table 14.3). In this chapter occurrence, distribution, disease signs, causal organism, favourable conditions for disease development, disease spread and crop losses of the major fungal, bacterial and viral diseases occurring in India are described.

Table 14.2. State wise area production and productivity of banana

State	Area (000HA)	Production (000MT)	Productivity(MT/HA)
Tamil Nadu	125.4	8253.0	65.8
Maharashtra	82.0	4303.0	52.5
Gujarat	64.7	3978	61.5
Andhra Pradesh	79.3	2774.8	35
Karnataka	111.8	2281.6	20.4
Madhya Pradesh	38.1	1719.6	45.2
Bihar	31.9	1517.1	47.6
Uttar pradesh	32.4	1346.1	41.5
West Bengal	42.0	1010.1	24.0
Assam	47.6	723.6	15.2
Others	175.3	1873.1	10.7
Total	**830.5**	**29779.9**	**35.9**

Table 14.3. Listing of important diseases affecting banana crop in India

Sl.No.	Name of the disease	*Causal organisam*
1	Fusarium wilt	*Fusarium oxysforum f.sp cuebense*
2	Black sigatoka	*Mycosphaerella fijiensis*
3	Yellow sigatoka	*M. musicola*
4	Septoria leaf spot	*M. eumusae*
5	Tip over or rhizome rot	*Erwinia caratovora*
6	Bunchy top disease	*Banana bunchy top virus*
7	Streak disease	*Banana streak virus*
8	Bract mosaic disease	*Banana bract mosaic virus*
9	Mosaic / Heart rot	*Cucumber mosaic virus*
10	Mild mosaic disease	*Banana mild mosaic virus*
11	Anthracnose	*Colletotrichum musarum*
12	Crown rot	*Colletotrichum musae, Fusarium spp. and B.theobromae*
13	Cigar end rot	*Verticiliumtheobromae, Verticilium.staphylidium, Trachysphaera.*
14	Botryodiplodia black tip disease	*Botryodiplodia theobromoece*
15	Pitting disease	*Pyricularia grisea*
16	Peduncle rot	*Colletotrichum gloeosporioides and Botryodiplodia theobromae.*
17	Deightoniella leaf and Fruit spot	*Deightoniella torulosa (Syd)*

1. Fungal diseases

1.1. Panama wilts of banana/Fusarium wilt: *Fusarium* wilt of banana is a major constraint in banana production worldwide. As a result of epidemics in export plantations, *Fusarium* wilt has been considered as the most destructive plant diseases in recorded history. Fusarium wilt of banana (*Musa* spp.) caused by *Fusarium oxysporum* f. sp. *cubense* (E.F. Smith) Snyder and Hansen is a destructive disease on many commercial cultivars in the world (Stover, 1962; Alves *et al.*, 1987; Sebasigari and Stover, 1988; Ploetz *et al.*, 1990; Thangavelu *et al.*, 2001).

1.1.1. Occurrence and distribution: Panama the place where this disease was recorded in epidemic form, hence the name. It was first reported in 1874 in Australia; later ten epidemics were noted in Central America (panama) in 1890. Fusarium wilt has been reported in all banana growing regions of the world except the South Pacific Islands, Somali land, and countries bordering on the Mediterranean Sea (Stover and Simmonds, 1987). During the 1950s the cultivation of the export-quality "Gros Michel" bananas was threatened almost to extinction due to the occurrence of fusarium wilt. Foc is considered one of the most important threats in Asia, Africa, Australia and tropical America (Hwang and Ko, 2004). In India this disease was first reported from Chinsurach, West Bengal in 1911 by Basu in cultivar Rasthali. Subsequently it was recorded in all other banana growing states of India and causes severe yield losses in most of the commercial cultivars grown (Sivamani and Gnanamanikkam, 1988; Lakshmanan *et al.*, 1987; Thangavelu *et al.*, 2001, 2011). Extensive survey conducted revealed that the incidence of the disease goes as high as 30 per cent in main crop and up to 85 per cent in ratooncrop. In Karnataka, the area under banana cultivation of var. Nanjangod Rasabale was reduced from 500 ha to less than 50 ha. (Narendrappa and Gowda, 1995). In Bihar, more than 55 % of the area are grown with susceptible variety were severely infected and the estimated yield reduction reported from these areas was 50–70%. In Tamil Nadu the disease severity was as high as 80-90 per cent. In Andhra Pradesh, the farmers abandoned the cultivation of the most susceptible var. Amritapani (Silk-AAB) for more than 15 years due to Fusarium wilt and since 2005, they started growing the var. Martamon (an eco- type of Rasthali received from West Bengal), since the Rasthali type fetches always high price in the local market. In this variety also the incidence was noted up to two per cent.

1.1.2. Symptoms: Fusarium wilt is a typical vascular disease causing disturbances of translocation. Fungus causes systemic foliage symptoms, which eventually lead to collapse of the crown and pseudostem. Generally symptom commences as a premature yellowing of the oldest leaves (Fig. 14.1) followed by necrosis and collapse close examination of the petioles of the lower leaves may reveal chlorotic steaks or patches and such leaves may collapse before chlorosis of the lamina the disease appears around 5th to 8th month after planting. Typically, diseased plants have few erect leaves with an increasing number hanging down beside the pseudostem. The unfurled heartleaf dies and only the pseudostem remains standing, which also collapse, induce course. When an infected plants pseudostem and rhizome are cut longitudinally the vascular tissue shows reddish to purplish brown discolouration. Inner surface of leaf sheaths showed brown flecks or spot. Splitting of pseudostem at the base is a common symptom in Rasthali, Karpuravalli and Ney Poovan. In Monthan, which is attacked by Race-2, conspicuous bright orange colour of older leaves is observed in severely affected plants where the progress of the disease is rapid, the whole plant dies within a span of one to two months after initial external symptom. Young suckers rarely show the symptoms. The affected plant will not yield bunch, however if the infection starts little later stage of the crop, it throw bunch, but there won't be any bunch development.

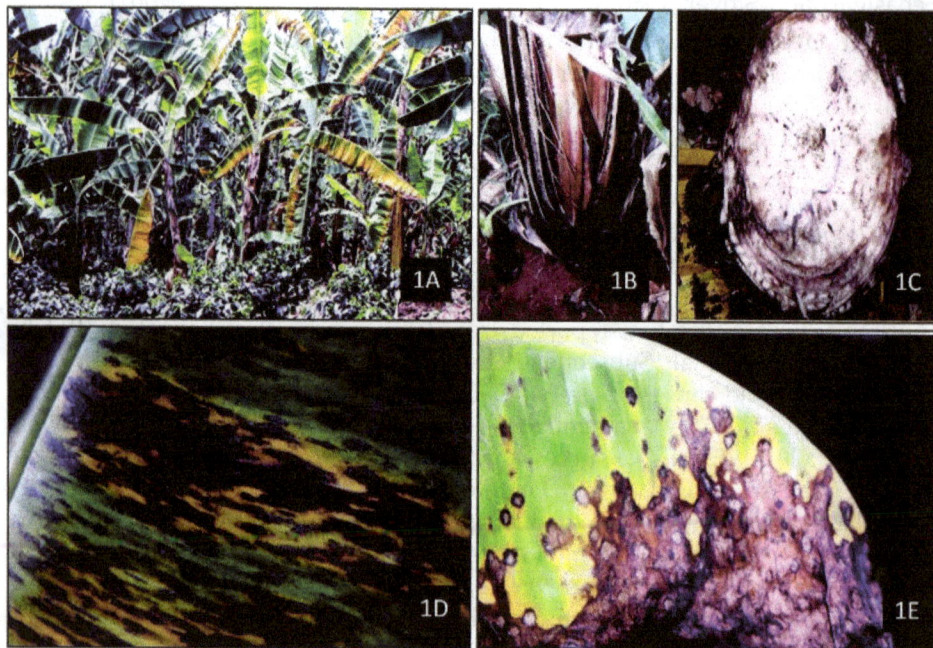

Fig. 14.1. A,B,C. Fusraium wilt in Hill Banana and D,E. Sigatoka disease

1.1.3. Causal organism: *Fusarium oxysporium* f.sp *cubense* is primarily a pathogen of *Musa spp.* This pathogen consider as a common soil-borne saprophyte and secondary invader of damaged plant root system. The fungus produces numerous micro and macro conidia and chalmydospores, which are characteristic of the genus. Sporodochia are tan to orange, and sclerotia are blue and submerged. Micro and macroconidia are produced on branched and unbranched monophialides. Microconidia are 5 to 16 × 2.4 to 3.5 μm, one- or two-celled, oval- to kidney-shaped, and are borne in false heads. Macroconidia are 27 to 55 × 3.3 to 5.5 μm, four- to eight-celled, and sickle-shaped with foot-shaped basal cells. Culture, invariably produce a reddish pigment and globose chlamydospores are produced in older culture. The pathogen survives in the soil and plant debris for long period. In stress or when the host is not found it can produce more number of resting spores called chlamydospores, which can survive in the soil for 15 to 18 years without germinating. When suitable host is present it will be induced to germinate causing the disease. Chlamydospores are not produced by isolates in vegetative compatibility group (VCG) 01214 of FOC has no known teleomorph. FOC has been classified into four physiological races based on pathogenicity to host cultivars in the field. FOC1 infects the cultivar Gros Michel; FOC2, 'Bluggoe'; FOC 3, *Heliconia* spp.; Laughe and FOC 4, Cavendish cultivars and all cultivars susceptible to Foc1 and FOC2 (Persley and Langhe, 1987). Earlier in the last century, FOC1 infection nearly destroyed the world's banana industry, which was based on the Gros Michel cultivar. Consequently, Gros Michel was replaced by Cavendish cultivars, which were resistant to FOC1. However, FOC4, which is capable of attacking Cavendish cultivars, was reported in Taiwan and Africa in 1967. To date, FOC4 has caused serious crop losses in Asia, Australia and Africa (Hwang and Ko, 2004). Thangavelu and Mustaffa (2010) reported the virulent strain of fusarium wilt pathogen (Race 1) infecting Cavendish group of banana in India.

1.1.4. Epidemiology: The speed of wilt in the field is influenced by soil and environmental conditions, being slowest on fertile soil and where cultural practices are good. Disease progress is rapidly expanding high level, the soil and cultural conditions are poorest. The time frame for first disease incidence to destruction of plant greatly depends upon initial inoculum level. It may vary from 2 to 20 years. Primary inoculum is from infected suckers and the pathogen survives in the soil in the form of chlamydospores for long period. Movement of infected plant material, contaminated soil, banana trash used for transport of harvested bunches are important methods of long distance dispersal. Intra field dispersal is favoured by surface irrigation and floodwater. The fungus gains entry to the plant through the roots. It then invades water conducting tissues of the corm and pseudostem reducing their efficiency. The planting of infected or contaminated corms is the main cause for the spread of the disease. Nematodes can help in entry of the pathogen. Nematode management is one of the precaution for avoiding the entry and spread of pathogen. Since infected rhizomes are often symptomless, they effectively spread the pathogen when used as seed pieces (Stover, 1962). The pathogen also moves in soil and running water, and on farm implements and machinery. It survives up to 30 years in the absence of banana, and non-host weed species that are infected by the pathogen are reservoirs of inoculum (Stover, 1962).

Soil populations are disturbed and decline rapidly in the absence of host. The pathogen doesn't colonized dead banana tissue in soil, nor does it grow through unsterilized soil. Survival is presumed to be by chlamydospore, which are produced abundantly in root tissues of diseased plants. Based on components of soil micro flora particularly bacteria inhibit growth of pathogen in soil and culture. Other factors namely soil texture, pH influencing pathogen behaviour in soil. Soil texture and organic matter increase the pathogen population in soil. Regarding soil pH the pathogen survive longer in light textured acid soil than in heavy alkaline soils. Certain crop residues may stimulate antagonistic micro flora and reduce survival of pathogen. It was reported that Fusarium has no collateral host, but the fungus spreads through the use of infected suckers from place to place.

Infection occurs principally through the small lateral or feeder roots. Wounding can only infect main roots. High inoculum potential is needed to initiate disease. First only few of the lateral roots infected then main roots and vascular tissue of rhizome infection reached. In fertile permeable soils being less favourable for disease initiation. Disease symptom develops most rapidly in conditions most suitable for banana growth and symptom appear 2-5 months after initial infection. The pathogen develops in the vascular system of the host plant by producing micro conidia, which are conveyed upwards in the transpiration stream. Production of toxins such as fusaric acid is associated with symptom development. These defence mechanism successfully block off infected xylem elements by production of gels and tyloses, and finally breakdown of the pathogen invades the xylem parenchyma. Resistance reactions appear to be similar in both susceptible and resistant cultivars. Poor drainage, acidic soil, high level of nitrogen in the soil also favours the disease development.

1.2. Sigatoka disease: Banana leaf streak disease (also known as black Sigatoka) is the most devastating disease affecting bananas (Marin *et al.*, 2003; Cordeiro *et al.*, 2004; Conde *et al.*, 2007). Black Sigatoka was originally identified in Fiji in 1963 (Rhodes, 1964) and now it is present in most of the banana growing areas worldwide (Arzanlou *et al.*, 2007). Yellow Sigatoka was first reported in Java (Indonesia) in 1902 (Zimmerman, 1902). The disease was subsequently reported in banana growing area. Black Sigatoka develops more rapidly, causes more damage and is more difficult to control than yellow Sigatoka (Simmonds, 1986; Johanson, 1993). Yellow sigatoka disease caused by *Mycosphaerella musicola* is well-known and wide spread. Later it has been reported to cause serious loss in many banana growing regions of the world. In India, yellow Sigatoka casued by *M.musicola* was reported in 1963 and it is present in most of the banana growing regions of India. In the recent past, virulent forms of other leaf spot diseases have been causing serious concern to the growers, as many cultivars known to have tolerance earlier, have fallen a prey to the new leaf spot. In the mid-1990s, a third *Mycosphaerella* sp., *M. eumusae*, was recognized as a new constituent of the Sigatoka complex of banana (Carlier *et al.*, 2000b, Crous and Mourichon, 2002). The presence of *M. eumusae*, causing septoria leaf spot disease in the Cavendish cv. Grande Naine in southern India has been confirmed (Selvarajan *et al.*, 2000: Carlier *et al.*, 2000a). Presently, *M. eumusae* is known from Southeast Asia and parts of Africa, where it affects cultivars that are highly resistant to both *M. musicola* and *M. fijiensis* (Jones, 2002). Leaf spot disease is prevalent in all banana-growing states of India, except in Jalgaon district of Maharastra. But recently, the incidence of leaf spot disease is rampant in Jalgaon, this occurrence probably due to change in climate, monocropping of Cavendish banana for long time and planting banana round the year in the region.

1.2.1. Symptoms: Black Sigatoka and yellow Sigatoka can cause indistinguishable symptoms depending on the cultivar infected, stage of the disease and the season of the year (Johanson, 1993). However, yellow Sigatoka is characterized by oval to round necrotic lesions, which first appear pale yellow on the lower surface of the leaf (Meredith and Lawrence, 1970). Initially pale yellow specks appear on the upper leaf surface; these lengthen to form yellow streaks and enlarge into mature spots. The Centre of the spots become brown and around spot yellow halo is seen. Spotting is common along the margins and at the apex of the leaf. The leaf spot is elliptical in shape. In severe conditions, the whole leaf dries from the tip. In case of Cavendish cultivars except 2-3 leaves all the leaves will dry out. If the weather is conducive, all the leaves will be spotted and eventually dries (Selvarajan *et al.*, 2000). With black Sigatoka, the first symptoms appear as dark brown specks on the lower surface of the leaf and causes reddish-brown streaks running parallel to the leaf veins, which aggregate to form larger, dark-brown to black compound streaks (Fig. 14.1 D, E). These streaks eventually form uniform or elliptical lesions that coalesce, form a water-soaked border with a yellow halo and, eventually, merge to cause extensive leaf necrosis. The disease does not kill the plants immediately, but weakens them by decreasing the photosynthetic capacity of leaves, causing a reduction in the quantity and quality of fruit, and inducing the premature ripening of fruit harvested from infected plants (Stover and Simmonds, 1987). The disease is influenced by rainfall, relative humidity and temperature. The incidence is more during north east and south west monsoon periods because of high humidity, intermittent rainfall and moderate temperature. The Sigatoka / Septoria leaf spot diseases were observed in four different commercial cultivars grown in Tamil Nadu (Selvarajan *et al.*, 2000). The symptom varied from variety to variety, depending upon the state of disease development, severity and duration of epidemic, the fruits on the bunch get affected. In severe cases, immature bunches fail to fill out. Normally 15-18 (functional) leaves are necessary at the time of shooting, for bunch development, but due to Sigatoka leaf spot it is difficult to maintain 15 leaves, if necessary control measures are not taken. The quality of banana is drastically reduced. Small fingers, premature ripening, peel splitting is commonly noticed. If bunches are harvested from affected orchards, during the transit the banana start ripening leading to damage of whole shipment. Due to abundant ethylene production in leaves affected with leaf spot cause premature ripening in banana in the field itself. Sufficient finger girth and length cannot be reached for export market. Uneven ripening leads to attack of many other postharvest pathogens.

1.2.2. Casual organism: Fungal identification has been based on sporulating structures, growth characteristics *in vitro* and symptoms in the field. *M. musicola* and *M. fijiensis* pathogens can produce indistinguishable symptoms in the field depending on the variety and environmental conditions (Stover and Simmonds, 1987; Johanson, 1993). The perithecia of the sexual stage of *M. fijiensis* and *M. musicola* (Fig. 14.2) are also very similar and are both found in the necrotic tissue of the lesion (Fullerton, 1994). *M. fijiensis* and *M. musicola* also exhibit similar growth characteristics in culture. Ascospores as well as conidia of both species germinate to produce two germ tubes, one on each end of the spore within 24 - 48 h (Meredith and Lawrence, 1969; Natural, 1985). Stover (1980) described two main cultural types produced by both *M. fijiensis* and *M. musicola in vitro*: a dark grey colony or grey brown with crenate edge and a pale grey or pink colony, where the dark grey colonies tend to become pink with age. It is therefore very difficult to accurately distinguish these two *Mycosphaerella* pathogens based on cultural characteristics. The major feature used to differentiate between *M. fijiensis* and *M. musicola* pathogens is the presence or absence of a scar at the base of the conidium. At the asexual stage of *M. musicola*, the conidiophores form on both the upper and lower surfaces of the leaf, but they are more abundant on the upper surface. Fullerton (1994) reported that conidiophores and conidia are more abundant on the lower surface of the leaf (in case of *M. musicola*) and on the upper surface (in case of *M. fijiensis*) of the leaf as a rapid guide to the identification of these *Mycosphaerella* species. With little modification, using the protocols of Natural (1990) and Carlier *et al.* (2000b), ascospores of *Mycosphaerella* spp. were isolated from samples collected during the survey. Selvarajan *et al.* (2000) has isolated the ascospores of the telemorph stage of Septoria pathogen. The Septoria pathogen produces conidia in pycnidia where as Sigatoka pathogens produce conidia on sporodochia.

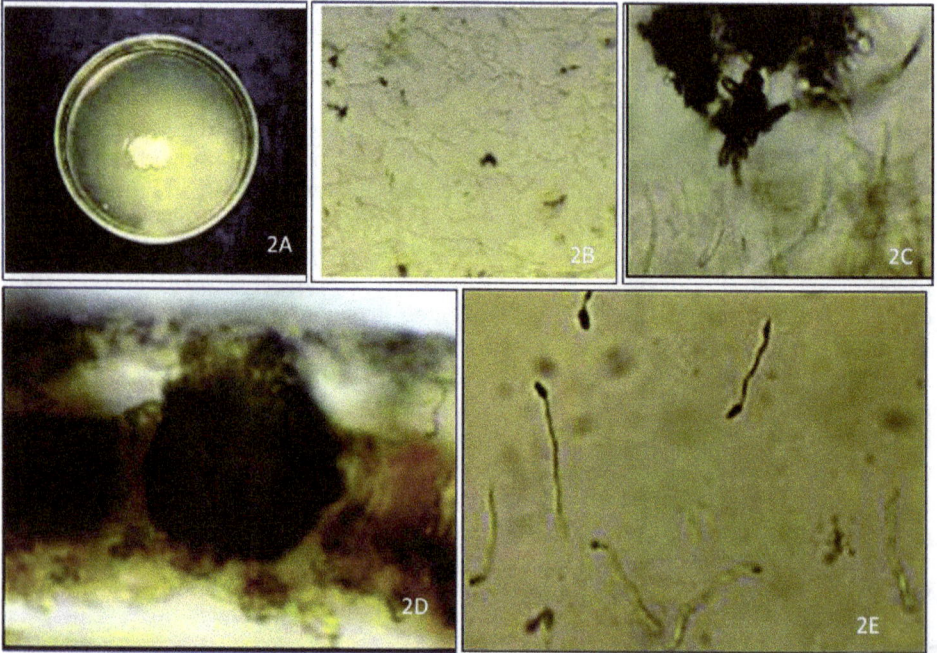

Fig. 14.2. Colonies of *Mycosphaerella musicola* after 3 months of culture on PDA

1.2.3. Epidemiology: Leaf spot disease development is strongly related to weather conditions and plant susceptibility. Under very favourable conditions in Costa Rica and with a susceptible host, incubation periods of *M. fijiensis* can be as short as 13 to 14 days, whereas during periods of unfavourable weather, the duration of the incubation period can extend up to 35 days (Marín *et al.*, 2003). During the rainy season, the incubation period was 14 days but in the dry season 24 days. Differences in the latent period from December 1993 to May 1995 for the susceptible cultivar Grande Nain, which is widely used for the fresh banana market, were observed in Guapiles, Costa Rica (Marín *et al.*, 2003).The latent period ranged from 25 days during the rainy season (June to December) to 70 days during the dry season at the same locality. When the weather is highly conducive for ascospore discharge and infection, many infections occur on the leaves. Under these conditions, leaves are rapidly and severely damaged. Black leaf streak disease and Sigatoka disease are disseminated locally by ascospores and conidia. Long distance spread is believed to be by the movement of germplasm (infected suckers, diseased leaves) and windborne ascospores. India, receives rain during two monsoon seasons. The south –west monsoon starts in June, (first week) ends in the month of September. The banana growing states receiving south west monsoon are Kerala, Karnataka, Maharashtra, parts of Andhra Pradesh, Madhya Pradesh and Gujarat and the North-east monsoon which starts in month of Oct and ends in Dec (second week) Tamil Nadu, parts of Andhra Pradesh, West Bengal and North

Eastern region are the parts receives N.E. monsoon rains. The rainy days which builds high humidity, reduces the temperatures favours disease development and spread. In tropical region, the spotting appears 8-10 days faster with black Sigatoka than Sigatoka (Stover, 1980). As the weather factors vary from place to place, the influence of each factor also vary or inducing disease incidence. Relative humidity and minimum temperature in the morning had positive correlation on disease incidence where as Relative humidity (evening) and maximum temperature had negative correlation. For every 1% rise in RH, there was an increase of disease index by 5.0 points in Coimbatore. Rise in 1°C temperature increased the Disease Development Time (DDT) by 6-14 days; similarly rise in RH decreased the DDT by 1.5 to 5 days (Selvarajan *et al.*, 2000). The rainy days, which build high humidity, reduces the temperatures favours disease development and spread Relative humidity (Morning) and minimum temperature have positive correlation on disease incidence whereas relative humidity (Evening) and maximum temperature had negative correlation. For every 1% rise in relative humidity 5.0 points increases in disease severity index in Coimbatore.

1.3. Pre-harvest and postharvest diseases

1.3.1. Crown rot: It occurs in all banana growing regions. It is the most serious postharvest problem, especially where dehandling and boxing of fruit is not carried out in modern centralised plants. In Asia, this disease is caused by *Colletotrichum musae, Fusarium spp.* and *B. theobromae.* Spores of these fungi invade fresh wound where the crown of the hand is cut from the stalk (Fig. 14.3A). Individual fingers

Fig. 14.3. A. Fruits of Cavendish cultivar affected by crown rot. B. Symptoms of anthracnose on finger of a Neypoovan cultivars. C.Cigar end rot. D.Pitting disease

may fall from the weakened crown if the rot penetrates deeply. Rotting spreads into crown during transportation. If the transit time exceeds 7 days high incidence of the disease can cause premature ripening. The pathogen can spread through the pedicel into the pulp causing fingers to seperate from hands. Disease stimulates ripening which is another cause of wastage in international trade.

The fungi causing crown rot are ubiquitous components of the microflora of banana plantations they live saprophytically in dead banana leaves, flower bracts, discarder fruits and bunch stems. Spores of pathogens are splashed on to fruit by rain or irrigation water and may remain viable for months in the field under extreme of relative humidity and temperature

1.3.2. Anthracnose: The causal agent is *Colletotrichum musae*. Black rot and ripe fruit rot are synonymous of anthracnose. Fruits are being liable to infection in the field, in local markets and in large consignments during ocean transport and especially during ripening. It attacks immature fruit and it may be present in very young fruit as invisible latent infections. The fungus reaches its greatest economic importance as a cause of wastage in scratched, blemished or otherwise wounded ripening and ripe fruit. If green bananas are inoculated before being placed in cold storage, sunken black diseased areas develop. On the other hand the more common aspect of Anthracnose, which becomes conspicuous as the fruit approaches final maturity, consists of more or less superficial brown diseased spots or patches (Fig. 14.3B) contrasting with the yellow skin. In a moist atmosphere, these spots or blemished areas become covered with the orange to salmon pink acervuli or conidial masses. On drying these look like small coral pustules.

1.3.3. Cigar end rot: It is caused by *Trachysphaera fructigera* and *Verticillium theobromae*. This disease occurs during rainfall. Immature fruits of Gros Michael and Cavendish varieties are more susceptible than mature fruits. Fungal infection of the perianth leads disease development. Black necrotic rot spreads slowly into the tip of the immature fingers upto 2cm. The emergence of conidiophores from the epidermis gives the necrotic tissue the appearance of the grey ash of a cigar end, hence the name. Infection generally begins from the tip of the finger (Fig. 14.3C) and slowly spreads backwards, causing greyishness of the skin. The infection of the fruit takes place through the pistillate end by air borned conidia. Necrosis spreads a short way up the finger and the necrtic tissue becomes corrugated and covered with greyish white fungal hyphae so that it looks like the ash at the end of a cigar. The occurrence of a dry rot of pulp is very characteristic of infection with V.theobrammae. There is a clear demarcation between the infected tip tissue and the rest of the finger. Cigar end rot is associated with periods of high humidity.

1.3.4. Pitting disease: The causal organism is *Pyricularia grisea*. It is characterized by round sunken pits of approximately 4-6 mm dia, which appears as infected fruit reaches maturity or after harvest. The sunken centre is surrounded by reddish-brown zone with a greenish, narrow water soaked halo. Smaller pits occur on finger stalks (Fig. 14.3D) and crown pads, which can lead to finger drop. The extend of pitting may increase considerably while fruit is in transit and during ripening cause serious post-harvest loss of quality.

1.3.5. Peduncle rot: Peduncle rot has become a serious threat to banana production in many banana growing regions of India. The problem is particularly more pronounced in regions where the temperature is high (above 35°C) and leaf spot diseases are severe. The incidence of the disease ranged from 5 to 25%. The upper portion of the peduncle is exposed to the hot sun, when the bunch emergence occurs during summer months and due to reduced functional leaves reduced due to leaf spot diseases. The exposed region of the peduncle become bleached and subsequently starts to rot. The rotting continues to spread up and down the peduncle resulting in reduced finger size and non-development or partial filling up of fingers (Fig. 14.4). When infection occurs at an early stage the disease can attain sufficient severity to spread into the fingers and results in the complete destruction of the bunch. This disease is caused by *Colletotrichum gloeosporioides* and *Botryodiplodia theobromae*.

Fig. 14.4. Peduncle rot

1.3.6. Botryodiplodia fruit rot: In India, reported the fungus as a causal agent for pseudostem rot and fruit rot. The causal organism *B. theobromae* is a wound parasite. It is often implicated in the more severe forms of main stalk rot; it may extend through the cushions and cause finger-stalk rot and dropping. It sometimes associated with fruit spots and blemishes. In the ripening room this pathogen may frequently be present as a tip rot. The infection, which originates in or immediately below the decayed perianth or style, spreads uniformly along the fruits causing progressive brownish-black discolouration and softening of the pulp. In fruit approaching maturity two third, even the whole finger may be affected, the flesh become pulpy and finally semiliquid, with a sweetish odour. The skin becomes soft black, wrinkled and encrusted with pycnidia.The fungus invades the fruit through wounds and the pulp turns rapidly into black watery mass. Surface growth of greyish black mycelia is characteristic of the disease. The infection originates in or immediately below the decayed perianth, stems end or style end, causing a progressive brownish black discolouration. The skin becomes black, soft and wrinkled. Later the skin gets encrusted with pycnidia. Maximum damage to fruits occurs at 25-30°C.

1.3.7. Deightoniella leaf and Fruit spot: The disease caused by *Deightoniella torulosa* (Syd.) is reported from Jamaica, Bermuda, Trinidad, Sri Lanka (Ceylon), India, Ghana etc. The disease appears to occur in three forms, which are black spot on leaf, fruit tip discolouration or black tip and fruit spot (speckle). It appears in the form of round, pin point, black spots on the main veins of the lamina in the proximity of leaf margin. Gradually these spots increase in size. They usually remain separated from the normal green leaf tissue by a narrow, yellow peripheral band. The large spots dry in the centre with pale brown areas, which extend to the edge of the leaf blade. In case of severe attack, petioles may be affected. The black tip below the perianth is seen advancing along the fruit. The disease areas may not have regular outlines, which get surrounded by a narrow, grey or yellowish margin. The intensity of spotting on fruit varies with age and position of fruit in the bunch. In some varieties nearly round spots with reddish brown or black centre and a darker green, water-soaked halo may be seen. Under the conditions of rapidly increasing temperature and decreasing relative humidity, there is a peak period of spore liberation, spores of the fungus can be trapped during June to November, particularly during wet spells. Water extracts of banana leaves and peel stimulate germination of spores. Germination is higher in the water film than at 100 % relative humidity. Infection requires dew or rain water and is more severe during wet weather. Isolates from fruit spot cause black spot on leaves and black tip on fruits. Spores are liberated during night show diurnal periodicity of spore liberation. In fruit spot, conidia germinate and produce 1 or 2 appressoria, which develop infection hyphae and then penetrate mechanically. The fungus can penetrate the unwounded leaf under undisturbed condition and remains long enough in a thin layer of water. However established its pathogenicity on leaves and fruits and found that injury and humidity are necessary for infection.

2. Bacterial disease

2.1. Tip over or rhizome rot: The disease caused by *Erwinia carotovora* (Jones) is widespread in banana growing areas of the world. Of late the Erwinia head rot is most commonly observed in banana plantations during hot summer months and also during monsoon seasons in some parts. This disease prominently observed in tissue culture plants at third to fifth month after planting and becoming economically very important. The disease is also noticed at the secondary hardening stage of tissue culture raised banana. In Rajamundry district of AP, and Theni district of TN the disease manifestation is seen even in later stage of plant growth. The pathogen is soil borne and enters the corm through wounds and also through decaying leaf sheath of suckers. In affected plants, the bunch fails to develop and fall apart leaving the rotten corm inside the soil. The affected plant shows severe splitting in pseudostem and being locally called *vedivazhai* (Split banana). The pathogen does not spread into the plant systemically through vascular bundles, rather it confine within the affected corm and can spread to corm of developing daughter suckers. Cultivars like Dwarf Cavendish, Grand Naine, Robusta, Nendran and Thella Chakkarakeli are most susceptible while other varieties also succumb to the disease.

Rotting of collar region is the commonest symptom of this disease (Fig. 14.5). The leaves of affected plants show epinasty and suddenly dry out. If the affected plants are pulled out, they topple from the collar region leaving the corms with their roots. Splitting of pseudostem is observed at late stage of infection in cultivars Robusta and Thella Chakkarakeli. If the affected plants are cut open at collar region, yellowish to reddish ooze can be seen. In early stage of infection, dark brown or yellow, water soaked areas are visible in the cortex area. In advanced stage the interior lesions may decay to such an extent as to form cavities surrounded by dark spongy tissues. The rot spreads radially towards the growing point through cortical tissues the disease. The rotten corm emits foul smell

Fig. 14.5. Erwinia rot

3. Viral diseases

3.1. Banana bunchy Top disease

3.1.1. Disease occurrence and distribution: Banana bunchy top disease (BBTD) is threatening the production in one quarter of the world's banana growing areas (Dale, 1987).Banana bunchy top disease (BBTD) caused by *banana bunchy top virus* (BBTV) is one of the most important and destructive viral disease in India. BBTD was first recorded in Fiji in 1889. It is believed that BBTD spread to Australia and Srilanka in 1913 through infected suckers from Fiji and later moved to Kerala, Southern India about 1940. Upon gaining entry in the south India, the virus spread to all the banana growing states of the country and probably it had later moved into Bangladesh. In India, BBTD was a major problem in Kerala, Andhra Pradesh, Tamil Nadu, Orissa, Maharastra, Gujarat, Bihar, Karnataka, West Bengal, Assam and Uttar Pradesh (Singh, 2003) and recently it has been recorded in North Eastern Hill states of India *viz.*, Nagaland, Meghalaya and Arunachal Pradesh. In lower Pulley hills of Tamil Nadu, India, a famous, unique flavoured elite dessert banana cultivar Virupakshi (Pome group, AAB) a choice banana cultivar registered under 'Geographical Indications' (GI) has been almost destroyed by the BBTD and the area under this banana has been reduced from 18,000 ha to 2,000 ha (Kesavamoorthy,1980). A loss of about Rs 40 million annually has been reported in Kerala state alone due to this disease. In Jalgaon district of Maharastra, India 17.16 million plants of Cavendish banana were affected with BBTV in 2008 and caused an estimated economic loss of around US$51 million (Selvarajan *et al.*, 2011). Survey in Kodur, Kadappa district of Andhra Pradesh during 2009, showed a BBTV incidence ranging 15-95% both in tissue culture plants and ratoon crop of conventional sucker grown orchards (Selvarajan *et al.*, 2011). The survey made during May 2009, in Lower Pulney hills (Kodaikanal) recorded 15.26 to 83.88% incidence in Hill banana (Selvarajan *et al.*, 2011).BBTV has been recorded during the bunch emergence stage of the crop leading to severe loss to the growers. Recent outbreaks of BBTV in 2011 in Theni district, TN lead to an infection of 0.3 million plants out of 3 million plants of both tissue culture and conventional sucker grown plants. In parts of Tamil Nadu and Kerala states this disease has caused enormous loss in early period of its introduction and even now the disease incidence is not uncommon. One of the reasons for the severity in hill regions is the disease inoculum and presence of the vector round the year where the crop is grown as perennially for 10 to 15 years. Among banana cultivars, Virupakshi (Hill banana),

Fig.14.6 A. Banana bunchy top disease; B, virion of BBTV; C, Banana black aphid (*Pentalonia nigronervosa*) vector of BBTV. D, Genome organization of BBTV infecting hill banana isolate

Robusta, Grande Naine, Nendran, Rasthali, Poovan, Ney Poovan, Monthan and Red Banana are severely affected by BBTD. There is no resistant gene source available in germplasm of bananas and plantains. Because of this direct influence on productivity, BBTV is considered one of the most economically destructive diseases of banana.

3.1.2. Symptoms: BBTV affected plants show intermittent dark green dots, dash, streaks of variable length like 'Morse code' pattern on leaf sheath, midrib, leaf veins and petioles of infected plants. Leaves produced are progressively shorter, brittle in texture, narrow and gives the appearance of bunchyness at the top (Fig. 14.6a), hence the name (Magee, 1927). Late infection of BBTV or in case of latency, the plant can throw bunch but the fingers never develop to maturity. Fruits of infected plants are malformed. In case Grand Naine, the BBTV affected plants throw bunch with extremely long or very short peduncle. Sometimes affected Grand Naine banana fingers appear like a non-Cavendish type. Marginal chlorosis and yellowing of whole leaf lamina resembling iron deficiency are common symptoms noticed in Lower Pulney Hills. Vein flecking symptoms in the lamina are also noticed. When infection occurs very late in the season the plant would show dark green streaks on the tip of the bracts of male flower bud. Sometimes the bracts tip of male bud is converted into green and leafy structure. Abaca bunchy top disease has been reported to cause by a strain of a BBTV also exhibit the similar symptoms in *Musa textiles* (Magee, 1927). Mild vein clearing and also blisters with dark green colour on mid rib and leaf sheath have also been recorded. In a recent survey at Kodur, AP symptoms induced by BBTV include floral greening, finger malformation, yellowing of newly emerging leaves, typical interrupted short dark green streaks in the mid rib and leaf lamina etc., were noticed.

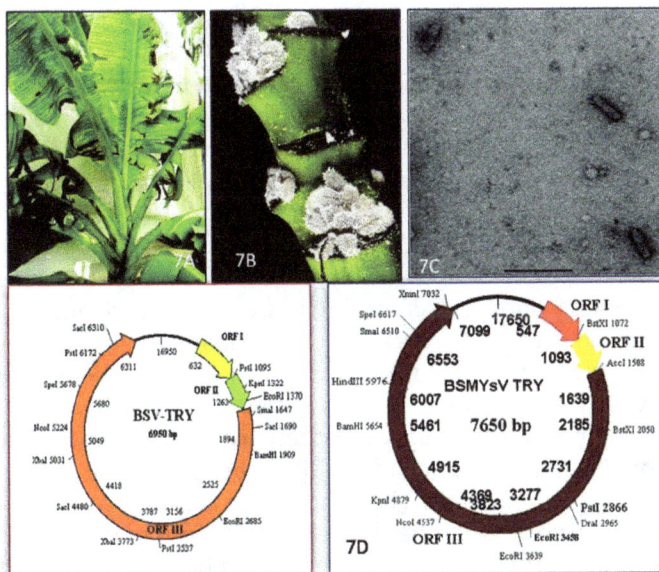

Fig. 14.7. Severe streak symptom of BSD in cv. Grand Nain

3.1.3. Causal organism: BBTV belongs to genus *Babuvirus* and family *Nanoviridae*. The size of the isometric virion (Fig. 14.6b) is 18-20 nm in diameter. BBTV is a multi-component virus and consists of at least six particles individually packed with six different circular single stranded DNA's (css DNA). Each genomic component (Fig. 14.6d) is approximately 1kb in size (BBTV DNA-R, -S, -M, -C, -N and -U) (Burns *et al.*, 1995; Harding *et al.*, 1993; Vetten *et al.*, 2005). Coat protein of BBTV is 19.6 kDa (Harding *et al.*, 1991; Burns *et al.*, 1995). BBTV-DNA R encodes a master replication initiation protein (Rep). The functions of DNA-U is unknown, while DNA-M,-C and -N encodes the movement protein, cell- cycle link (clink) protein and nuclear shuttle protein (NSP) respectively (Wanitchakorn *et al.*, 1997; 2000). Burns *et al.* (1995) identified two regions viz the major common region and a stem loop common region that are highly conserved between all six components (BBTV DNA-1 to 6) from Australian BBTV isolates. Each component had a structure similar to that of BBTV DNA-1 in that each component had one large ORF in the virion sense (except for BBTV DNA-2) with an associated potential TATA box and polyadenylation signal(s) and also a sequence 5' of the ORF capable of forming a stable stem-loop structure. The stem-loop common region (CR-SL) extends up to 25 nucleotides at 5' of the stem-loop sequence and up to 13 nucleotides 3' of the stem-loop sequence. The major common region (CR-M) of a length between 65 and 92 nucleotides located at 5' of the CR-SL. In India, complete genome of three BBTV isolates from Lucknow (UP), Bhagalpur (Bihar) and Lower Pulney Hills (TN) have been sequenced and characterized and found that all belongs to South Pacific group (Selvarajan *et al.*, 2011; Nazrul Islam *et al.*, 2010). Complete genome sequences of these Indian isolates showed high degree of similarity with the corresponding sequences of BBTV isolates originating from Fiji, Egypt, Pakistan, and Australia. An analysis of the cp sequences of 16 Indian isolates with distinct geographical origins revealed that they belong to the South Pacific group, except the isolates from Shervroy and Kodaikanal hills, TN (Selvarajan *et al.*, 2011).

3.1.4. Epidemiology: Bunchy top primarily spread through the infected planting material such as suckers, corms, corm bits and also through the tissue culture plants. The secondary transmission is through banana aphids. BBTV is transmitted by banana black aphid, *Pentalonia nigronervosa* in a persistent manner (Fig. 14.6c). The nymphal stage of the aphids is more efficient in transmission. No other aphids are known to transmit the disease. The aphids are usually found clustered around the unfurled heart-leaf and the sheathing leaf base of petioles which are ideal locations for feeding and protection. They are also found on the base of the pseudostem and on very young suckers. The aphids flourish throughout the year in hills, but are more numerous during the rainy season. Both winged and wingless individuals occur in a normal aphid colony. Banana aphids produce large quantities of "honey-dew" which attracts ants. The presence of ants is a good indication of the presence of aphids on a banana plant. BBTV, being obligate it can survive only in the live host or infected plants. The aphids once acquire the virus it can remain infective for 13 to 20 days. BBTV has been found to express visual symptoms only 23 to 25 days after inoculation but the virus could be detected early from young roots or cortex tissue even before the symptom expression (Nancy, 2005). It has been found that the optimum temperature range for acquisition of virus by the vector was 25°C to 27°C. Minimum inoculation feeding of 3 hr was required to successfully acquire the BBTV by the vector. Menon and Christudas (1967) reported the life history and population dynamics of *Pentalonia nigronervosa* in Kerala state in India. It was observed that the aphid bred throughout the year in banana but was more common during July. Menon and Christudas (1967) stated that climatic conditions existing during the summer months and rainy weather are unfavourable for the banana aphid in Kerala state. Samraj *et al.* (1970) have reported that a minimum time of 5 days and maximum 10-15 days is required for the down ward movement of the virus after inoculation with the aphid and this might change depending upon the vigour of the plant.

3.2. Banana streak disease

3.2.1. Disease occurrence and distribution: Banana Streak Virus Disease (BSD) has become a major threat to banana production, international exchange of germplasm and crop improvement programmes. *Banana streak virus* (BSV) is the causal agent of streak disease and is the most widely distributed disease in all producing countries (Hull *et al.*, 2000). This disease was first reported in Ivory Coast in 1966 (Lassoudiere, 1974) but the virus involved with the disease was not isolated until 1986. BSV was first identified on variety 'Poyo' of Cavendish subgroup (AAA) of banana by Lockhart (1986) in Morocco. In India, BSV was first reported from NRC Banana in 1996 (Jones, 2000; Singh *et al.*, 2000). Though the Mysore group of bananas was suspected for a virus infection even before 40 years (Wardlaw, 1972), but it remained unnoticed by Musa pathologists as the symptoms produced suspected to be the genetic characters of Mysore group. A yield loss of 49.48% has been recorded in cv. Poovan (Mysore, AAB) due to BSV (Selvarajan *et al.*, 1997; Thangavelu *et al.*, 2000). Banana streak has also been reported from Rasthali (sub group Silk, AAB) and Grande Naine (sub group Cavendish, AAA) and *Musa zebrina*. BSV severely infects Mysore group of banana namely Poovan (Syn: Palayankodan, Mysore, Champa, Alpon, Lalvelchi, Karpura Chakerakeli), Red banana, Robusta, Nendran and also in Cavendish group (Basrai, Grand Nain) and *Musa zebrina*. Many wild bananas also found to have BSV genome as integrant in their genomes but very rarely get infected. Sometimes in the inter-specific hybrids, the infection occurs by activation of endogenous pararetroviral sequence integrated into Musa genome. Such activation may occur during tissue culture and hybridization process. BSV exist in the host without producing any symptoms for long periods and in addition the integration of more than a full-length genome of virus in the host chromosome has caused problems to quarantine authorities and organizations involved in international exchange of *Musa* germplasm.

3.2.2. Symptoms: The most characteristic symptom of disease is chlorotic and necrotic streaks on leaves. Initially small dots with golden yellow colour develops, later it extend to form long streaks. The chlorotic streaks (Fig. 14.7a) become necrotic giving a blackish appearance on lamina. Necrotic streaks are also observed on midrib, petiole and pseudostem. Bunch choking, abortion of bunch and seediness in fingers are observed in infected plants. Sometimes diseased plants are stunted and fruit get distorted with a thinner peel and bunches are small in size. On some occasions, rotting of heart leaf and eventually the plant dies. Necrosis of cigar leaf and death of entire plant has been recorded in plantain hybrids in Nigeria (Harper and Hull, 1998). Leaf stripping symptoms are observed in infected cvs. Poovan, Grand Nain and Robusta. In Trichy, bunches bear a female phase and after a short male phase again a female phase has been observed in BSV infected in cv. Poovan. Emergence of bunch by piercing out from the side of the pseudostem has frequently been observed in cv. Poovan. In cv. Rasthali, the affected plant bears small bunches with reduced finger size. Fruit malformation and seediness in cv. Poovan are associated with the BSV infection. Pseudostem splitting and peel splitting are also caused by BSV (Lockhart and Jones, 2000a). Depending on infecting BSV species, highly susceptible banana cultivars can develop more severe symptoms, such as pseudostem splitting and necrosis, eventually leading to the death of infected plants. BSV has a very restricted host range. BSV has been found naturally in many *Musa* spp and Ensete.

3.2.3. Casual organisms: BSV is a pararetrovirus belonging to the genus, *Badnavirus* in the family *Caulimoviridae*. The virions are non-enveloped, bacilliform in shape (Fig. 14.7c) with an average size of 130–150 nm X 30 nm and contain a circular dsDNA as its genome (~ 7-8 kbp) (Lockhart and Olszewski, 1993). Harper and Hull (1998) cloned and sequenced the genome of a Nigerian isolate of BSV and shown to comprise 7389 bp and organized in a manner characteristic of badnavirus named as *Banana streak Obino l'Ewai virus* (BSOLV). The virus codes for three proteins from three well defined ORF's. The first two proteins functions are unknown. The ORF III polyprotein is most characterized part in the genus *badnavirus*. BSV was believed to be a single virus species with many strains and isolates having genomic and serological heterogeneity, but, now ten BSV isolates were classified as independent species of the genus *Badnavirus*. They are *Banana streak OL virus* (BSOLV), *Banana streak GF virus* (BSGFV), *Banana streak MY virus* (BSMYV), *Banana streak VN virus* (BSVNV), *Banana streak IM virus* (BSIMV), *Banana streak CA virus* (BSCAV), *Banana streak UA virus* (BSUAV), *Banana streak UI virus* (BSUIV), *Banana streak UL virus*(BSULV), *Banana streak UM virus* (BSUMV) (James *et al.*,2011). In India, complete genome of two BSV species (Fig. 14.7d) from banana cv. Mysore has been cloned and sequenced (Selvarajan *et al.*, 2007b). The BSOLV-TRY isolate found to be 6950 bp with three ORF's. However, BSVOLV-TRY isolate, a deletion of 450 bp in the ORF III of published BSVOL sequence has been observed. BSOLV-TRY was closely related to BSV-GD and there was 100 % identity between the amino acid sequences. The complete nucleotide sequence of BSMYV infecting CV Poovan named as BSMYV-TRY, was cloned and sequenced. The length of the sequence was 7650 bp. BSMYV-TRY was closely related to BSMYV (Australian isolate), and there was 99 % identity between the amino acid sequences. Lafleur *et al.* (1996) first time reported that portions of banana streak badna virus genome are integrated in the genome of its host *Musa*. 'B genome' containing banana cultivars and wild accessions has been known to have integration of viral genome in their genomes. Most of the cultivated bananas are manmade selections derived from inter specific natural hybridization containing accuminata (A) genome and balbisiana (B) genome. Integrated badnavirus sequences, termed endogenous pararetroviruses (EPRVs), are known to occur within the banana genome. Although some EPRV sequences show homology to the genomes of recognized or tentative

BSVs including BSMyV, BSGFV, BSOLV and BSImV, many others have no known episomal counterpart (Geering *et al.*, 2005). Two types of integrated BSV sequences are known to occur in banana. The first type contains the majority of banana EPRVs and comprises incomplete virus genomes which are incapable of causing infections. The second type of integrated sequences, known as endogenous activatable BSVs (eaBSVs), consist of the entire genome of characterized episomal BSVs which exist as multiple non-contiguous regions of the virus DNA combined with host-genomic sequences. Under certain stress conditions, particularly tissue culture and hybridisation, recombination events occur in the integrated sequences allowing the reconstituted viral genome to be activated thereby resulting in episomal infections. Integration of this viral genome into the host Musa causes problems in safe movement and also breeding new varieties for other biotic stresses and yield. Tissue culture process trigger, excision of integrated viral sequence into episomal virus especially in bispecific clones having diploid or triploid genomes. Selvarajan *et al.* (2006) reported that the integration of BSV genome in *Musa* germplasm accessions having one or more balbisiana (B) genome as their constituent. The copy number of integrated BSOLV has been found to be one based on the real time PCR assay for the different banana cultivars / wild banana (Selvarajan, Unpublished)

3.2.4. Epidemiology: The temperature plays a major role in symptom expression. At ambient conditions the expression of symptoms by the virus is very mild and plants kept at 22° C express more typical symptoms (Selvarajan and Sathiamoorthy, 2000; Dahal *et al.* 1998). BSV is primarily transmitted through planting material like suckers, corms, tissue culture raised plants and secondarily it is transmitted in a semi-persistent manner by several species of mealy bugs (Dahal *et al.*, 1999). *Ferrisia virgata* (striped mealy bug) (Fig. 14.7b) reported to transmit BSV from banana to banana (Selvarajan *et al.*, 2006). BSV does not get transmitted by mechanical sap inoculation (Lockhart, 1995). Like other badnaviruses, BSV also shown to be seed transmitted. BSV is not known to be transmitted mechanically by sap or by tools used for inter-cultivation operations. Seed/vegetative transmission is the primary mode of transmission. BSV is also transmitted to new hybrids through the activation of integrants present in the parents having B genomes. Meyer *et al.* (2008) reported that transmission of activated episomal Banana *streak OL (badna) virus* (BSOLV) to cv. Williams banana (*Musa* sp.) by three mealy bug species viz., *Dysmicoccus brevipes*, *Planococcus citri*, and *P. ficus*.

3.3. Banana bract mosaic disease

3.3.1. Disease occurrence and distribution: Banana bract mosaic disease (BBrMD) is becoming serious threat to banana and plantain production. BBrMD caused by *banana bract mosaic virus* (BBrMV) was and first reported in 1979 in the Philippines at Davao on the island of Mindanao (Thomas and Magnaye, 1996). It has subsequently been shown to be widespread throughout the banana growing areas (Diekmann and Putter, 1996). In India, BBrMD was first reported in French plantain cv. Nendran as a Kokkan disease of unknown etiology in Kerala by Samraj *et al.* (1966). Later the causal agent of Kokkan disease has been authentically confirmed as BBrMV (Singh *et al.*, 2000). Now BBrMV is reported to occur in several banana growing states in India viz., Kerala, Tamil Nadu, Karnataka, Andhra Pradesh and Maharastra (Anitha Cherian *et al.*, 2002; Kiranmai *et al.*, 2005; Singh *et al.*, 2000; Selvarajan and Jeyabaskaran,2006). In India, banana bract mosaic disease is mostly present in states of Tamil Nadu and Kerala where French plantain cultivar 'Nendran' is particularly susceptible. Around 40% reduction of bunch weight of kokkan diseased plant over healthy was recorded in Kerala (Anitha Cherian *et al.*, 2002). The virus infects cvs. Nendran, Poovan, Robusta, Ney Poovan, Rasthali, Red Banana, Karpooravalli and Monthan are also affected but the yield loss is minimal. In Tamil Nadu, the per cent incidence of BBrMD in different cultivars ranged from 0.5% to 56.8 % and the infected plants showed significant reduction in height, girth, leaf area, finger weight, and girth over healthy plants (Selvarajan and Singh, 1997). Based on three years of survey, per cent infection and yield loss assessment studies, an extrapolation on the loss due to BBrMV was made in cultivar Nendran. An amount of 387 million per annum has been the loss due to BBrMV (Kokkan) in cultivar Nendran has been estimated by NRCB (Selvarajan *et al.*, 1997). The yield loss assessment study due to BBrMV was done in 1995-96 by NRCB showed that 68.34, 50.00 and 46.34 per cent reduction in bunch weight in cultivars Nendran, Poovan and Ney Poovan over uninfected healthy plants (Thangavelu *et al.*, 2000). The yield loss caused by banana bract mosaic viral disease in Kerala was 52 and 70 per cent in cvs. Nendran and Robusta (Anitha Cherian *et al.*, 2002). Selvarajan and Jeyabaskaran (2006) reported that the average yield reduction in cv Nendran due to BBrMV was 30% and the yield loss varied from place to place depending on soil fertility.

3.3.2. Symptoms: BBrMV infected plants develop distinct, dark coloured, broad streaks on the bracts of the inflorescence (Fig. 14.8a). The characteristic mosaic symptoms on the flower bracts give the disease its common name. Necrotic streaks on fingers, leaf, pseudostem and mid rib are also recorded due to BBrMD. In Nendran, the leaf orientation changes in such a way giving the appearance of 'Travelers palm' plant (Balakrishnan *et al.* 1996). Bunches with unusually very long or very short peduncle, chocking of bunches, raised corky growth on peduncle are also observed (Selvarajan and Jeyabaskaran, 2006). Though this disease bears the name of bract mosaic and reported that this disease could only be diagnosed when symptom appears on bract, in Thiruvananthapuram district of Kerala, farmers have named the disease as Pola roga, means 'disease of pseudostem' in cultivar Nendran (Thangavelu *et al.*, 2000). Singh (2003) reported that the bract of the infected plants exhibited spindle shaped discontinuous dark red streaks. Dark red to purple mosaic streaks were also observed on the pseudostem after the removal of leaf sheath. The emerging suckers were deeply pigmented. Foliar symptoms appeared as chlorotic streaks parallel to veins and petioles.

Fig. 14.8. Typical mosaic symptom of Banana bract mosaic disease in bract

3.3.3. Casual organism: BBrMV belongs to the family *Potyviridae* and the genus *potyvirus*. BBrMV has flexuous (Fig. 14.8c) filamentous particles (660-760 x 12 nm) with a single stranded positive sense RNA genome (Thomas *et al.*, 1997). Rodoni *et al.* (1997) characterized BBrMV in India and the sequence comparisons of the coat-protein-coding and 3' untranslated regions revealed that the Indian isolates of BBrMV had greater than 96.6 and 97.2% homology with a Philippines isolate nucleotide and amino acid levels, respectively. Rodoni *et al.* (1999) sequenced the entire coat protein (CP)-coding region and 162 nucleotides of the 3' untranslated region (UTR) of nine different isolates of BBrMV from five different countries. A variability of between 0.3% and 5.6%, and 0.3% and 4.3%, was observed at the nucleotide and amino acid levels, respectively. Phylogenetic analysis of the BBrMV isolates did not reveal any relationship between the geographic locations of the isolates. Sankaralingam *et al.* (2006) amplified, cloned and sequenced 1062 nucleotides (nt) including 900 nt of the CP coding region and 162 nt from the 3' UTR from BBrMV infected banana cv. Ney Poovan (AB group) from Coimbatore. Sequence analysis of this complete CP gene showed a variability of 1.0- 4.6% at nucleic acid level and 0.7-2.0% at the amino acid level with other Indian isolates and 4.5-5.3% at nucleotide and 1.4-3.0% at amino acid variability level with Southeast Asian isolates respectively. The complete genome sequence of a BBrMV isolate from the Philippines (BBrMV-PHI) was 9711 nucleotide long excluding the 3' terminal poly (A) tail (Ha *et al.*, 2008). Recently a complete genome of the BBrMV-TRY isolate has been characterized at NRC Banana. The genome consists of a single large open reading frame (ORF) of 9378 nucleotides. When compared with BBrMV-PHI isolate, BBrMV-TRY had 94% nucleotide sequence identity and ten mature proteins had amino acid sequence identities ranging from 88-98 % (Balasubramanian and Selvarajan, 2012)

3.3.4. Epidemiology: BBrMV is primarily transmitted through vegetative planting material including suckers and tissue cultured plantlets. The virus is non–persistently transmitted through several aphid species viz., *Pentanlonia nigronervosa, Rhopalosiphum maidis, Aphis gossypii* (Magnaye and Espino, 1990; Munez, 1992) and in addition the cowpea aphid *Aphis craccivora* (Fig. 14.8b) has also been reported to transmit the disease (Selvarajan *et al.*, 2006). Attempts made to transmit BBrMV by mechanical sap inoculation were unsuccessful (Munez, 1992; Magnaye and Espino, 1990; Diekmann and Putter, 1996).

3.4. Banana Mosaic

3.4.1. Disease occurrence and distribution: Banana mosaic caused by *Cucumber mosaic virus* (CMV) is one of the common viral diseases affecting the bananas and plantains and occurs throughout the world (Lockhart and Jones, 2000b). Previously this disease was called in various names including infectious chlorosis, heart rot, sheath rot, cucumber mosaic and banana mosaic (Stover, 1972). Banana mosaic was first reported in Australia by Magee in 1930 (Magee, 1940), and later it was found in South and Central America, the Caribbean, India and the Philippines. In India, it was reported in Maharastra in 1943 by Kamat and Patel. Now this disease is known to occur in almost all the banana growing states of the country. It has reported a yield loss of 3 million rupees per annum in Gujarat state. Rao (1980) reported the incidence of banana mosaic disease in cultivars Robusta, Dwarf Cavendish and Rasthali from 2 to 7 per cent in parts of Andhra Pradesh and Tamil Nadu. An outbreak of banana mosaic was recorded in Poovan in Tamil Nadu during 1986 and 1988 (Kathirvel *et al.* 1986; Mohan and Laksmanan 1988). An epidemic of CMV in tissue culture plants of Grand Nain banana was recorded in Jalgaon, Maharastra during 2008-2010 (Selvarajan, unpublished). Severe, mild and heart rot strains have been reported in India. The heart rot of banana in Maharastra was reported by Mali and Rajegore in 1980.

3.4.2. Symptoms: Banana mosaic or infectious chlorosis is characterized by a range of symptoms from diffused foliar mosaic severe chlorosis, chlorotic streaking or flecking, stripes, line patterns, ring spots, leaf curling, distortion, rosette appearance of leaf arrangement to stunting of plant (Fig. 14.8c). Symptoms have often been confused with those of BSV, as happen with BSV infection in banana, CMV infected plant also show symptoms sporadically on few leaves.

3.4.3. Casual organism: *Cucumber mosaic virus* (CMV) belongs to the genus, *Cucucmovirus* and family, *Bromoviridae* (Roossinck *et al.*, 1999). The virus particles (Fig. 8d) are isometric in shape measures 29 nm in diameter and each particle composed of 180 subunits. CMV is a multi-component virus, its genome is single stranded tripartite positive sense RNA's (RNA 1, RNA 2 and RNA 3) and an additional sub-genomic RNA (RNA 4) derived from RNA 3 (Hubili and Francki, 1974; Paden and Symons, 1973). RNA 1, RNA 2 encode 1a, 2a proteins involved in virus replication, while as RNA 3a encode 3a movement protein (MP) and 3b expressed from RNA 4 coat protein (CP) with 5′ cap structures and 3′ conserved regions. Like most other plant viruses with divided genomes, the genomic RNA's are packaged in separate particles. Several strains of CMV have been classified using serology, host range, peptide mapping and nucleic acid hybridization. Numerous strains of CMV have been classified into subgroups I and II based on serological properties and nucleotide sequence homology (Palukaitis *et al.*, 1992). The subgroup I has been further divided into two groups (IA and IB) by phylogenetic analyses (Roossinck *et al.*, 1999). One of the subgroup (IB) is restricted to Asia, whereas the other two subgroups (IA and II) distributed a worldwide. Isolates of CMV from banana (CMV-B) have been identified as belonging to subgroup I, which includes most of the CMV isolates from the tropics. Coat protein gene of CMV isolate infecting banana in India has been sequenced and compared (Selvarajan *et al.*, 2007; Shahanavaj Khan *et al.* 2011) and the sequence analysis showed 93%-98% (at nucleotide) and 94%-99% (at amino acid) sequence identity and phylogenetic analysis revealed that all banana infecting isolates belong to subgroup IB.

3.4.4. Epidemiology: Primarily this virus is transmitted through infected suckers and it is also acquired from a wide range of host plants growing near banana fields through aphid vectors. The Aphids, *Aphis gossypii, A.craccivora, Rhopalosiphum maidis, R.purnifolia, Myzus persicae* and *Macrosiphum pisi* have been reported to carry and spread this virus disease (Magee, 1940; Waite, 1960; Capoor and Varma, 1968; Mali and Rajagore, 1980; Rao, 1980). Depends upon the virus strain and temperature the symptom will change. In some varieties, high temperature may suppress symptoms and it will be more severe when temperature falls below 24°C.

REFERENCES

Alves, E. J., Shepherd, K., and Dantas, J. L. L.1987. Cultivation of bananas and plantain in Brazil and needs for improvement. In: Banana and Plantain Breeding Strategies, ACIAR Proceedings . 21: 187.

Anita Cherian, K., Rema Menon., Suma, A., Shakunthala Nair., and Sudheesh, M. V. 2002. Impact of Banana bract mosaic diseases on the yield of commercial banana varieties of Kerala. Global Conference on Banana and Plantain 155.

Arzanlou, M., Abeln, E. C. A., Kema G. H. J., Waalwijk, C., Carlier, J., deVries, I., Guzmán, M., and Crous, P. W. 2007. Molecular diagnostics for the Sigatoka disease complex of banana. Phytopathoogy. 97: 1112-1118.

Balakrishnan, S., Gokulapalan, C., and Paul, S. 1996. A widespread banana malady in Kerala, India... Infomusa, 5: 28-29.

Balasubramanian. V., and Selvarajan, R. 2012. Complete genome sequence of a banana bract mosaic virus isolate infecting the French plantain cv. Nendran in India. Archives of Virology, 157: 397-400.

Burns, T.M., Harding, R.M., and Dale, J. L. 1995. The genome organization of banana bunchy top virus: analysis of six ssDNA components. J. Gen. Virol.76: 1471-1482.

Capoor, S. P., and Varma, P. M. 1968. Banana Mosaic in the Deccan. Indian Phytopath. Soc. Bull. 4: 11-14.

Carlier, J., Foure, F., Gauhl, F., Jones, D. R., Lepoivre, P., Mourichon, X., Pasberg-

Gauhl, C., and Romero, R. A. 2000a. Black leaf streak. Disease of Banana, Abaca and Enset. 37-47.

Carlier, J., Zapater, M. F., Lapeyre, F., Jones, D. R., and Mourichon, X. 2000b. Septoria leaf spot of banana: A newly discovered disease caused by*Mycosphaerella eumusae* (anamorph *Septoria eumusae*). Phytopathology 90: 884-890.

Conde-Ferraez, L., Waalwijk, C., Canto-Canché, B. B., Kema, G. H. J., Crous, P. W., James, A. C., and Abeln., E. C. A. 2007. Isolation and characterization of the mating type locus of *Mycosphaerella fijiensis*, the causal agent of black leaf streak disease of banana. Mol. Plant Ecol. 8: 111-120.

Cordeiro, Z. J. M., Matos A. P. M., Silva, S. O., Gasparotto, L., and Cavalcante, M. J. B. 2004. Impact and Management of Black Sigatoka in Brasil. XVI Reunião Internacional ACORBAT, Joinville, 63-69.

Crous, P. W., and Mourichon, X. 2002. *Mycosphaerella eumusae* and itsanamorph *Pseudocercospora eumusae* spp. nov.: causal agent of eumusaeleaf spot disease of banana. Sydowia 54: 35-43.

Dahal, G., Hughes, J. d' A., Thottappilly, G., and Lockhart, B. E. L. 1998. Effect of temperature on symptom expression and reliability of banana streak badnavirus detection in naturally infected plantain and banana (*Musa* spp.). Plant Disease, 82: 16-21.

Dahal, G., Gauhl, F., Pasberg-Gauhl, C., Hughes, J. A., Thottappilly, G., and Lockhart, B. E. L. 1999. Evaluation of micropropagated plantain and banana (Musa spp.) for banana streak badnavirus incidence under field and screen house conditions in Nigeria. Ann. Appl. Biol. 134: 181-191.

Dale, J. L. 1987. Banana bunchy top: an economically important tropical plant virus disease. Advances in Virus Research 33: 301-325.

Diekmann, M., and Putter, C. A. J. 1996. FAO/IPGRI Technical Guidelines for the Safe Movement of Germplasm. Musa, 2nd edn, 28 .

Fullerton, R. A. 1994 Sigatoka leaf diseases. Compendium of Tropical Fruit Diseases Ploetz *et al* . American Phytopathological Society. 12-14.

Garrett, S. D. 1950. Biology of Root-Infecting Fungi. Cambridge University Press.

Geering, A. D. W., Olszewski, N. E., Harper, G, Lockhart, B. E. L., Hull, R., and Thomas, J. E. 2005 Banana contains a diverse array of endogenous badnaviruses. J Gen Virol. 86: 511-520.

Ha, C., Coombs, S., Revil, P. A., Harding, R. M, Vu, M., and Dale, J. L. 2008. Design and application of two novel degenerate primer pairs for the detection and complete genomic characterization of potyviruses. Arch Virol 153: 25-36.

Harding, R. M., Burns, T. M., and Dale, J. L. 1991. Virus-like particles associated with banana bunchy top disease contain small single-stranded DNA. J. Gen. Virol,. 72: 225-230.

Harding, R. M., Burns, T. M., Hafner, G., Dietzgen, R. G., and Dale, J. L. 1993.Nucleotide sequence of one component of the banana bunchy top virusgenome contains the putative replicase gene. J. Gen. Virol., 74: 323-328.

Harper, G., and Hull, R. 1998. Cloning and sequence analysis of banana streak virus DNA. Virus Genes, 17: 271-278.

Hwang, S. C., and Ko, W. H. 2004. Cavendish banana cultivars resistant to fusarium wilt acquired through somaclonal variation in Taiwan. Plant Disease 88: 580–588.

Hubili, N., and Francki, R. I. B. 1974. Comparative studies on tomato aspermy and cucumber mosaic viruses. III. Further studies on relationship and construction of a virus from part of the two viral genomes. Virology 61: 443-449.

Hull, R., Harper, G., and Lockhart, B. E. L. 2000. Viral sequences integrated into plant genomes. Trends in Plant Science 5: 362-365.

James, A. P., Geijskes, R. J., Dale, J. L., and Harding, R. M. 2011. Molecular characterization of six badnavirus species associated with leaf streak disease of banana in East Africa. Annals of Applied Biology.

Johanson, A. 1993. Molecular methods for the identification and detection of the *Mycosphaerella* species that cause Sigatoka leaf spots of banana and plantains. PhD Thesis, University of Reading, England.

Jones, R. D. 2002. The distribution and importance of the *Mycosphaerella* leaf spot diseases of banana. Proc. 2nd Int. Workshop on *Mycosphaerella* Leaf Spot Disease of Bananas. 25-42.

Jones, D. R. 2000. Diseases of Banana, Abaca and Enset, CAB Publishing, UK, 560.

Kamat, M. N. and Patel, K. 1951. Notes on two important plant diseases in Bombay. Stat. Bull. Pl. Prot., New Delhi, 3: 16.

Kathirvel, A. K., Sathiamoorthy, S., Baskaran, T. L., and Letchoumanae, S. 1986. Outbreak of banana mosaic in Trichy District, TNAU News Letter, 15: 3.

Kesavamoorthy, R. C. 1980. Radical changes in ecosystem in the Pulney hills. Proc 13th National Seminar on Banana production Technology, 23-28.

Kiranmai, G., Lavakumar, P., Hema, M., Venkatramana, M., Kirshna prasadji, J., Mandheva Rao., and Sreenivasalu, P. 2005. Partial characterization of a potyvirus causing bract mosaic of banana in Andhra Pradesh. Indian Journal of Virology, 16, 7-11.

Lakshmanan, P., Selvaraj, P., and Mohan, S. 1987. Efciency of different methods for the control of Panama disease. Tropical Pest Management 33: 373–376.

LaFleur, D. A., Lockhart, B.E.L., and Olszewski, N. E. 1996. Portions of the banana streak badnavirus genome are integrated in the genome of its host. Phytopathology, 86-100.

Lassoudière, A. 1974. La mosaïque dite "a tirets" du bananier "Poyo" en Côte d'Ivoire. Fruits, 29: 349–357.

Lockhart, B. E. L. 1986. Purification and serology of a bacilliform virus associated with banana streak disease. Phytopathology, 76: 995–999.

Lockhart, B. E. L. 1995. Banana streak badnavirus infection in Musa: epidemiology, diagnosis and control. Food and Fertilizer Technology Center Technical Bulletin, 143: 1-11.

Lockhart, B. E. L., and Olszewski, N. E. 1993. Serological and genomic heterogeneity of banana streak badnavirus: implications for virus detection in *Musa* germplasm. In Breeding banana and plantain for resistance to diseases and pests, (Eds J. Ganry). 105-113.

Lockhart, B. E. L., and Jones, D. R. 2000a. Banana streak. In Diseases ofbanana, abaca and enset 263-274.

Lockhart, B. E. L., and Jones, D. R. 2000b. Banana Mosaic. In: Diseases of Banana, Abaca and Enset . pp.256–263.

Magee, C. J. P. 1927. Investigation on the bunchy top disease of banana. Council for Scientific and Industrial Research, Melbourne, Australia .86.

Magee, C. J. P. 1940. Transmission ofinfectious chlorosis or hearts rot of banana and its relationship to cucumber mosaic. Journal of Australian Institute of Agriculture. 6: 44-47.

Magnaye, L. V., and Espino, R. R. C. 1990. Note: Banana bract mosaic, a new disease of banana I. Symptomatology. The Philippine Agriculturist, 73, 55-59.

Mali, V. R., and Rajegore, S. B. 1980. A cucumber mosaic virus of banana in India. Phytpathologische Zeitschrift, 98: 127-136.

Marin, D. H., Romero, R. A., Guzman, M., and Sutton, T. B. 2003. BlackSigatoka an increasing threat to banana cultivation. Plant Dis. 87: 208-222.

Menon, M. R., and Christudas, S. P. 1967. Studies on the population of aphid pentalonia nigronervosa Coq. On banana plants in Kerala. Agriculture Research Journal of Kerala, 5: 84-86.

Meredith, D. S., and Lawrence, J. S. 1969. Black leaf streak disease of bananas (*Mycosphaerella fijiensis*); symptoms of the disease in Hawaii and notes on the conidial state of the causal fungus. Transactions of the British Mycological Society 52: 459-476

Meredith, D. S., and Lawrence, J. S. 1970. Black leaf streak disease of bananas (*Mycosphaerella fijiensis*): Susceptibility of cultivars. Tropical Agriculture Trinidad 47: 375-387.

Meyer, J. B., Kasdorf, G. G. F., Nel, L. H., and Pietersen, G. 2008. Transmission of activatedepisomal *Banana streak OL (badna)virus* (BSOLV) to cv. Williams banana (*Musa* sp.) by three mealybug species. Plant Dis. 92: 1158-1163.

Mohan, S., and Lakshmanan, P. 1988. Outbreak of cucumber mosaic virus on *Musa* sp.in Tamil Nadu, India. Phytoparasitica, 16: 281-282.

Munez, A. R. 1992. Symptomatology, transmission and purification of banana bract mosaic virus (BBMV) in 'giant cavendish' banana. In Faculty of Graduate School, 1 - 57.

Narendrappa, T., and Gowda, B. J. 1995. Integrated management of Panama wilt on banana cv. Nanjangud Rasabale. *Curr. Res.*, University of Agricultural Sciences, Bangalore 24: 53-55.

Nazrul, I., Naqvi, A. R., Arif, T. J., and Qazi, M., and Rizwanul, H. 2010. Genetic Diversity and Possible Evidence of Recombination among Banana Bunchy Top Virus (BBTV) Isolates. International Research Journal of Microbiology 1: 001-012

Natural, M. P. 1985. An update on the development of an in vitro screening procedure for resistance to Sigatoka leaf spot diseases of bananas. INIBAP. 208 –230.

Natural, M. P. 1990. An update on the development of an in vitro screening procedure for resistance to sigatoka leaf disease of banana in Proceedings of an international workshop on 'sigatoka leaf spot diseases of bananas' (R.A. Fullerton and R.H. Stover Eds.). 208-230.

Nancy, F. S. J. 2005. Movement of Banana Bunchy Top Virus and its relationship with its aphid vector (*Pentalonia nigronervosa*) M.Sc. thesis, Bharathidasan University, Tiruchirapalli, Tamil Nadu, and India.

Palukaitis, P., Roossinck., M. J., Dietzgen, R. G., and Francki, R. I. B. 1992. *Cucumber mosaic virus*. Advances in Virus Research, 41: 281-348.

Paden, K. W. C., and Symons, R. H. 1973.Cucumber mosaic virus contains a functionally divided genome. Virology, 53: 487-492.

Persley, G. J., and Langhe, E. A. D. 1987. Summary of discussions and recommendations. in Banana and Plantain Breeding Strategies (G.J.Persley & E.A. De Langhe, eds.).9-17.

Ploetz, R. C., Herbert, J., Sebasigari, K., Hernandez, J. H., Pegg, K. G., Ventura, J. A., and Mayato L.S. 1990. Importances of Fusarium wilt in different banana-growing regions. In: Ploetz RC, ed. FusariumWilt of Bananas. St Paul, MN, USA: APS Press, 9–26.

Rao, D. G. 1980. Studies on a new strain of banana mosaic virus in South India. In: Proc. National Seminar on Banana Production Technology. C. R.Muthukrishnan and J. B. M. AbdulKhader (Eds.). TamilnaduAgriculturual University, Coimbatore, India. pp.155-159.

Rhodes, P. L. 1964. A new banana disease in Fiji. Commonwealth Phytopathological News 10: 38-41.

Rodoni, B. C., Ahlawat, Y. S., Varma, A., Dale, J. L. and Harding, R. M. 1997. Identification and characterization of banana bract mosaic virus in India. Plant disease, 81: 669- 672.

Rodoni, B. C., Dale, J. L., and Harding, R. M. 1999. Characterization and expression of the coat protein-coding region of banana bract mosaic potyvirus, development of diagnostic assays and detection of the virus in banana plants from five countries in Southeast Asia. Arch. Viro., 144: 1725-1737.

Roossinck, M. J., Bujarski, J., Ding. S., Hajimorad, W. R., Hanad, K., Scott, S., and Tousignant, M. 1999. In: Family *Bromoviridae*(ed). Virus taxonomy-eight report of theinternational Committee on Taxonomy of Viruses. Academic Press. San Diego,Calif. 923-935.

Samraj, J., Menon, M. R., and Christudas, S. P. 1970. The movement of banana bunchy top virus in plant. Agriculture Research Journal of Kerala, 8: 106-108.

Samraj, K. S., Menon, M. R., and Christudas, S. P. 1966. Kokkan a new disease of banana. Agriculture Research Journal of Kerala, 4: 116.

Sankaralingam, A., Baranwal, V. K., Ahlawat, Y.S., Renuka Devi, P., and Ramaiah, M. 2006. RT-PCR detection and molecular characterization of Banana bract mosaic virus from the pseudostem and bract of banana. Archives of Phytopathology and Plant Protection, 39: 273-281.

Sebasigari, K., and Stover, R. H. 1988. Banana Diseases and Pests in East Africa. Report of a survey in November 1987.INIBAP, Montpellier, France.

Selvarajan, R., and Singh, H. P. 1997. Occurrence, geographical distribution and Electron microscopy of BBrMV in India. In the proceedings of International Conference on Integrated disease management for sustainable agriculture held at New Delhi on November, 11-15.

Selvarajan, R., and Sathiamoorthy, S. 2000. Influence of temperature and rainfall on symptom expression of banana streak virus in cultivar Poovan. In symposium on "Emerging trends in plant disease management" Organised by IPS, Southern zone.

Selvarajan, R., Uma, S., and Sathiamoorthy, S. 2000. Etiology and survey of banana leaf spot diseases in India. In the proceedings of 3rd RAC of Asia and Pacific (ASPNET / INIBAP) held at Bangkok, Thailand.

Selvarajan, R., Balasubramanian, V., and Sathiamoorthy, S. 2006. Vector transmission of banana bract mosaic and banana streak viruses in India. In Abstracts of XVI Annual convention and international symposium on "Management of vector – borne viruses, ICRISAT, 7-10th February, 2006. 110.

Selvarajan, R., and Jeyabaskaran, K. J. 2006. Effect of Banana bract mosaic virus (BBrMV) on growth and yield of cultivar Nendran (Plantain, AAB). Indian Phytopathology, 59: 496-500.

Selvarajan, R., Mary Sheeba, M., Balasubramanian, V., and Mustaffa, M. M. 2007. Molecular Characterization and Expression of Coat Protein Gene of Cucumber Mosaic Virus infecting Banana -TRY isolate. In. National Conference on Banana: "Production and utilization of banana for economic livelihood and nutritional security" held at Trichy, Tamil Nadu.

Selvarajan. R., Mary Sheeba, M., Balasubramanian. V., Rajmohan. R., Lakshmi Dhevi, N., and Sasireka, T. 2011. Molecular characterization of geographically Different *Banana Bunchy Top Virus* (BBTV) Isolates in India. Indian Journal of Virology, 21: 110-116.

Shahanavaj, K., Arif, T. J., Bushra, A., Qazi, M., and Rizwanul, H. 2011. Coat Protein Gene based Characterization of *Cucumber Mosaic Virus* Isolates Infecting Banana in India. Journal of Phytopathology, 3: 94-101.

Simmonds, N. W. 1986. Banannas *Musa* cvs. In: Simmonds N. W. ed. Breeding for Durable Resistance. FAO Plant Production and Protection. 17 – 24.

Singh, S. J., Selvarajan, R., and Singh, H. P. 2000. Detection of bract mosaic virus (kokkan disease) by electron microscopy and serology. In Banana-Improvement, Production and Utilization (Eds. H.P.Singh and K.L.Chadha). Proceedings of the Conference on Challenges for Banana Production and Utilization in 21st century, AIPUB, NRCB, Trichy, India. 381-383.

Singh, S. J. 2003.Viral diseases of Banana, 1st eds. Kalyani publishers, Ludhiana, India.

Sivamani, E., and Gnanamanickam, S. S. 1988. Biological control of *Fusarium oxysporum* f. sp. *cubense* in banana by inoculation with Pseudomonas fluorescens. Plant Soil 107: 3–9.

Stover, R. H. 1972. In: Banana,Plantain and Abaca Diseases. Commonwealth Mycological Institute (ed). Kew Survey, England. 316.

Stover, R. H. 1962. Studies on Fusarium wilt of bananas. IX. Competitive saprophytic ability of *F. oxysporum* f. *cubense*. Can. J. of Botany 40: 1473-1481.

Stover, R. H., and Simmonds, N. W. 1987. Bananas. Third Edition. Longman Scientific & Technical. Longmans, London, United Kingdom..467 .

Stover, R. H. 1980. Sigatoka leaf spot of bananas and plantains. Plant Disease 64: 750-755.

Thangavelu, R., Sundararaju, P., Sathiamoorthy, S., Raghuchander, T., Velazhahan, R., Nakkeeran S., and Palaniswami, A. 2001. Status of Fusarium wilt of banana in India. In 'Banana Fusarium wiltmanagement: towards sustainable cultivation'. (Eds AB Molina, NH Nikmasdek, KW Liew) 58–63.

Thangavelu, R., and Mustaffa, M. M. 2010. First report on the occurrence of a virulent strain of Fusarium wilt pathogen (Race- 1) infecting Cavendish (AAA) group of bananas in India. Plant Disease 94.

Thangavelu, R., Muthu Kumar, K., Ganga Devi, P., and Mustaffa, M. M. 2011. Genetic Diversity of *Fusarium oxysporum* f.sp. *cubense* Isolates (Foc) of India by Inter Simple Sequence Repeats (ISSR) Analysis. Molecular Biotechnology.

Thangavelu, R., Selvarajan, R., and Singh, H. P. 2000. Status of banana streak virus and banana bract mosaic virus diseases in India. In Banana: Improvement, Production and utilization. (Eds. H.P.Singh and K.L.Chadha,) Proceedings of the Conference on Challenges for Banana Production and Utilization in 21st century, AIPUB, NRCB, Trichy, India. 364-376.

Thomas, J. E., and Magnaye, L. V. 1996. Banana bract mosaic disease. Musa Disease Fact Sheet, 1.

Thomas, J. E., Geering, A. D. W., Gambley, C. F., Kessling, A. F., and White, M. 1997. Purification, properties and diagnosis of banana bract mosaic potyvirus and its distinction from abaca mosaic potyvirus. Phytopathology, 87: 698-705.

Vetten, H. J., Chu, P. W., Dale, J. L., Harding, R. M., Hu, J., Katul, L., Kojima, M., Randles, J. W., Sano, Y., and Thomas, J. E. 2005. In Virus Taxonomy: Eighth Report of the International Committee on Taxonomy of Viruses, (ed. by C.M. Fauquet, M.A. Mayo, J. Maniloff, U. Desselberger, L.A. Ball) Academic Press, San Diego, London. 343–352.

Waite, B. H. 1960. Virus diseases of bananas in Central America. Proc. Caribb.Reg.Am.Soc.Horti.Sci.4: 26-30.

Wardlaw, C. W. 1972. Banana diseases, including plantains and abaca (2nd ed.). Longmans, Green. UK. 878.

Wanitchakorn, R., Hafner, G. J., Harding, R. M., and Dale, J. L. 2000.Functional analysis of proteins encoded by *banana bunchy top virus* DNA-4 to-6. Journal of General Virology, 81: 299-306.

Wanitchakorn, R., Harding, R. M., and Dale, J. L. 1997. *Banana bunchy top virus* DNA-3 encodes the viral coat protein. Archives of Virology, 142: 1673- 1680.

Zimmerman, A. 1902. Uber einige tropische Kulturpflanzen beobachte Pilze. II.Zbl. Bakt., II, 8: 216-221.

Diseases of Field and Horticultural Crops
Editor-in-Chief: P. Chowdappa
Published by: Daya Publishing House, New Delhi

15

Diseases of Cardamom

C.N. Biju, R. Praveena and P.Chowdappa

From the dawn of human civilization, spices were sought after as eagerly as that of gold and precious stones. Discovery of the spice land was one of the major objectives of majority of circumnavigations and adventurous explorations that the era of Renaissance had witnessed. One such navigational pursuit undertaken by the Portuguese sailors in search of the renowned "Land of Spices" reached the coastal regions of the erstwhile princely state of Malabar in South India. The Portuguese subsequently established trade links with the merchants of Malabar and the spices including black pepper, cardamom, ginger and cinnamon were the most preferred commodities for export. The Arab merchants, who established trade relations with the Malabar regions, were the pioneers in spreading cardamom to ancient and medieval Greece and Rome, from which it spread to other Mediterranean and West European countries (Pushpangadan, 2001).

Cardamom (*Elettaria cardamomum* Maton), otherwise known as Malabar cardamom, true cardamom or small cardamom, adorns the unique position of the "Queen of spices" because of its very pleasant, mild aroma and taste. Cardamom is a perennial, herbaceous, rhizomatous monocot, belonging to the family Zingiberaceae. It is a native of the biodiversity rich moist evergreen forests of the Western Ghats of Southern India. The cardamom of commerce is the dried ripe fruits (capsules) of cardamom plant. It is grown extensively in the mountainous tracts of South India at elevations of 800–1300 m above MSL mainly as an under crop in the forest lands. The status of cardamom, earlier considered as a forest produce was shifted to a commercial commodity with the intervention of British planters, who started organized cultivation of this valuable crop (Ravindran ,2002).

India and Guatemala are the major players in world economy of cardamom. Other small producers are Tanzania, Sri Lanka, Papua New Guinea, Honduras, Costa Rica, El Salvador, Thailand and Vietnam. India accounted for nearly 65 per cent of the world production in the early 1970s, but only 28 per cent in 1997–98. Guatemala on the other hand, stepped up its production from the middle of 1960s contributing 21.5 per cent of world production in the early 70s, but now her share is more than 65 per cent. In Indian cultivation of cardamom is mainly concentrated in three states; Kerala, Karnataka and Tamil Nadu. Among these, Kerala accounts for 59 per cent of area with a production of 70 per cent and Karnataka has 34 per cent area and contributes 23 per cent to the total production, while Tamil Nadu has 7 per cent area as well as production. Majority of the cardamom growing areas in Kerala are located in the districts of Idukki, Wayanad and Palakkad. In Karnataka, cardamom is grown in Kodagu, Chikmagalur and Hassan districts and to a lesser extent in North Canara .Whereas, in Tamil Nadu, cardamom based cropping system is mainly confined to certain pockets of Pulney and Kodai hills . In India, cardamom is considered as a small holders' crop, approximately 40,000 holdings spreading over an area of 80,000 ha (George and John, 1998).

In India the area under cardamom has come down over the last one decade from 1,05,000 ha during 1987–88 to 69,820 ha during 1997–98, while production has gone up from 3200 tons during 1987–88 to 9290 tons in 1999–2000. During the period, the productivity has increased from almost 47 kg/ha to 173 kg/ha.

Among the factors contributing to the decline in both production and productivity of cardamom in India , the following plays a significant role:

1. Recurring climatic vagaries, especially crop losses incurred due to drought conditions which is a resultant of non adoption of irrigation during summer months.

2. Maintaining senile plants and aversion towards periodical rejuvenation of the plantation with improved high yielding varieties.

3. Deforestation and consequent adverse changes in the prevailing ecological conditions prevailing in cardamom based cropping systems.

4. Replacing cardamom with other remunerative crops in the traditional cardamom growing regions.

5. Failure in the timely adoption of adequate pest and disease management measures.

In spite of the above limiting factors cardamom production has registered an increase both in production and productivity during the recent past. The shift in the production pattern was predominantly due to the use of high-yielding varieties with desirable traits, better adoption of various crop production technologies and better awareness of the different pest and disease management strategies among the farming communities. However, concerted efforts are highly imperative to alleviate various production constraints, to achieve sustainable yield levels and also to regain the lost glory as a leader in cardamom production. In order to achieve a quantum jump in productivity, a thorough knowledge on various disease of the crop is highly essential in formulating a holistic approach to manage the disease effectively, economically and eco friendly.

Cardamom is affected by a number of diseases incited by fungi, bacteria, viruses and nematodes in nurseries as well as in main plantations. Among the fungal diseases, capsule rot (Azhukal), rhizome rot/ clump rot, leaf blight (Chenthal) and among viral diseases, katte/ mosaic, Chlorotic streak are wide spread, causing considerable crop loss in nurseries and main plantations. Damages inflicted by nematodes in the nurseries and main fields also results in considerable crop damage (Thomas and Bhai, 2002). Major diseases such as the rots, leaf blights and nematode infestation are often wide spread and lead to crop losses while crop losses due to minor disease are negligible. Disease alone can cause up to 50 % crop loss if not managed properly. The diseases of cardamom incited by various pathogens are presented in Table 15.1.

Table 15.1. Diseases of cardamom

Sl.No.	Disease	Causal Organism
Fungal	**Diseases**	
1.	Capsule rot (Azhukal)	*Phytophthora nicotianae* var. *nicotianae*
2.	Rhizome rot/ clump rot	*Rhizoctonia solani, Pythium vexans, Fusarium oxysporum*
3.	Leaf Blight (Chenthal)	*Colletotrichum gloeosporioides*
4.	Damping off	*Rhizoctonia solani,Pythium vexans*
5.	Seedling rot or clump rot	*Rhizoctonia solani, Pythium vexans Fusarium oxysporum, Meloidogyne incognita*
6.	Leaf blotch	*Phaeodactylium alpiniae*
7.	*Phytophthora* leaf blight	*Phytophthora meadii*
8.	Leaf rust	*Phakopsora elettariae (Uredo elettariae)*
9.	*Sphaceloma* leaf spot	*Sphaceloma cardamomi*
10.	*Glomerella* leaf spot	*Glomerella cingulata*
11.	*Cercospora* leaf spot	*Cercospora zingiberi*
12.	*Phaeotrichoconis* leaf spot	*Phaeotrichoconis crotolariae*
13.	*Ceriospora* leaf spot	*Ceriospora elettariae*
14.	Nursery leaf spots	*Phyllosticta elettariae*
	Phyllosticta leaf spot	*Colletotrichum gloeosporioides*
	Colletotrichum leaf spot	
15.	Sooty mould	*Trichosporiopsis* sp
16.	Capsule tip rot	*Rhizoctonia solani*
17.	*Fusarium* capsule disease	*Fusarium moniliforme*
18.	Capsule ring spot	*Marasmius* sp
19.	Stem lodging	*Fusarium oxysporum*
Bacterial	**Diseases**	
20.	*Erwinia* rot	*Erwinia chrysanthemi*
21.	Capsule canker	*Xanthomonas* sp.
Viral	**Diseases**	
22.	Mosaic or *Katte* Disease	*Cardamom mosaic virus*
23.	Chlorotic streak	*Banana bract mosaic virus*
24.	Cardamom Vein Clearing or *Kokke Kandu* Disease	*Potyvirus*
25.	Cardamom Necrosis or Nilgiri Necrosis Disease	*Carlavirus*
26.	Infectious Variegation Disease	*Virus*
27.	Nematode diseases	*Meloidogyne incognita, Pratylenchus coffeae* and *Radopholus similis*

1. Capsule rot (Azhukal)

1.1. Occurrence and distribution: Capsule rot, popularly known as Azhukal (rotting in Malayalam) is perhaps the most wide spread and devastating disease of cardamom. Menon *et al.* (1972) reported the disease for the first time from plantations of Idukki district in Kerala state. Capsule rot is a major problem affecting cardamom cultivation in Idukki and Wyanad districts of Kerala and Anamalai hills in Tamil Nadu (Thomas *et al.*, 1989). However, the disease is not observed in low rainfall areas in Tamil Nadu. Cardamom plantations in Karnataka state receive good monsoon rains but this disease rarely occurs in this geographical locality. Nambiar and Sarma (1976), who studied the disease earlier, have reported a crop loss of 30 per cent. However later it has shown that as high as 40 per cent crop loss can occur in severely disease affected plantations.

1.2. Symptoms: The first visible symptoms appear as discoloured water soaked lesions on young leaves and capsules which subsequently enlarge resulting in the decay of the affected portions. The infection occurs on capsules and tender leaves either simultaneously or first on capsules followed by foliar infection (Thomas *et al.*, 1991). When the pathogen infects the foliage, water soaked lesions appear on the leaf tips or margins which later enlarges leading to the coalescing of adjacent lesions to form large patches. As a result of infection, immature unopened leaves fail to unfurl. As the disease advances, the lesion turn necrotic, leaves decay, shrivels and finally gives a shredded appearance. On the capsules, the symptoms develops as water soaked discoloured areas which later turn brownish (Fig. 15.1) and subsequently such capsules decay and drops off prematurely(Thomas *et al.*, 1991; Thomas *et al.*, 1989; Thomas and Bhai, 2002) . The rotten capsules emanate a foul smell. The pathogen infects the capsules of all ages however; young capsules are seriously affected by the disease. From the capsules, infection spreads to panicle and whole inflorescence is affected (Alagianagalingam and Kandaswamy, 1981).

Fig. 15.1. Capsule rot – Symptoms on capsules

During favourable conditions, the disease becomes more severe and the infection extends to other plant parts like panicles, tender shoots and even to rhizomes and roots. In severe disease conditions, whole pseudostem and panicle decays completely. In such cases, rotting extends to underground parts like rhizomes and roots resulting in the rotting and collapse of the entire plant.

1.3. Causal Organism: Earlier studies indicated that the disease is caused by *Phytophthora palmivora* . The fungi *Pythium vexans* and *Fusarium* sp. were also found to be associated with the disease (Menon *et al.*, 1972; Thankamma and Pillai, 1973; Radha and Joseph, 1974). Cardamom isolate of P. *palmivora* was found to be infective to cocoa, coconut, arecanut, black pepper and rubber (Manomohanan and Abi, 1984). Later, *Phytophthora nicotianae* var. *nicotianae* was identified as the causative organism of the disease (Nair, 1979). Wild *Colocasia* plants are found to be a collateral host of *Phytophthora nicotianae* var. *nicotianae* (Nair, 1979).

1.4. Epidemiology: The disease makes its appearance after the onset of South-West monsoon and the disease incidence is directly correlated to high and continuous rainfall during the monsoon seasons. The disease development in main fields is favoured by environmental factors such as high soil moisture levels (34.3 - 37.6%), low temperature (20.4 – 21.3⁰ C), high relative humidity (83 – 90.6 %) and high rainfall (320 – 400 mm) which prevails during the months of June to August (Nair and Menon, 1980). *Phytophthora* survives in soil in the form of hyphae and sporangia for 4-6 weeks while as chlamydospores, it can survive as long as 48 weeks in moist soils. Presence of high level of inoculum in the soil, thick shade, closer spacing, high soil moisture and water logging are the other factors which pre-disposes the plants to the disease in main plantations.

2. Rhizome rot/ clump rot

2.1. Occurrence and distribution: Rhizome rot, also called as clump rot is an important disease in plantations throughout the cardamom growing regions of the world. The disease is widely distributed in the cardamom plantations of Kerala and Karnataka and in heavy rainfall areas of Tamil Nadu such as the Anamalai hills. In severely affected areas as much as 20 per cent disease incidence was recorded. This disease was first reported by Park in 1937. Later, Subba Rao (1938) described the disease as clump rot disease.

2.2. Symptoms: The first visible symptom is the development of pale yellow colour in the foliage, which leads to the premature death of older leaves. The leaves also show wilting symptoms. The collar portion of aerial shoots in the infected plants become brittle and the tiller breaks at slight disturbance. Symptoms of rotting develop at the collar region, which becomes soft and turns brown in colour. In the advanced stages of disease development all the affected aerial shoots fall off from the base (Fig. 15.2a and 15.2b). The rotting extends to the panicles, young shoots, rhizomes and roots also resulting in total destruction of crop.

Fig. 15.2 a. Rhizome rot – Advanced stage

Fig. 15.2 b. Rhizome rot – close-up view

2.3. Causal Organism: Subba Rao (1938) reported that rhizome rot is caused by *Rhizoctonia solani* Kuhn in association with a nematode. Later, Ramakrishnan (1949) reported, *Pythium vexans* de Bary as the causal agent of rhizome rot. It was also reported that *Fusarium oxysporum* is also occasionally found to cause rhizome and root rot infections.

2.4. Epidemiology: Prolonged dampness is conducive for disease development (Venugopal *et al.*, 2006).The disease makes its appearance during the south – west monsoon period by about middle of June in the cardamom plantations. Falling of shoots resulting from rhizome rot infection becomes severe during July- August months. Presence of inocula in the soil and plant debris, overcrowding of plants and thick shade are congenial conditions for disease development.

3. Leaf blight (Chenthal)

3.1. Occurrence and distribution: Leaf blight disease popularly known as Chenthal has been reported in cardamom plantations from Idukki district of Kerala state. Though it was reported as a minor diseases of limited spread, presently the situation is alarming as the disease is spreading to newer areas and is becoming a major problem.

3.2. Symptoms: The characteristic symptoms of leaf blight are rectangular lesions on the foliage which later elongates to form parallel streaks (Fig. 15.3a). In the later stages, the lesions become yellowish-brown to orangish-red in colour and the central portions become necrotic. In the advanced stage of disease development, more number of lesions develops on older leaves. These lesions subsequently coalesce together resulting in the drying of the affected area. The disease affected gardens present a burnt appearance (Fig. 15.3b). Flowers produced after disease incidence fail to form capsules (George and Jayashankar, 1979; Thomas and Bhai, 2002).

Fig. 15.3 a. Leaf blight infested field

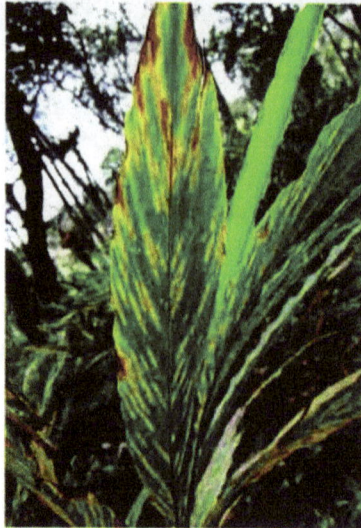

Fig. 15.3 b. Affected leaf showing typical symptoms

3. Causal Organism: The causal agent of chenthal disease was earlier reported to be *Corynebacterium* sp (George and Jayasankar, 1977). Later, detailed investigations on symptomatology, etiology and management clearly indicated that, the disease is caused by *Colletotrichum gloeosporioides* (Penz.) Penz and Sacc (Govindaraju *et al.*, 1996).

3.4. Epidemiology: The disease though prevalent through out the year in the plantations, often assumes severity during post monsoon period. The disease spread is faster in partially deforested areas and less shaded plantations.

4. Mosaic/ Katte/ Marble Disease

4.1. Occurrence and distribution: 'Katte' or mosaic or marble disease of cardamom has been responsible for low yields and rapid decline of cardamom in India and elsewhere. In South India, *katte* is widely distributed in all cardamom growing tracts with an incidence ranging from 0.01-99 per cent (Mayne 1951, Venugopal, 2002, Biju *et al.*, 2010). Loss in yield due to the disease depends on growth stage at the time of infection. In early infected plants, loss will be almost total. While, late infection results a gradual decline in productivity. Crop losses of 10-60, 26-91 and 82-92 per cent were reported under cardamom-areca mixed cropping in first, second and third years of production, respectively. In monocrop conditions, a yield reduction of 38.62 and 68.7 per cent were reported for the first, second and third year of infection (Venugopal and Naidu, 1987; Venugopal, 1995). In general, total decline of plants occur within 3-5 years of infection.

4.2. Symptoms: On the young leaves, prominent, discontinuous yellowish stripes running out from the midrib to the margin are noticed (Fig. 15.4a). The size of leaves gets reduced progressively; the plant looses vigour and becomes stunted. Later, the mottling manifests on the leaf sheaths also. Young plants when infected rarely produce capsules. Whereas older plants may produce a few lean crops. The disease gradually spreads to all tillers in a clump (Fig. 15.4b) and in advanced stages; the affected plants produce shorter and slender tillers with few short panicles. The incidence in a new plantation is usually low but it increases steadily and in the course of 3-4 years the incidence may reach 70 per cent or even higher.

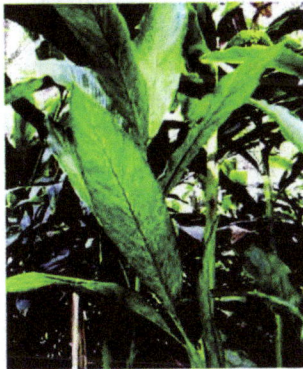

Fig. 15.4 a. Katte symptoms on leaves

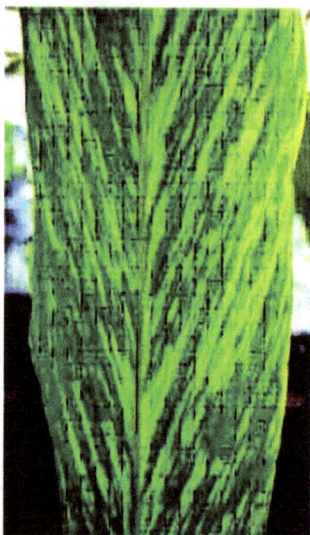

Fig. 15.4 b. Katte symptoms on tillers

4.3. Causal organism: Involvement of a flexuous rod shaped virus with a length of 650 nm and 10-12 nm wide belonging to the genus *Potyvirus* has been reported as the causal virus of *katte* disease (Naidu *et al.*, 1985). Further, the presence of pinwheel type of inclusion bodies in the cytoplasm of infected plants supports its inclusion in the genus, *Potyvirus* (Gonsalves *et al.*, 1986). Based on the coat protein and 3′ untranslated region (3′ UTR) sequence studies of a Karnataka (Sakleshpur) isolate, the *katte* virus (*Cardamom mosaic virus* -CdMV) has been placed as a new member of the genus, *Macluravirus* of the family Potyviridae (Jacob and Usha, 2001). Further, sequencing coat protein and 3′ UTR regions of biologically distinct isolates of CdMV from different geographical locations of Karnataka and Kerala revealed the existence of different strains of the virus. The nucleotide sequence of strains varied mainly at the N-terminal region of coat protein. Subsequently, procedures for total RNA isolation and RT-PCR based detection of the virus using primers designed for the conserved region of the coat protein was developed and validated by testing more than 50 field samples collected from different geographical regions of the Karnataka and Kerala (Biju *et al.*, 2010).

4.4. Epidemiology: Primary sources of infection includes: infected clones, nearby infected plantations and volunteers (Venugopal, 2002). In plantations, primary spread occurs at random due to the activity of viruliferous alate forms of vector. Secondary spread within the plantation is mainly internal and rate of spread is very low and concentrated within 40 m radius (Deshpande *et al.*, 1972; Naidu and Venugopal, 1989). However, rate of disease spread is very fast and natural infection may reach 83 per cent within six months of planting in areca-cardamom mixed cropping system. (Venugopal, 2002).

The first experimental transmission of *katte* disease in India was obtained with banana aphid (*Pentalonia nigronervosa* Coq) (Uppal *et al.*, 1945). So far 13 aphid species were reported to transmit the disease (Rao and Naidu, 1974). It was earlier thought that the aphids found on banana and cardamom were the same, but later it was found that *P. nigronervosa* f. *typica* breeds on *Musa* and related genera, while *P. nigronervosa* f. *caladii* breeds on cardamom, colocasia and caladium (Siddappaji and Reddy, 1972; Venugopal, 1995). Recently, based on DNA sequence of mitochondrial cytochrome oxidase subunit I and nuclear gene elongation factor 1α as well as morphometric differences, *P. nigronervosa* f. *caladii* has been restored to full species status and designated as *Pentalonia caladii*. Incubation period for the virus within the plant vary from 20-114 days and expression is directly influenced by growth of plants. Young seedlings at 3-4 leaf stage express the symptoms within 15-20 days of inoculation while grown up plants take 30-40 days for symptom expression (Naidu and Venugopal, 1989).

5. Chlorotic Streak

5.1. Occurrence and distribution: In a survey of 77 cardamom plantations in 49 locations of Kerala, Karnataka and Tamil Nadu states of India, incidence of the disease was found to be 0-15 per cent (Siljo *et al.*, 2012).

5.2. Symptoms: disease is characterized by the presence of continuous or discontinuous spindle shaped yellow or light green intravenous streaks along the veins and midrib (Fig. 15.5). The disease initially manifests as spindle shaped chlorotic streaks, intravenously and on the midrib region. These streaks coalesced together and imparted yellow or light green colour to the veins. Discontinuous spindle shaped mottling on the pseudostem and petioles are other associated symptoms. In severe cases, tillering of the affected plants was suppressed. The distinguishing feature of the disease was the formation intravenous chlorotic streaks and the name.

Fig. 15.5. Chlorotic streak symptoms

5.3. Causal Organism: Leaf dip electron microscopy of infected leas samples revealed the presence of flexuous virions resembling *Potyvirus*. Cloning and sequencing of coat protein gene from different geographical isolates showed an identity of >94% with *Banana bract mosaic virus* (BBrMV) isolates indicating that causal virus is a strain of BBrMV (Siljo *et al.*, 2012). Multiple sequence alignment and phylogenetic analyses showed high sequence conservation among BBrMV isolates irrespective of host and geographical origin. Subsequently a procedure of total RNA isolation from cardamom plants and RT-PCR for the detection using primers designed for the conserved coat protein region of the virus was developed (Siljo *et al.*, 2012).

6. Cardamom Vein Clearing or Kokke Kandu Disease

This disease is confined in few endemic pockets of Karnataka. The disease is reported from Kodagu, Hassan, Chickmagalur, Shimoga and North Canara districts of Karnataka. This disease may occur either singly or mixed infections with katte. Unlike katte disease, the plants affected with this disease decline rapidly with yield reduction upto 62-84% in the first year of peak crop (Anonymous, 1995; 1996; 1997). The affected plants become stunted and perish within 1-2 years of infection. Several thousands of hectares of cardamom plantations in the Hongadahalla Zone of Hassan district and arecanut based mixed crop in North Canara district of Karnataka have become uneconomical due to the infection of mosaic and *kokke kandu*.

6.1. Symptoms: The first visible symptoms of the disease include continuous or discontinuous vein clearing (Fig. 15.6a). Later rosetting, loosening of leaf sheath and shredding of leaves were seen. New leaves get entangled in the older leaves, resulting in the formation of hook-like tiller (Fig. 15.6b) and hence the name *kokke kandu*. Leaf sheaths of the infected plants exhibit mottling symptoms. Light green patches with shallow grooves are seen on immature capsules. Cracking of fruits and partial sterility of seeds are also associated with the disease. (Venugopal, 2002).

Fig. **15.6 a.** Kokke kandu - Continuous or discontinuous clearing of the veins

Fig. 15.6 b. Kokke kandu - Hook-like tiller

6.2. Causal organism: The exact etiology of the disease is yet to be established. Diseased cardamom samples showed positive serological relationship with potyviruses such as *Peanut mottle virus, Sugarcane stripe virus* and Indian and Guatemalan mosaic virus isolates indicating the possible involvement of a virus belonging to the genus, *Potyvirus.*

6.3. Epidemiology: The virus is not transmitted through seed, soil or sap. The disease is transmitted through cardamom aphid, *P. caladii* in a semi persistent manner. Incubation period for the virus within the plant ranges from 22-128 days. Primary sources of inoculum includes, infected planting material and disease affected plantations. Primary spread in new plantations occurs due to the activity of incoming alate viruliferous vectors. Random spread was reported in new plantations located up to 2000 m away from the affected plantations (Venugopal, 2002). Secondary spread within the plantations is mediated by alate forms of aphid and the rate of spread varies from 1.3–8.5 per cent per year.

7. Cardamom Necrosis or Nilgiri Necrosis Disease

The disease was first reported from Nilgiris, Tamil Nadu. The disease is known to occur in few pockets in Kerala and Tamil Nadu. In general, incidence of the disease is found to be low (0.1-1 per cent), although higher incidence (upto 80 per cent) was recorded from certain areas of Nilgiris (Venugopal, 2002).

7.1. Symptoms: Initially, the symptoms manifests on young leaves as whitish-yellowish, continuous or broken streaks, proceeding from midrib to leaf margins.

In advanced stages of infection, these streaks turn reddish-brown (Fig. 15.7) and results in shredding of the le aves along the streaks. Leaves are reduced in size with distorted margins. Early infected plants produce only few panicles and capsules and in advanced stages of infection, tillers are highly stunted and fail to bear panicles. All the types of cardamom cultivars are susceptible to the disease. Studies carried out in disease affected plantations indicated 55 per cent reduction in yield in early infected plants and total yield loss in late infected plants .

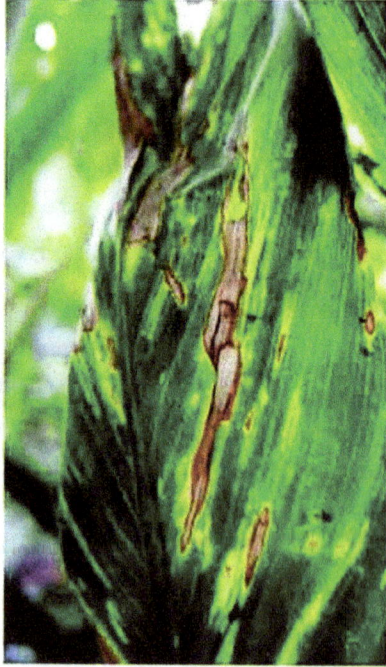

Fig. 15.7. Nilgiri necrosis - necrotic patches on the leaves

7.2. Causal organism: Association of flexuous particles of 570-700nm long and 10-12nm wide were seen in leaf dip preparations of infected leaf tissues and it belongs to genus, *Carlavirus*.

7.3. Epidemiology: The disease is not transmitted through seed, soil, or sap/ mechanically. It is mainly transmitted through infected planting materials originating from affected plantations. Aphids (*P. caladii*), thrips (*Sciothrips cardamomi*) and whitefly (*Dialeurodes cardamomi*) failed to transmit the disease under green house conditions. Infected rhizomes/seedlings raised from diseased nurseries are the primary sources of inoculum. Monitoring new infections at regular intervals in affected plantations revealed that the spread of the disease is mainly internal and new infections occur in a centrifugal fashion from the source of inoculum. Most of the infections occurred within 10-15 m radius from the sources of inoculum and the number of new infections decreased as the distance increased. The pattern of spread is similar to that of katte disease. The rate of spread of the disease was low *i.e.*, 3.3 per cent for the period of one year.

8. Nematode diseases

8.1. Occurrence and distribution: Nematode diseases are major problems in nurseries as well as in main plantations in India. Among the 28 genera of various plant parasitic nematodes reported from cardamom soils, the root knot nematode, *Meloidogyne incognita* causes severe damage to the cardamom in nurseries and also in plantations (Ali, 1983). In cardamom nurseries, nematodes cause significant reduction (more than 50 per cent) in seed germination and the infested seedling fail to establish on transplantation (Ali and Koshy 1982).

8.2. Symptoms: The galls are seen only in young seedlings but not on roots of mature plants in the plantation (Eapen 1992) Aerial symptoms induced by root knot nematode includes, stunting of the plants, reduced tillering, rosetting and narrowing of the leaves (Eapen 1994) Symptoms manifested on foliage include yellow banding on the leaf blades (Fig. 15.8) and drying of leaf tips and margins. Flowering is normally delayed in the affected plants and immature fruits drop resulting in yield reduction. Underground symptoms develop on the roots in the form of prominent galls while the tender roots exhibit spherical or ovoid swellings.

Fig. 15.8. Nematodes - yellow banding on the leaf blades

8.3. Causal organism: Root knot nematodes (*M. incognita*) are commonly found in almost all plantations and nurseries while, the root lesion nematode, *Pratylenchus coffeae* and the burrowing nematode, *Radopholus similis* are noticed in plantations where cardamom is grown along with other component crops. Micro plot studies under simulated field conditions showed 46.6 per cent yield loss in cardamom, at an initial inoculum level of 4 nematodes/ 100 cm^3 soil (Eapen 1994).

8.4. Epidemiology: Root knot nematode population is high in cardamom soils during September – January *i.e* post-monsoon period (Eapen, 1993). These fluctuations in nematode population are generally influenced by rainfall and its subsequent effect on soil moisture, soil temperature and root regeneration of the host plant. Nematode multiplication in cardamom soils are favoured by heavy shade, moisture level of the soil and warm humid weather (Ramana and Eapen, 1992).

9. Damping off: Wilson *et al.,* (1979) observed the incidence of damping off in young seedlings at the age of 1-6 months. The initial disease symptoms are noticed on leaves, which show slight paleness and yellowing of leaves at the top. Gradually yellowing spreads into leaf blades and leaf sheaths followed by withering of plants. In primary nursery, infected seedlings collapse at collar region and die in patches. Overcrowding of seedlings and excess soil moisture are the predisposing factors of this disease.The causal organism of damping off was identified as *Rhizoctonia solani* (Wilson *et al.,* 1979) and *Pythium vexans* (Nambiar *et al.,* 1975).

10. Seedling rot or clump rot: Usually the disease is observed in nurseries when the seedlings attain an age of 6 – 12 months .In nurseries, the disease incidence varies from 10-60 per cent. Ali and Venugopal (1993) reported the incidence of this disease as high as 64.5 per cent in some nurseries in Karnataka.

In grown up seedlings the infection starts from collar and spreads into rhizome which first becomes discoloured and decay. The symptoms are characterized by wilting and dropping of leaves. The pseudostems and rhizomes of grown up plants when infected becomes soft, ultimately resulting in the death of the clump. The *Rhizoctonia* infection is indicated in the form of brownish discolouration in the collar, whereas pinkish discolouration and soft decay is the symptom of *Pythium* infection. As the infection advances, the young tillers fall off and the entire seedling collapse (Fig. 15.9a and 15.9b) (Thomas and Bhai, 2002).

Fig. 15.9 a. Damping off- Collapsed seedling

Fig. 15.9 b. Damping off- affected nursery

Soil borne pathogenic fungi *R. solani* and *P. vexans* are the causal organisms of seedling rot. Siddaramaiah (1988) reported that *Fusarium oxysporum* also causes seed and seedling rot. Ali and Venugopal (1993) have reported the association of root knot nematode, *Meloidogyne incognita*, along with *R. solani* and *P.vexans*. The disease is often seen during rainy season in overcrowded nurseries and also in nurseries with impeded drainage. Excessive soil moisture and lack of proper drainage in the nursery are the pre disposing factors for the infection by *P. vexans* whereas, damping off caused by *R. solani* appears when warm temperature prevails.

REFERENCES

Alagianagalingam, M. N., and Kandaswamy, T. K. 1981.Control of capsule rot and rhizome rot of cardamom (*Elettaria cardamomum* Maton). Madras Agric. J. 68: 554-557.

Ali, S. S. 1983. Nematode problems in cardamom and their control measures. Sixth Workshop of All India Coordinated Spices and Cashew nut Improvement Project, Calicut, November 10 – 13, 1983.

Ali, S. S., and Koshy, P. K. 1982. Occurrence of root knot nematodes in cardamom plantations of Kerala. Nematol. Medit. 10: 107-110.

Ali, S. S., and Venugopal, M. N. 1993. Prevalence of damping off and rhizome rot disease in nematode infected cardamom nurseries in Karnataka. Curr Nematol. 4(1): 19 – 24.

Anonymous. 1995. Annual Report 1994-95, Indian Institute of Spices Research, Calicut, Kerala, India.

Anonymous. 1996. Annual Report 1995-96. Indian Institute of Spices Research, Calicut, Kerala, India.

Anonymous. 1997. Annual Report 1996-97. Indian Institute of Spices Research, Calicut, Kerala, India.

Biju, C. N., Siljo, A., and Bhat, A. I. 2010. Survey and RT-PCR based detection of Cardamom mosaic virus affecting small cardamom in India. Indian J. of Virol. 21: 148–150.

Deshpande, R. S., Siddappaji, C., and Viswanath, S. 1972. Epidemiological studies on katte disease of cardamom (*Elettaria cardamomum* Maton). Mysore J. Agric. Sci. 6: 4-9.

Eapen, S. J. 1992. Influence of plant age on root knot nematode development in cardamom. Nematol. Medit. 20: 193-295.

Eapen, S. J. 1993. Seasonal variations of root knot nematode population in cardamom plantation. Indian J. Nematol. 23: 63-68.

Eapen, S. J. 1994. Pathogenicity of root knot nematode on small cardamom (*Elettaria cardamomum* Maton). Indian J. Nematol. 24: 31-37.

George, M., and Jayasankar, N. P. 1977. Control of Chenthal (Bacterial Blight) disease of cardamom with penicillin. Curr. Sci. 46: 237.

George, M., and Jayasankar, N. P. 1979. Distribution and factors influencing chenthal disease of cardamom. In proceedings of PLACROSYM-II, CPCRI, Kasargod, 343-347.

George, C. K., and John, K. 1998. Future of cardamom industry in India. Spice India. 11(4): 20-24.

Gonsalves, D., Trujillo, E., and Hoch, H. C. 1986. Purification and some properties of a virus associated with cardamom mosaic, a new member of Potyvirus group. Plant Dis. 70: 65-69.

Govindaraju, C., Thomas, J., and Sudharsan, M.R. 1996. Chenthal disease of cardamom caused by *Colletotrichum gloeosporioides* Penz. and its management. In: N.M. Mathew and C.K. Jacob (Eds.) Developments in Plantation Crop Research, Allied Publ., New Delhi, 255 – 259.

Jacob, T., and Usha, R. 2001. 3' - terminal sequence analysis of the RNA genome of the Indian isolate of cardamom mosaic virus, a new member of the genus Macluravirus of Potyviridae. Virus Genes. 23: 81-88.

Manomohanan, T. P., and Abi, C. 1984. *Elettaria cardamomum*, A new host for *Phytophthora palmivora* (Butler). In: Proc. of PLACROSYM –VI, CPCRI, Kasaragod, pp. 133 – 137.

Mayne, W. W. 1951. Report on cardamom cultivation in South India. ICAR Bull., 50. Indian Council of Agricultural Research, New Delhi. p. 62.

Menon, M. R., Sajoo, B. V., Ramakrishnan, C. K., and Remadevi, L. 1972. A new Phytophthora disease of cardamom (*Elettaria cardamomum* Maton). Curr. Sci. 41: 231.

Naidu, R., Venugopal, M. N., and Rajan, P. 1985. Investigations on strainal variation, epidemiology and characterization of 'katte' virus agent of small cardamom. Final Report of Research Project, Central Plantation Crops Research Institute, Kasargode, Kerala, India.

Naidu, R., and Venugopal, M. N. 1989. Epidemiology of katte virus disease of small cardamomi. Foci of primary disease entry, patterns and gradients of disease entry and spread. J. Plantation Crops, 16 (Suppl.). 267-271.

Nair, R. R. 1979. Investigations on fungal diseases of cardamom. Ph.D Thesis, Kerala Agricultural University, Vellanikkara, Thrissur, p. 161.

Nair, R. R., and Menon, M. R. 1980. Azhukal disease of cardamom. In: K.K.N. Nambiar (Ed.) Proceedings on the workshop on Phytophthora Diseases of Tropical Cultivated Plants, CPCRI, Kasaragod, pp. 24 – 33.

Nambiar, K. K. N., and Sarma, Y. R. 1976. Capsule rot of cardamom *Pythium vexans* de Bary as a causal agent. J. of Plantation Crops. 4: 21–22.

Nambiar, M. C., Pillai, G. B., and Nambiar, K. K. N. 1975. Diseases and pests of cardamom – A resume of research in India. Pesticides Annual, 1975.

Park, M. 1937. The seed treatment of ginger. Trop. Agric. 84: 3-7.

Pushpangadan, P. 2001. The legendary history of spices In: Spices Indica-Silver Jubilee Souvenir (Eds.) Sarma, Y.R.,Sasikumar, A. and Chempakam,B. Indian Institute of Spices Research, Calicut . pp 76-88.

Radha, K., and Joseph, T. 1974. Investigations on the bud rot disease (*Phytophthora palmivora* Butl.) of coconut. Final report of PL. 480 1968 – 1973, p. 30, CPCRI, Kayamkulam.

Ramakrishnan, T. S. 1949. The occurrence of *Pythium vexans* de Bary in South India. Indian Phytopath. 2: 27-30.

Ramana, K. V., and Eapen, S. J. 1992. Plant parasitic nematodes of black pepper and cardamom and their management. In: Proc. of National Seminar on Black pepper and Cardamom, 17th – 18th May, 1992, Calicut, Kerala, pp. 43 – 47.

Rao, D. G., and Naidu, R. 1974. Additional vectors of katte disease of small cardamom. Indian J. Hort. 31: 380 - 381.

Ravindran, P. N. 2002. Introduction. In: Cardamom. The genus *Elettaria*. (Eds.) Ravindran, P.N. and Madhusoodanan, K.J. Taylor and Francis, London and NewYork . pp 1-10.

Siljo, A., Bhat, A. I., Biju, C. N., and Venugopal, M. N. 2012. Occurrence of Banana bract mosaic virus on cardamom. *Phytoparasitica*. 40: 77–85.

Siddappaji, C., and Reddy, D. N. R. N. 1972. A note on the occurrence of the aphid (*Pentalonia nigronervosa* f. *caladii* Van der Goot (Aphididae: Hemiptera) on cardamom (*Elettaria cardamomum* Maton). Mysore J. Agric. Sci. 6: 192-195.

Siddaramaiah, A. L. 1988. Seed rot and seedling – wilt a new disease of cardamom. Curr. Res. 17 (3): 34–35.

Subba Rao, M. K. 1938. Report of the Mycologist 1937-1938. Admn. Rept .Tea Sci. Dept. United. Plant. Assoc. S. India, 1937-1938. pp. 28-42.

Thankamma, L., and Pillai, P. N. R. (1973) Fruit rot and leaf rot diseases of cardamom in India. F.A.O. Plant Prot. Bull. 21: 83–84.

Thomas, J., Suseela Bhai, R., and Naidu, R. 1989. Comparative efficacy of fungicides against *Phytophthora* rot of small cardamom. Pesticides. 40 – 42.

Thomas, J., Suseela Bhai, R., and Naidu, R. 1991. Capsule rot disease of cardamom *Elettaria cardamomum* (Maton) and its control. J. Plantation Crops. 18 (Suppl) pp. 264 – 268.

Thomas. J., and Suseela Bhai, R. 2002. Diseases of cardamom (fungal, bacterial and nematode diseases). In: Cardamom. The genus *Elettaria*. (Eds.) Ravindran, P.N. and Madhusoodanan, K.J. Taylor and Francis, London and NewYork . pp 160 – 179.

Uppal, B. N., Varma, P. M., and Capoor, S. P. 1945. A mosaic disease of cardamom. Curr. Sci. 14: 208-209.

Venugopal, M. N. 1995) Viral diseases of cardamom (*Elettaria cardamomum* Maton) and their management. J. Spices and Aromatic Crops. 4: 32-39.

Venugopal, M. N., and Naidu, R. 1987. Effect of natural infection of *katte* mosaic disease on yield of cardamom - a case study. In: Proc. PLACROSYM - VI, Oxford and IBH, New Delhi, pp. 115 - 119.

Venugopal, M. N., Prasath, D., and Mulge, R. 2006. IISR Avinash – a rhizome rot resistant and high yielding variety of cardamom (*Elettaria cardamomum* Maton). J. Spices and Aromatic Crops. 15(1): 14–18.

Venugopal, M. N. 2002. Viral diseases of cardamom. In: Cardamom- The genus *Elettaria*. (Eds.) Ravindran, P.N. and Madhusoodanan, K.J. Taylor and Francis, New York, pp. 143-159.

Wilson, K. I., Sasi P. S., and Rajagopalan, B. 1979. Damping off of cardamom caused by *Rhizoctonia solani* Kuhn. Curr. Sci.48: 364.

Diseases of Field and Horticultural Crops
Editor-in-Chief: **P. Chowdappa**
Published by: **Daya Publishing House, New Delhi**

Pages **475-489**

16

Crop losses in Cucurbits by Diseases

M. Loganathan, V. Venkataravanappa, A.B. Rai and P. Chowdappa

Cucurbits are vegetables belonging to family cucurbitaceae, comprising of 118 genera and 825 species. Cucurbits shares 5.6 % of the total vegetable production in India which includes both major and minor cucurbits like cucumber (*Cucumis sativus*), bitter gourd (*Momordica charantia*), bottle gourd (*Lagenaria siceraria*), watermelon (*Citrullus lanatus*), melon (*Cucumis melo*), long/serpent melon (*Cucumis melo* var. *flexuosus*), snapmelon (*Cucumis melo* var. *momordica*), ridge gourd (*Luffa acutangula*), sponge gourd (*Luffa cylindrical*), pumpkin (*Cucurbita moschata*), summer squash (*Cucurbita pepo*),winter squash (*Cucurbita maxima*), ash gourd (*Benincasa hispida*), pointed gourd (*Trichosanthes dioica*), ivy or scarlet gourd (*Coccinia cordifolia* syn. *C. indica*), round melon (*Praecitrullus fistulosuos*) and sweet gourd (*Momordica cochinchinensis*) (Rai *et al.* 2008). Cultivation of cucurbits is severely hampered by both biotic and abiotic factors. Among the biotic factors the diseases caused by pathogens are the most important as they accounted for 30-50% production loss in vegetables (Khan, 1999; Hossain *et al.* 2010).

Major diseases of cucurbits

Cucurbits are known to attack by different kinds of pathogens such as fungi, bacteria and viruses and they causes mildew, leaf spot, blight, wilt etc (Table 16.1). Among the fungal diseases, mildews are more destructing and distributed on large number of cucurbits (Zitter *et al.* 1996). Since these diseases occurs on the foliage they cause severe yield losses in short period of time

Cucurbits are susceptible to many viruses and some major viruses infecting cucurbits reported from India are *Cucumber green mottle mosaic virus* (Raychaudhuri and Varma, 1978), *Cucumber mosaic virus* (Vani, 1987), *Watermelon mosaic virus* (Raychaudhuri and Varma, 1975), *Watermelon bud necrosis virus* (Jain *et al.*, 1998; Mandal *et al.*, 2003), *Squash mosaic virus* (SqMV), *Cucumber mosaic virus* (CMV), *Watermelon mosaic virus 2* (WMV-2), *Papaya ringspot virus-W* (formerly, *Watermelon*

Table 16.1. Major diseases of cucurbits caused by fungal and bacterial pathogens

Disease	Causal organisms	Host	Reference
Downy mildew	*Pseudoperonospora cubensis*	Cucumber, gourds, squash,pumpkin, melons	Zitter *et al.*, 1996; Ferguson *et al* 2009
Powdery mildew	*Podosphaera xanthii* *Erysiphe cichoracearum*	Cucumber, gourds, squash, pumpkin, musk melon	Zitter and Kyle 1992; Zitter *et al* 1996
Vascular wilt	*Fusarium oxysporum*	Bottle gourd, cucumber, Muskmelon and water melon	Palodhi and Sen 1980; Radhakrishnan and Sen 1985
Blight	*Phytopthora capsici*	Cucumber, melons, pumpkins, squash	Babadoost 2000; Lee *et al* 2001; Tian and Babadoost 2004
Alternaria leaf spot	*Alternaria cucumerina*	Cucumber, watermelon, Muskmelon, pumpkin	Jackson 1959; Schenk 1968; Ibrahim *et al* 1975;Cohen and Rotem,1987; Latin 1992;Rotem 1994
Anthracnose	*Colletotrichum orbiculare* (Berk. And Mont.)	Cucumber, watermelon, Muskmelon, and bottle gourd	Goode, 1958; Hadwiger and Hall, 1963; Sen *et al* 1999.
Gummy stem blight	*Didymella byroniae* (Auersw.) Rehm (Ana.: *Phoma cucurbitacearum* (Fr.) Sacc.)	All cucurbits	St Amand and Wehner 1991; Keinath *et al* 1995
Angular leaf spot	*Pseudomonas syringae* pv. *lachrymans* (Smith and Bryan)	Cucumber, gherkin, muskmelon, pumpkin, squash, and watermelon	Jindal 1994

mosaic virus 1), *Zucchini yellow mosaic virus* (ZYMV), *Tobacco ringspot virus* and *Tomato ringspot virus*. Apart from these viruses, begomoviruses are emerging as serious problems in many cucurbit crops (Varma and Malathi, 2003). The genus *Begomovirus* under family *Geminivirideae*, consists of a group of viruses exclusively transmitted by whitefly *Bemisia tabaci* and infects dicotyledonous plants are reported in many cucurbits such as cucumber, bitter gourd, pointed gourd, bottle gourd, muskmelon, sponge gourd, and winter squash (Raj and Singh, 1996; Varma and Giri, 1998; Singh *et al.*, 2001; Khan *et al.*, 2002). Recently, three different begomoviruses infecting different cucurbits have been documented: *Squash leaf curl China virus* in pumpkin (Muniyappa *et al.*, 2003), and *Tomato leaf curl New Delhi virus* in sponge gourd (Sohrab *et al.*, 2003) and chayote (Mandal *et al.*, 2004) and *Tomato leaf curl palampur virus* in pumpkin (Namrata *et al.*, 2009). Some of the viruses are seed transmitted, which includes *Cucumber mosaic virus, Squash mosaic virus* and *Cucumber green mottle mosaic virus* (Kang *et al.* 2010). These seed transmitted viruses are widely prevalent in almost all the cucurbits growing states of India and causing huge yield loss to the crops. Different viruses infecting various cucurbits reported from India have been well documented (Table 16.2; Fig. 16.1-16.8)

Table 16.2. List of viral diseases reported in different cucurbits

Disease	Virus	Group	Transmission	Reference
Ashgourd				
Mosaic disease	PVY	potyvirus	Aphid	Bhargava and Bhargava, 1977
Necrosis	GBNV	Tospovirus	Thrips	Singh *et al.*, 2003
Bittergourd				
Mosaic disease	CMV	Cucumovirus	Aphid	Nagarajan and Ramakrishnan 1971, Takami *et al.*, 2006
	TMV	Tobamovirus	Mechanical	Vasudeva and Lal 1943
	PRSVW	Poty virus	Aphid/mechanical	Chin and Ahmad, 2007, Tomar and Jitendra, 2001
Yellow vein mosaic	ToLCNDV	Begomoviruses	Whitefly	Rajinimala *et al.*, 2005, Tahir and Haider, 2005 Tiwari *et al.*, 2010b,
	ICMV	Bogomoviruses	Whitefly	Rajinimala and Rabindran, 2007
Leaf curl disease	PepLCBDV	Begomoviruses	Whitefly	Raj *et al.*, 2010
Cucumber				
Mosaic Disease	CGMV	Tobamovirus	Mechanical	Vasudeva and Nariani 1952
	CMV	Cucumovirus	Aphid/ mechanical	Horvath, 1979, Mukhopadayay and Saha, 1968, Sharma and Chohan 1973
		Poty virus	Aphid/mechanical	Bhargava and Bhargava, 1977
Cucumber	GBNV	Tospovirus	Thrips/ mechanical	Singh *et al.*,2003
necrosis disease	MeYSV	Tospovirus	Thrips/ mechanical	Laxmi Devi *et al.* 2010
Musk melon				
Mosaic	TMV	Tobamovirus	Mechanical	Nariani *et al.*, 1977
	PVY	Poty virus	Aphid/mechanical	Singh *et al.*,2003
Necrosis	GBNV	Tospovirus	Thrips/ mechanical	Singh and Rajverma, 2003
Mosaic disease	ToLCNDV	Begomovirus	Whitefly	Varma and Giri, 1998
Pumpkin				
Pumpkin enation mosaic	PVY	Poty virus	Aphid/mechanical	Ghosh and mukhopadayaya, 1971, Bhargava and Bhargava, 1977
Pumpkin necrosis	GBNV	Tospovirus	Thrips/ mechanical	Singh *et al.*,2003
Pumpkin vein banding	TMV	Tobamovirus	Mechanical	Shankar and Nariani, 1974
Pumpkin yellow mosaic disease	ToLCNDV	Begomoviruses	Whitefly	Maruthi *et al.*, 2006, Varma, 1955
	ToLCPaV	Begomoviruses	Whitefly	Namrata *et al.*, 2009
	SLCCV	Begomoviruses	Whitefly	Muniyappa *et al.*, 2003

Contd...

Contd...

Disease	Virus	Group	Transmission	Reference
Ridge gourd				
Ridge gourd	CMV	Cucumovirus	Aphid/mechanical	Goel and Verma, 1973
mosaic	PVY	Poty virus	Aphid/mechanical	Kride and Lokhande 1996
	TMV	Tobamovirus	Mechanical	Singh *et al.*, 1999
Yellow mosaic and leaf curling disease	ToLCNDV	Begomoviruses	Whitefly	Tiwari *et al.*, 2012b
Snake gourd				
Snake gourd	CMV	Cucumovirus	Aphid/mechanical	Pillai,1971
mosaic	PVY	Poty virus	Aphid/mechanical	Nagarajan and Ramakrishnan 1971
	TMV	Tobamovirus	Mechanical	Venkatsubbiah and Saigopal,1997
Watermelon				
Watermelon bud necrosis	WBNV	Tospovirus	Thrips/ mechanical	Singh and Reddy 1991.
Watermelon mosaic	PVY	Potyvirus	Aphid/mechanical	Vasudeva and Pavgi, 1945, Singh *et al.*, 2003
Bottle gourd				
Bottle gourd	PRSV	Potyvirus	Aphid/mechanical	Mantri *et al.*,2004
mosaic disease	CMV	Cucumovirus	Aphid/mechanical	Vasudeva and Lal., 1943
	CGMV	Tobamovirus	Seed borne/mechanical	Raychaudhuri and Varma., 1978
	PVY	Potyvirus	Aphid/mechanical	Mantri *et al.*,2004
	PRSV-W	Potyvirus	Aphid/mechanical	Bhargava and Bhargava 1977
	ZYVMV	Potyvirus	Aphid/mechanical	Verma *et al.*, 2004
Chlorotic Curly Stunt disease	ToLCNDV	Begomoviruses	Whitefly	Sohrab *et al.*, 2010
Chayote *(Sechium edule)*	ToLCNDV	Begomoviruses	Whitefly	Mandal *et al.*, 2004
Sponge gourd				
Mosaic disease	ToLCNDV	Begomoviruses	Whitefly	Sohrab *et al.*, 2003, Tiwari *et al.*, 2012a
	PRSV	Potyvirus	Aphid/mechanical	Verma *et al.*, 2006
Winter and summer Squash				
Yellow mosaic	SLCCV	Begomoviruses	Whitefly	Saritha *et al.*, 2011
disease	ToLCPaV	Begomoviruses	Whitefly	Tiwari *et al.*, 2010a

Potato virus Y (PVY); *Ground nut bud necrosis virus* (GBNV); *Cucumber mosaic virus* (CMV), *Tobacco mosaic virus* (TMV), *Tomato leaf curl New Delhi virus* (ToLCNDV), Tomato leaf curl palampur virus (ToLCPaV), *Indian cassava mosaic virus* (ICMV); *Pepper leaf curl Bangaladesh virus* (PepLCBDV); *Cucumber green mottle virus* (CGMV); *Melon yellow spot virus* (MeYSV); *Papaya ring spot virus-W* (PRSV-W); *Squash leaf curl China virus* (SLCCNV); *Water melon bud necrosis virus* (WBNV); *Zucchini yellow mosaic virus* (ZYMV)

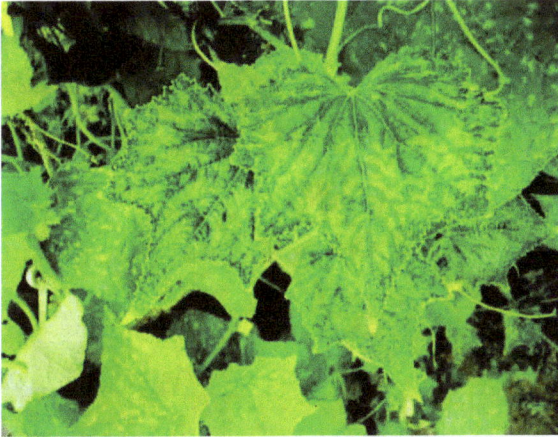

Fig. 16.1. CMV on cucumber

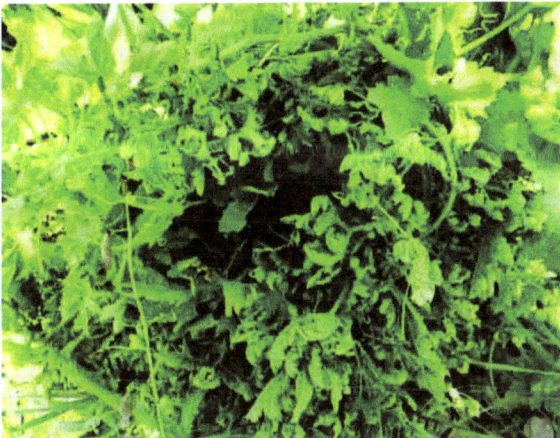

Fig. 16.2. Bittergourd leaf curl disease

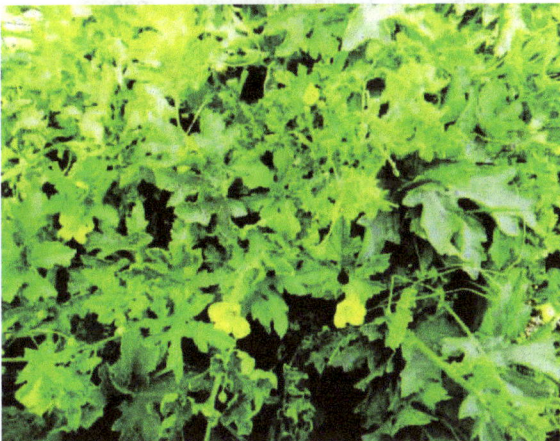

Fig. 16.3 Bittergourd mosaic caused by begomoviruses

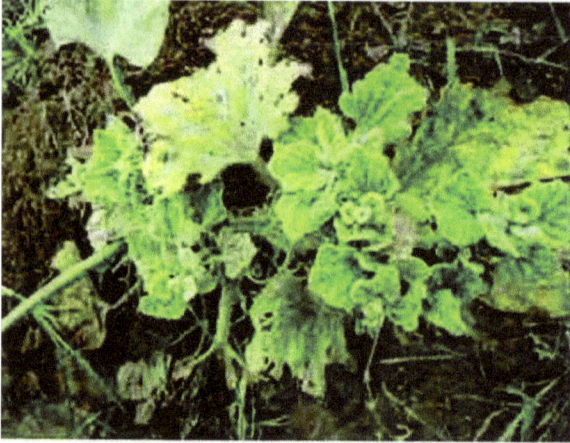

Fig. 16.4. Bottle gourd chlorotic curly stunt disease

Fig. 16.5. Bottle gourd yellow mosaic

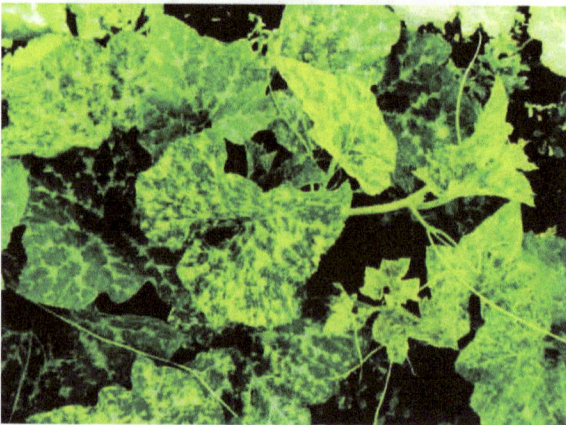

Fig. 16.6. Pumpkin yellowvein mosaic disease caused by virus complex

Fig. 16.7. Ridgegourd yellow mosaic and leaf curl complex

Fig. 16.8. Squash yellow mosaic disease

Losses due to major diseases

Information on yield loss in cucurbits due to diseases is scanty. However the based of available information disease wise yield loss has bee been documented. Diseases *viz.*, powdery mildew, downy mildew, scab or gummosis (*Cladosporium cucumerinum*), blotch or target spot (*Corynespora cassiicola*), angular leaf spot and *Fusarium* wilt are the major threats to cucumber cultivation (Pivovarov 1984; Lebeda 1986; Lower and Edwards 1986; Yurina 1987; Yurina *et al.* 1998) and they reported to cause 50-70% yield loss. *Phytopthora* blight could cause 100% yield loss in severely infected field of cucumber, pumpkin, squash and watermelon (Sherf and MacNab, 1986; Babadoost 2004). Warm temperature and tropical climate occurrence of downy mildew is severe. In India the disease is reported from all the parts of the country and it is more severe in cucumber, muskmelon, watermelon, sponge gourd and ridge gourd, and less severe in bottle gourd and pumpkin. Early infection of downy mildew in cucurbits could cause 61 per cent yield loss (Sen *et al.* 1999) and also it is having potentiality to cause 100% yield loss (Hausbeck and Cortright, 2009; Holmes *et al.*, 2006; Lebeda and Urban, 2007). In USA it is estimated to cause loss of more than $246.2 million per annum (Anonymous, 2009). Kuepper (2003) reported this disease is most destructive to cucumber and cantaloupe than others.

Powdery mildew is found severe in cucumber, muskmelon, bottle gourd, squash, and pumpkins and rarely on watermelon. The disease is severe under warm weather with moisture (Sen *et al.* 1999). It is one of the major causes of yield loss in cucurbits throughout the world (Cohen *et al.*, 2004; Køístková *et al.*, 2009). Losses due to powdery mildew disease are reduced photosynthetic rate leads to reduction in plant growth, premature foliage loss and yield loss (Mossler and Nesheim 2005). In honeydew melon (*Cucumis melo* L.) and muskmelon due to the disease on foliage, the sugar content of fruit has been reduced resulted in reduction in marketable fruit quality. Mossler and Nesheim (2005) reported that yield loss is directly proportionate to severity of the disease and the length of time under which the plants are infected. The disease is estimated to affect 70 per cent of crop area (Hector *et al.* 2006). Fusarium wilt at early stage causes damping off hence the losses start at seedling stage however flowering and fruiting stages are more vulnerable. Among the cucurbits, bottle gourd, cucumber, muskmelon and watermelons are highly susceptible to wilt and reported up to 80 per cent yield loss (Sen *et al.* 1999). Anthracnose disease is severe in humid region grown water melon, muskmelon, bottle gourd and cucumber while squash and pumpkins are less affected (Sen *at al.* 1999). Gummy stem blight can attack both glasshouse and field grown cucurbits and it causes both pre-harvest and post harvest loss (Zitter and Kyle,1992). Angular leaf spot is found severe in cooler regions (Jindal, 1994). Leaf spot disease especially *Alternaria* is reported to cause huge yield loss in pumpkins and watermelon estimated as 80% and 88 % respectively (Bhargava and Singh, 1985).

Plant viruses also cause considerable damage to various cucurbits. Nearly, 193 viruses infecting cucurbits have been found major a limiting factor for the production of cucurbits in worldwide including India (Lovisolo, 1980; Provvidenti, 1996) and they reported to cause massive damage to crops even up to 100% yield loss. The important viral diseases of cucurbits and their extend of damage have been well documented (Table 16.3).

Table 16.3. Viral disease incidence on different cucurbits

Crop	Virus	Incidence (%)	Reference
Cucucmber	Cucumber Green mottle virus	10.0-20.0	Rani *et al.*, 1971
Muskmelon	CMV	20.0-45.0	Nariani *et al.*, 1977, Bandoppadhyay and Mukhopaddhay, 1977
Squash	PotY	42.3-46.5 80.0-85.0	Reddy and Nariani 1963 Bonsal *et al.*, 1990
Snake gourd and bottle gourd	PotY	85.0-100.0	Rani *et al.*, 1971
Cucurbits	Cucumber mosaic virus Squash mosaic virus Cucumber green mottle mosaic virus	6.28-24.59%	Kang *et al.*, 2010

Conclusion

In general cucurbits are known to attack by 201 pathogens/strains indicating that the crops are having very rare or no chance to escape from the diseases as they are reported to occur starts from germination to harvesting period of crops and also to cause approximately 100% yield loss. The present report indicates that strategy for the management of multiple diseases through biocontrol agents or resistant cultivars will have new scope for its cultivation.

REFERENCES

Anonymous, 2009. National online statistics. US Dept. Agric. Natl. Agric. Stat. Serv. http: //www.nass.usda.gov.

Babadoost, M. 2000. Outbreak of *Phytophthora* foliar blight and fruit rot in processing pumpkin fields in Illinois. Plant Dis. 84: 1345.

Babadoost, M. 2004. *Phytophthora* blight of cucurbits. Department of Crop Sciences University of Illinois at Urbana-Champaign. Report on Plant Disease No 945, p 1-3.

Bandoppadhyay, S. R., and Mukhopaddhay, S. 1977. Incidence of virus disease of muskmelon (*Cucumism melo* L. var. *reticulatus*) in west Bengal. Curr. Sci.46: 566-567.

Bhargava., and Bhargava. 1977. Cucurbit Mosaic viruses in Gorakhpur. Indian J. Agric. Sci. 47: 1-5.

Bhargava, A. K., and Singh, R. D. 1985. Comparative study of *Alternaria* blight, losses and causal organisms of cucurbits in Rajasthan. Indian J. Mycol. Plant Pathol. 15: 150-154.

Bonsal, R.D., Singh, P., and Cheema, S. S. 1990. Characterization of the viruses associated with mosaic disease of summer squash. Ann.Biol. 6: 81-86.

Chin, M., and Ahmad, M. H. 2007. *Momordica charantia* is a weed host reservoir for *Papaya ring spot virus type P* in Jamaica. Plant Dis. 91: 15-18.

Cohen, R., Burger, Y., and Katzir, N. 2004. Monitoring physiological races of Podosphaera xanthii (syn. Sphaerotheca fuliginea), the causal agent of powdery mildew in cucurbits: factors affecting race identification and the importance for research and commerce. Phytoparasitica, 32: 174-183.

Cohen, Y., and Rotem, J. 1987. Sporulation of foliar pathogens. In: Fungal infection of plants (edited by G. F. Pegg and P. G. Ayres), Cambridge University Press, Cambridge, pp314-333.

Ferguson, G., Cerkauskas, R., and Celetti, M. 2009. Downy Mildew of Greenhouse cucumber. Fact Sheet, Ministery of Agriculture, Food and Rural Affairs, Ontario, P1-3.

Ghosh, S. K., and Mukhopadhyay, S. 1971. Viruses of pumpkin (*Cucurbita moschata*) in West Bengal. Phytopath. Z. 94: 172-184.

Goel, R. K., and verma, J. P. 1973. Mosaic disease of Ridgegourd (*Luffa acutangula*. Roxb.) In Haryana. HAU J. Res. Hissar. 3: 135-144.

Goode, M. G. 1958. Physiological specialization in *Colletotrichum lagenarium*. Phytopathology. 48: 79-82.

Hadwiger L. A., and Hall, C. V. 1963. A biochemical study of the host parasite relationship between *Colletotrichum lagenarium* and cucurbit hosts. Proc. Am. Soc. Hortic. Sci. 82: 378-387.

Hausbeck, M. K., and Cortright, B. D. 2009. Evaluation of fungicides for control of downy mildew of pickling cucumber. 2007. Plant. Dis. Manag. Rep. 3: V112.

Hector, G., Nuñez-Palenius, Hopkins, D., and Cantliffe, D. J. 2006. Powdery Mildew of Cucurbits in Florida. Publication No: HS1067, The Horticultural Sciences Department, Florida Cooperative Extension Service, Institute of Food and Agricultural Sciences, University of Florida. http: //edis.ifas.ufl.edu.

Holmes, G., Wehner, T., and Thornton, A. 2006. An old enemy re-emerges. Am. Vegetable Grow. Feb, 14–15.

Horvath. 1979. New artificial hosts and non-hosts of plant viruses and their role in the identification and separation of virus's X. Cucumovirus group: Cucumber mosaic virus. Acta Phytopath. Acd. Sci Hung. 14: 285-295.

Hossain, M. T., Hossain, S. M. M., Bakr, M. A., Rahman, A. K. M. M., and Uddin, S. N. 2010. Survey on major diseases of vegetable and fruit crops in Chittagong region. Bangladesh J. Agril. Res. 35(3): 423-429.

Ibrahim, A. N., Abdel-Hak, T. M., and Mahrous, M. M. 1975. Survival of *Alternaria cucumerina*, the causal organism of leaf spot disease of cucurbits. Acta Phytopathol. Acad. Sci. Hung. 10: 309-313.

Jackson, C. R. 1959. Symptoms and host-parasite relations of the Alternaria leaf spot disease of cucurbits", Phytopathol. 49: 731-733.

Jain, R. K., Pappu, H. R., Krishna Reddy, M., and Vani, A. 1998. Watermelon bud necrosis tospovirus is a distinct virus species belonging to serogroup IV. Arch. Virol. 143: 1637-1644.

Jindal, K. K. 1994. Occurrence of angular leaf spot bacterium *Pseudomonas syringae* pv. *lachrymas* on cucumber plants. Plant Dis. Res. 9: 66-67.

Kang, S. S., Inder, K. S., Abhishek, S., and Sandhu P. S. 2010. Characterization of seed-borne viruses associated with cucurbits in Punjab. Plant Dis. Res. 25(2): 162-165.

Keinath, A. P., Faenham, M. W., and Zitter, T. A. 1995. Morphological, pathological and genetic differentiation of *Didymella bryoniae* and *Phoma* spp., isolated from cucurbits. Phytopathol. 85: 364-369.

Khan, J. A., Siddiqui, M. R., and Singh, B. P. 2002. Association of *Begomovirus* with bitter melon in India. Plant Dis. 86: 328.

Khan, N. U. 1999. Studies on epidemiology, seed-borne nature and management of Phomopsis fruit rot of brinjal. An MS thesis submitted to the Department of Plant Pathology, Bangladesh Agricultural University, Mymensingh. pp. 25-40.

Kride, C. V., and Lokhande, N. M. 1996. Biological reservoirs of tomato spotted wilt virus in Nara prefecture. Annuals of Phytopath. Soc. of Japan. 50: 541-544.

Køístková, E., Lebeda, A., and Sedláková, B. 2009. Species spectra, distribution and host range of cucurbit powdery mildews in the Czech Republic, and in some other European and Middle Eastern countries. Phytoparasitica. 37: 337-350.

Kuepper, G. 2003. Downy mildew control in Cucurbits. http://www.attra.ncat.org/attra-pub/PDF/downymil.pdf

Latin, R. X. 1992. Modeling the relationship between *Alternaria* leaf blight and yield loss in muskmelon. Plant Dis. 76: 1013-1017.

Laxmi Devi, V., Krishna Reddy, M., Samuel, D. K., and Jalali, S. 2010. Molecular identification of a new tospovirus infecting cucumber in India. Conference on whitefly and thrips transmitted viruses held at University of Delhi - south campus New Delhi during August 27-28, P 61.

Lebeda, A. 1986. *Pseudoperonospora cubensis. In* Methods of testing vegetable crops for resistance to plant pathogens. (Lebeda A, ed) VHJ Sempra, Research Institute of Vegetable Crops, Olomouc (CZ), pp 81-85

Lebeda, A., and Urban, J. 2007. Temporal changes in pathogenicity and fungicide resistance in *Pseudoperonospora cubensis* populations. Acta Hortic. 731, 327–336.

Lee, B. K., Kim, B. S., Chang, S. W., and Hwany, B. K. 2001. Aggressiveness of isolates of *Phytophthora capsici* from pumpkin and pepper. Plant Dis. 85: 797-800.

Lovisolo, O. 1980. Viruses and viroid diseases of cucurbits. Acta Hortic. 88: 33-82.

Lower, R., and Edwards, M. D. 1986. Cucumber breeding. *In* Breeding vegetable crops (Bassett MJ, ed) Avi Publishing Co, Westport (CT, USA) pp 173-203.

Mandal, B., Mandal, S., Sohrab, S. S., Pun, K. B., and Varma. A. 2004. A new yellow mosaic disease of chayote in India. Plant Pathol. 53: 797.

Mandal, B., Chaudhary, V., and Jain, R. K. 2003. First report of Natural infection of Luffa acutangula by Watermelon bud necrosis virus in India. Plant Dis. 87: 598.

Mantri, N. L., Kitkatu, A. S., Misal, M. B. and Ravi, K. S. 2004. First report of Papaya ring spot virus-W in bottle gourd *(Lagenaria siceraria)* from India. New Dis Rep. 10: 35.

Maruthi, M. N., Rekha A. R., Alam, S. N., Kader, K. A, Cork, A., and Colvin, J. 2006. A novel begomovirus with distinct genomic and phenotypic features infects tomato in Bangladesh. Plant Pathology. 55,290.

Mossler, M. A., and Nesheim O. N. 2005. Florida Crop/Pest Management Profile: Squash. Electronic Data Information Source of UF/IFAS Extension (EDIS). CIR 1265. February, 3, 2005. http: //edis.ifas.ufl.edu/.

Mukhopadayay, S., and Saha, K. 1968. Transmission of cucumis virus (Cucumber mosaic virus) through seeds of *Cucurbita maxima* L. Sci. and Cult. 34: 436-437

Muniyappa, V., Maruthi, M. N., Babitha, C. R., Colvin, J., Briddon, R. W., and Rangaswamy, K. T. 2003. Characterization of Pumpkin yellow vein mosaic virus from India. Ann. App. Biol. 142: 323-331.

Nagarajan, K., and Ramakrishnan, K. 1971. Studies on cucurbit viruses in Madras state-1 A new virus disease in bitter gourd. Pro. Ind. Acd. Sci., sect. B. 73: 30-35

Namrata, J., Saritha, R. K., Datta, D., Singh, M., Dubey, R. S., Rai, A. B., and Rai, M. 2009. Molecular Characterization of Tomato leaf curl Palampur virus and Pepper leaf curl betasatellite Naturally Infecting Pumpkin (*Cucurbita moschata*) in India. Indian J Virol. 21(2): 128-132.

Nariani, T. K., Vishwantha, S. M., Raychaudhuri, S. P., and Moharir, A. V. 1977. Studies on a mosaic disease of muskmelon. Curr. Sci. 46: 47-48.

Palodhi, P. R., and Sen, B. 1980. *Fusarium moniliforme Schle.* and *F. moniliforme var.* subglutinans Wr. and Reink., causing wilt in cucurbits. Indian Phytopath. 33: 474-475.

Pillai, N. G. 1971. A mosaic disease of Snake gourd (*Trichosanthes Anguina* L.,). Sci & Cult. 37: 46-47

Pivovarov, V. 1984. Cucumber breeding for resistance to false mildew (*Pseudoperonospora cubensis* R) using various ecological geographic zones. In: Cucumis and Melon. Proceedings of the 3th EUCARPIA meeting on breeding of cucumber and melon, Plovdiv (BG), pp 85-86.

Provvidenti, R. 1996. Diseases caused by viruses. In: Zitter T.A., Hopkins D.L., Thomas C.E. (eds.). Compendium of cucurbit diseases, pp. 37-45. APS Press, St. Paul, Minn.

Radhakrishnan, P., and Sen, B. 1985. Efficacy of different methods of inoculation of *Fusarium oxysporum* and *Fusarium solani* for inducing wilt in muskmelon. Indian Phytopath. 38: 70-73.

Rai, M., Pandey, S., and Kumar, S. 2008. Cucurbit research in India: a retrospect Cucurbitaceae 2008, Proceedings of the IXth EUCARPIA meeting on genetics and breeding of Cucurbitaceae (Pitrat M, ed), INRA, Avignon (France), May 21-24, 2008.

Raj, S. K., and Singh, B. P. 1996. Association of geminiviral infection with yellow mosaic disease of *Cucumis sativus*: Diagnosis by nucleic acid probes. Indian J. Exp. Biol. 34: 603-605.

Raj, S. K., Snehi, S. K., Khan, M. S., Tiwari, A. K., and Rao, G. P. 2010. First report of Pepper leaf curl Bangladesh virus strain associated with bitter gourd (*Momordica charantia* L.) yellow mosaic disease in India. Australasian Plant Disease Notes. 5: 14–16.

Rajinimala, N., and Rabindran, R. 2007. First report of Indian cassava mosaic virus on bittergourd (*Momordica charantia*) in Tamil Nadu, India. Australasian Plant Disease Notes 2, 81–82.

Rajinimala, N., Rabindran, R., Ramiah, M., and Kamlakhan, A. 2005. Virus vector relationship of Bitter gourd yellow mosaic virus and whitefly Bemisia tabaci germ. Acta Phytopathologica et Entomologica Hungrica. 40: 23-30.

Rani, S., Verma, G. S., and verma, H. N. 1971. Epidemiology of virus diseases of cucurbits. Proc. Indian Natl.Sci. Acad.37: 345-351.

Raychaudhuri, M., and Varma, A. 1978. Mosaic disease of Muskmelon caused by minor variant of cucumber green mottle mosaic virus. Phytopath. Z. 93: 120-125.

Raychaudhuri, M., and Varma, A. 1975. Virus disease of cucurbits in Delhi. Proc. 62nd Indian Sci. Cong. Part III, P. 74.

Reddy, K. S. C., and Nariani, T. K. 1963. Studies on mosaic disease of vegetable marrow. India Phytopath.16: 260-267.

Rotem, J. 1994. "The genus Alternaria: biology, epidemiology, and pathogenicity", APS Press. St. Paul, Minnesota, p 326.

Saritha, R. K., Bag, T. K., Loganathan, M., Rai, A. B., and Rai, M. 2011. First report of Squash leaf curl china virus causing mosaic symptoms on summer squash (*Cucurbita pepo*) grown in Varanasi district of India. Archives of Phytopathology and Plant Protection 44(2): 179-185.

Schenk, N. C. 1968. Incidence of airborne fungus spores over watermelon fields in Florida. Phytopathology. 58: 91-94.

Sen, B., Majumder, S., and Kumar, S. 1999. Fungal and bacterial diseases of cucurbits. In: Diseases of Horticultural Crops-Vegetables, Ornamentals and Mushrooms (L.R. Verma and R.C. Sharma), Indus Publishing Co., New Delhi.

Shankar, G., and Nariani, T. K. 1974. A mosaic disease of watermelon (*Citullus Vulgaris* Schrad.). Curr. Sci. 43: 281-282.

Sharma, Y. R., and Chohan, J. S. 1973. Transmission of cucumis virus 1 and 3 through seeds of cucurbits. Indian phytopath. 26: 529-598.

Sherf, A. F., and MacNab, A. A. 1986. Fusarium wilt of muskmelon. In: Vegetable diseases and their control, 2nd ed. Wiley, New York, pp 334–337.

Singh, R., Raj, S.K., and Chanadra, G. 2001. Association of monopartite Begomovirus with yellow mosaic disease of pumpkin in India. Plant Dis. 85. 1029.

Singh, R. P., Mohan, J., and Singh, D. P. 1999. Assessment of losses and management of Ridge gourd mosaic virus. Pl Dis Res. 14: 134-138.

Singh, S. J., and Rajverma. 2003. Muskmelon necrosis –A new tospovirus disease in India. Indian J. Mycol. Plant Pathol. 65: 4-5.

Singh, S. J., and Reddy, M. K. 1991). Watermelon bud necrosis –A new viral disease. IIHR annual report 1991/1992.

Singh, S. J., Verma, R., Ahlawat, Y. S., Singh, R. K., Prakash, S., and Pant, R. P. 2003. Natural occurrence of a yellow mosaic disease on zucchini in India caused by a potyvirus. Indian Phytopath. 56: 244-249.

Sohrab, S. S., Mandal B., Ali A., and Varma A. 2010. Chlorotic Curly Stunt: A Severe Begomovirus Disease of Bottle Gourd in Northern India. Indian J. Virol. 21(1): 56-63.

Sohrab, S. S., Mandal, B., Pant, R. P., and Varma, A. 2003. First report of association of Tomato leaf curl New Delhi virus with yellow mosaic disease of *Luffa cylindrica* in India. Plant Dis. 87: 1148.

St Amand, P. C., and Wehner, T. C. 1991. Crop loss to 14 diseases of cucumber in North Carolina from 1983 to 1988. Cucurb. Genet. Coop. 14: 15-17.

Tahir, M., and Haider, M. S. 2005. First report of Tomato leaf curl New Delhi virus Infecting bitter gourd in Pakistan. Plant Pathol. 54: 807.

Takami, K., Okubo, H., Yamasaki, S., Takeshita, M., and Takanami, Y. 2006. A cucumber mosaic virus isolated from *M. charantia* L. J. Gen. Plant Pathol. 72: 391-392.

Tian, D., and Babadoost, M. 2004. Host range of *Phytophthora capsici* from pumpkin and Pathogenicity of isolates. Plant Dis. 88: 485-489.

Tiwari, A. K., Snehi, S. K., Singh, R., Raj, S. K., Rao, G. P., and Sharma, P. K. 2012a. Molecular identification and genetic diversity among six Begomovirus isolates affecting cultivation of cucurbitaceous crops in Uttar Pradesh, India. Archives of Phytopathology and Plant Protection. 45 (1): 62-72.

Tiwari, A. K., Mall, S., Khan, M. S., Snehi, S. K., Sharma, P. K., Rao, G. P., and Raj, S. K. 2010a. Detection and identification of Tomato leaf curl Palampur virus infecting *Cucurbita pepo* in India. Guanaxi Agric. Sci. 41: 1291-1.295.

Tiwari, A. K., Sharma, P. K., Khan, M. S., Snehi, S. K., Raj, S. K., Rao, G. P. 2010b. Molecular detection and identification of Tomato leaf curl New Delhi virus isolate causing yellow mosaic disease in Bitter gourd (*Momordica charantia*), a medicinally important plant in India. Med. Plants. 2(2): 117-123.

Tiwari, A. K., Snehi, S. K., Khan, M. S., Raj, S. K., Sharma, P. K., and Rao, G. P. 2012b. Molecular detection and identification of Tomato leaf curl New Delhi virus associated with yellow mosaic and leaf curling disease of *Luffa cylindrica* crops in India, Indian Phytopathol. 65(1): 80-84.

Tomar, S. P. S., and Jitendra, M. 2001. Bitter gourd mosaic virus in weed host. Journal of Living world. 8: 24-27.

Vani, S. 1987. Studies on viral disease of muskmelon and watermelon. Ph.D.Thesis. P.G. School, IARI, New Delhi.

Varma, A., and Giri, B. K. 1998. Virus disease of cucurbits in India. In: Cucurbits, (Eds, N.M. Nayar and T.A. More). Oxford and IBH Publishing House Pvt. Ltd., New Delhi.

Varma, A., and Malathi, V. G. 2003. Emerging geminivirus problem: A serious threat to sustainable crop production. Ann. App. Biol. 142: 145-164.

Varma, P. M. 1955. Ability of the whitefly to carry more than one virus simultaneously. Curr. Sci. 24: 317-318.

Vasudeva R. S., and Lal, T. B. 1943. A mosaic disease of bottle gourd. Indian J. Agric. Sci. 13: 182-191.

Vasudeva, R. S, and Nariani, T. K. 1952. Host range of bottle gourd mosaic virus and its inactivation by plant extracts. Phytopathology 42: 149-152.

Vasudeva, R. S., and Pavgi, M. S. 1945. Seed transmission of melon mosaic virus. Curr. Sci. 14: 271-272.

Venkatsubbiah, K., and Sai Gopal, D. V. R. 1997. Characterization and identification of a tobamoviruses infecting Snakegourd (*Trichosanthes anguina* Linn.) In Andhra Pradesh. Indian J. Virol. 13: 153-157.

Verma, R., Ahlawat, Y. S., Tomer, S. P. S., and Pant, R. P. 2004. First report of Zucchini yellow mosaic virus (ZYMV) in bottle gourd in India. Plant Dis.88: 426.

Verma, R., Baranwal, V. K., Prakeash, S., Tomer, S. P. S., Pant, R. P., and Ahlawat. Y. S. 2006. First report of papaya ringspot virus W in Sponge Gourd from India Plant Disease 90: 7.

Yurina, O., Pivovarov, V., and Balashova, N. 1998. Cucumber breeding for disease resistance. *In:* Breeding and seed production of *Cucurbitaceae* crops in Russia (Pivovarov V, ed), Moscow Press (RU), pp 248-252.

Yurina, O. 1987. Hereditary and methodological aspects of cucumber breeding for the complex disease resistance. *In* Vegetable Crop Breeding, (Sichev S, ed), VNIISSOK, Moscow Press (RU), pp. 67-73.

Zitter, T. A., and Kyle, M. M. 1992. Impact of powdery mildew and gummy stem blight on collapse of pumpkins (*Cucurbita pepo* L.). Cucurb. Genet. Coop. 15: 93-96.

Zitter, T. A., Hopkins, D. L., and Thomas C. E. 1996. Compendium of cucurbit diseases. Amer. Phytopathol. Soc., St. Paul, Minn.

Diseases of Field and Horticultural Crops
Editor-in-Chief: P. Chowdappa
Published by: Daya Publishing House, New Delhi

Pages **476-505**

17

Diseases of Carnation

Nakkeeran, S., Dinesh, D., Renukadevi, P., Jawaharlal, M. and P. Chowdappa

Floriculture is a fast emerging venture throughout the world. In India the area under traditional flowers during 2007 was 73,536 ha under open condition. At present more than 1000 ha is cultivated under protected condition with the production of 3, 65,668 MT. The major states carrying out floriculture business in India are Karnataka (20,780 ha), Tamil Nadu (17,750 ha) and West Bengal (13,750 ha) (Patil *et al.*, 2004). National income generated from this industry is about Rs.500 crores per annum including traditional flowers (jasmine, crossandra, tuberose *etc.,*) and modern cut flowers (rose, chrysanthemum, carnation, gerbera, lillium, anthurium *etc.,*). Global trade in cut flowers is estimated as 40 billion US dollars. Indian floriculture industry is growing at a compounded annual growth rate (CGAR) of 25 per cent over the past decade. Indian flower export markets are estimated at 11 billion US dollars at present and expected to grow upto 20 billion US dollars by 2020 (Gian Aggarwal, 2011). The awareness on the usage of cut flowers for various occasions has raised the demand for flowers in the market. The production of cut flowers has gone upto 6,667 million stems (2011) from 2,071 million stems (2007) and this is due to the improvement in the standard of living and quality of life which ultimately increases the growth of domestic and export markets (Jafar Naqvi, 2011).

Carnation (*Dianthus caryophyllus* L.), a "Divine Flower", native of Mediterranean region belonging to the family Caryophyllaceae, is an introduced cut flower crop in India and adapts well to the regions having mild climatic conditions like Nilgiris, Kodaikanal, Yercaud, Bangalore, Pune and Shimla. Carnation cultivation is of late picking up and provides very good remuneration to the growers. Carnation holds an esteemed position among the cut flowers in the global floriculture trade and is gaining increasing popularity among the flower growers and traders. India has a vast potential to grow good quality carnations. The climatic conditions prevailing in Nilgiris, Yercaud Kodaikanal (Tamil Nadu) are most favourable for growing carnation, chrysanthemum, lillium, gerbera and anthurium. Continuous cultivation of carnation, chrysanthemum and exploitation of soil under protected cultivation, indiscriminate use of fungicides, and availability of different cultivars makes the carnation cultivation to be highly susceptible for the attack of *Fusarium* wilt caused by *Fusarium oxysporum* Schlechtend: Fr. f.sp. *dianthi* (Prill & Delacr.) W.C. Snyd. & H.N.Hans. Production systems of carnation in Nilgiris are managed intensively under monoculture for 2 years. Hence there is no sufficient gap between one crop to the other. The soil is continuously exploited, which deteriorate the soil health. The major impediment in the cultivation of carnation is due to the infection of soil borne and foliar diseases (Table 17.1) causing substantial yield loss, leading to deterioration in quality and quantity of the marketable blooms.

Table 17.1. Diseases of carnation

S.No	Disease	Causal Organism
Bacterial diseases		
1	Bacterial leaf spot	*Burkholderia andropogonis*
2	Bacterial wilt	*Burkholderia caryophylli*
3	Bacterial slow wilt	*Erwinia chrysanthemi*
Fungal diseases		
1	Alternaria leaf blight	*Alternaria dianthi*
2	Calyx rot	*Stemphylium botryosum*
		Pleospora tarda [teleomorph]
3	Downy mildew	*Peronospora dianthicola*
4	Fairy-ring leaf spot	*Mycosphaerella dianthi*
		Cladosporium echinulatum [anamorph]
5	*Fusarium* wilt	*Fusarium oxysporum* f.sp. *dianthi*
6	Gray mold	*Botrytis cinerea*
		Botryotinia fuckeliana [teleomorph]
7	*Pythium* root rot	*Pythium* spp.
		Pythium ultimum
8	*Phytophthora* rot	*Phytophthora nicotianae* var. *parasitica*
9	*Rhizoctonia* stem rot	*Rhizoctonia solani*
		Thanatephorus cucumeris [teleomorph]
10	*Sclerotinia* stem rot	*Sclerotinia sclerotiorum*
11	Southern blight	*Sclerotium rolfsii*
12	Rust	*Uromyces dianthi*
Viral diseases		
1	Carnation etched ring	genus *Cauliovirus*, Carnation etched ring virus (CERV)
2	Carnation latent	genus *Carlavirus*, Carnation latent virus (CLV)
3	Carnation mottle	genus *Carmovirus*, Carnation mottle virus (CarMV)
4	Carnation necrotic fleck & Carnation streak	genus *Closterovirus*, Carnation necrotic fleck virus (CNFV)
5	Carnation ring spot	genus *Dianthovirus*, Carnation ringspot virus (CRSV)
6	Carnation vein mottle	genus *Potyvirus*, Carnation vein mottle virus (CVMV)

1. Bacterial Diseases

1.1. Bacterial Wilt (*Burkholderia caryophylli* (Burkholder) Yabuuchi or *Pseudomonas caryophylli* (Burkholder) Starr & Burkholder and *Erwinia chrysanthemi* pv. *dianthicola* Burkholder *et al.*)

Burkholderia caryophylli infected plants exhibit grayish green foliage followed by sudden wilting and death of the plants. Roots of the affected plants rot and emit bad odour. Affected vascular tissues turn yellowish to brown. Basal portion of the affected stems crack (Fig. 17.1a, 1b). Plants infected by *E. chrysanthemi pv. dianthicola* results in slow wilting. Affected plants are stunted. Infected root gets rotten and the root system as a whole get reduced. Vascular bundles turn brown and the basal portion of the stem gets necrotized, leading to death of the plant (Fig. 17.2).

Fig. 17.1a. Yellowing of leaves   **Fig.17.1b.** Stem splitting

Fig. 17.2. Rotting and death of plant

1.2. Bacterial leaf spot/blight (*Burkholderia andropogonis* (Smith) Gillis Syn; *Pseudomonas andropogonis* (Smith) Stapp & P. *woodsii* (Smith) Stevens): Bacterial leaf spot initially appear as small, water soaked, yellow specks. During severe stages of infection, spots coalesce and cause extensive blighting. Diatloff and Rochecouste (1991) explained that, the extent of spread varies depending upon the climatic conditions like temperature and rainfall. Based on it, the disease development occurs either as slow peripheral spread, rapid extensive spread and declining vertical spread.

2. Fungal Diseases

2.1. Stem and Root Rots: *Phytophthora nicotianae.* parasitica (Breda de Haan var. parasitica (Dastur) Waterhouse). Plants affected by stem and root rot are characterized with wilting symptoms. Examination of the infected plants expresses the presence of brown discolouration on the collar region followed by the plant death (Fig. 17.3). *P. nicotianae var. parasitica* associated with other hosts and *P. capsici* infecting chillies can also infect carnation (Ann *et al.*, 1990). Pathogen grows between 10-35°C. But temperature of 27°C is optimum for the growth of pathogen. But, disease development is more rapid at 5-35°C than at 15-20°C (Ann *et al.*, 1990). The other pathogens like *P. capsici* and *P. cryptogea* were also found to be associated with stem and root rot infected samples.

Fig. 17.3. Stem rot

2.2 *Sclerotinia* stem rot (*Sclerotinia sclerotiorum*): Stem rot affected plants wilt gradually and the wilting is not observed on one side as *Fusarium* wilt. The leaves become straw coloured. The affected stems are hollow. Longitudinal section of the affected stem reveals the presence of brown coloured sclerotia in irregular shape. In the affected plants, vascular system turns brown (Fig 17.4a, b, c, d. Soil and infected plants serves as the source of primary inoculum. Raising beans as alternate crop also spreads the disease. Besides cool soil temperatures also favours the disease development. Soil drenching with biocontrol agents like *Bacillus subtili*s, *B. amyloliquefaciens, B. cereus* controls the disease development and spread of the disease. Besides soil drenching with fungicides like axoxystrobin, tebuconazole+ trifloxystrobin and carbendazim suppress the disease incidence.

Fig. 17.4a. Sclerotinia infected plant

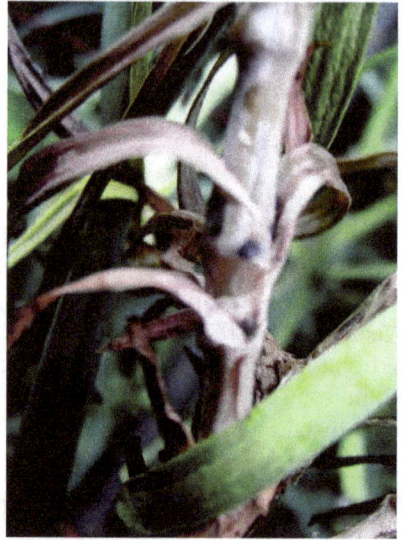

Fig. 17.4b. sclerotia on stem

Fig. 17.4c. Stem and root rot advanced stage

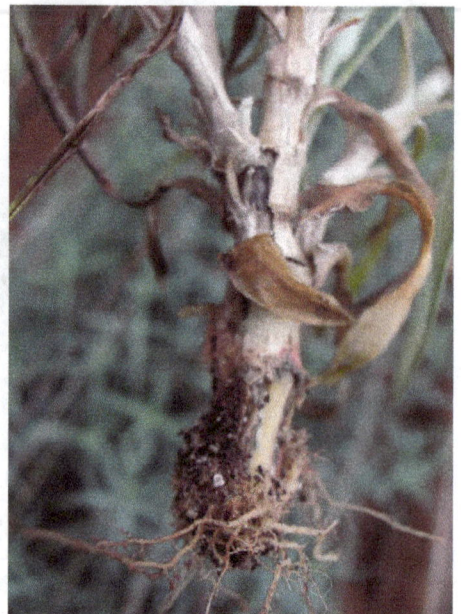

Fig. 17.4d. Sclerotinia stem rot and death of the plant

2.3. Collar rot/ Stem rot/ Root rot (*Rhizoctonia solani* Kuhn Tel.: *Thanatephorus cucumeris* (Frank) Donk): *Rhizoctonia solani* infect carnation plants at or just below the soil level. Affected plants are stunted. Lesions develop on the stem and the stem breaks off. Subsequently, the entire plant wilts and dies (Fig 17.5a, 17.5b). Dry and humid weather spreads the disease faster. But, the degree of susceptibility decreases with plant age. Disease spread is faster at 5-30ºC, but symptom does not express when the soil temperature is below <20ºC.

Fig. 17.5a. Collar rot   **Fig. 17.5b.** Lesions on stem

2.4. Fusarium stub dieback/Fusarium basal rot (*Fusarium roseum, Fusarium avenacearum, F. culmorum, F. graminearum*): Pathogen gains entry in to the plant after the harvest of the first flush. Sometimes, infection occurs after pinching of the shoots. After gaining entry, the pathogen kills the stubs of mature plants and girdles the main branches. Further the infection spreads down the roots, leading to the breakdown of main stem, causing wilting and death of branches without discolouration of the vascular tissues (Fig. 17.6). Sometimes the upper part of the

Fig. 17.6. Stub die back

plant is also infected and dieback occurs. The disease is severe in unmaintained polyhouses and in the gardens with high flower yield (Arbelaez, 1987).The injuries due to continuous harvesting of the cuttings favour the spread and entry of the pathogen. The pathogen survives on the infected stumbs and main stems. Further the ascospore released from perithecia also spreads the disease. Warm, humid conditions, high N fertilization and high N: K ratio also makes the plant susceptible for the infection. Use of clean cuttings and rooting medium and spraying the cuttings in rooting medium with thiophanate-methyl controls the disease.

2.5. Sclerotium root rot/ basal rot (*Sclerotium rolfsii* Sacc. Tel.: Athelia rolfsii (Curzi) Tu & Kimbrough): Leaves of the carnation plants infected by *Sclerotium rolfsii* turns pale green and dry. Infected stems and roots, rot and lead to the death of the plants. Infected plants exhibit the presence of white, cottony growth of mycelium at the collar region of the stem leading to rotting of the stems at the soil level. During severe cases of infection, rotting of stem starts and the pathogen spread to the leaf. Subsequently, it leads to rotting of the stem and death of the plants. During the advanced stages of infection, on the soil level and at the basal portion of the stem white to dark brown spherical sclerotial bodies were noticed. Pathogen survives as sclerotial bodies in the soil and on the crop residues. Allternate wetting and drying favours the disease spread.

2.6. *Pythium* root rot/foot rot (*Pythium vexans* de Bary, *P. irregulare, P. aphanidermatum* (Edson) Fitzp.): Pathogen attacks at the collar region and results in gradual discolouration followed by drying of leaves from bottom to upward. The infection results in rotting of stem at soil level resulting in death of the plant. Increased soil moisture coupled with increased relative humidity and the presence of crop residues spreads the disease (Fig 17.7).

Fig. 17.7. *Phythium* root rot

2.7. Wilt (*Fusarium oxysporum* f. sp. *dianthi):*

2.7.1. Crop loss: In our study, the wilt incidence was observed in both seedlings (30 days old) and older plants. The symptoms associated with wilt in the seedling stage include, yellowing of lower most leaves, and it subsequently spread to entire plant. Affected leaves droop and finally wilt. The longitudinal section of the infected roots and stems exhibit discolouration of the vascular tissues. However, if the infection occurs during later stages of the crop growth (70 days after planting), yellowing of the leaves are noticed on few shoots. Later the affected leaves droop down and dry. Subsequently, the infection gradually spread to other shoots. It leads to wilting of the entire plant within 3 to 4 weeks after infection. Examination of the infected plants shows the presence of vascular discolouration in roots and stems (Fig 8a, b, c, d). The wilt incidence in carnation varies depending upon the cultivar, monocropping and stage of infection. The incidence of wilt in the susceptible carnation cultivar Fantasia ranged from 22 to 50% (Ben-Yephet *et al.,* 1994). The occurrence of fusarial wilt in different cultivars of carnation varied from 40-79% (Katoch, 1999). Carnation varieties were surveyed for the incidence of *Fusarium* wilt in Nilgiris district of Tamil Nadu, India. Ten varieties of white, red, yellow, 9 light pink, 7 dark pink and 9 bicoloured carnation varieties were surveyed for the occurrence of *Fusarium* wilt in Nilgiris, Coonoor and Kothagiri of Nilgiris district in Tamil Nadu, India. Commercially cultivated white coloured varieties at Nilgiris, Coonoor and Kothagiri are White Dona (RR), Hunza (RRR), Praga (RR), Baltico (R), Dovar (RR), Lisa (RR), White liberty (R), Madame collete (RR), Randal (RRR) and Snowstorm (RRR). The survey results revealed that, the average wilt incidence in white coloured carnation at Nilgiris district was 31.51%. Among the ten different white carnation varieties cultivated in Nilgiris, the minimum incidence of 22.0%, 20.2% and 21.0% was observed in the variety Randal at Nilgiris, Coonoor and Kothagiri respectively. However, the maximum wilt incidence of 48.2%, 47.8% and 32.24% was observed in Baltico variety cultivated at Nilgiris, Coonoor and Kothagiri respectively (Table 17.2).

Fig. 17.8a. Yellowing of leaf

Fig. 17.8b. *Fusarium* wilt affected plants

Fig. 17.8c. Wilted plant

Fig. 17.8d. Wilt severity in polyhouse

Table 17.2. Survey for the occurrence of Fusarium wilt in White colour carnation variety

S.No	Variety	Colour	Company	Per centage of wilt incidence			
				Nilgiris	Coonoor	Kothagiri	Mean
1	White dona(RR)	White	Florence flora	40	32	28.8	33.60
2	Hunza(RRR)	White	Florence flora	25	28.2	22.2	25.13
3	Praga(RR)	White	Florence flora	22.8	31.2	32.2	28.73
4	Baltico(R)	White	Barberet & Blanc	48.2	47.8	32.4	44.46
5	Dovar(RR)	White	Barberet & Blanc	28.2	34.2	34.8	32.40
6	Lisa(RR)	White	Hilverda kooij	34	30	31.8	31.93
7	White Liberty(R)	White	Hilverda kooij	47.4	28.8	30.4	35.53
8	Madame collete(RR)	White	selecta	38	36	32.2	35.40
9	Randal(RRR)	White	selecta	22	20.2	21.0	23.53
10	Snow Storm(RRR)	White	selecta	24	28.2	28.4	24.40
	Mean			32.96	32.16	29.42	31.51

The red coloured varieties cultivated in Nilgiris, Coonoor and Kothagiri are, Big Red (RRRR), Pintado (RRR), Tuareg (RRR), Turbo (RR), Domingo (RRRR), Master (RRR), Eskimo(R), Gaudina (RRR), Don Pedro (RR) and Aicardi (R). The survey for the Fusarium wilt of red coloured carnation cultivated at Nilgiris, indicated that, maximum incidence of 49.8% was observed in the variety Gaudina, while the minimum incidence of 18%, was observed in the variety Bigred. However, the maximum incidence of 47% was noticed in the variety Domingo and the minimum incidence of 20.2% wilt was observed in the variety Big Red cultivated at Coonoor. Besides, survey for the incidence of wilt in Kothagiri, reflected that, the maximum incidence was associated with the variety Gaudina (50.4%) and the minimum incidence of 18% was observed in the variety Master (Table 17.3).

Table 17.3. Survey for the occurrence of Fusarium wilt in Red colour carnation variety

S.No	Variety	Colour	Company	Per centage of wilt incidence			
				Nilgiris	Coonoor	Kothagiri	Mean
1	Big Red(RRRR)	Red	Florence flora	18	20.2	20.8	19.66
2	Pintado(RRR)	Red	Florence flora	30.4	28.4	30.4	29.73
3	Tuareg(RRR)	Red	Florence flora	32	24	20	25.33
4	Turbo(RR)	Red	Florence flora	42	38	39.4	39.80
5	Domingo(RRRR)	Red	Barberet & Blanc	38	47	28.2	37.73
6	Master(RRR)	Red	Barberet & Blanc	36	42	18	32.00
7	Eskimo(R)	Red	Hilverda kooij	47.4	30	24.5	33.96
8	Gaudina(RRR)	Red	Hilverda kooij	49.8	32.4	50.4	44.20
9	Don Pedro(RR)	Red	selecta	38	30.4	28.8	32.40
10	Aicardi(R)	Red	selecta	35.2	28	31.2	31.46
	Mean			36.68	32.04	29.17	32.63

In Nilgiris, Coonoor and Kothagiri 10 yellow coloured carnations are commercially cultivated. They are, Elisir (RR), Soto (RR), Luna (RRRR), Kiro (RRR), Cameron (RR), Harvey (R), Liberty (R), L.E.D (RRR), Victoria (RR) and Hermes (RRR). The mean average wilt incidence observed in Nilgiris district was 26.35%. Association of Fusarium wilt on yellow coloured carnation cultivated at Nilgiris, revealed that, maximum incidence of 38.0% was observed in the variety Harvey, while the minimum incidence of 18.0%, was observed in the variety Luna. However, the maximum incidence of 32% was noticed in the variety Victoria and the minimum incidence of 18.0% wilt was observed in the variety Kiro cultivated at Coonoor. Similarly, survey for the incidence of wilt in Kothagiri reflected that, the maximum incidence of wilt was observed in the variety Kiro (36.2%) and the minimum incidence of 16.8% was observed in the variety Hermes (Table 17.4).

Table 17.4. Survey for the occurrence of *Fusarium* wilt in Yellow colour carnation variety

| S.No | Variety | Colour | Company | Per centage of wilt incidence | | | |
				Nilgiris	Coonoor	Kothagiri	Mean
1	Elisir(RR)	Yellow	Florence flora	32	30	24.4	28.80
2	Soto(RR)	Yellow	Florence flora	28	24	18.8	23.60
3	Luna(RRRR)	Yellow	Florence flora	18	19.8	22.4	20.06
4	Kiro(RRR)	Yellow	Barberet & Blanc	34.2	18	36.2	29.46
5	Cameron(RR)	Yellow	Hilverda kooij	28	24.6	18.2	23.60
6	Harvey(R)	Yellow	Hilverda kooij	38	29.6	20.4	29.33
7	Liberty(R)	Yellow	Hilverda kooij	36	31.2	28.6	31.93
8	L.E.D(RRR)	Yellow	selecta	19	26.2	28.3	24.16
9	Victoria(RR)	Yellow	selecta	28	32	32.4	30.80
10	Hermes(RRR)	Yellow	selecta	30.4	18.2	16.8	21.80
	Mean			29.06	25.36	24.65	26.35

Table 17.5. Survey for the occurrence of *Fusarium* wilt in Light pink colour carnation variety

| S.No | Variety | Colour | Company | Per centage of wilt incidence | | | |
				Nilgiris	Coonoor	Kothagiri	Mean
1	Big mama(RRR)	Light pink	Florence flora	28	28.8	34.2	30.33
2	Haris(RRR)	Light pink	Florence flora	24.8	28.6	24.8	26.06
3	Paloma(RR)	Light pink	Florence flora	38	32	30.8	33.60
4	Navona(RR)	Light pink	Barberet & Blanc	37.4	28.8	26.2	30.80
5	Pink dovar(RR)	Light pink	Barberet & Blanc	36.4	20	28.4	28.26
6	Charment(RR)	Light pink	Hilverda kooij	28.4	20.4	26.8	25.20
7	Kleos(RR)	Light pink	Selecta	38.4	36.2	28	34.20
8	Candy pink(RRR)	Light pink	Selecta	28	34.4	28	30.13
9	Bicocca(RRR)	Light pink	Selecta	18	28.2	26.4	24.20
	Mean			27.74	25.74	25.36	26.28

Light Pink coloured carnation varieties like, Big mama(RRR), Haris(RRR), Paloma (RR), Navona (RR), Pink Dovar (RR), Charment (RR), Kleos (RR), Candy pink (RRR) and Bicocca (RRR) are cultivated at Nilgiris district under protected condition. The mean average wilt incidence observed in Nilgiris district was 26.28%. Incidence of Fusarium wilt on light pink carnation variety cultivated at Nilgiris, revealed that, maximum incidence of 37.4% was observed in the variety Navona, while the minimum incidence of 18.0%, was observed in the variety Bicocca. But, the maximum incidence of 36.2% was noticed in the variety Kleos and the minimum incidence of 20.0% was observed in the variety Pink Dovar cultivated at Coonoor. Besides, survey for the incidence of wilt in Kothagiri, revealed that, the maximum incidence of 34.2% wilt was observed in the variety Big mama and the minimum incidence of 24.8% was observed in the variety Haris (Table 17.5).

Dark Pink coloured carnation varieties like, Dona (RR), Golem (RRRR), Dumas (RRRR), Bizet (RR), Farida (R), Jurano (RR) and Cinderella (RR) are cultivated at Nilgiris district in aerodynamic poly house under protected condition. Incidence of Fusarium wilt on dark pink coloured carnation variety cultivated at Nilgiris, had the maximum incidence of 48.0% in the variety Farida, while the minimum incidence of 28.0%, was observed in the variety Jurano. Similarly, the maximum incidence of 38.0% was recorded in the variety Farida and the minimum incidence of 12.0% was observed in the variety Dumas cultivated at Coonoor.

In addition, survey for the incidence of wilt in Kothagiri, revealed that, the maximum incidence of 38.0% was observed in the variety Jurano and the minimum incidence of 14.0% was observed in the variety Bizet and the mean average wilt incidence recorded in Dark Pink coloured carnation varieties in the Nilgiris district was 19.68% (Table 17.6).

Table 17.6. Survey for the occurrence of *Fusarium* wilt in Dark pink coloir carnation variety

S.No	Variety	Colour	Company	Per centage of wilt incidence			
				Nilgiris	Coonoor	Kothagiri	Mean
1	Dona(RR)	Dark pink flora	Florence	34.4	32	22.8	29.73
2	Golem(RRRR)	Dark pink flora	Florence	34.8	18	16.2	23.00
3	Dumas(RRRR)	Dark pink Blanc	Barberet &	32.2	12	16	20.06
4	Bizet(RR)	Dark pink kooij	Hilverda	38	20	14	24.00
5	Farida(R)	Dark pink kooij	Hilverda	48	38	26	37.33
6	Jurano(RR)	Dark pink kooij	Hilverda	28	38	38	34.66
7	Cinderella(RR)	Dark pink	selecta	32	24	28.2	28.06
	Mean			24.74	18.2	16.12	19.68

In Nilgiris, Coonoor and Kothagiri 9 bicoloured carnations are commercially grown. They are happy golem (RRRR), Folgore (RRRR), Diana (RRRR), Falicon (RRR), Solar (RR), Readez-vouz (RRR), Yellow viana (RRR), Alibaba (R) and Felica (RR). The mean average wilt incidence in bicoloured carnation varieties observed in Nilgiris district was 18.58%. The highest incidence of wilt was recorded in the bicoloured variety Felica, in irrespective of places like Nilgiris (38.0%), Coonoor (38.40%) and Kothagiri (38.2%). Similarly the variety, Folgone recorded the minimum incidence of wilt in Nilgiris, Coonoor and Kothagiri accounting for 14.0%, 12.0% and 8.0% respectively (Table 17.7).

Table 17.7. Survey for the occurrence of Fusarium wilt in Bicolour carnation variety

S.No	Variety	Colour	Company	Per centage of wilt incidence			
				Nilgiris	Coonoor	Kothagiri	Mean
1	Happy Golem(RRRR)	Bicolour	Florence flora	18	12	10	13.33
2	Folgore(RRRR)	Bicolour	Florence flora	14	12	8	11.33
3	Diana(RRRR)	Bicolour	Florence flora	18	12.2	14.6	14.93
4	Falicon(RRR)	Bicolour	Barberet & Blanc	14	12.8	18.2	15.00
5	Solar(RR)	Bicolour	Barberet & Blanc	32	30.4	30.2	30.86
6	Rendez-vouz(RRR)	Bicolour	Hilverda kooij	18	16.2	17.2	17.13
7	Yellow viana(RRR)	Bicolour	Hilverda kooij	18.8	16.8	11.2	15.60
8	Alibaba(R)	Bicolour	selecta	38	30	26.2	31.40
9	Felica(RR)	Bicolour	selecta	32	38.4	38.2	36.20
	Mean			20.28	18.08	17.38	18.58

2.7.2. Symptoms: The symptoms of carnation wilt (*Fusarium oxysporum* f. sp. *dianthi*) include yellowing of leaves, withering of leaves at the basal portion, and yellowing of midribs. The infected leaves turn chlorotic and finally wilt. During certain occasions, a portion of the plant wilts and subsequently spread to all the portions of the plant (Sohi, 1992). *Fusarium oxysporum* f. sp. *dianthi* infect vascular tissues and impairs absorption of water and nutrients, followed by chlorosis and wilting of lower leaves and shoots. Initially, these symptoms were observed on one side of the plant. Affected stems get shriveled, the leaves attached to the shriveled stem withers completely. But, the leaves in the uninfected shoots were green for a short period and the infection spreads to other shoots rapidly and leads to death of the plant. The wilted plants are stunted, yellow and dry often with hollow stem (Hood and Stewart, 1957; Baayen and Maat, 1987). Vascular browning of stem was also associated with the disease.

2.7.3. Causal organism: *Fusarium oxysporum* f. sp. *dianthi the* causal agent of vascular wilt on carnation is highly diverse in nature. Eight physiological races were reported in Italy. Races 1 and 8 are associated with the Mediterranean carnation ecotypes, which exist in Italy, France, and Spain (Garibaldi, 1983). But Race 2 is found in all carnation growing areas across the World. Race 3 of *F. oxysporum* f. sp. *dianthi* was classified as *F. redolens* f. sp. *dianthi* Race 3 (Baayen *et al.*, 1997). Race 4 is found to be associated with American carnation cultivars in the United States (Baayen *et al.*, 1997), Italy (Garibaldi *et al.*, 1986), Israel (Ben-Yephet *et al.*, 1992), Spain and Colombia (Baayen *et al.*, 1997). Races 5, 6, and 7 were reported by Garibaldi *et al.* (1986) on diseased carnations from Great Britain, France, and the Netherlands. Three new races (Daboussi and Langin., 1994) were reported to occur on diseased carnations from Australia (race 9 - Kalc Wright *et al.*, 1996) and the Netherlands (races 10 and 11 - Baayen *et al.*, 1997). Booth (1971) described that *F. oxysporum* f. sp.*dianthi* was characterized by the presence of microconidia, macroconidia and chlamydospores. The Microconidia were abundant, hyaline, continuous or single septate, ovoid to ovate and measured 4.0-8.5 x 2.0-3.5 μm. Macroconidia were sparse, fusoid and variable, 3 septate or rarely 4-5 septate and measured 18.5-28.0 x 3.0-4.5 μm. Chlamydospores were hyaline, usually vacuolated and spherical, measured 3.5-8.0 μm in diameter. The culture of the fungus on potato dextrose agar produced pinkish to purple coloured mycelium.

The pathogen *F. oxysporum* f .sp. *gladioli* produced white to peach pale salmon or purple mycelium. Microconidia were abundant, hyaline, and ovoid to ovate. Macroconidia were scarce, often lacking and variable, three septate. The chlamydospores were hyaline, usually vacuolate and spherical. Smith *et al.* (1988) reported that, *F. oxysporum* exhibits varying cultural morphology on Potato Dextrose Agar (PDA). The aerial mycelium first appears white and then may change to a variety of colours, ranging from violet to dark purple according to the strain of *F. oxysporum*. Nakkeeran *et al* (2012) observed *F. oxysporum* f.sp. *dianthi*, under ESEM at 7000x magnification. It was characterized with the chain of ovoid shaped microconidia, fusoid shaped macroconidia. Besides spherical shaped, intercalary chlamydospores were also observed under 1200x magnification.

2.7.4. Epidemiology: Pathogen development is favoured by acidic conditions (Orozco-de-Amezquita *et al.*, 1993). Soil pH 6.0, 60 per cent soil moisture and soil temperature of 30°C favour highest wilt incidence, while 15°C soil temperature, 4.0 pH and 10 per cent soil moisture did not favour disease development. The most severe wilt epidemics developed at low radiation intensities (200-300μE/m²/s) and at temperature 25-26°C. But, plants remained symptomless at higher intensities of solar radiation (>1000μE/m²/s) and temperature <18°C. The wilt pathogen also spread by water and infected cuttings. The spread is faster after colonization of the soil by the pathogen.

2.8. Alternaria leaf spot/blight (*Alternaria dianthi* Stevens & Hall and *A. dianthicola* Neergaard) **:** Infections are generally observed on lower surface of the leaf and subsequently spread to upper surface. The disease appears during rooting time. The symptoms appear during December- January and become severe during July – October. Initial symptoms are characterized with small purple lesions and later turn as grayish-brown spots. (Fig 17.9). During favourable conditions, lesions enlarge and merge together and results in blighting of leaves. Lesions also spread from the leaf to the stems and lead to girdling of the stem (Arbelaez, 1987).The optimum temperature and pH for the establishment of host parasite relationship by *A. dianthi* is 25-30ºC and pH 6 for vegetative growth. But the temperature of 20-25ºC favours spore germination. Disease severity is correlated with the minimum temperature. Delayed planting also aggravates the disease severity. Pathogen survives in infected plants and debris. Existence of moist conditions for 8 to 10 hours required for infection. The disease spreads through airborne spores. Good air circulation and low humidity reduce the disease severity. Foliar application with fungicides like iprodione, difenaconazole, mancozeb, azoxystrobin or chlorothalonil controls the disease.

Fig. 17.9. Symptoms of *Alternaria* infection

2.9. Gray mold (*Botrytis cinerea*): The pathogen infects flowers. Symptoms are characterized with the presence of brown to grey coloured spots with full of sporulation. During humid weather, the infection spread to entire flower and thus not found suitable for marketing. The sporulation of the pathogen is observed as woolly gray fungal spore mass (Fig. 17.10 a, b, c, d). Pathogen survives in the plant debris. Low temperature, high moisture and increased humidity and airborne spores spread the disease. Maintenance of low humidity, horizontal air movement, removal of old flowers in the garden prevents the disease outbreak and spread. Foliar spray with iprodione, mancozeb, iprovalicarb, triazole groups helps in suppressing the disease development.

Fig. 17.10a. Gray mold on leaf

Fig. 17.10b. Gray mold on stem

Fig. 17.10c. Gray mold in flower

Fig. 17.10d. Grey mold severity

2.10. **Fairy ring leaf spot** (*Cladosporium echin*ulatum (Berk.) De Vries Tel.: *Mycospharella dianthi* (Burt) Jorst.): Leaf spot is characterized with pin head like necrotic tan spots on the leaf and leaf sheath. The margin of the spots is purple to dark purplish. During favourable conditions, the spots enlarge as circular – oval spots with grey centre. Conidiophores and conidia develop in the spots. Dark spores form in spots. This brownish growth appears as dull and dark bands, giving the name 'fairy ring' spot to the disease. During severe cases the lesions coalesce and the affected leaves get blighted. Blighted leaves dry and defoliate. During favourable environmental conditions, lesions spread on to petals as oval shaped tan spots characterized with grey centre. Later minute black fructification appears at the centre of the lesions (Fig. 17.11 a, b, c, d). Subsequently, infection spread to calyx, petals and stems. Infected flowers are not suited for marketing (Arbelaez, 1999). Pathogen survives on the infected plants and debris. Wet weather, moisture adhering on the leaf coupled with increased relative humidity favours the disease spread. Good air circulation, low humidity and leaves free from moisture reduce the disease severity. Foliar application of difenaconazole, axoxystrobin, propiconazole, mancozeb, iprovalicarb or probineb reduces the disease severity.

Fig. 17.11a. Mycosphaerella lesion on leaf

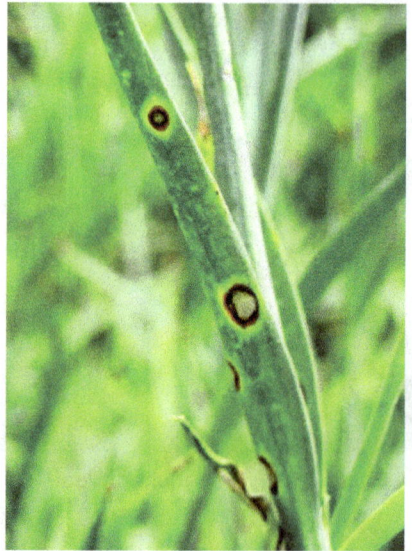

Fig. 17.11b. Mycosphaerella lesions on leaf

Fig. 17.11c. Mycosphaerella leaf blight

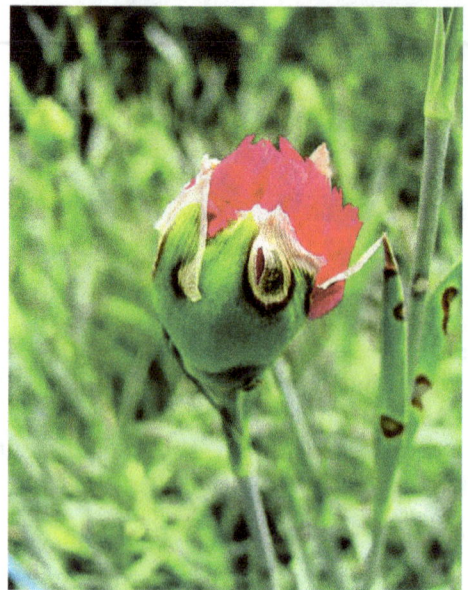

Fig. 17.11d. Mycosphaerella spots on flowers

2.11. Rust (*Uromyces dianthi* (Pers.) Niessl): Rust symptoms are initially characterized with light green coloured lesions. On the lesions small brown pustules with powdery mass of brown spores develop on the leaves. During severe cases of infection, the lesions also spread to the stems. Due to high disease intensity, plants wither and affect the growth and flower quality. Severe infection leads to drying of leaves and stems. It in turn affects the flower quality and yield (Fig 17.12 a, b, c, d). The pathogen is air borne in nature and urediniospores remain viable for 6 months. The pathogen is more prevalent in areas of high precipitation irrespective of plant age (Arbelaez, 1987). Low temperature coupled with high humidity favours the disease development and spread. Rust development is also negatively correlated with maximum temperature while positively correlated with minimum temperature. Foliar application with myclobutanil, krexoxym methyl, tebuconazole, mancozeb, azoxystrobin or iprovalicarb suppresses rust. Besides cultivation of resistant cultivars also reduce the disease incidence.

Fig. 17.12a. Initial infection

Fig. 17.12b. Severity on leaf

Fig. 17.12c. Drying of leaf

Fig. 17.12d. Rust pustules on stem

3. Viral Diseases

Cultivation of carnation is hampered by the infection of viral diseases. Viruses affecting carnation cultivation are, Carnation mottle virus (genus Carmovirus, CarMV), Carnation vein mottle virus (genus Potyvirus, CaVMV), Carnation ring spot virus (genus Dianthovirus, CaRSV), Carnation etched ring virus (genus

Caulimovirus, CaERV), Carnation latent virus (genus Carlavirus, CaLV) and Carnation necrotic fleck virus (genus Closterovirus, CaNFV) are most serious. Plants infected by these viruses show wide variety of mosaic symptoms like mottling, chlorotic mottle or green vein mottle, necrotic flecks, rings, line patterens 'etched', whitish or necrotic streaks etc. Most of the viruses are transmitted either by sap (mechanically) or by the aphid (*Myzus persicae*).

4. Yield loss in carnation due to diseases

The major yield limiting factors in the cultivation of high valued carnation crop include diseases like Fusarium wilt, stem and root rot, stub rot and rust diseases. The disease is severe in states like Tamil Nadu, Andhra Pradesh, Karnataka and Himachal Pradesh. Continuous exploitation of the soil under protected cultivation, leads to depletion of soil health. Application of soil fumigants like basamid, kills all the beneficial microbes. Cultivation of carnation in soils of poor health leads to the severity of soil borne diseases and foliar diseases. The intensity of loss depends on the stage of infection of the crop and the choice of the variety. If wilt infection coincides during Ist flush of the crop and yield loss of 70 to 80% occurs. But, if the infection coincides with second flush, an average yield loss of 30 to 40% occurs.

Besides, infection by grey mold pathogen on flowers results in deterioration of the flower quality. The colour of the flower gets changed into black colour due to mould infection. Sometimes incorporation of the diseased carnation stems along with *Mycosphaerella dianthi* infected leaf spreads the infection onto the flowers during transport of the produce. It leads to complete disposal of the respective containers with severe infection.

Studies on the monetary yield loss due to wilt infection over an area of $1000m^2$ under protected cultivation reflect that a farmer get an net income of Rs 16.8 Lakhs under disease free condition. But if the wilt incidence occurs during the second flush farmer losses an income of Rs 6.72 Lakhs/$1000m^2$. On an average of 33 crores is lost through wilt and stub rot infection in Tamil Nadu during one cropping season (Two years). This not only leads to heavy monetary loss, besides the standard of living of the cut flower growers are also affected. During severe cases of infection, farmers withheld the cultivation of cut flowers like carnation and go for the cultivation of crops like beans which is not economical to be grown under protected cultivation.

5. Integrated Management of carnation disease

5.1 Cultural Practices: Crop sanitation is one of the basic steps used for the reduction of inoculum pressure and crop susceptibility. Removal of Fusarium wilt, rust, bacterial wilt infected plants and their debris reduces the primary inoculum load and maintains at minimum level. Deep summer ploughing, followed by soil solarization reduce the soil-borne inoculums of the pathogens. Bio-fumigation of soil with cucurbit crop residues also reduces the primary inoculum load in the soil. Manual removal of rust and leaf spot infected leaves and infected debris reduces the leaf spots and rust pathogens in carnations (Arbelaez, 1987). Removal of the left out carnation stumps after harvest and burning the residues avoid the secondary spread of the carnation diseases. Similarly, removal of weeds in the carnation poly houses alters the microclimate by lowering the soil moisture and humidity and thereby prevents the occurrence of foliar diseases like rust of carnation (Vide Sharma and Sharma, 2008).

5.2. Physical methods/Soil solarization: Soil solarization for a period of seven weeks significantly reduced the inoculum potential and inoculum pressure of *F. oxysporum f.sp. dianthi* (Elena and Tjamos, 1997) and *P. nicotianae var. parasitica* (Garibaldi and Tamietti, 1989). Steaming before planting also suppressed the occurrence of vascular pathogens (Arbelaez, 1988). Improving the horizontal air movement and air circulation inside the poly house reduce the humidity and leaf moisture. It thereby reduces the incidence of foliar diseases of carnation.

5.3. Crop rotation: Inoculum density of soil-borne pathogens can be reduced through crop rotation. As most of these pathogens have wide host range, care should be taken while selecting crops for rotation. *P. nicotianae var. parasitica* and *P. capsici* from other hosts have been reported to be pathogenic to carnation (Ann *et al.*, 1990). Similarly, *P. cryptogea* has also been isolated from carnations. However, crop rotation is not practically viable, since farmers are not willing to raise a non-remunerative non host crops.

5.4. Adjustment of planting date: The choice of planting in relation to crop disease has one principal aim to reduce to a minimum the period over which pathogen meets the susceptible stage of the host. This can be exploited for disease management by altering the date of planting. The incidence of *Alternaria* leaf spot can be reduced by altering the time of planting of carnation in such a way that the crop does not coincide with the stage of susceptibility. Since, carnation is a two year old crop, adjusting the planting time will not have a major impact in avoiding the disease development.

5.5. Nutrient management: Maintenance of soil nutrition at optimum level maintains the plant health appropriately. But, application of excess doses of nitrogen in carnation results in lush growth and high relative humidity which predisposes the crop for the attack of Fusarium stub dieback (Sharma and Sharma, 2008). Soil application of organic amendments that induce nitrogen deficiency reduced the incidence of carnation wilt. (Filippi and Bagnoli, 1992). The number of colonies of *F. oxysporum f.sp. dianthi* in the soil decreased following treatments with potassium nitrate and potassium sulfate as they increased soil pH (Burbano *et al.*, 1990).Application of fresh, non-decomposed organic matter resulted in the acidic pH and increased the incidence of Fusarium wilt. Best control of wilt of carnation was noticed by planting the cuttings in the top soil of the container media amended with manures with neutrl pH (Duskova and Kovacikova, 1989). Application of vermicompost immediately after planting of carnation significantly suppressed the spread of wilt and reduced pathogen propagules by 50 per cent (Sharma and Sharma, 2008).

5.6. Regulatory methods: Growers on their own interest import planting materials from other countries and there by introduce new diseases like crown gall due to the negligence of the post entry quarantine measures. The incorporation of

infected cuttings from different countries has been highlighted as major cause of
F. oxysporum f.sp. dianthi and *Phialophora cinerescens* in Colombia, which have
caused big losses in production of carnation flowers (Arbelaez, 1987, 1999). Lack
of stringent quarantine measures, has resulted in the spread of several races of
FOD across the globe. The most virulent race FOD2 is prevailing in carnation
growing areas of the world. The cyst nematode *Heterodera trifoli* was introduced
by importing carnation seedlings from Israel during 1984 (*Arbelaez et al., 1984*).
Hence, management of carnation diseases has to be viewed seriously by
strengthening both pre and post entry quarantines while importing the carnation
cuttings. All the quarantine stations should be well equipped with latest technologies
and man power.

5.7. Chemical Control: Application of soil fumigants like methyl bromide,
chloropicrin, metham-sodium, dazomet, formaldehyde, etc along with steaming
reduced the incidence of vascular wilt (Arbelaez, 1988; Orozco-de-Amezquita *et al.*,
1993; Navas-Becerra *et al.*, 2000). Similarly application of dazomet followed by soil
drenching with water and covering the soil with poly propylene bags completely
suppressed the incidence of vascular wilt diseases in carnation (Personal
Communication). Vascular wilts and Rhizoctonia root rot was controlled by
drenching/soil mixing of benomyl, carbendazim, thiophanate methyl. Application
of strobilurins (azoxystrobin, kresoxim-methyl and trifloxystrobin) was effective in
inhibiting *F. oxysporum f.sp. dianthi*, *R. solani* and *P. nicotianae var. parasitica* when
applied as soil drenching or mixing at transplanting (Gullino *et al.*, 2000).
Rhizoctonia root rot can also be controlled by dipping the cuttings in captaf (0.3%)
or carbendazim (0.1%) for 30 minutes before planting. Treatment with propiconazole
(Tilt) provided moderate control of Fusarium wilt (while 100 per cent survival of
sclerotium root rot infected plants was obtained with tebuconazole (0.1%) drench
(Sharma and Sharma, 2008).

Use of broad spectrum protectant fungicides (captafol, mancozeb, zineb,
chlorothalonil) gave good control of Alternaria leaf blight and cladosporium leaf
spot. Sprays of systemic fungicides (difenoconazole, hexaconazole, bitertanol,
oxycarboxin, myclobutanil) were quite effective against most of the foliar pathogens
(Ferrin and Rohde, 1991). The combination or alternate use of both protective and
systemic fungicides could be the best strategy for the control of these diseases. The
applications of streptomycin sulfate, oxytetracycline and fosetyl-Al have been
advocated for the control of bacterial leaf spot (Trujillo and Nagata, 1994). Many
triazole compounds have good fungicidal and plant growth regulating activities.
Triazole compounds with 1, 3-dioxolanes have both preventive and control activities
against several plant diseases (Zhao *et al.*, 1995). Dipping of rooted cuttings and
soil drenching with tebuconazole + trifloxystrobin @1g/litre was effective in reducing
the *Fusarium* wilt (8.33%), but the number of shoots (3.40), plant height (61.00 cm)
and flower yield (141 numbers/ m²) was lower than all the fungicides tested in
the study. The treated seedlings were stunted and it took a minimum of 30 days
to restore the normal growth. It indicated that, tebuconazole + trifloxystrobin acted
as a growth retardant at early stage of application. However, application of the
same fungicide at same concentration after three months of planting does not have
any impact on retarding the growth of the plant. Propiconazole and difenoconazole
has been used as the most efficient triazole fungicide in the control of some
common plant diseases. Thirteen novel triazole analogs of difenoconazole containing

1, 3-dioxolane rings have been synthesized, and they express plant-growth regulatory activity similar to those of a difenoconazole (Xu *et al.*, 2004).

Delivering of difenoconazole through root dipping and soil drenching @ 2ml/litre recorded 54.4 per cent reduction of wilt incidence over control. The flower yield increased upto 93.4% over control. The flower yield was around 241.8 numbers/m² area, which was only 125.0 numbers/m². Besides the mean number of shoots, plant height, root length and dry weight of the root were 7.5 numbers, 95.9 cm, 24.2cm and 3.62g respectively. Days taken for flower bud initiation, bud opening, duration of flowering, length of flower stalk and number of flower buds was 102.3 days, 142.3 days, 57.2 days, 94.53 cm and 7.80 numbers of flower buds was noticed respectively. It differed significantly from all the other treatments, indicating that earliness was induced in the treated seedlings rather than the other fungicides. It was followed by dipping of rooted cuttings and soil drenching with azoxystrobin @ 2ml/litre (Table 17.7 and 17.8). As reported by Xu and his co-workers during 2004, very good growth promotional activities were observed in the growth parameters of carnation and the flower yield (Nakkeeran *et al.*, 2012). Sebastian *et al.* (2002) reported enhanced chlorophyll synthesis in *Dianthus caryophyllus* treated with palobutrazol.

Table 17.7. Effect of fungicides on number of shoot, root length, plant height, wilt incidence and yield in carnation variety Gaudina Red

Treatments	No.of Shoots 4MAP	Root length (cm) 4MAP	Plant height (cm) 7MAP	Wilt incidence 7MAP	Flower yield/m² 7MAP
T1- Root dipping and soil drenching with carbend-azim @ 2.0g/litre	7.30[b]	16.40[e]	84.70[d]	23.61[e]	173.00[d]
T2- Root dipping and soil drenching with Azoxys-trobin @ 2.0ml/litre	6.50[c]	21.20[c]	90.00[c]	16.60[c]	195.00[b]
T3- Root dipping and soil drenching with Pyraclostrobin @ 2.0g/litre	7.50[a]	21.40[c]	92.30[b]	25.00[f]	202.50[b]
T4- Root dipping and soil drenching with Tebuco-nazole + Trifloxystrobin @ 1.0 g/litre	3.40[e]	23.80[b]	61.00[f]	8.33[a]	141.90[e]
T5- Root dipping and soil drenching with Difenoco-nazole @ 2.0 ml/litre	7.50[a]	24.20[a]	95.90[a]	13.81[b]	241.80[a]
T6- Root dipping and soil drenching with Iprovalicarb + Probineb @ 2 g/litre	6.60[b]	16.80[d]	92.00[b]	22.22[d]	184.80[c]
T7- Untreated control	4.20[d]	16.00[f]	70.80[e]	30.50[g]	125.00[f]

*Values are mean of three replications

Means followed by a common letter are not significantly different at 5% level by DMRT

Table 17.8. Effect of fungicides on Days taken for flower bud initiation, Days taken for flower bud opening, Duration of Flowering, Length of flower stalk, Bud circumference (at paint brush stage) in carnation variety Gaudina Red

Treatments	Days taken for flower bud initiation	Days taken for flower bud opening	Duration of flowering	Length of flower stalk (cm)	Bud circumference (at paint brush stage) (cm)
T1- Root dipping and soil drenching with carbendazim@@ 2.0g/litre	104.0[b]	153.20[b]	64.00[b]	90.57[c]	6.97[c]
T2 -Root dipping and soil drenching with Azoxystrobin @ 2.0ml/litre	108.7[c]	162.50[c]	62.70[b]	93.73[b]	7.53[a]
T3 -Root dipping and soil drenching with Pyraclostrobin @ 2.0g/litre	108.0[c]	162.60[c]	69.20[c]	93.47[b]	6.27[f]
T4- Root dipping and soil drenching with Tebuconazole + Trifloxystrobin @ 1.0 g/litre	127.0[e]	217.00[f]	93.00[f]	73.53[e]	6.00[g]
T5- Root dipping and soil drenching with Difenoconazole @ 2.0 ml/litre	102.3[a]	142.30[a]	57.20[a]	94.53[a]	6.80[d]
T6- Root dipping and soil drenching with Iprovalicarb + Probineb @ 2 g/litre	115.3[d]	185.30[d]	73.70[d]	91.20[c]	6.53[e]
T7- Untreated control	129.3[f]	211.80[e]	88.70[e]	86.27[d]	7.20[b]

*Values are mean of three replications
Means followed by a common letter are not significantly different at 5% level by DMRT

5.8. Biological control: Management of plant diseases aims in the rational use of fungicides, bactericides, biocontrol agents and the application of non-chemical methods that cause less impact to the environment. Biocontrol agents like, *Bacillus* spp., *Pseudomonas* spp., and *Trichoderma* spp., complement chemical pesticides for the successful control of several plant diseases. *Bacillus* species govern the interest of Microbiologist and Plant Pathologist all over the universe, owing to its safety, and ubiquitous distribution in diverse habitats, endospore development and long term survival. Besides, the extracellular compounds released by Bacillus species in to the rhizosphere region, it also promote plant growth and suppress plant diseases by the way of producing antimicrobial substances, plant hormones and aid in the uptake of nutrient existing in the soil. Several strains of *Bacillus* control plant diseases through antibiosis, ISR and competition for nutrient sources and space, which could be explored for the management of diseases of high valued flower crops grown under protected cultivation. Origin of antimicrobial peptides (AMPs) from *Bacillus* spp. plays a vital role in biocontrol of plant pathogens responsible for foliar, soil, and postharvest diseases (Montesinos, 2007) and in plant growth promotion (Kloepper *et al.*, 2004; Idris *et al.*, 2007). Cyclic lipopeptides based AMPs include compounds like fengycin, iturin, bacillomycin, and surfactin with intense surfactant activities and broad spectrum of antimicrobial action against several plant pathogens (Stein, 2005). Apart from lipopeptides, bacilysin a dipeptide based compound produced by *B. amyloliquefaciens* besides subtilin a lantibiotic compound produced by *B. subtilis* is also effective for the management of plant pathogens (Lee and Kim, 2011). *B. subtilis* isolate, BS2 effective against fusarium wilt of carnation has four antibiotic biosynthetic genes such as bacillomycin, fengycin, iturin A and surfactin. *B. amyloliquefaciens* isolate BSC7 has iturin and surfactin genes. *B. cereus* isolate BSC11 has iturin and surfactin genes (Nakkeeran *et al.*, 2012). Biocontrol potential of plant pathogens by different strains of *Bacillus* has been linked to the presence of AMP biosynthetic genes *bmyB, fenD,ituC, srfAA*, and *srfAB* (Gonzalez-Sanchez *et al.*, 2010) which can be well exploited for the management of diseases of flower crops due to the availability of conducive environment for the multiplication of introduce organism under protected cultivation. The simultaneous production of different AMPs with broad range of antagonistic activity and the occurrence of antibiotics as a mixture of bacillomycin, fengycin, and iturin A by *B. subtilis* has been related to the effective control of *Podosphaera fusca* in cucurbits (Romero *et al.*, 2007). In addition, combined production of bacilysin, iturin, and mersacidin by *B. subtilis* ME488 effectively suppressed *Fusarium* wilt of cucumber and blight of pepper caused by *Phytophthora* (Chung *et al.*, 2008). Accordingly, strains of *Bacillus* with multiple AMP biosynthetic genes are most effective at inhibiting the growth of *Rhizoctonia solani* and *Pythium ultimum* than other *Bacillus* isolates that lack one or more of those markers (Joshi and McSpadden-Gardener, 2006).

Similarly, delivering of *B. subtilis* isolate BS2 through root dipping and soil drenching @ 5ml/Litre recorded 1% wilt incidence with flower yield of 244.80 numbers over an area of 1m^2 and was not found to differ significantly, from the consortia of *Bacillus* spp., comprising of isolates *B. subtilis* BS2 + *B. amyloliquefaciens* BSC7 + *B. cereus* BSC11 based liquid formulation (Nakkeeran *et al.*, 2012). Some of *B. subtilis* and *B. amyloliquefaciens* strains produce surfactin, iturin A and fengycin. They showed synergistic effects on antifungal activity, which was higher than only iturin A production (Athukorala *et al.*, 2009; Arrebola *et al.*, 2010; Chen *et al.*, 2009). Likewise, in our study, *B. subtilis* isolate BS2, had bacillomycin, surfactin, fengycin and iturin genes. Besides, *B. amyloliquefaciens* isolate BSC 7 had both surfactin and iturin genes, which might be responsible for the better performance under field conditions to control vascular wilt of carnation. However, in soil drenching with consortia, significant difference in the flower yield was observed with the isolate BS2. The flower yield increased upto 7.3%, which was 262.80 numbers, which indicate that, plant growth promotion was observed to a maximum level compared to the application of BS2 alone to the rhizosphere. Besides, the population buildup of *Bacillus* was the maximum in the application of consortia, which might be due to the response of quorum sensing between the bacterial cells. Similarly, Kloepper *et al.* (2004) and Idris *et al.* (2007) reported that, antimicrobial peptides (AMPs) from *Bacillus* spp. play a vital role in biocontrol of plant pathogens responsible for foliar, soil, and postharvest diseases and in plant growth promotion. Bacteria associated in the plants along with rhizosphere promote plant growth under field conditions (Kloepper, 1992) as reflected by the increase in population dynamics of *Bacillus* spp., in the rhizosphere of the present study. The versatile and diversified bacterial antagonists namely, *Pseudomonas* and *Bacillus* species in the natural environment are diversified and may produce novel beneficial metabolites that promote plant growth and yield (Sharma and Kaur, 2010). Comparison of number of shoots and root length indicated that, shoot length was the maximum in the seedlings treated with *P. aeruginosa* isolate P1 and was followed by the application of consortia comprising of BS2+BSC7+BSC11. However, it was superior over isolate P1 in increasing the flower yield. But the plant height and length of flower stalk in the consortia comprising of isolates BS2+BSC7+BSC11 was the maximum than *P. aeruginosa* isolate P1. Comparison of duration of harvesting of flowers was retained from 42 to 55 days either between individual or combined application of *Bacillus* isolates. But, the duration extended upto 88 days in untreated control, which indicated the induction of earliness in flowering among different *Bacillus* isolates. Similarly, the bud circumference, in the carnation rooted cuttings treated with consortia of Bacillus isolates with different antibiotic genes was maximum than untreated control, indicating that *Bacillus* spp., increased the flower size due to growth promotion and disease reduction. Production of surfactin is widespread among *B. subtilis* and *B. amyloliquefaciens* which aid in cell attachment and detachment to surfaces during the formation of biofilm and in swarming motility (Raaijmakers *et al.*, 2010). Similarly, in our study, combined application of Bacillus species with antibiotic genes including surfactin, might have helped in the attachment of all the strains of *Bacillus* and resulted in the multiplication of the isolates in rhizosphere leading to the biofilm formation, thus suppress *F. oxysporum* f.sp. *dianthi* in carnation and increase plant growth promotion and soil health (Table 17.9 and 17.10). Similarly, bacyllomicin, produced by *B. amyloliquefaciens* FZB42 along with fengycin exhibited strong antifungal activity against *Fusarium oxysporum* (Koumoutsi *et al.*, 2004). Delivering of fungal biocontrol agent *Trichoderma viride* 85/1 and *T. harzianum* 658 through root dip in carnation promoted plant growth and controlled *F. oxysporum* f.sp. *dianthi* (Manka *et al.*, 1989). Mixing of oospores of *Pythium oligandrum* with peat (100 oospores/gram of peat) 10 days before carnation planting inhibited wilt pathogen (Sharma and Sharma, 2008). Soil application of *T. harzianum*, effectively reduced the incidence of *Phytophthora* root rot of carnation.

Table 17.9. Effect of PGPR on number of shoots, root length, plant height, wilt incidence and flower yield in carnation variety Gaudina Red

Treatments	No.of Shoots/ plant 4MAP	Root length (cm) 4MAP	Plant height (cm) 7MAP	Wilt incidence	Flower yield/ m^2
T$_1$-Root dipping and soil drenching of BS2@5ml/lt	6.80^b	26.40^b	98.20^a	1.00^a	244.80^b
T$_2$-Root dipping and soil drenching of BSC7 @5ml/lt	7.20^{ab}	14.20^e	90.40^c	5.50^c	227.80^c
T$_3$-Root dipping and soil drenching of BSC11@ 5ml/lt	6.40^b	16.00^d	86.50^d	8.30^d	204.60^e
T$_4$-Root dipping and soil drenching of BSC7+ BSC11@5ml/lt	7.30^{ab}	23.40^c	97.20^b	2.70^b	245.03^b
T$_5$-Root dipping and soil drenching of BS2+BSC7 +BSC11@5ml/lt	7.40^{ab}	23.20^c	98.40^a	1.00^a	262.80^a
T$_6$-Root dipping and soil drenching of P1 @5ml/lt	7.70^a	30.00^a	90.57^c	11.10^e	224.00^d
T$_7$-Root dipping and soil drenching of TV1 @5g/lt	5.60^c	14.00^e	72.80^f	16.60^f	174.00^g
T$_8$-Root dipping and soil drenching of carbendazim@ 2g/ml	7.30^{ab}	16.40^d	84.70^e	23.60^g	200.77^f
T$_9$-Uninoculated control	4.20^d	12.80^f	70.8^g	30.50^h	125.03^h

*Values are mean of three replications
Means followed by a common letter are not significantly different at 5% level by DMRT

Table 17.10. Effect of PGPR on Days taken for flower bud initiation, Days taken for flower bud opening, Duration of Flowering, Length of flower stalk, Bud circumference (at paint brush stage) in carnation variety Gaudina Red

Treatments	Days taken for flower bud initiation	Days taken for flower bud opening	Duration of Flowering	Length of flower stalk (cm)	Bud circumference (at paint brush stage) (cm)
T$_1$-Root dipping and soil drenching of 2 BS @5ml/lt	92.00 ^{a}	119.90 ^{a}	42.80 ^{a}	98.00 ^{a}	7.03^b
T$_2$-Root dipping and soil drenching of BSC7@5ml/lt	96.50 ^{b}	126.80 ^{b}	42.00 ^{a}	92.00 ^{c}	7.05 ^{b}
T$_3$-Root dipping and soil drenching of BSC11@5ml/lt	96.00 ^{b}	142.00 ^{d}	55.40^b	80.40 ^{e}	6.97 ^{bc}
T$_4$-Root dipping and soil drenching of BSC7+BSC11@5ml/lt	93.10 ^{a}	127.20 ^{b}	44.03 ^{a}	96.30 ^{b}	7.25 ^{a}
T$_5$-Root dipping and soil drenching of BS2+BSC7+BSC11@5ml/lt	93.20 ^{a}	121.50 ^{a}	42.80 ^{a}	99.50 ^{a}	7.20 ^{a}
T$_6$-Root dipping and soil drenching of P1 @5ml/lt	96.50 ^{b}	136.60 ^{c}	46.60 ^{a}	81.90 ^{e}	6.90 ^{bc}
T$_7$-Root dipping and soil drenching of TV1 @5g/lt	118.00 ^{d}	198.20 ^{f}	80.40 ^{d}	90.50 ^{d}	6.63 ^{d}
T$_8$-Root dipping and soil drenching of carbendazim@ 2g/ml	104.20 ^{c}	153.20 ^{e}	64.00 ^{c}	92.80 ^{c}	6.60 ^{d}
T$_9$-Untreated control	129.30 ^{e}	211.80 ^{g}	88.70 ^{e}	68.33 ^{f}	6.00 ^{e}

*Values are mean of three replications
Means followed by a common letter are not significantly different at 5% level by DMRT

REFERENCES

Ann, P.J., Kunimoto, R., and Ko, W. H. 1990. Phytophthora wilt of carnation in Taiwan and Hawaii. Plant Protection Bull. 32: 145-157.

Arbelaez, G. 1987. Fungal and bacterial diseases on carnation in Colombia. Acta Hort. 216: 151-157.

Arbelaez, G. 1988. Primer curso internacional sobre patógenos vasculares del clavel. Enfermedades vasculares del clavel en Colombia: aspectos históricos y situación actual. Acta Hort. 216: 77-84.

Arbelaez, G. 1999. Overview of the cut flowers pathology in Colombia. Acta Hort. 482: 91-96.

Arbeláez, G., Granada, E., and Acosta, A. 1984. El nemátodo *Heterodera trifolii* G.; una nueva enfermedad del clavel en Colombia. Agronomía Colombiana. 2: 97-100.

Arrebola, E., Jacobs, R., and Korsten, L. 2010. Iturin A is the principal inhibitor in the biocontrol activity of *Bacillus amyloliquefaciens* PPCB004 against postharvest fungal pathogens. J. Appl. Microbiol. 108: 386-395.

Athukorala, S. N. P, Fernando, W. G. D., and Rashid, K. Y. 2009. Identification of antifungal antibiotics of *Bacillus* species isolated from different microhabitats using polymerase chain reaction and MALDI-TOF mass spectrometry. Can. J. Microbiol. 55: 1021-1032.

Baayen, R. P., and Maat, W. 1987. Passive transport of microconidia of *Fusarium oxysporum* f. sp. *dianthi* in carnation after root inoculation. Netherlands J. Plant Pathol. 94: 273-288

Baayen, R. P., Van Dreven, F., Krijger, M. C., and Waalwijk, C. 1997. Genetic diversity in *Fusarium oxysporum* f. sp. *dianthi* and *Fusarium redolens* f. sp. *dianthi*. Eur. J. Plant Pathol. 103: 395-408.

Ben-Yephet, Y., Reuven, M., Lampel, M., Nitzani, Y., and Mor, Y. 1992. *Fusarium oxysporum* f. sp. *dianthi* races in carnation(Abstr.). Phytoparasitic. 20: 225.

Ben-Yephet, Y., Reuven, M., and Genizi, A. 1994. Effects of inoculum depth and density on Fusarium wilt in carnations. Phytopathology. 84: 1393-1398.

Booth, C. 1971. The genus *Fusarium*. Common wealth Mycological Institute, Kew Surrey, UK, pp.142-143.

Burbano, L. E., Erazo, A., Amezquita, M., and Granda, E. 1990. Effect of nitrogen fertilization on the incidence of Fusarium oxysporum f.sp. dianthi and Heterodera trifollii in carnation. Agronomia Colombiana. 7: 61-69.

Chen, X. H, Koumoutsi, A., Scholz, R., Schneider, K., Vater, J., Sussmuth, R., Pie, l. J., and Borriss, R. 2009. Genome analysis of *Bacillus amyloliquefaciens* FZB42 reveals its potential for biocontrol of plant pathogens. J. Biotechnol. 140: 27-37.

Chung, S., Kong, H., Buyer, J. S., Lakshman, D. K., Lydon, J., Kim, S. D., and Roberts, D. P. 2008. Isolation and partial characterization of *Bacillus subtilis* ME488 for suppression of soil borne pathogens of cucumber and pepper. Appl. Microbiol. Biotechnol. 80: 115-123.

Daboussi, M. J., and Langin, T. 1994. Transposable elements in the fungal plant pathogen *Fusarium oxysporum*. Genetica. 93: 49-59.

Diatloff, A., and Rochecouste, J. 1991. The pattern of spread of bacterial leaf spot of carnations in a commercial field crop. Aust. Plant Pathol. 20(1): 27-30.

Duskova, E., and Kovacikova, E. 1989. Occurrence of Fusarium wilt depending on substrate composition. Sbornik UVTIZ, Zahradnictvi. 16: 47-53

Elena, K., and Tjamos, E. C. 1997. Soil solarization for the control of Fusarium wilt of greenhouse carnation. Phytopathologia Mediterranea. 36: 87-93.

Ferrin, D. M., and Rohde, R.G. 1991. Tests compare fungicides for control of rust on greenhouse carnations. Cal. Ag. 45: 16-17.

Filippi, C., and Bagnoli, G. 1992. A relation between nitrogen deficiency and protective effect against tracheo-fusariosis (*Fusarium oxysporum f.sp. dianthi*) in carnation plants. Zentralblatt fur Mikrobiologie. 147: 345-350

Garibaldi, A. 1983. Resistenza di cultivar di garofano nei confronti di otto patotipi di *Fusarium oxysporum* f. sp. *dianthi* (Prill. *et* Del.) Snyd. *Et* Hans. (in Italian) *Riv.* Ortoflorofrutti. Ital. 67: 261-270.

Garibaldi, A., and Tamietti, G. 1989. Solar heating: recent results obtained in northern Italy. Acta Hort. 255: 125-130.

Garibaldi, A., Lento, G., and Rossi, G. 1986. *Fusarium oxysporum* f. sp. *dianthi* in Liguria. Indagine sulla diffusione dei diversi patotipi nelle colture di garofano. (in Italian) Panorama Floricolo. 11: 1-4.

Gian Aggarwal. 2011. Indian Greenhouse Industry. Floriculture Today. 16 (1): 30-31.

Gonzalez - Sanchez, M. A., Perez-Jimenez, R. M., Pliego, C., Ramos, C., De Vicente, A., and Cazorla, F. M. 2010. Biocontrol bacteria selected by a direct plant protection strategy against avocado white root rot show antagonism as a prevalent trait. J. Appl. Microbiol. 109: 65-78.

Gullino, M. L., Gilardi, G., and Garibaldi, A. 2000. Activity of strobilurins against three soil-borne pathogens of carnation. Gent. 65(2B): 733-737.

Hood, J. R., and Stewart, R. N. 1957. Factors affecting symptom expression in fusarium wilt of dianthus. Phytopathology. 47: 173-178.

Idris, E. E., Iglesias, D. J., Talon, M., and Borriss, R. 2007. Tryptophan-dependent production of indole-3-acetic acid (IAA) affects level of plant growth promotion by *Bacillus amyloliquefaciens* FZB42. Mol. Plant Microbe. Interact. 20: 619-626.

Jafar Naqvi. 2011. Editorial. Floriculture Today. 16(2): 8.

Joshi, R., and McSpadden-Gardener, B. B. 2006. Identification and characterization of novel genetic markers associated with biological control activities in *Bacillus subtilis*. Phytopathology. 96: 145-154.

Katoch, R. 1999. Studies on Fusarium wilt of Carnation. M.Sc. Thesis, Univ. Horti and Forestry, Nauni, Solan, (India). pp. 53.

Kloepper, J. W. 1992. Plant growth-promoting rhizobacteria as biological control agents. In: Soil microbial ecology: applications in agricultural and environmental management. Ed. F. B. Metting, Jr. pp. 255–274.

Kloepper, J. W., Ryu, C. M., and Zhang, S. 2004. Induced systemic resistance and promotion of plant growth by *Bacillus* spp. Phytopathology. 94: 1259-1266.

Koumoutsi, A., Chen, X. H., Vater, J., and Borriss, R. 2004. DegU and YczE positively regulate the synthesis of bacillomycin by *Bacillus amyloliquefaciens* strain FZB42. Appl. Environ. Microbiol. 73: 6953–6964.

Lee, H., and Kim, H. Y. 2011. Lantibiotics, Class I bacteriocins from the genus *Bacillus*. *J. Microbiol.* Biotechnol. 21: 229–235.

Manka, M., and Fruzynska-Jozwiak, D. 1989. An attempt of biological control of glasshouse carnation Fusarium wilt. Acta Hort. 25: 255-259.

Montesinos, E. 2007. Antimicrobial peptides and plant disease control. FEMS Microbiol. Lett. 270: 1-11

Nakkeeran, S., Indhumathi, T., Dinesh, D., Jawaharlal, M., Renukadevi, P., Chandrasekar, G., and Jonathan, E. I. 2012. Effect of liquid formulations of Bacillus spp., Pseudomonas aeruginosa and talc based formulation of Trichoderma viride for the management of carnation wilt induced by Fusarium oxysporum f.sp. dianthi under protected cultivation. In, Second International Symposium on Biopesticides and Ecotoxicological Network (ISBioPEN), September 24-26, 2012, Bangkok, Thailand, organized by Departmetn of Zoology, Faculty of Science, Kasetsart University, Bangkok. P- 48.

Navas-Becerra, J. A., Vela-Delgado, M. D., Lopez-Rodringuez, M., Basallote-Ureba, M. J., Prados-Ligero, A., Zea-Bonilla, T., Lopez-Herrera, C. J., and Melero- Vara, J. M. 2000. Comparative effectiveness of several methods of soil disinfestation to control Fusarium wilt of carnation. Acta Hort. 532: 247-251.

Orozco de Amezauita, M., Garces Degranada, E., and Arbelaez. Y. G. 1993. Efecto de diferentes niveles de nitrógeno, potasio Y pHen el desarrollo de *Fusarium oxysporum* f. sp. *dlanthl*, agente causal del marchitamiento vascular del clavel. Agronomía Colombiana. 10: 90-102.

Patil, R. T., Reddy, B. S., Jholgiker, P., and Kulkurni, B. S. 2004. Correlation studies in carnation. J. of .Orn.Horti. 7(3-4): 7-10.

Raaijmakers, J. M., De Bruijn, I., Nybroe, O. and Ongena, M. 2010. Natural functions of lipopeptides from *Bacillus* and *Pseudomonas*: more than surfactants and antibiotics. FEMS Microbiol. Rev. 34: 1037-1062.

Romero, D., De Vicente, A., Rakotoaly, R. H., Dufour, S. E., Veening, J. W., Arrebola, E., Cazorla, F. M., Kuipers, O. P., Paquot, M., and Perez-Garcia, A. 2007. The iturin and fengycin families of lipopeptides are key factors in antagonism of *Bacillus subtilis* toward *Podosphaera fusca*. Mol. Plant-Microbe. Interact. 20: 430-440.

Sebastian, B., Alberto, G., Emillo, A. C., Jose, A. F., and Juan, A. F. 2002. Growth development and colour response of potted Dianthus caryophyllus to paclobutrazol treatment. Sci. hort. 1767: 17.

Sharma., and Kaur. 2010. Antimicrobial activities of rhizobacterial strains of *Pseudomonas* and *Bacillus strains* isolated from rhizosphere soil of carnation (*Dianthus caryophyllus* cv. Sunrise). Indian J. Microbiol. 50: 229-232.

Sharma, S., and Sharma, N. 2008. Carnation diseases and their management – A review. 2008. Agric. Rev. 29: 11-20.

Smith, I. M., Dunez, J., Phillips, D. H., Lelliot, R. A., and Archer, S. A. 1988. European Handbook of Plant Diseases. Blackwell Scientific Publications, Oxford, UK, pp. 583.

Sohi, H. S. 1992. Diseases of ornamental plants in India. Publication and information division, Indian Council of Agricultural Research, Krishi Anusandhan Bhavan, New Delhi, India. pp. 46.

Stein, T. 2005. *Bacillus subtilis* antibiotics: structures, syntheses and specific functions. Mol. Microbiol. 56: 845-857.

Trujillo, E. E., and Nagata, N. M. 1994. Bacterial Blight of Carnation Caused by Pseudomonas woodsii and Susceptibility of Carnation Cultivars. Plant Dis. 78: 91-94.

Xu, S. S., Friesen, T. L., and Mujeeb-Kazi, A. 2004. Seedling resistance to tan spot and *Stagonospora nodorum* blotch in synthetic hexaploid wheats. Crop Sci. 44: 2238-2245.

Zhao, D., Seip, H. M., Zhao, D., and Zhang, D. 1995. Pattern and cause of acidic deposition in the Chongqing region, Sichuan Province, China. Water, Air and Soil Pollution. 77: 27-48.

Diseases of Field and Horticultural Crops
Editor-in-Chief: **P. Chowdappa**
Published by: **Daya Publishing House, New Delhi**

Pages **506-527**

18

Diseases of Chrysanthemum

S. Nakkeeran., R. Dheepa., P. Renukadevi, M. Jawaharlal and P. Chowdappa

Chrysanthemum (*Chrysanthemum morifolium*), the "Queen of East" is a leading commercial flower crop, grown for cut and loose flowers. The Genus Chrysanthemum belongs to the family Asteraceae and comprises around 200 species. It is grown globally, owing to its beauty and economic value. Apart from ornamental values, the flowers of 52 varieties of chrysanthemum are edible. Globally, chrysanthemum is cultivated over an area of 3,05,105 ha (Bhattacharjee and De, 2003).

In India, chrysanthemum is commercially cultivated in Bangalore (Karnataka), Pune (Maharashtra), Kolkata (West Bengal), Tamil Nadu, Punjab, Rajasthan, Gujarat, Delhi, Shimla and Solan. The area under flower crops in Tamil Nadu is around 20,274 ha with annual production of 1, 61,655 tonnes with 9 tonnes productivity per ha. In Tamil Nadu, chrysanthemum is grown over an area of 1724 ha with the production and productivity of 15516 tonnes and 9 tonnes per ha respectively. The potential areas of cut chrysanthemum in Tamil Nadu are Nilgiris, Yercaud, Coimbatore and Kodaikanal. Cultivation of cut chrysanthemum is slowly picking up at Yercaud and Nilgiris under protected cultivation. The crop duration is around 70 to 120 days. The seedlings of cut chrysanthemum are imported from other countries and cultivated commercially. It has resulted in the outbreak of white rust, stem blight, wilt and crown gall which are the major determinants of yield reduction (Table 18.1).

1. Foliar Diseases

1.1. *Septoria* leaf spot (*Septoria chrysanthemi*): Leaf spot is widely distributed in all chrysanthemum growing tracts of the World. The disease has wide geographical distribution in Asia and European continents. Disease is prevalent in United States since 1891. The quality and marketability of the blooms grown as

Table 18.1. Diseases of chrysanthemum

S.No	Name of the disease	Causal organism
	A.Fungal Diseases	
1	*Alternaria* leaf spot	*Alternaria* sp.
2	*Septoria* leaf spot	*Septoria chrysanthemi*
		Bipolaris setaria (Sawada) Shoemaker
3	*Bipolaris* leaf spot	*Cochliobolus setariae* (Ito & Kuribayashi in Ito) Drechs. ex Dastur [teleomorph]
4	*Cercospora* leaf spot	*Cercospora chrysanthemi* Heald & F. A. Wolf
5	Charcoal stem rot	*Macrophomina phaseolina* (Tassi) Goidanich
6	*Cylindrosporium* leaf spot	*Cylindrosporium chrysanthemi* Ellis & Dearn.
7	*Fusarium* wilt	*Fusarium oxysporum* Schlechtend.: Fr. f. sp. *chrysanthemi* G. M. Armstrong *et al.* or
		F. oxysporum Schlechtend.: Fr. f. sp. *tracheiphilum*(E. F. Sm.) W.C. Snyder & H. N. Hans. (race 1)
8	Gray mold	*Botrytis cinerea* Pers.: Fr.
		Botryotinia fuckeliana (de Bary) Whetzel [teleomorph]
9	*Itersonilia* petal blight	*Itersonilia perplexans* Derx
10	*Phymatotrichum* root rot	*Phymatotrichum omnivorum* Duggar
11	Powdery mildew	*Erysiphe cichoracearum*
12	Ray blight	*Didymella ligulicola* (K. Baker *et al.*) Arx in E. Müller & Arx
		Phoma chrysanthemi Voglino [anamorph]
		= *Ascochyta chrysanthemi* F. Stevens
13	Ray speck	*Stemphylium lycopersici* (Enjoji) W. Yamamoto
		= *S. floridanum* Hannon & Weber
14	*Rhizoctonia* stem rot	*Rhizoctonia solani* Kühn
		Thanatephora cucumeris (A. B. Frank) Donk [teleomorph]
15	Root rot	*Phoma chrysanthemicola* Hollos
16	Rust	*Puccinia tanaceti* DC.
17	*Sclerotinia* rot	*Sclerotinia sclerotiorum* (Lib.) de Bary
18	*Pythium* root rot	*Pythium* spp.
		P. ultimum Trow
19	Southern blight	*Sclerotium rolfsii* Sacc.
20	Stem rot	*Fusarium solani* (Mart.) Sacc.
		Nectria haematococca Berk & Broome [teleomorph]
21	*Verticillium* wilt	*Verticillium albo-atrum* Reinke & Berthier
		V. dahliae Kleb.
22	*Fusarium* wilt	*Fusarium oxysporum* f. sp. *chrysanthemi* and *F. oxysporum*.f. sp. *tracheiphilum*
23	White rust	*Puccinia horiana* Henn.
	B.Bacterial Diseases	
24	Bacterial blight	*Erwinia chrysanthemi* Burkholder *et al.*
25	Bacterial leaf spot Marginal necrosis, Water soaked spots, blighting.	*Pseudomonas cichorii* (Swingle) Stapp
26	Crown gall	*Agrobacterium tumefaciens* (Smith & Townsend) Conn
27	Stem necrosis	*Pseudomonas cichorii* (Swingle) Stapp
28	Fasciation	*Rhodococcus fascians* (Tilford) Goodfellow
		= *Corynebacterium fascians* (Tilford) Dowson
	C.Viral diseases	
29	Aspermy & Chrysanthemum mosaic	Genus *Cucumovirus, Tomatoaspermy virus* (TAV)
30	Spotted wilt	Genus *Tospovirus, Tomato spotted wilt virus* (TSWV)
31	Chrysanthemum chlorotic mottle	Chrysanthemum chlorotic mottle viroid
32	Chrysanthemum stunt	Chrysanthemum stunt viroid
	D.Phytoplasmal diseases	
33	Aster yellows	
	E.Nematodes	
34	Bud and leaf	*Aphelenchoides ritzemabosi* (Schwartz) Steiner & Buhrer
35	Lesion	*Pratylenchus pratensis* (de Man) Filipjev
36	Root-knot	*Meloidogyne* spp.

cut flower is severely affected due to the infection of septoria leaf spot (Wadel and Webber, 1963). Initially the spots appear on the lower leaves and progress upward.The leaf spots are characterized with brown irregular spots surrounded by yellow halo. The spots coalesce into large necrotic areas. During severe cases of infection the entire leaf dry and persist on the leaf and lead to death of the plants (Fig. 18.1). Pathogen infects during moist weather with increased relative humidity. The infection is faster in the presence of the moisture adhering on to leaf. Pathogen produces pycnidia on the affected leaf surface. The conidia are disseminated to the neighbouring plants and within the crop canopy through rain splash. The pathogen germinates; germtube enters through stomatal opening and establishes the infection. The fungus overwinters upto 4 months on the leaf debris and serves as the source of inoculum for further spread of the disease (Alferi, 1968). Foliar spraying with mancozeb @ 0.25% at weekly intervals or azoxystrobin @ 0.1% reduces the disease incidence. Besides, during the rainy season and humid weather spraying on alternate days is found to be effective. Removal of infected crop debris and safe disposal helps in reducing the further spread of the disease. Care should be taken to always keep the leaf surface free from moisture. Other fungicides like chlorothalonil, myclobutanil, propiconazole, or thiophanate methyl may also be applied depending on the disease severity.

Fig. 18.1 Septoria leaf spot

1.2. *Alternaria* Leaf Spot (*Alternaria* spp.): Infections are noticed on the lower most leaves. The symptoms are characterized with necrotic spots on the leaves. Initially the spots are yellowish, and later turn dark brown and black. Prematured defoliation of the affected leaves are also noticed. Sporulation of the pathogen can be noticed as a white mass of spores on the affected areas during humid conditions (Fig. 18.2a, b).Pathogen survives on the infected crop debris. Over head irrigation helps in the spread of the conidia. Disease can be managed by removing the infected leaves and dead plant debris. Pathogen overwinters as spores in the affected debris. During severe conditions, foliar application of fungicides like chlorothalonil or mancozeb or myclobutanil or propiconazole or thiophanate methyl at periodical intervals, helps in the management of the disease (Dreistadt, 2001).

Fig. 18.2 *Alternaria* leaf spot.

1.3. Powdery mildew (*Erysiphe cichoracearum***)** : Disease is characterized by the colonization of the pathogen on the upper surface of the leaf as a white to ash-gray powdery growth. Infections start as small spots and rapidly expand to encompass the entire bud or flower. During severe conditions the powdery growth also spreads to the stems. Infected foliages are puckered and losses its shape and get distorted. Severely infected leaves will shrivel and get defoliated (http: // urbanext.illinois.edu/focus/per_chrysanthemum.cfm). The disease is serious during hot, humid weather. Compared to the other fungal diseases, free water is not a prerequisite for the infection by powdery mildew. Increased relative humidity coupled with dry weather favours the disease development. Rain splash or splashing water on to the crop raised under protected cultivation spreads the disease. Besides, fallen crop residues serves as the inoculum source for the subsequent spread of the pathogen (http: //www.clemson.edu/extension/hgic/ pests/plant_pests/flowers/hgic2101.html; (http: //www.apsnet.org/publications/ commonnames/Pages/Chrysanthemum.aspx).Damage created by powdery mildew pathogen can be avoided by increasing the spacing between the plants. Adoption of 15cm x15cm and 20cm x 20cm reduce the humidity rather than the crop raised with a spacing of 10 cm x 10cm. It increases air circulation and reduces the buildup of humidity and arrest the survival rate of the pathogen. Foliar application of fungicides after the initial appearance of the sign and symptom reduces the disease severity. Foliar application of any one of the fungicides like azoxystrobin (0.15%), pyraclostrobin (0.1%), tebuconazole (0.1%), probineb (0.25%) and mancozeb (0.25%), at weekly intervals helps in better management of the disease (Personal Communication).

1.4. Gray mold (*Botrytis cinerea)*** : Pathogen infects leaves, stem and petals. Disease appears as brown water soaked spots on leaf, petal or stem. Brown cankerous growth appears on the stem. During cloudy, humid weather coupled with increased relative humidity, the infected portions will be covered with gray to brown, woolly powdery masses of spores. Senescing tissues are most susceptible. Severe infection leads to rotting of lower most leaves and the pathogen gain entry into the affected stem tissues and girdle the plant. The yield, quality and marketability of the chrysanthemum crop raised under protected cultivation are severely affected (http://www.apsnet.org/publications/commonnames/Pages/Chrysanthemum.aspx). Conidia on the infected plant debris serve as the inoculum source for the secondary spread of the disease. During adverse conditions pathogen survive as resting structures in the form of sclerotia and spread the disease to the subsequent crop. Closer spacing and dense crop canopy with poor air circulation and high relative humidity favours the disease spread (Dreistadt, 2001). Removal of affected, senescing leaves and good sanitation reduces the disease severity. Care should be taken to keep the leaf surface free from the moisture. Avoiding closer spacing and excess watering on to the flowers also reduce the disease severity. Proper ventilation with improved air circulation and heating reduces the humidity and prevents the disease spread. During severe cases of infection foliar spray with fungicides immediately after initial appearance of the disease with fungicides like mancozeb, chlorothalonil, trifloxystrobin, difenaconazole, iprodione, azoxystrobin, and thiophanate methyl gives better control of the disease. Based on the disease severity the frequency of application of the fungicide has to be standardized. Crop rotation with non-host crops also suppresses the disease spread (http://extension.psu.edu/plant-disease-factsheets/all-fact-sheets/chrysanthemum-diseases).

1.5. Rusts (*Puccinia chrysanthemi*): Rust is mostly predominant in summer season. Disease is characterized with the presence of dirty brown small pustules on the lower surface of the leaf. Correspondingly on the upper surface yellowish green spots can be noticed. It is mostly seen in the crops raised under field conditions but not noticed under greenhouse condition. During conducive stage all the pustules coalesce and damage most of the leaves leading to defoliation (Dreistadt, 2001). Powdery mass of uredospores released from the infected living plant tissues serve as the inoculum source to spread the disease. Low temperature, high relative humidity coupled with the prevalence of moisture on the leaf surface is essential for further spread of the disease. Free air circulation, non-availability of moisture on the leaf surface prevents the establishment of host pathogen interaction. Periodical removal of the infected leaves, improved ventilation also helps in reducing the disease severity. During severe cases of infection foliar spray with fungicides like mancozeb (0.2%) or probineb (0.25%) or tebuconazole (0.1%) or triadimefon (0.1%) spray either at weekly or fortnightly intervals based on the disease severity helps in the management of the disease.

1.6. White Rust (*Puccinia horiana*)

1.6.1. Symptoms: Following infection, pale-green to yellow spots, up-to 5 mm in diameter, develop on the upper surface of the leaves. The centres of these spots become brown and necrotic with aging. On the corresponding lower surface, raised, buff or pinkish, waxy pustules (telia) are found. As the spots on the upper surface become sunken, these pustules become quite prominent and turn whitish when basidiospores are produced. Telia are occasionally found on the upper leaf surface. Severely attacked leaves wilt, hang down the stem and gradually dry up completely (Fig. 18.3 a, b, c, d, e, f, g). On bracts and stems, sori sometimes develop when crops are excessively affected (Fig. 18.4a, 18.4b). On flowers, Infection has been recorded as necrotic flecking with occasional pustules (Dickens, 1970).

Fig. 18.3a. Yelllow circular lesions

Fig. 18.3b. Rust infected foliage

Fig. 18.3c. Necrotic spots on rust infected leaves

Fig. 18.3d. Severe infection of rust on upper surface

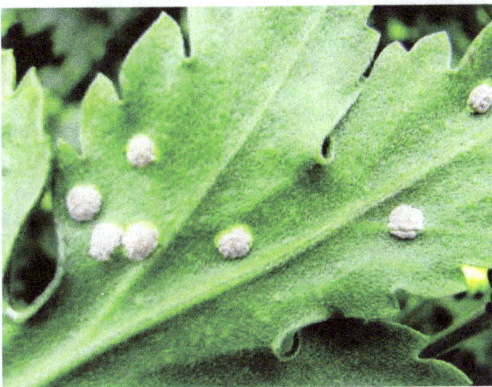

Fig. 18.3e Rust pustules on lower leaf surface

Fig. 18.3f. Leaf colonized with pustules

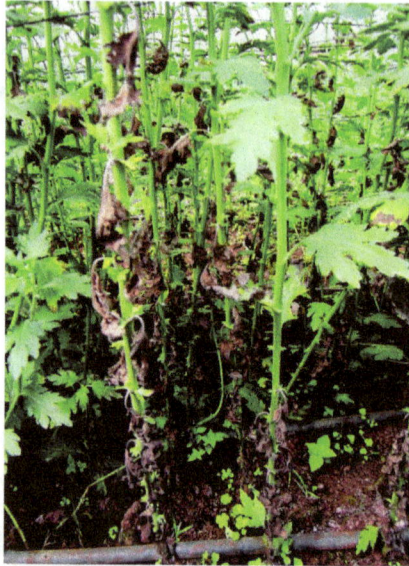

Fig. 18.3g. Drying of rusted leaves

Fig. 18.4a Rust pustules on stem

Fig. 18.4b Rust pustules on stem

Table 18.2. List of susceptible varieties in Tamil Nadu

S. No	Variety	White rust	Crown gall
1	Bonfire yellow	Susceptible	Nil
2	Bonfire orange	Susceptible	Nil
3	Kuga Bright Orange	Susceptible	Nil
4	Vanity Pink	Susceptible	Susceptible
5	Saffin Pink	Susceptible	Susceptible
6	Brighton Pink	Susceptible	Nil
7	Terror Red	Resistant	Nil
8	Punch White	Susceptible	Susceptible

1.6.2. Causal organism: Telia hypophyllous, rarely epiphyllous, compact, pinkish-buff to white, 2-4 mm in diameter (Fig. 18.5a, 18.5b). Teliospores on pedicels are upto 45 μm long; pale-yellow, oblong to oblong-clavate, slightly constricted, 30-45 x 13-17 μm, with thin walls, 1-2 μm thick at sides, thicker commonly 4-9 μm at apex. Germination of teliospores can be observed *in situ*. Basidiospores hyaline, slightly curved, broadly ellipsoid to fusiform, 7-14 x 5-9 μm. (Hennings, 1901; Baker, 1967; Firman & Martin, 1968).

| Fig. 18.5a. Teleospores | Fig. 18.5b. Teliospore |

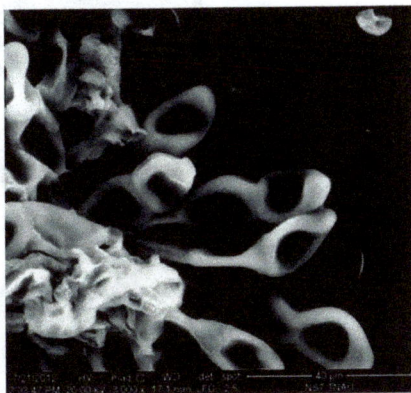

1.6.3. Epidemiology: New infections are initiated by basidiospores released from pustules during periods of high relative humidity (96 to 100%) when temperatures are between 40 and 73°F (optimum 63°F). Spores landing on a plant surface can germinate and penetrate within two hours at optimum temperatures (63 to 75°F). A film of free water is required for infection. For 5 to 14 days after infection the fungus grows within the plant as a latent infection, after which chlorotic (yellow) spots appear on upper surface of leaf, and ultimately pustules, appear. Teliospores produced in pustules remain attached to leaves, germinating to produce the next generation of basidiospores when conditions of temperature and humidity are favourable. Long distance dispersal of white rust depends on movement of infected plant material. Because, cuttings may not display symptoms for as long as two weeks after infection, apparently healthy cuttings are not guarantee of safety. This is the basis for the requirement for six-month post entry quarantine for imported cuttings.

Basidiospores can be carried to a short distances by wind currents. These spores are so short-lived that even under ideal conditions (100% humidity, cool temperatures) they only survive long enough to be carried a short distance before dying due to desiccation. Inspite of their very brief lives, basidiospores are responsible for the explosive character of local epidemics when conditions are right. Infested debris carrying viable teliospores may also play a vital role in dispersal. White rust survives for extended periods in association with host plant tissue. Basidiospores (the airborne spores) are thin walled and subjected to rapid desiccation. They are thus extremely short-lived, surviving only 5 minutes at 80% relative humidity and less than one hour at 90% relative humidity. Teliospores (the spores that remain attached to the leaf) associated with dried plant debris survive for at most two weeks when buried in air dried soil (Fig. 18.6).

Fig. 18.6. Rust infected host debris

1.6.4. Host range: The host range of white rust extends to a number of species of chrysanthemum and their close relatives. Since 1990 white rust has been detected in the United States in commercial and residential plantings of the florist's chrysanthemum and chrysanthemum *(Chrysanthemum x morifolium* Ramat. or *Dendranthema x morifolium* (Ramat.) Tzvelev, the perennial garden plant *Chrysanthemum pacificum* Nakai (synonym *Ajania pacifica* (Nakai) K. Bremer & Humphries), and also the Nippon or Montauk daisy (*Nipponanthemum nipponicum* (Franch. ex Maxim.) Kitam. Some cultivars of chrysanthemum appear to be resistant to some races of white rust. When exposed to other races, however, they may be quite susceptible. Thus, cultivar selection is not a reliable approach for the control of this disease.

1.6.5. Management: Exclusion is our first and best means of defense against CWR. Use of disease free cuttings prevents the disease spread. Cut chrysanthemums should not be mixed with growing plants. This means do not have them in the same facility or within 400 m (preferably >700m). Cut chrysanthemums, especially from off shore countries where CWR is established, are a major threat to growing plants because they can bring inoculum and does not show symptoms. Cultural control also helps in the suppression of white rust. Adoption of increased plant spacing 20cm × 20cm and avoiding over head irrigation followed by the removal of leaves from basal portion of the soil to a height of 30 cm increases the air circulation and reduces the relative humidity. This in turn prevents the establishment of the rust pathogen (Fig. 18.7). Removal of the rust pustules infected leaves reduce the inoculums pressure. In addition to exclusion, eradication and cultural control principles, a fungicide program should be used for protection against CWR. Some fungicides are better for CWR eradication while others are better for protectants.

Fig. 18.7. Removal of basal leaflet for rust

Azole fungicides are effective against CWR for the past 19 years (Wojdyla, 2002). Ammerman *et al.* (1992) and Brunelli *et al.* (1996) reported that natural products like strobilurin A, strobilurin B, oudemansin A and myxothiazol; inhibit mitochondrial respiration by blocking electron transfer between cytochrome C1. Azoxystrobin showed excellent effectiveness in the control of *P. horiana.* (Wojdyla 1999; Wojdyla and Oriikowski 1999). Benodanil, bitertanol, oxycarboxin, propiconazole, triadimefon and triforine was effective for the control of chrysanthemum white rust (Grouet & Allaire, 1973; Dirkse, 1980; Krebs, 1985; Dickens, 1990). Rattink *et al* (1985) reported that active ingredients found useful include oxycarboxin, triforine, benodanil, triadimefon, diclobutrazol, dibitertanol and propiconazole.

Myclobutanil, which acts as a sterol biosynthesis inhibitor, can be used to eradicate early infections of CWR. Routine use of this fungicide on chrysanthemums is not recommended. It is not as effective as some other fungicides as a protectant for CWR and resistance to this class of pesticides is a high risk. Discussion with cut chrysanthemum growers in Tamil Nadu revealed that, fungicides like krexoxym methyl, azoxystrobin, tebuconazole, propiconazole, hexaconazole, iprodione, thiophanate methyl, mancozeb and carboxin are used for the management of CWR. The most difficulty of controlling *P. horiana* on chrysanthemum is experienced during the late summer and autumn when temperature and moisture are favourable for fungus development. In these periods chemical control is necessary to prevent heavy losses (Baker, 1967; Firman and Martin, 1968). Foliar application of triforine seems to be one of the most useful one in chrysanthemum protection against CWR (Wojdyla and Orlikowski, 1981). Oxycarboxin was reported to be very effective in the control of CWR (Gjaerum, 1979 and Dirkse, 1980).

1.7. Bacterial leaf spot (*Pseudomonas cichorii*): Bacterial leaf spots are characterized with the presence of black to dark brown spots or blotches surrounded with yellow halo. The infection is noticed initially on the lower surface of the leaf. Subsequently it spreads to the upper portion of the plant and later on to the flowers. The spots enlarge and become irregular in shape. Infected leaves get necrotized and dry. The affected portion turns brittle and crack. Since the growth of the bacteria is limited along the veins, the lesions are angular in shape. It is followed by wilting of leaves and by death of the affected plants (Fig. 18.8). Bacteria persist in or on infected plants, crop debris, infected seed, contaminated soil, and infested pots and tools. Use of disease free seeds or disease free planting material prevents the primary spread of the disease. Cultivation of resistant varieties followed by good sanitation, and avoiding overhead irrigation prevents the disease spread. Removal of infected plants and neighbouring plants helps in preventing the subsequent spread of the disease. Foliar application of copper fungicides and bacterial antibiotics suppress the disease spread.

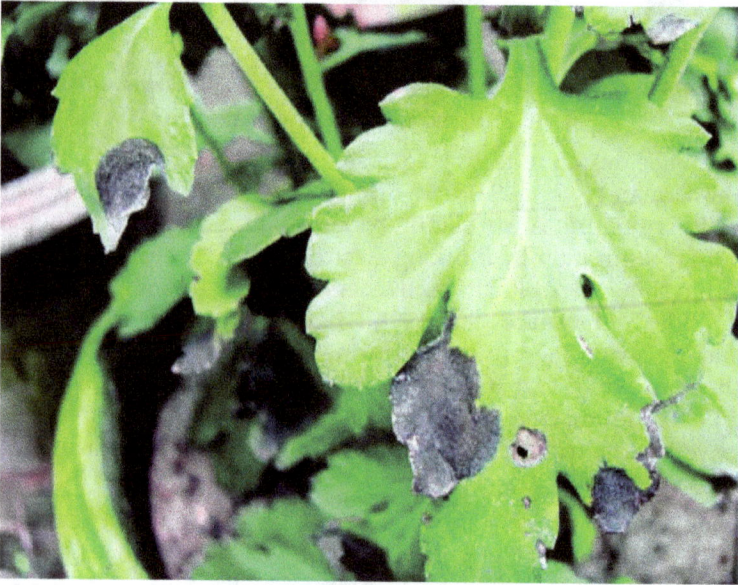

Fig. 18.8. Bacterial leaf spot

1.8. Bacterial blight (*Erwinia chrysanthemi*): Disease is characterized by the presence of water soaked lesions on leaves and extends to stems, leading to the rotting of the infected cuttings. Affected plants exhibit decay at the base of the stem. Infected mother plants in the nursery or the cuttings raised under greenhouse turn dark brown and collapse. In few cases marginal leaf scorching can also be noticed. The infected cuttings will be surviving without expressing the symptoms. Established plants in the greenhouse infected with the pathogen wilt during day time and recover during night. Subsequently it leads to death of the seedlings (Fig 18.9 a, b). Pathogen survives in crop debris and in the soil. Pathogen is favoured by high soil moisture, temperature and humidity. Over head irrigation spreads the pathogen through water splashes to the neighboring plants. Infected tools, soils adhering to the shoes and equipments facilitate the pathogen spread (Dreistadt, 2001). Use of disease free cuttings prevents the establishment of the disease. Dipping the cuttings in the antibiotic solution (Agrimycin/ Streptomycin) for 4 hours avoid the establishment of the pathogen in the nurseries. In the greenhouses, soil in which diseased plants grew should be pasteurized. Reduced moisture in the soil surface, avoiding over head irrigation, reduction in humidity, increased air circulation, avoiding the presence of moisture or wetting of the foliage, increased air circulation, rouging of the infected plants and safe disposal by imposing proper sanitary measures will help in the management of bacterial infection by *Erwinina chryanthemi* (Dreistadt, 2001).

Fig. 18.9a. Erwinia Stem blight-wilting

Fig. 18.9b. Stem blight infected stem

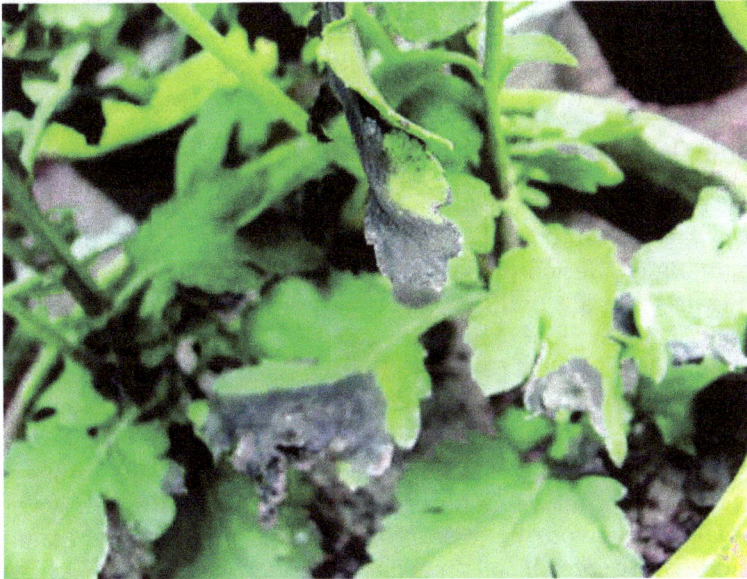

Fig. 18.10. Nematode

1.9. Foliar nematodes (*Aphelenchoides ritzema-bosi*) : Plants infected with nematodes exhibit yellow to brown, angular spots on lower leaves and progress to the upper portion of the plant. Lesions on the leaves coalesce to cover the entire leaf leading to withering and defoliation of the leaves (Fig 18.10). Severe infestation leads to death of the plants. Nematodes are slender, unsegmented microscopic roundworms, live in the infested plant debris fallen on to the soil. Nematodes swim in a film of water on the leaves and spread to uninfected leaves. Foliar nematodes overwinter in the soil, in infested plant material and swim up in the film of water on the plants and enter leaves through the stomata. Nematodes are dormant and survive for a year in fallen leaves. Avoiding wetting of foliage and over head irrigation prevents the spread of the disease. Rouging of the infected plants in the nursery prevents the spread of the disease in the greenhouse (Dreistadt, 2001).

1.10. Virus, Viroid and Phytoplasmal diseases: Chrysanthemums are susceptible to viruses like Chrysanthemum Mosaic Virus, Necrotic Spot Virus, Tomato Aspermy Virus and TOSPO Virus. Besides Viroid diseases include Chrysanthemum chlorotic mottle viroid and Chrysanthemum Stunt Viroid. Besides Aster Yellows caused by Pytoplasmas is also another disease limiting the productivity of the crop. Symptoms of virus (viroid) infected plants include stunting and formation of dense rosettes. Infected flowers are small, distorted or exhibit streaking and colour break. Symptoms on the leaf include yellowing, appearance of ring spots, lines, mottling, mosaics, vein clearing, distortion, crinkling, wilt and defoliation of the leaf. Aster Yellows results in chlorotic foliage, stunting, upright yellow shoots with few or no flowers, flower distortion and failure to colour. Sucking insects such as aphids, thrips and leafhoppers play a vital role in the spread of the disease.Use of disease free, certified seedlings prevents the disease spread. Removal of weeds that serve as alternate host and foliar application of insecticides minimize the viral diseases (Dreistadt, 2001).

2. Vascular Wilt diseases

2.1. Fusarium Wilt (*Fusarium oxysporum* f. sp. *chrysanthemi* and *F. oxysporum*.f. sp. *tracheiphilum*): Symptoms of *Fusarium* wilt are characterized with yellowing of foliage, stunting of affected plants, drooping of leaves followed by wilting of the affected plants. Generally, wilting often progress along one side of plant. Affected plants seem to look like that of the plants suffering from water stress. The affected plant wilts and dies. Infected plants are characterized with the presence of reddish brown discolouration of the vascular system (Fig. 18.11). Pathogen survives in the soil and the affected plant debris. Survive for longer period in the soil through chlamydospores. The pathogen spread in contaminated soil and through infected cuttings. Similarly warm temperatures, high relative humidity, over watering, and poor drainage also favour the disease spread. Cultivation of resistant varieties prevents the disease spread. Soil drenching with biocontrol agents like *Trichoderma viride, Pseudomonas fluorescens, Bacillus amyloliquefaciens* and *B. subtilis* also controls the disease.

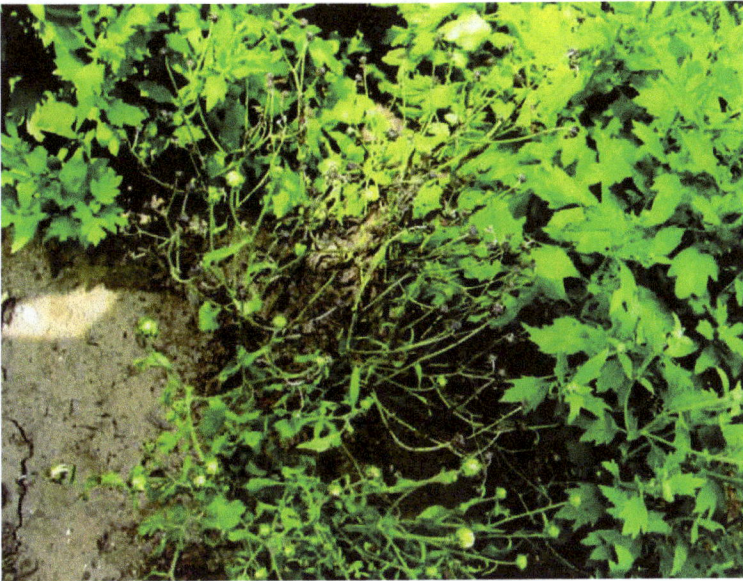

Fig. 18.11. Fusarium wilt

2.2. Verticillium Wilt (*Verticillium alboatrum*): *Verticillium* wilt often appears only after the initiation of blossom buds. Young vigorous plants do not express the wilt symptom. In the affected plants, lower most leaves turn yellow and wilt. Leaves wither and remain to be attached to the plant. Dark coloured streaks were observed in the vascular system (Dreistadt, 2001). Cool weather followed by hot weather is highly favourable for the occurrence and spread of the disease. Cultivation of resistant varieties prevents the disease spread. Soil drenching with biocontrol agents like *Trichoderma viride, Pseudomonas fluorescens, Bacillus amyloliquefaciens* and *B. subtilis* also controls the disease.

3. Diseases of the Flowers

3.1. Ray Blight (*Mycosphaerella ligulicola*): The ray flowers (marginal flowers of an inflorescence) are infected and leads to blackish rot of ray florets and deformation of blooms, so that the blooms are deformed and one-sided. Infection extends in to the floral stalks. From the flower, infection spreads to lower leaves and stem. Severe infection leads to blasting of buds. Lower leaves and stems also rot and foliage may distort or die on one side of stem. Pathogen persists in plant debris and spores are spread by wind and water. The disease is favoured by overhead irrigation or rain. Usage of disease free cuttings and avoiding wetting of foliage helps in preventing the disease (Fig. 18.12). Foliar application with fungicides like chlorothalonil, mancozeb, myclobutanil, propiconazole, or thiophanate methyl aids in the management of disease (Dreistadt, 2001).

3.2. Ray Speck (*Stemphylium lycopersici*): Pathogen infection on florets causes small, necrotic, light brown to dark brown lesions which may coalesce and cause blossom death (Fig. 18.13). The pathogen is favoured by wet conditions and temperatures between 60° to 85° F. Rogue out and dispose infected plants. Provide good air circulation and avoiding overcrowding of plants, overhead irrigation and maintaining the leaf surface under dry conditions controls the disease.

Fig. 18.12 Ray blight

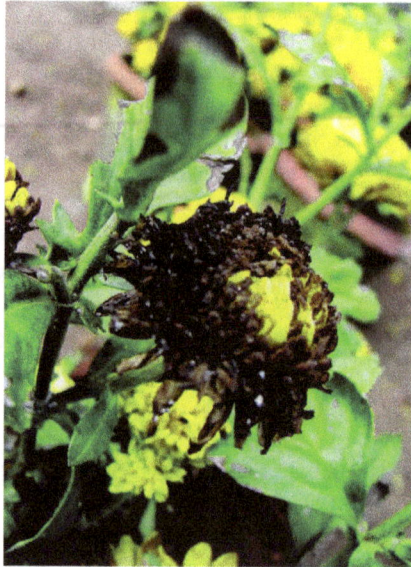

Fig. 18.13. Ray speck

4. Diseases of Roots and Crowns

4.1. Sclerotinia rot/ Cottony rot (*Sclerotinia sclerotiorum*): Infected plants are characterized with rotting of stems. Foliage turn pale green, droop and wilt. During severe cases flowers are also infected and the symptoms appear as that of gray mold. Rotted stem tissues will be colonized with white cottony fungal mass. Examination of the splitted stems reveals the presence of black coloured irregular shaped sclerotia (Dreistadt, 2001). Pathogen survives as sclerotia in soil. Ascospores produced from over wintering sclerotia are airborne and infect the petals of the flowers and dead tissues. Disease spread is favoured by high humidity coupled with increased soil moisture. Crop rotation with non-host crops followed by soil drenching with fungicides like carbendazim or trifloxystrobin+tebuconazole or difenconazole reduce the inoculum level and controls the disease. Foliar spray with any one of the fungicides like tebuconazole, azoxystrobin, difenaconazole, thiophanate methyl controls the infection by ascospres on the petals.

4.2. *Pythium* **root and stem rot (***Pythium* **spp.,):** Necrotic black lesions occur near the soil level and leads to girdling of the stem at the collar region. Later infection spreads to root followed by rotting of the roots and death of the plant. Disease is severe in nurseries and in the transplanted stem cuttings in the green house (Fig. 18.14 a, b).Pathogen is soil borne. The increased soil moisture and poor drainage spreads the pathogen. Soil drenching with metalaxyl compounds controls the disease. Besides application of Plant growth promoting rhizobacteria to soil along with compost helps in controlling the disease.

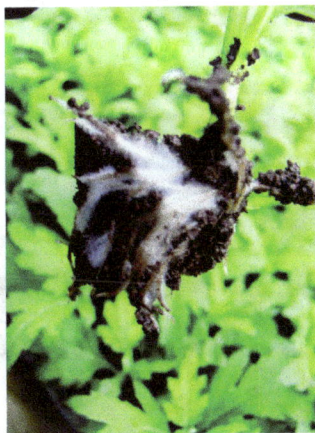

Fig. 18.14a. *Phytium* root rot   **Fig. 18.14b.** *Phythium* root rot affected field

Fig. 18.15 *Rhizoctonia* root rot

4.3. Rhizoctonia stem rot (*Rhizoctonia solani*): Pathogen infects at the collar region. Necrotic lesion appear on the collar region and leads to shriveling of the collar region followed by the death of the plant. Under warm humid conditions cottony growth of the mycelium can be observed on the affected areas (Fig 18.15). Pathogen is soil borne and survives as sclerotial bodies in the soil in the crop debris. Warm humid temperature is highly conducive for the disease spread. Soil drenching with fungicides and biocontrol agents controls the disease.

4.4. Crown gall (*Agrobacterium tumefaciens*): Crown gall is caused by a soil-borne bacterium A. *tumefaciens*. Over 600 plant species in more than 90 families can be infected. Common hosts include aster, chrysanthemum, cydonia, daisy, malus, marigold, prunus, pyrus, roses, and willows. Galls may develop on the crown, roots or, in some cases, on the aerial shoots and branches of infected plants. Galls are usually soft, spongy and white at first, but later turn hard and brown. They range in size from a few millimeters to several centimeters in diameter. Infected plants often first show symptoms of nutrient deficiency, such as yellowing or discolouration of leaves, followed by a general decline and stunting (Fig 18.16 a, b, c). A large gall at the crown may be more damaging than several smaller galls on roots or stems, since it interferes with the main vascular system of the plant (Dreistadt, 2001).

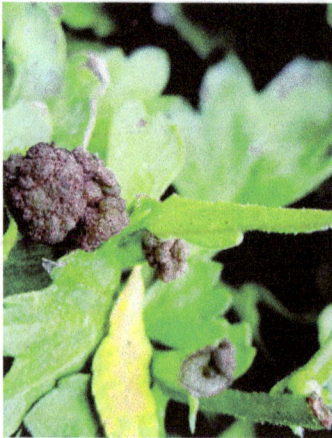

Fig. 18.16a. Crown gall on cut ends

Fig. 18.16b. Crown gall on leaves

Fig. 18.16c. Crown gall on infected plants

Agrobacterium tumefaciens is an aerobic gram negative, rod shaped and motile non sporing bacteria. The motility is due to the flexuous peritrichous flagella on the cell body. It has circular chromosome with two plasmids, one linear and the other one circular. It is a soil inhabiting bacterium found mainly in the rhizosphere and known for its disease inducing ability on many plants. The diseases is characterized by the proliferation of cells on many parts of susceptible plants especially at the stem base and roots resulting in big swellings. The cause of the swellings is the presence of plasmids that have the ability to induce tumours when taken up by a susceptible host. The plasmids containing DNA with the ability to cause cells to proliferate and form tumours are known as Ti (Tumour inducing) plasmids. They are approximately 200 kilobases (Kado 2002). *A. tumefaciens* is a rhizoplane bacterium, Gram-negative, strictly aerobic, bacilliform rods measuring 1 x 3 µm, and whose nutritional requirements are non-fastidious. The rods bear flagella that are arranged subpolarly around the cylindrical circumference of the cell, referred as circumthecal flagellation.

There are no effective chemical to control crown gall. Although antibiotics and copper bactericides are able to kill the bacterium on contact, they do not penetrate the plants and therefore fail to come into contact with bacteria residing systemically. In general, bacterial diseases of plants are very difficult to control owing to the lack of effective chemicals. Kerr discovered non-pathogenic strains of *A. radiobacter* having the ability to compete with pathogenic strains for food and space in mixed inoculations, preventing the pathogenic bacterium from becoming established (Farrand 1990). This strain is marketed in many parts of the world and it is used by suspending the bacterial cells in water, then dipping seeds, seedlings or cuttings in this suspension before planting. One disadvantage is that it acts only as a preventative treatment, and not curative. K84 inhibits strains of *A. tumefaciens* containing a nopaline-type *Ti* plasmid which are also the most common strains attacking horticultural crops. Some resistance to *A. radiobacter* strain K84 has been reported to occur due to the transfer of the genes responsible for conferring immunity to agrocin 84 antibiotics. A genetic modification by deletion of the gene responsible for exchange of DNA material among bacteria has resulted to the development of a new strain, *A radiobacter*, strain K1026 in Australia. *A. radiobacter* strain K84 as a potential biocontrol agent against crown gall in Commercial Flower Farms in Ethiopia showed that there were no significant differences (P < 0.05) in the number of galls between treated and untreated plots at one flower farm, while significant differences (P > 0.05) were observed in two farms (Derso and Yalemtesfa 2011).

Plant nursery disease management programs should be designed to maintain healthy plants. The first steps in management program are preventative, starting with clean plant production nursery area. Proper site selection of plant nurseries is very critical. Plant nurseries should be cleared of from previous crop debris prior to establishing a new crop. Weed control in and around plant nursery areas eliminates alternate hosts. The selection of pest and disease free seeds, plants and cuttings is very important. All plants brought into the nursery should be carefully inspected and infected planting material should be discarded. The key to produce clean planting material includes monitoring, scouting and recording of pests and disease. Other cultural considerations include: avoid damaging roots and stems whenever handling the plants. Use clean tools or hands during pruning and disinfect regularly and rogue all infected plants. It is also important to treat irrigation water, particularly recycled water to kill microorganisms. Some growers use pest-resistant or tolerant plants to reduce the need for pesticides.

Plants infected should be rouged as a matter of routine practice (Sigee, 1993) to reduce infection of *A. tumefaciens*. Incorporation of high organic matter content in crop land and regulating nutrients to promote uptake of copper also reduce crown gall. Water can harbour *A. tumefaciens* and should therefore be treated before use for irrigation. A water source situated close to an infected nursery could become contaminated and thus serve as a source of inoculum for further infection. Since the pathogen moves systemically in plants, it is important that tools used to cut plants should be disinfected between cuts to avoid its spread from one plant to another. Some farmers in Tanzania have also resorted to the use of olive oil as a control measure. The olive oil is applied on the galls to prevent the further growth.

A number of pest control products have been tested locally for the control and management of crown gall. Bronopol has both bactericidal and bacteristatic effects. It acts by oxidation of Mercapto group of bacterial enzymes causing inhibition of dehydrogenase activity, resulting to irreversible membrane damage. It acts as an imunomodulator by modifying the immune systems of plants. It mimics the natural Systemic Activated Resistance (SAR) by changing the contents of the phenols, proteins, nitrogen and certain enzymes and makes the plants resist bacterial attack. It was introduced in 1964 as a preservative for cosmetics, pharmaceutical preparations and industrial water systems for control of bacterial growth.

5. Role of plant quarantine in disease management : The major diseases that limit the productivity of chrysanthemum are white rust caused by *Puccina horiana* and crown gall incited by *Agrobacterium tumefaciens*. These diseases were not found to exist earlier in India. White rust originated in East Asia having first been reported in China and Japan in 1895. From there it has spread to all continents with its first sighting in the UK recorded in 1963, although it did not become established in Great Britain until 1988 and Northern Ireland in 1990. Chrysanthemum (*Chrysanthemum morifolium*) is the only commonly cultivated species that is susceptible to white rust infection. Chrysanthemum White Rust (CWR) is widespread and considered endemic in parts of Europe and South America and other areas of the world. Any introduction of this pathogen from offshore is a significant threat to the floriculture industry.

Chrysanthemum white rust, caused by the fungus *P. horiana*, is presently classified as a quarantine significant pathogen in the U.S. and Australia. This rust has been described as the most serious disease of greenhouse produced chrysanthemums because infected plants are unmarketable resulting in large economic losses. *Puccinia horiana* is autoecious rust, native to Asia. Chrysanthemum white rust was introduced into England from Japan in 1963. For more than twenty years an eradication campaign and quarantine measures were strictly enforced to prevent movement of the pathogen. These measures were ultimately unsuccessful and in 1989 the quarantine was lifted. Chrysanthemum white rust is now endemic in England. The rust has also become endemic in the Netherlands, which exports almost half of their chrysanthemum cuttings and flowers. Colombia, which is the second largest flower exporter behind the Netherlands, sends 97% of its total chrysanthemum exports to the U.S. White rust have been reported in Colombia since the late 1980s and eradication efforts have been in place to remove the pathogen from export producing areas. If white rust were to be detected on imported plant material from Colombia, all U.S. imports would be stopped, resulting in enormous financial losses for Columbian producers and U.S. distributors. A strict eradication and control campaign has been implemented in Columbia to keep all chrysanthemum exports free of *P. horiana*.

This campaign has been funded by emergency funds obtained from Colombian growers and financial backing from the flower industry. Isolated outbreaks of white rust have occurred in the 1990s in New Jersey, Pennsylvania, Washington, and Oregon. The discovery of white rust on chrysanthemum plants in production areas of California in 1992 prompted a reevaluation of the eradication program in place for control of the disease. Weekly sprays of triazole or strobilurin fungicides such as azoxystrobin, hexaconazole, myclobutanil, and propiconazole were found to be suitable regulatory treatments for exclusion and eradication of this pathogen. However, in 2001 isolates of white rust, insensitive to both the triazole and strobilurin classes of chemicals were noticed in England.

The white rust and crown gall were not known to be present earlier in India. But, the interest among the growers to cultivate high yielding varieties has resulted in the import of seedlings from other countries like Netherlands and Malaysia. The plants infected with white rust remains as a symptom less carrier for 2 weeks. Hence, if it is not properly quarantined, white rust will spread rapidly in greenhouse and nursery environments, resulting in severe losses. As a quarantine action pest, detection of white rust requires national and state regulatory action. The prevention and control of white rust depends on effective plant quarantine laws, healthy planting material, management of humidity, irrigation, and the proper selection and use of fungicides. Hence, Government should give utmost care to screen the importing planting materials to ascertain the disease free nature of the crop. Proper embargo facilities should be made available. Awareness program on the threat of the introduced diseases should be made aware among the Inter and Intra state quarantine ports in India. Since, crown gall pathogen has more than 600 host species; it is very difficult to manage the crop free from disease. All the quarantine centers should be well equipped with latest molecular diagnostic techniques for the effective prevention of quarantined diseases.

6. Impact of diseases on yield loss: Severity of diseases like white rust and crown gall depends upon the season and variety. Cultivation of cut chrysanthemum during summer does not have maximum threat on the yield and quality loss of the flowers. The severity of white rust is found to be more only during winter months. Closer spacing of 15cm x 15cm, cultivation of susceptible varieties, increased relative humidity coupled with low temperatures increase the white rust severity. Expression of rust pustules and scorching on all the leaves, reduce the flower size and yield. Though the plant has the potentiality to flower even in the presence of rust infection, the leaves which are having the rust pustules are not preferred for marketing. Hence, there is every possibility for 100 per cent yield loss. Based on our experience with white rust, the severely infected plants in several green houses were totally uprooted and burnt, which incurred a total loss to the farming community. In this juncture, exclusion of the propagating material infected by quarantined diseases must be viewed very seriously to avoid the yield loss.

7. Researchable and Policy Issues: The level of awareness among policy makers about the significance of Inter and Intra level quarantines should be increased to prevent the introduction of new pathogens into the country.

Since, the import of apparently healthy seedlings does not mimic the freeness from pathogen attack (symptom less carrier), molecular detection technique for the quarantined diseases and pathogen has to be well established so as to facilitate the prevention of pathogen entry in to the country.

- The research on management of white rust and crown gall has to be strengthened.
- The impact of seasonal effect, varieties, location specificity and their interaction on disease severity and yield loss has to be documented.
- An IDM module for the management of major yield limiting diseases has to be well documented.

REFERENCES

Alfieri, S. A. 1968. Septoria leaf spot of chrysanthemum. Plant Pathology Circular No. 73. Florida Department of Agriculture, Division of Plant Industry.

Ammerman, E., Lorenz, G., and Schelberger, K. 1992. BAS 490 F-a broad spectrum fungicide with a new mode of action. Brighton Crop protection conference – Pest and Diseases. 403-410.

Baker, J. J. 1967. Chrysanthemum white rust in England and Wales 1963-66. Plant Pathology. 16: 162-166.

Bhattacharjee, S. K., and De, L. C. 2003. Dried flowers and plant parts. In: Advanced Commercial floriculture. Avishkar Publishers, Jaipur, pp 162-173.

Brunelli, A., Minuto, G., Monchiero, M., and Gulino, M. L. 1996. Efficacy of strobilurin derivatives against grape powdery mildew in northern Italy. Brighton crop protection conference - Pest and Disease: 137-142.

Derso, E., and Yalemtesfa, B. 2011. Evaluation of *Agrobacterium radiobacter* strain K84 as potential biocontrol agent against Crown gall (*Agrobacterium tumefaciens*) in Commercial Flower Farms under Greenhouse Conditions. Proceedings of Video conference on global competitiveness of the flower industry in the East African region. Kenya Development Learning Centre (KDLC) 07th June 2011.

Dickens, J. S. W. 1970. Infection of chrysanthemum flowers by white rust (*Puccinia horiana*). Plant Pathology. 19: 122-124.

Dickens, J. S. W. 1990. Studies on the chemical control of chrysanthemum white rust caused by *Puccinia horiana*. Plant Pathology. 39: 434-442.

Dirkse F. B.1980. Bestrijding van Japanase roest. Vakblad voor de Bloemisterij 34: 30-31.

Dreistadt, S. H. 2001. Integrated Pest Management for Floriculture and Nurseries. University of California Division of Agriculture and Natural Resources. Publication 3402.

Farrand, S. K. 1990. *Agrobacterium radiobacter* K84: a model biocontrol system. pp. 679-691 in, New Directions in Biological Control: Alternatives for Suppressing Agricultural Pests and Diseases. (Publ. Alan R Liss Inc.)

Firman, I. D., and Martin P. H. 1968. White rust of chrysanthemum. Ann. Appl. Biol., 62: 429–442.

Gjaerum, H. B. 1979. Rust fungi on glasshouse crops. Forsk.Fors. I Landbruket. 30: 91-109.

Grouet, D., and Allaire, L. 1973. Efficacite de l'oxycarboxine et de la triforine contre la rouille blanche du chrysantheme (*Puccinia horiana* P. Henn.). Phyt. Phytopharm. 22: 177-188.

Hennings, P. 1901. Quelques nouvelles rouilles japonaises. Hedwigia. 40: 25-26.

http: //extension.psu.edu/plant-disease-factsheets/hrysanthemum-diseases

http: //urbanext.illinois.edu/focus/per_chrysanthemum.cfm

http://www.apsnet.org/publications/commonnames/Pages/Chrysanthemum. aspx

http: //www.clemson.edu/extension/hgic/pests/plant pests/flowers/ hgic2101.html

Kado, C. I. 2002. Crown gall. Plant health instructor pp. 7

Krebs, K. E. 1985. Chrysanthemum white rust can be controlled. Gb+Gw. 85(3): 69-73.

Rattink, H., Zamorski, C., and Dil, M. C. 1985. Spread and control of white rust (Puccinia horiana) on chrysanthemums on artificial substrate. Med. Fac. Landbouww. Univ. Gent 50(3b): 1243-1249.

Sigee, D. C. 1993. Biology of Agrobacterium and infection behaviour, Bacterial Plant Pathology: Cell and Molecular Aspects. Cambridge University Press.

Wadel, H. T., and Weber, G. F. 1963. Physiology and pathology of Septoria species on Chrysanthemum. Mycologia, 55: 442-452.

Wojdy³a, A. T. 1999. Susceptibility of chrysanthemum cultivars to *Puccinia horiana*. Folia Horticulturae 11(2): 115–122.

Wojdyyla, A. T. 2002. Azoles in the control of *Puccinia horiana*. J. Plant Protection Res. 42: 261–270.

Wojdyyla, A. T., and Orlikowski, L. B. 1999. Strobilurin compounds in control of rust, powdery mildew and black spot on some ornamental plants. Med. Fac. Landbouww. Univ. Gent. 64: 539-545.

Subject index

F

G

P